Lecture Notes in Mathematics

1928

Editors:
J.-M. Morel, Cachan
F. Takens, Groningen
B. Teissier, Paris

T0215917

Jakob Jonsson

Simplicial Complexes
of Graphs

 Springer

Jakob Jonsson
Department of Mathematics
KTH
Lindstedtsvägen 25
10044 Stockholm
Sweden
jakobj@math.kth.se

ISBN 978-3-540-75858-7   ISBN 978-3-540-75859-4 (eBook)

DOI 10.1007/978-3-540-75859-4

Lecture Notes in Mathematics ISSN print edition: 0075-8434
                              ISSN electronic edition: 1617-9692

Library of Congress Control Number: 2007937408

Mathematics Subject Classification (2000): 05E25, 55U10, 06A11

Cover design: *design & production* GmbH, Heidelberg

Printed on acid-free paper

9 8 7 6 5 4 3 2 1

springer.com

Preface

This book is a revised version of my 2005 thesis [71] for the degree of Doctor of Philosophy at the Royal Institute of Technology (KTH) in Stockholm. The whole idea of writing a monograph about graph complexes is due to Professor Anders Björner, my scientific advisor. I am deeply grateful for all his comments, remarks, and suggestions during the writing of the thesis and for his very careful reading of the manuscript.

I spent the first years of my academic career at the Department of Mathematics at Stockholm University with Professor Svante Linusson as my advisor. He is the one to get credit for introducing me to the field of graph complexes and also for explaining the fundamentals of discrete Morse theory, the most important tool in this book. Most of the work presented in Chapters 17 and 20 was carried out under the inspiring supervision of Linusson.

The opponent (critical examiner) of my thesis defense was Professor John Shareshian; the examination committee consisted of Professor Boris Shapiro, Professor Richard Stanley, and Professor Michelle Wachs. I am grateful for their valuable feedback that was of great help to me when working on this revision.

The work of transforming the thesis into a book took place at the Technische Universität Berlin and the Massachusetts Institute of Technology. I thank Björner and Professor Günter Ziegler for encouraging me to submit the manuscript to Springer.

Some chapters in this book appear in revised form as journal papers: Chapters 4, 17, and 20 are revised versions of a paper published in the Journal of Combinatorial Theory, Series A [67]. Chapter 5 is a revised version of a paper published in the Electronic Journal of Combinatorics [70]. Chapter 26 is a revised version of a paper published in the SIAM Journal of Discrete Mathematics [72]. I am grateful to several anonymous referees and editors representing these journals, and also to anonymous referees representing the FPSAC conference, who all provided helpful comments and suggestions.

In addition, I thank two anonymous reviewers for this series for providing several useful comments on the manuscript and the editors at Springer for showing patience and being of great help during the preparation of the manuscript.

Finally, and most importantly, I thank family and friends for endless support.

For the reader's convenience, let me list the major revisions compared to the thesis version of 2005:

- Chapter 1 has been extended with a more thorough discussion about applications of graph complexes to problems in other areas of mathematics.
- Recent results about the matching complex M_n and the chessboard complex $M_{m,n}$ have been incorporated into Sections 11.2.3 and 11.3.2.
- Section 15.4 has been updated with a more precise statement about the Euler characteristic of the complex $DGr_{n,p}$ of digraphs that are graded modulo p and a shorter proof of a formula for the Euler characteristic of $DGr_n = DGr_{n,n+1}$.
- Section 16.3 has been updated with a proof that the complex NXM_n of noncrossing matchings is semi-nonevasive.
- Section 18.5 is new and contains a brief discussion about the complex of disconnected hypergraphs.
- Section 19.4 is new and contains a generalization of the complex NC_n^2 of not 2-connected graphs along with yet another method for computing the homotopy type of NC_n^2. The theory in this section is applied in Section 22.2, which is also new and contains a discussion about the complex $DNSC_n^2$ of not strongly 2-connected digraphs.
- At the end of Section 23.3, we discuss a recent observation due to Shareshian and Wachs [121] about a connection between the complex NEC_{kp+1}^p of not p-edge-connected graphs on $kp + 1$ vertices and the poset $\Pi_{kp+1}^{1 \bmod p}$ of set partitions on $kp + 1$ elements in which the size of each part is congruent to 1·modulo p.

Cambridge, MA,   *Jakob Jonsson*
March 2007

Summary. Let G be a finite graph with vertex set V and edge set E. A *graph complex* on G is an abstract simplicial complex consisting of subsets of E. In particular, we may interpret such a complex as a family of subgraphs of G. The subject of this book is the topology of graph complexes, the emphasis being placed on homology, homotopy type, connectivity degree, Cohen-Macaulayness, and Euler characteristic.

We are particularly interested in the case that G is the complete graph on V. *Monotone graph properties* are complexes on such a graph satisfying the additional condition that they are invariant under permutations of V. Some well-studied monotone graph properties that we discuss in this book are complexes of matchings, forests, bipartite graphs, disconnected graphs, and not 2-connected graphs. We present new results about several other monotone graph properties, including complexes of not 3-connected graphs and graphs not coverable by p vertices.

Imagining the vertices as the corners of a regular polygon, we obtain another important class consisting of those graph complexes that are invariant under the natural action of the dihedral group on this polygon. The most famous example is the associahedron, whose faces are graphs without crossings inside the polygon. Restricting to matchings, forests, or bipartite graphs, we obtain other interesting complexes of noncrossing graphs. We also examine a certain "dihedral" variant of connectivity.

The third class to be examined is the class of digraph complexes. Some well-studied examples are complexes of acyclic digraphs and not strongly connected digraphs. We present new results about a few other digraph complexes, including complexes of graded digraphs and non-spanning digraphs.

Many of our proofs are based on Robin Forman's discrete version of Morse theory. As a byproduct, this book provides a loosely defined toolbox for attacking problems in topological combinatorics via discrete Morse theory. In terms of simplicity and power, arguably the most efficient tool is Forman's divide and conquer approach via decision trees, which we successfully apply to a large number of graph and digraph complexes.

Contents

Part I

Introduction and Basic Concepts

1

Introduction and Overview

This book focuses on families of graphs on a fixed vertex set. We are particularly interested in *graph complexes*, which are graph families closed under deletion of edges. Equivalently, a graph complex Δ has the property that if $G \in \Delta$ and e is an edge in G, then the graph obtained from G by removing e is also in Δ. Since the vertex set is fixed, we may identify each graph in Δ with its edge set and hence interpret Δ as a simplicial complex. In particular, we may realize Δ as a geometric object and hence analyze its topology. Indeed, this is the main purpose of the book.

Fig. 1.1. Δ contains all graphs isomorphic to one of the four illustrated graphs.

As an example, consider the simplicial complex Δ of graphs G on the vertex set $\{1, 2, 3, 4\}$ with the property that some vertex is contained in all edges in G. This means that G is isomorphic to one of the graphs in Figure 1.1. Denoting the edge between i and j as ij, we obtain that

$$\Delta = \{\emptyset, \{12\}, \{13\}, \{14\}, \{23\}, \{24\}, \{34\}, \{12, 13\}, \{12, 14\}, \{13, 14\},$$
$$\{12, 23\}, \{12, 24\}, \{23, 24\}, \{13, 23\}, \{13, 34\}, \{23, 34\}, \{14, 24\},$$
$$\{14, 34\}, \{24, 34\}, \{12, 13, 14\}, \{12, 23, 24\}, \{13, 23, 34\}, \{14, 24, 34\}\}.$$

See Figure 1.2 for a geometric realization of Δ. It is easy to see that Δ is homotopy equivalent to a one-point wedge of three circles.

Monotone Graph Properties

In the above example, note that a given graph G belongs to Δ if and only if all graphs isomorphic to G belong to Δ. Equivalently, Δ is invariant under the

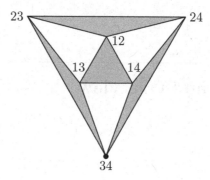

Fig. 1.2. Geometric realization of the complex Δ.

action of the symmetric group on the underlying vertex set. A family of graphs satisfying this condition is a *graph property*. We will be mainly concerned with graph properties that are also graph complexes, hence closed under deletion of edges. We refer to such graph properties as *monotone* graph properties.

In this book, we discuss and analyze the topology of several monotone graph properties, some examples being matchings, forests, bipartite graphs, non-Hamiltonian graphs, and not k-connected graphs; see Chapter 7 for a summary. Some results are our own, whereas others are due to other authors. We restrict our attention to topological and enumerative properties of the complexes and do not consider representation-theoretic aspects of the theory.

Remark. Some authors define monotone graph properties to be graph properties closed under *addition* of edges. While such graph properties are not simplicial complexes, they are quotient complexes of simplicial complexes and hence realizable as geometric objects; see Sections 3.2 and 3.5.

Other Graph Complexes

Monotone graph properties are not the only interesting graph complexes. For example, for any monotone graph property Δ and any graph G, one may consider the subcomplex $\Delta(G)$ consisting of all graphs in Δ that are also subgraphs of G; this is the induced subcomplex of Δ on G. In some situations, $\Delta(G)$ is interesting in its own right; we would claim that this is the case for complexes of matchings, forests, and disconnected graphs. In other situations, $\Delta(G)$ is of use in the analysis of the larger complex Δ; one example is the complex of bipartite graphs.

With graph properties being invariant under the action of the symmetric group, a natural generalization would be to replace the symmetric group with a smaller group. In this book, we concentrate on the dihedral group D_n. This group acts in a natural manner on the family of graphs on the vertex set $\{1, \ldots, n\}$: Represent the vertices as points evenly distributed in a clockwise manner around a unit circle and identify a given edge with the line segment between the two points representing the endpoints of the edge. We refer to this representation of a graph as the *polygon representation*; the vertices are the corners in a regular polygon. The action of the dihedral group consists of rotations and reflections, and combinations thereof, of this polygon. The associahedron is probably the most well-studied graph complex with a natural dihedral action. Some other interesting "dihedral" graph complexes are complexes of noncrossing matchings, noncrossing forests, and graphs with a disconnected polygon representation. See Chapter 8 for more information.

Finally, we mention complexes of directed graphs; we refer to such complexes as *digraph complexes*. Some important examples are complexes of directed forests and acyclic digraphs. We also discuss some directed variants of the property of being bipartite and the property of being disconnected. See Chapter 9 for an overview.

Remark. As is obvious from the discussion in this section, our graph complexes are completely unrelated to Kontsevich's graph complexes [83, 85].

Discrete Morse Theory

The most important tool in our analysis is Robin Forman's discrete version of Morse theory [48, 49]. As we describe in more detail in Chapter 4, one may view discrete Morse theory as a generalization of the concept of collapsibility. A complex Δ is collapsible to a smaller complex Σ if we can transform Δ into Σ via a sequence of elementary collapses. An elementary collapse is a homotopy-preserving operation in which we remove a maximal face τ along with a codimension one subface σ such that the resulting complex remains a simplicial complex (i.e., closed under deletion of elements).

To better understand the generalization, we first interpret a collapse as a giant one-step operation in which we perform all elementary collapses *at once*, rather than one by one. This way, a collapse from a complex Δ to a subcomplex Σ boils down to a partial matching on Δ such that Σ is exactly the family of unmatched faces. Dropping the condition that the unmatched faces must form a simplicial complex, we obtain discrete Morse theory.

More precisely, under certain conditions on a given matching – similar to the ones that we would need on a matching corresponding to an ordinary collapse – Forman demonstrated how to build a cell complex homotopy equivalent to Δ using the unmatched faces as building blocks. Indeed, this very construction is the main result of discrete Morse theory. As an immediate corollary of Forman's construction, we obtain upper bounds on the Betti numbers.

Remark. We should mention that some aspects of the above interpretation of discrete Morse theory are due to Chari [32]. In addition, while we discuss only simplicial complexes, Forman considered a much more general class of cell complexes.

Divide and Conquer

One of the more typical ways of applying discrete Morse theory in practice is to partition the complex under consideration into smaller subfamilies and then define a matching on each subfamily. Babson et al. [3, Lemma 3.6] provided a very early application of this divide and conquer approach in their proof that a certain complex related to the complex of not 2-connected graphs is collapsible.[1] For arguably the very first full-fledged application of discrete Morse theory, we refer to Shareshian [118], another paper about complexes of not 2-connected graphs. See Chapter 19 for more discussion. Finally, we mention Forman's divide and conquer approach via decision trees [50], which we discuss in detail in Chapter 5.

1.1 Motivation and Background

Many graph complexes are beautiful objects with a rich topological structure and may hence be considered as interesting in their own right. Nevertheless, some background seems to be in place, particularly since this area of research to some extent emerged from developments in other fields. In this section, we provide a random selection of prominent examples, referring the reader to the literature and later chapters of this book for details.

1.1.1 Quillen Complexes

Let G be a finite group and let p be a prime. Brown [22, 23] and Quillen [110] studied topological properties of two complexes known as the *Brown complex* and the *Quillen complex*. The Brown complex is the order complex $\Delta(S_p(G))$ of the poset $S_p(G)$ of nontrivial p-subgroups of a finite group G. The Quillen complex $\Delta(A_p(G))$ is the order complex of the subposet $A_p(G)$ of $S_p(G)$ consisting of all nontrivial elementary abelian p-subgroups of G. Quillen demonstrated that $\Delta(S_p(G))$ and $\Delta(A_p(G))$ have the same homotopy type.

For $G = \mathfrak{S}_n$ and $p = 2$, it turns out that one may deduce information about the Quillen complex by examining the matching complex M_n; see Chapter 11. Specifically, one may identify the barycentric subdivision of M_n with the order complex of the poset of nontrivial abelian subgroups of $\mathfrak{S}_n$ generated by transpositions. Examining the natural inclusion map from this complex to $\Delta(A_p(G))$ and using the fact that M_n is simply connected for $n \geq 8$ (see

---

[1] The paper [3] was published in 1999, but the authors announced their results already two years earlier.

Corollary 11.2), Ksontini [88, 89] was able to deduce that $\Delta(A_2(\mathfrak{S}_n))$ is simply connected for $n \geq 8$. Some further detailed analysis yielded simple connectivity also for $n = 7$. Bouc [21] initiated the study of M_n motivated by the Quillen complex.

For odd primes p, there is a similar connection between a relative of the complex HM_n^p of p-hypergraph matchings (see Chapter 11) and $\Delta(A_p(\mathfrak{S}_n))$. This relative differs from HM_n^p in that we have $(p-1)!$ disjoint copies of each hyperedge in the complete p-hypergraph. Specifically, one may identify the face poset of this complex with the poset $T_p(\mathfrak{S}_n)$ of nontrivial elementary abelian p-subgroups of $\mathfrak{S}_n$ generated by p-cycles. Using properties of $T_p(\mathfrak{S}_n)$, Ksontini [88, 89] demonstrated that $\Delta(A_p(\mathfrak{S}_n))$ is simply connected if and only if $3p + 2 \leq n < p^2$ or $n \geq p^2 + p$. Shareshian [119] built on this work, providing a concrete description of the homotopy type of $\Delta(A_p(\mathfrak{S}_n))$ in terms of that of $\Delta(T_p(\mathfrak{S}_n))$ whenever $p^2 + p \leq n < 2p^2$. Using a computer calculation of the homology of HM_{13}^3 carried out by J.-G. Dumas, Shareshian also demonstrated that $\tilde{H}_2(\Delta(A_3(\mathfrak{S}_{13})); \mathbb{Z})$ contains 2-torsion, thereby providing the first known example of a Quillen complex with nonfree integral homology.

1.1.2 Minimal Free Resolutions of Certain Semigroup Algebras

An interesting connection between ring theory and topological combinatorics is given by a well-known correspondence between semigroup algebras over semigroup rings and certain associated simplicial complexes. This correspondence was exploited by Reiner and Roberts [111], and subsequently by Dong [36], who were led to study complexes of graphs with a bounded vertex degree; see Chapter 12.

To explain the correspondence, let $n \geq 1$. For a sequence $\lambda = (\lambda_1, \ldots, \lambda_n)$ of nonnegative integers, let $\overline{\mathsf{BD}}_n^\lambda$ be the simplicial complex of simple graphs with loops allowed on the vertex set $\{1, \ldots, n\}$ such that the degree of the vertex i is at most λ_i for $1 \leq i \leq n$; see Chapter 12 for the exact definition. If all λ_i are equal to one, then we obtain the matching complex M_n; see Chapter 11.

Let $\mathbb{F}$ be a field and consider the polynomial rings $A = \mathbb{F}[\{z_{ij} : 1 \leq i \leq j \leq n\}]$ and $\mathbb{F}[\mathbf{x}] = \mathbb{F}[x_1, \ldots, x_n]$. Defining $\phi(z_{ij}) = x_i x_j$, we obtain an A-algebra structure on $\mathbb{F}[\mathbf{x}]$. The *second Veronese subalgebra* $\mathrm{Ver}(n, 2, 0)$ is the subalgebra $\phi(A)$ of $\mathbb{F}[\mathbf{x}]$.

By a well-known theorem (e.g., see Stanley [132, Th. 7.9] and Reiner and Roberts [111, Prop. 3.2]), we have that

$$\dim_{\mathbb{F}} \mathrm{Tor}_i^A(\mathrm{Ver}(n, 2, 0), \mathbb{F}) = \sum_\lambda \dim_{\mathbb{F}} \tilde{H}_{i-1}(\overline{\mathsf{BD}}_n^\lambda; \mathbb{F}),$$

where the sum is over all sequences $\lambda = (\lambda_1, \ldots, \lambda_n)$ such that $\sum_{i=1}^n \lambda_i$ is even. One easily checks that $\overline{\mathsf{BD}}_n^\lambda$ has vanishing reduced homology for all but finitely many λ; hence the sum makes sense.

There is also a bipartite variant of this construction involving the so-called Segre algebra. In this case, a chessboard variant of $\overline{BD}_n^{\lambda}$ is of importance; see Reiner and Roberts [111] for details.

1.1.3 Lie Algebras

Complexes of graphs of bounded degree also appear in the analysis of the homology of the free two-step nilpotent Lie algebra. See Józefiak and Weyman [77] and Sigg [123] for details and for information about the representation-theoretic aspects of the theory.

Let $n \geq 1$ and let $\{e_1, \ldots, e_n\}$ denote the standard basis of complex n-space $\mathbb{C}^n$. Define $\mathcal{L}(n) = \mathbb{C}^n \oplus (\mathbb{C}^n \wedge \mathbb{C}^n)$; this is the *free two-step nilpotent complex Lie algebra* of rank n, where the Lie bracket is defined on basis elements by $[e_i, e_j] = e_i \wedge e_j$ and zero otherwise. Identifying e_i with the vertex i and $e_i \wedge e_j = -e_j \wedge e_i$ with the edge ij for $i < j$, we obtain that a basis for $\mathcal{L}(n)$ is given by the union of the set of vertices and the set of edges of the complete graph K_n.

The homology of a Lie algebra $\mathcal{A}$ with trivial coefficients is defined to be the homology of the exterior algebra complex $(\Lambda^* \mathcal{A}, \delta)$, where

$$\delta(x_1 \wedge \cdots \wedge x_p) = \sum_{i<j} (-1)^{i+j+1} [x_i, x_j] \wedge x_1 \wedge \cdots \wedge \hat{x}_i \wedge \cdots \wedge \hat{x}_j \wedge \cdots \wedge x_p;$$

$\hat{x}_i$ denotes deletion. It is easy to see that $(\Lambda^* \mathcal{L}(n), \delta)$ splits into many small pieces. Specifically, for a basis element $x = a_1 b_1 \wedge a_2 b_2 \wedge \cdots \wedge a_r b_r \wedge c_1 \wedge c_2 \wedge \cdots \wedge c_s$, we define the weight $\gamma(x)$ to be the vector $(\gamma_1, \ldots, \gamma_n)$ with the property that γ_i is the number of occurrences of the vertex i in x. For example, $\gamma(13 \wedge 26 \wedge 1 \wedge 2 \wedge 4) = (2, 2, 1, 1, 0, 1)$. The boundary operator δ preserves the weight, which implies that we obtain a natural decomposition

$$(\Lambda^* \mathcal{L}(n), \delta) \cong \bigoplus_{\gamma} ((\Lambda^* \mathcal{L}(n))_{\gamma}, \delta),$$

where $(\Lambda^* \mathcal{L}(n))_{\gamma}$ is generated by all basis elements with weight γ.

Let Σ_n^{γ} be the quotient complex of loop-free graphs on n vertices with the property that the degree of the vertex i is either $\gamma_i - 1$ or γ_i for each i. The complex homology, and hence cohomology, of Σ_n^{γ} coincides with the homology of the complex $\overline{BD}_n^{\gamma}$ (see previous section for definition). An easy way to prove this is to use the construction in the proof of Proposition 12.16 and apply Lemma 3.16.

Now, we may define a homomorphism φ from $(\Lambda^i \mathcal{L}(n))_{\gamma}$ to the cochain group $\tilde{C}^{m-i-1}(\Sigma_n^{\gamma}; \mathbb{C})$ by mapping $a_1 b_1 \wedge a_2 b_2 \wedge \cdots \wedge a_r b_r \wedge c_1 \wedge c_2 \wedge \cdots \wedge c_s$ to $a_1 b_1 \wedge a_2 b_2 \wedge \cdots \wedge a_r b_r$; $m = 2r + s$ and $i = r + s$. While φ is a group isomorphism, it is not the case that $\delta(\varphi(x)) = \varphi(\delta(x))$. Still, as Dong and Wachs demonstrated [37, Sec. 4], the homology group of degree i of $((\Lambda^* \mathcal{L}(n))_{\gamma}, \delta)$ is indeed isomorphic to the cohomology group of degree $m - i - 1$ of $\overline{BD}_n^{\gamma}$.

For the special case $\gamma = (1, \ldots, 1)$, meaning that $\overline{\mathsf{BD}}_n^\gamma = \mathsf{M}_n$, we may tweak φ by defining

$$\varphi(a_1 b_1 \wedge a_2 b_2 \wedge \cdots \wedge a_r b_r \wedge c_1 \wedge c_2 \wedge \cdots \wedge c_s) = \operatorname{sgn}(\pi) \cdot a_1 b_1 \wedge a_2 b_2 \wedge \cdots \wedge a_r b_r,$$

where π is the permutation $\begin{pmatrix} 1 & 2 & 3 & 4 & \cdots & 2r-1 & 2r & 2r+1 & 2r+2 & \cdots & n \\ a_1 & b_1 & a_2 & b_2 & \cdots & a_r & b_r & c_1 & c_2 & \cdots & c_s \end{pmatrix}$; $2r + s = n$. It is easy to check that φ does satisfy $\delta(\varphi(x)) = \varphi(\delta(x))$ this time. For the general case, we refer to Dong and Wachs [37, Prop. 4.4].

1.1.4 Disconnected k-hypergraphs and Subspace Arrangements

The complex $\mathsf{HNC}_{n,k}$ of disconnected k-hypergraphs (see Section 18.5) is closely related to the lattice of set partitions in which each set either is a singleton set or has size at least k. Björner and Welker [16] studied this lattice to derive information about certain subspace arrangements. Note that $k = 2$ yields the complex NC_n of disconnected graphs discussed in Section 18.1.

Let $2 \le k \le n$. For any indices $1 \le i_1 < i_2 < \cdots < i_k \le n$, let $H_{i_1, \ldots, i_k}$ be the subspace of $\mathbb{R}^n$ consisting of all points $(x_1, \ldots, x_n)$ satisfying $x_{i_1} = \cdots = x_{i_k}$. Define $\mathcal{A}_{n,k}^{\mathbb{R}}$ as the arrangement in $\mathbb{R}^n$ consisting of all such subspaces. Moreover, define

$$V_{n,k}^{\mathbb{R}} = \bigcup_{i_1, \ldots, i_k} H_{i_1, \ldots, i_k}$$

and $M_{n,k}^{\mathbb{R}} = \mathbb{R}^n \setminus V_{n,k}^{\mathbb{R}}$.

The intersection lattice $L_{\mathcal{A}}$ of a subspace arrangement $\mathcal{A}$ is the set of all intersections $K_1 \cap \cdots \cap K_r$ of subspaces $K_1, \ldots, K_r \in \mathcal{A}$ ordered by reverse inclusion. The minimal element $\hat{0}$ in $L_{\mathcal{A}}$ is the full space $\mathbb{R}^n$ corresponding to the "void" intersection (i.e., $r = 0$). Write $\Pi_{n,k}^{\mathbb{R}} = L_{\mathcal{A}_{n,k}^{\mathbb{R}}}$. By a theorem due to Goresky and McPherson [54], we have that

$$\tilde{H}^i(M_{n,k}^{\mathbb{R}}; \mathbb{Z}) \cong \bigoplus_{\mathcal{U} \in \Pi_{n,k}^{\mathbb{R}} \setminus \hat{0}} \tilde{H}_{n-\dim(\mathcal{U})-i-2}(\Delta(\Pi_{n,k}^{\mathbb{R}}(\hat{0}, \mathcal{U})); \mathbb{Z}),$$

where $\Pi_{n,k}^{\mathbb{R}}(\hat{0}, \mathcal{U})$ is the subposet consisting of all elements $\mathcal{U}'$ such that $\hat{0} < \mathcal{U}' < \mathcal{U}$ and $\Delta(P)$ is the face poset of P; see Section 2.3.8.

It is easy to see that the elements of $\Pi_{n,k}^{\mathbb{R}}$ are in bijection with partitions of $[n]$ such that each set either is a singleton set or has size at least k. As a consequence, $\Delta(\Pi_{n,k}^{\mathbb{R}}(\hat{0}, \hat{1}))$ has the same homotopy type as the complex $\mathsf{HNC}_{n,k}$ of disconnected k-hypergraphs. This follows from the fact that we may define a closure operator on the face poset of $\mathsf{HNC}_{n,k}$ by mapping any given hypergraph H to the hypergraph obtained by adding all hyperedges that do not reduce the number of connected components in H. The image of this map turns out to be isomorphic to $\Pi_{n,k}^{\mathbb{R}}(\hat{0}, \hat{1})$, which implies that the two complexes are homotopy equivalent; apply Lemma 6.1. A similar examination yields that the suspension of $\Delta(\Pi_{n,k}^{\mathbb{R}}(\hat{0}, \mathcal{U}))$ is homotopy equivalent to a join consisting of one copy of $\mathsf{HC}_{m,k}$ for each set in $\mathcal{U}$ of size m, where $\mathsf{HC}_{m,k}$ is the quotient complex of connected k-hypergraphs on m vertices.

Björner and Welker [16] proved that each $\Delta(\Pi_{n,k}^{\mathbb{R}})$, and hence each $\mathsf{HNC}_{n,k}$ and $\mathsf{HC}_{n,k}$, is homotopy equivalent to a wedge of spheres in various dimensions. In particular, all homology is free, and we may hence easily deduce the cohomology of $M_{n,k}^{\mathbb{R}}$ from the homology of $\mathsf{HNC}_{m,k}$ for $1 \le m \le n$.

1.1.5 Cohomology of Spaces of Knots

Complexes of connected and 2-connected graphs appear in Vassiliev's analysis of the cohomology groups of certain spaces of knots [141, 142, 143]. Below, we provide a heuristic and simplified description of the construction; for a more accurate and detailed description, we refer to Vassiliev's work.

Let $n \ge 3$ and let $\mathcal{K}$ be the space of all smooth maps ϕ from the real line $\mathbb{R}$ into $\mathbb{R}^n$ such that ϕ coincides with the natural embedding $x \mapsto (x, 0, 0)$ for sufficiently large $|x|$. $\phi \in \mathcal{K}$ is a (non-compact) *knot* if ϕ is an embedding, meaning that ϕ is injective and has no local singularities $\phi'(x) = 0$. The *discriminant* of $\mathcal{K}$ is the subset Σ of all non-knots of $\mathcal{K}$. Two knots are considered equivalent if they lie in the same connected component of $\mathcal{K} \setminus \Sigma$.

Define

$$\Psi = \{(x, y) \in \mathbb{R}^2 : x \le y\}.$$

The *resolution* σ of Σ is defined roughly in the following manner; we refer to Vassiliev [141] for details. Let I be a "generic" embedding of Ψ in $\mathbb{R}^N$, where N is extremely large but finite. For a map $\phi \in \Sigma$, we say that ϕ *respects* a point $(x, y) \in \Psi$ if either $x \ne y$ and $\phi(x) = \phi(y)$ (an intersection) or $x = y$ and $\phi'(x) = 0$ (a cusp). ϕ respects a set $X \subset \Psi$ if ϕ respects each point in X. Let $\Delta(X)$ be the convex hull of $I(X)$ and define $\Delta(\phi) = \Delta(X_\phi)$, where X_ϕ is the set of all points (x, y) respected by ϕ. Using certain approximations, one may assume that X_ϕ is finite and that $\Delta(\phi)$ is a finite-dimensional simplex whose vertex set coincides with the point set $I(X_\phi)$.

Define

$$\sigma = \bigcup_{\phi \in \Sigma} \{\phi\} \times \Delta(\phi) \subseteq \mathcal{K} \times \mathbb{R}^N.$$

Vassiliev [141] proved that the Borel-Moore homology of Σ coincides with the homology of σ and that a duality argument yields a correspondence between this homology and the cohomology of $\mathcal{K} \setminus \Sigma$.

For any finite set $X \subset \Psi$, the family of maps ϕ respecting X forms an affine subspace of $\mathcal{K}$ of codimension a multiple cn of n for some integer c. The value c is the *complexity* $\xi(X)$ of X. To compute the homology of σ, Vassiliev [143] forms a filtration

$$\sigma_1 \subset \sigma_2 \subset \sigma_3 \subset \cdots,$$

where σ_i consists of all $\{\phi\} \times \Delta(X)$ such that $\xi(X) \leq i$. By a theorem due to Kontsevich [84], the spectral sequence associated with this filtration degenerates already at the first term.

For a finite set $X \subseteq \Psi$, form a graph G_X with one vertex for each x appearing in X and with an edge between x and y whenever $(x, y) \in \Psi$; if $(x, x) \in \Psi$, then we add a loop at x. The complexity of X satisfies the identity $\xi(X) = v(G_X) + \ell(G_X) - c(G_X)$, where $v(G_X)$, $\ell(G_X)$, and $c(G_X)$ denote the number of vertices, the number of loops, and the number of connected components, respectively, of G_X. Let Y be a set such that the graph obtained from G_Y by removing all loops has the property that each connected component is a clique. Define $\Gamma(Y) = \{X \subseteq Y : \xi(X) = \xi(Y)\}$. It is straightforward to check that $\Gamma(Y)$ is a join of quotient complexes of the form C_r, where C_r is the quotient complex of connected graphs on a vertex set of size r; see Chapter 18. This observation is of use in the analysis of $\sigma_i \setminus \sigma_{i-1}$, where $i = \xi(Y)$.

To proceed further, Vassiliev considers yet another filtration

$$\Phi_1^i \subset \Phi_2^i \subset \cdots \subset \Phi_{i-1}^i$$

of the relevant term $\sigma_i \setminus \sigma_{i-1}$ from the first filtration for each i. Define $\alpha(X) = v(G_X) - c(G_X) - b(G_X)$, where $b(G_X)$ is the number of 2-connected components in the graph obtained from G_X by removing all loops. We define Φ_j to be the union of all $\{\phi\} \times \Delta(X) \subset \sigma$ such that $\alpha(X) \leq j$. Write $\Phi_j^i = \Phi_j \cap (\sigma_i \setminus \sigma_{i-1})$.

We say that X is *block-closed* if each 2-connected component of G_X is a clique. For a set X, we let $\overline{X}$ be the block-closed set obtained from X by adding (x, y) whenever x and y belong to the same 2-connected component of G_X. For a block-closed set Y and a subset X of Y, it is easy to see that $\xi(X) = \xi(Y)$ and $\alpha(X) = \alpha(Y)$ if and only if $\overline{X} = Y$. This implies that $\{X \subseteq Y : \xi(X) = \xi(Y) = i, \alpha(X) = \alpha(Y) = j\}$ is a join of quotient complexes of the form C_r^2, where C_r^2 is the quotient complex of 2-connected graphs on a vertex set of size r; see Chapter 19. Using this fact and properties of C_r^2, one may obtain useful information about the homology of $\Phi_j^i \setminus \Phi_{j-1}^i$.

As a side note, we observe that $\xi(X) + \alpha(X) = 2v(G_X) + \ell(G_X) - 2c(G_X) - b(G_X)$. Whenever G_X is block-closed and loop-free, this is the rank of G_X in the lattice $\Pi_{n,2}$ of block-closed graphs; see Theorem 19.2.

1.1.6 Determinantal Ideals

The famous theory of Hochster, Reisner, and Stanley provides a fundamental link between ring theory and topology of simplicial complexes [63, 113, 132]. Specifically, there is a natural correspondence between simplicial complexes and ideals generated by square-free monomials, and several of the most fundamental ring-theoretic concepts turn out to admit elegant interpretations in terms of simplicial topology. Dimension, multiplicity, depth, and Cohen-Macaulayness are a few examples; see Section 3.8 for some more information.

For a particularly fruitful application of this interaction, let us discuss determinantal ideals; see Bruns and Conca [25] for a survey. In such ideals, each variable is indexed by a position in a certain matrix, which means that we may interpret each variable as an edge in a bipartite graph (or a directed edge in a digraph). Herzog and Trung [62] showed how to transform determinantal ideals into ideals generated by square-free monomials and analyzed the corresponding simplicial complexes, which are effectively graph complexes, to establish results about the multiplicity and Cohen-Macaulayness of the original determinantal ideals.

To describe the construction, we let $M = (X_{ij} : 1 \leq i \leq r, 1 \leq j \leq s)$ be a generic $r \times s$ matrix. Let $\mathbb{F}$ be a field and let $k \geq 2$. We define $D_{r,s,k}$ to be the ideal in $\mathbb{F}[\{X_{ij} : 1 \leq i \leq r, 1 \leq j \leq s\}]$ generated by all $k \times k$ minors of M. Pick any total order $\geq$ on the set of variables such that $X_{ij} \geq X_{kl}$ whenever $i \leq k$ and $j \leq l$ and extend this order to a total order of all monomials using lexicographic order, arranging the variables in each monomial in decreasing order. For example, $X_{11}X_{21}^2 = X_{11}X_{21}X_{21} \geq X_{11}X_{21}X_{22}$, because the monomials coincide on the first two positions, and the variable on the third position in the first monomial is X_{21}, which is greater than the variable X_{22} on the third position in the second monomial.

For any given element p in $D_{r,s,k}$, the *leading monomial* in p is the largest monomial with a nonzero coefficient in p. The *initial ideal* $I_{r,s,k}$ of $D_{r,s,k}$ is the ideal generated by all leading monomials of elements in $D_{r,s,k}$. Herzog and Trung [62] demonstrated that $I_{r,s,k}$ is the ideal generated by all monomials of the form $X_{i_1 j_1} \cdots X_{i_k j_k}$, where $i_1 < \cdots < i_k$ and $j_1 < \cdots < j_k$. In particular, $I_{r,s,k}$ is the Stanley-Reisner ideal of the simplicial complex on the vertex set $[1, r] \times [1, s]$ such that

$$\{\{i_1 j_1, \ldots, i_k j_k\} : i_1 < \cdots < i_k \text{ and } j_1 < \cdots < j_k\}$$

is the family of minimal nonfaces. By a theorem due to Björner [7], this complex is shellable. For $k = 2$, the complex is of importance in our analysis of the homology of complexes of (not) 3-connected graphs; see Section 20.3.

1.1.7 Other Examples

Chessboard complexes, i.e., matching complexes on complete bipartite graphs, first appeared in Garst's analysis of *Tits coset complexes* [53]. Let G be a group and let $G_1, \ldots, G_m$ be subgroups of G. The maximal faces of the Tits coset complex $\Delta(G; G_1, \ldots, G_m)$ are sets of the form

$$\{gG_1, gG_2, \ldots, gG_m\},$$

where $g \in G$. Choosing $G = \mathfrak{S}_n$ and $G_i = \{\sigma : \sigma(i) = i\}$, we obtain a complex isomorphic to the chessboard complex $\mathsf{M}_{m,n}$.

Chessboard complexes also appeared in the analysis of halving hyperplanes in a paper by Živaljević and Vrećica [153]. Given a finite point set $S \subset \mathbb{R}^d$

in general position, such a hyperplane is the affine hull of d of the points and divides the set of remaining points into two sets of equal cardinality. The chessboard complex comes into play in the authors' solution to the problem of finding the maximum number of halving hyperplanes, where the maximum is taken over all point sets S in $\mathbb{R}^d$ with a given cardinality for a fixed d.

In their analysis of certain graph coloring problems, Babson and Kozlov encountered a spherical cell complex [5, §4.2]. The face poset of this complex turns out to be closely related to the complex of bipartite digraphs; such digraphs have the property that each vertex has either zero out-degree or zero in-degree (see Section 15.3).

Björner and Welker [17] discovered an intriguing relationship between the poset of all posets on a fixed vertex set and certain complexes of acyclic digraphs (see Section 15.2) and not strongly connected digraphs (see Section 22.1). There are many other examples of natural interactions between graph complexes and posets. For example, Sundaram [137] examined the lattice of set partitions in which each set has size at most k; this lattice corresponds to the complex of graphs in which each connected component contains at most k vertices (see Section 18.2). In Section 1.1.4, we discussed the correspondence between another lattice and the complex of disconnected k-hypergraphs.

1.1.8 Links to Graph Theory

Not surprisingly, a successful analysis of the topology of a monotone graph property often relies on applying the appropriate graph-theoretical results about the given property. Maybe the most prominent example appears in the work of Linusson, Shareshian, and Welker [95], who applied the Gallai-Edmonds structure theorem (see Lovász and Plummer [97]) in their analysis of complexes of graphs with bounded matching size. See Chapter 24 for more information about their work.

Another example appears in Chapter 26, where we apply results of Hajnal [58] and Berge [6] to analyze the homotopy type of complexes of graphs admitting a small vertex cover; a vertex cover of a graph is a vertex set such that every edge in the graph contains some vertex from the set.

For yet another example, we may mention our work on non-Hamiltonian graphs in Chapter 17. Using the fact that the Petersen graph is cubical, 3-connected, and edge-maximal among non-Hamiltonian graphs, we obtain a nontrivial upper bound on the connectivity degree of the complex of non-Hamiltonian graphs on ten vertices. The discovery of a larger class of graphs with this property would be a big leap forward in the analysis of complexes of non-Hamiltonian graphs.

Alas, we know very little about the existence of results in the other direction, i.e., proofs of nontrivial graph-theoretical theorems based on topological properties of certain graph complexes. On a more general level however, topology surely has proved to be a fundamental tool in graph theory and combinatorics; see Björner [9] for a survey of some of the most celebrated examples and Babson and Kozlov [5] for a very recent application of topology to graph coloring problems.

1.1.9 Complexity Theory and Evasiveness

Several of the monotone graph properties discussed in the book – the properties of having a bounded covering number, not containing a Hamiltonian cycle, and being t-colorable for $t \geq 3$ – correspond in a natural manner to NP-complete problems; see Section 26.8 for some discussion. A potentially interesting area of research would be to examine whether information about the homology of a monotone graph property can tell us anything useful about the corresponding decision problem.

One of the most fundamental classes of simplicial complexes is the class of contractible simplicial complexes. Two important subclasses are the ones consisting of *collapsible* complexes and *nonevasive* complexes, respectively. A complex is collapsible if it is collapsible to a single point. Nonevasive complexes are collapsible complexes with additional structure and are of some importance in the complexity theory of decision trees. See Section 3.4 and Chapter 5 for details.

Karp's famous *evasiveness conjecture* states that there are no nonevasive monotone graph properties except for the void complex and the full simplex. Kahn, Saks, and Sturtevant [78] settled Karp's conjecture in the case that the underlying vertex set is of cardinality a prime power. The proof of Kahn et al. relied on the observation that nonevasive complexes are contractible and hence $\mathbb{Z}$-acyclic. Specifically, they demonstrated that a (nontrivial) monotone graph property cannot be $\mathbb{Z}$-acyclic if the cardinality of the underlying vertex set is a prime power. For other cardinalities exceeding six, it is not known whether there are $\mathbb{Z}$-acyclic monotone graph properties. See Chakrabarti, Khot, and Shi [28] for some recent progress on Karp's conjecture and Yao [149] for the case of monotone bipartite graph properties.

In this context, it is worth mentioning that there are indeed plenty of $\mathbb{Z}$-acyclic and contractible simplicial complexes with a vertex-transitive automorphism group; see the work of Lutz [98]. Moreover, there exists at least one nontrivial $\mathbb{Q}$-acyclic monotone graph property; see Section 5.5. In fact, we would not be surprised if a $\mathbb{Z}$-acyclic or contractible monotone graph property turned out to exist. Note however that it may well be the case that such a complex is not nonevasive or even collapsible and hence does not provide a counterexample to Karp's conjecture.

1.2 Overview

This book is divided into seven parts. The first part consists of this introduction and two chapters listing the basic concepts to be used in the book. In the second part, we present our main proof techniques, most notably discrete Morse theory and decision trees. The third part provides an overview of the complexes to be examined in the last four parts. These complexes appear in parts IV, V, VI, or VII depending on whether they are defined in terms of vertex degree, cycles and crossings, connectivity, or cliques and stable sets, respectively.

Below, we present a rough summary of the book.

Part I – Introduction and Basic Concepts

We give an introduction to the subject and introduce basic and fundamental concepts in graph theory and topology.

Chapter 1 – Introduction and Overview. This is the present chapter and contains an overview of the book.

Chapter 2 – Abstract Graphs and Set Systems. We introduce basic concepts and definitions about graphs, posets, simplicial complexes, and matroids.

Chapter 3 – Simplicial Topology. We provide a summary of the most important concepts and results about the homology and homotopy type of simplicial complexes. We also discuss some important classes of simplicial complexes, including contractible and shellable complexes.

Part II – Tools

We describe the different techniques that we use in later parts to examine the topology and Euler characteristic of different simplicial complexes.

Chapter 4 – Discrete Morse Theory. We present a simplicial variant of Forman's discrete Morse theory [49]. The greater part of this chapter is a revised version of two sections in a published paper [67].

Chapter 5 – Decision Trees. We consider topological aspects of decision trees on simplicial complexes, concentrating on how to use decision trees as a tool in topological combinatorics. This chapter relies heavily on work by Forman [49, 50] and Welker [146] and is a revised version of a published paper [70].

Chapter 6 – Miscellaneous Results. We present miscellaneous results about posets, depth, vertex-decomposability, and enumeration.

Part III – Overview of Results

We give an overview of the complexes to be analyzed in the last four parts. We also present a very sketchy summary of the main theorems about these complexes.

Chapter 7 – Graph Properties. We discuss monotone graph properties.

Chapter 8 – Dihedral Graph Properties. We discuss monotone dihedral graph properties.

Chapter 9 – Digraph Properties. We discuss monotone digraph properties.

Chapter 10 – Main Goals and Proof Techniques. We discuss our main goals – homotopy type, homology, connectivity degree, depth, and Euler characteristic – and give some hints about the most important proof techniques.

Part IV – Vertex Degree

We consider complexes defined in terms of vertex degree.

Chapter 11 – Matchings. We examine complexes of matchings; a matching is a graph in which each vertex has degree at most one. We list some of the main results in the area, focusing on achievements by Bouc [21] and Shareshian and Wachs [122].

Chapter 12 – Graphs of Bounded Degree. We deal with complexes defined in terms of more general bounds on the vertex degree, discussing both graphs without loops and graphs with loops. The latter case is particularly well-studied, and we provide a summary of results from the literature, emphasizing on the work of Dong [36].

Part V – Cycles and Crossings

We consider complexes of graphs and digraphs avoiding cycles or crossings and some variants and combinations thereof.

Chapter 13 – Forests and Matroids. We look at the complex of forests (i.e., cycle-free graphs) in a matroid-theoretic setting and generalize the independence complex of a matroid to a larger class of well-behaved complexes.

Chapter 14 – Bipartite Graphs. We discuss complexes of bipartite graphs, reviewing results due to Chari [31] and Linusson and Shareshian [94]. The greater part of the chapter is devoted to graphs that admit an "unbalanced" bipartition.

Chapter 15 – Directed Variants of Forests and Bipartite Graphs. We examine some directed variants of the properties of being a forest, acyclic, or bipartite. The most important examples are directed forests and acyclic digraphs studied by Kozlov [86] and by Björner and Welker [17] and Hultman [64], respectively.

Chapter 16 – Noncrossing Graphs. This chapter deals with dihedral graph complexes defined in terms of avoiding crossing edges. The most important example is the associahedron, which we discuss in the context of the work of Lee [90]. The remainder of the chapter is devoted to complexes of noncrossing matchings, forests, and bipartite graphs.

Chapter 17 – Non-Hamiltonian Graphs. We consider the complex of non-Hamiltonian graphs. The chapter is a revised version of two sections in a published paper [67].

Part VI – Connectivity

We consider graph and digraph complexes defined in terms of connectivity.

Chapter 18 – Disconnected Graphs. We examine the property of being disconnected. We also examine some related properties defined in terms of restrictions on the size of the connected components. One important example, studied by Sundaram [137], is the complex of graphs in which each component has at most k vertices. Section 18.3 is a revised and extended version of a section in a published paper [70].

Chapter 19 – Not 2-connected Graphs. The emphasis is on the important complex of not 2-connected graphs. We summarize the main results of Babson, Björner, Linusson, Shareshian, and Welker [3], Turchin [139], and Shareshian [118, 117].

Chapter 20 – Not 3-connected Graphs and Beyond. The greater part of this chapter is devoted to the complex of not 3-connected graphs. Not much is known about general k-connected graphs, but we give a short summary of some results at the end of the chapter. The first two sections in the chapter constitute a revised version of two sections in a published paper [67].

Chapter 21 – Dihedral Variants of k-connected Graphs. We present some dihedral variants of connectivity.

Chapter 22 – Directed Variants of Connected Graphs. We proceed with a few directed variants of connectivity, most importantly strongly connected graphs, summarizing results due to Björner and Welker [17] and Hultman [64].

Chapter 23 – Not 2-edge-connected Graphs. We consider edge connectivity, focusing on the complex of not 2-edge-connected graphs. We also review some results about complexes of factor-critical graphs due to Linusson, Shareshian, and Welker [95].

Part VII – Cliques and Stable Sets

We focus on complexes defined in terms of cliques and stable stets.

Chapter 24 – Graphs Avoiding k-matchings. We discuss complexes of graphs that do not contain a matching (i.e., a union of two-cliques) of a specified size, summarizing the results of Linusson, Shareshian, and Welker [95].

Chapter 25 – t-colorable Graphs. We proceed with complexes of graphs admitting a vertex coloring with a specified number of colors (i.e., a partition of the vertex set into a specified number of stable sets). The results of this chapter are due to Linusson and Shareshian [94].

Chapter 26 – Graphs and Hypergraphs with Bounded Covering Number. We elaborate on complexes of graphs admitting a vertex cover of a specified size. For certain parameter choices, we obtain the Alexander dual of the complex of triangle-free graphs. This chapter is a revised version of a published paper [72].

2

Abstract Graphs and Set Systems

We introduce basic concepts and notation related to graphs, posets, abstract simplicial complexes, and matroids. In Section 2.1, we discuss graphs, digraphs, and hypergraphs. Section 2.2 is devoted to posets and lattices. We proceed with abstract simplicial complexes in Section 2.3 and conclude the chapter with some matroid theory in Section 2.4 and a few words about integer partitions in Section 2.5.

Basic Notation

In the below definitions, n and k are nonnegative integers, x is a real number, and S is a finite set.

$|x|$ is the absolute value of x; $|x| = x$ if $x \geq 0$ and $|x| = -x$ if $x < 0$. $\lfloor x \rfloor$ is the largest integer less than or equal to x, whereas $\lceil x \rceil$ is the smallest integer greater than or equal to x. For $n \geq 1$ and every integer a, $a \bmod n$ is the unique integer b in the set $\{0, \ldots, n-1\}$ such that $(b-a)/n$ is a integer.

$\mathbb{Q}$ and $\mathbb{R}$ are the fields of rational and real numbers, respectively, whereas $\mathbb{Z}$ is the ring of integers. Define $\mathbb{Z}_n = \mathbb{Z}/n\mathbb{Z}$; this is the ring of integers modulo n. If n is a prime, then $\mathbb{Z}_n$ is a field.

We denote the empty set by $\emptyset$. 2^S is the family of all subsets of the set S, including S itself and $\emptyset$. $|S|$ is the cardinality (size) of the set S. Let $\binom{S}{k}$ be the family of all subsets T of S satisfying $|T| = k$; clearly, $|\binom{S}{k}| = \binom{|S|}{k}$. $\mathfrak{S}_S$ denotes the symmetric group on the set S, i.e., the group of permutations (bijections) $\pi : S \to S$. Multiplication is defined by $(\pi\pi')(x) = \pi(\pi'(x))$. Finally, we define $[k, n] = \{m \in \mathbb{Z} : k \leq m \leq n\}$ and $[n] = [1, n] = \{1, \ldots, n\}$.

2.1 Graphs, Hypergraphs, and Digraphs

We present standard graph-theoretic concepts.

2.1.1 Graphs

A (simple) *graph* $G = (V, E)$ consists of a finite set V of *vertices* and a family E of subsets of V of size two called *edges*; $E \subseteq \binom{V}{2}$. An edge should be thought of as a line connecting the two vertices in it. A graph being *simple* means that there is at most one edge between any two vertices; E is not a multiset. The edge between the two vertices a and b is denoted as ab or $\{a, b\}$. Two vertices a and b are *adjacent* in G if $ab \in E$.

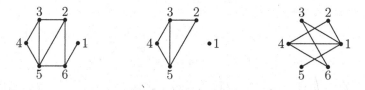

Fig. 2.1. The graph $G = ([6], \{16, 23, 25, 26, 34, 35, 45, 56\})$ to the left, the induced subgraph $G([5])$ in the middle, and the complement of G to the right. We have that $N_G(6) = \{1, 2, 5\}$ and $\deg_G(6) = 3$. The vertex set $\{1, 2, 4\}$ is a stable set in G, whereas $\{2, 3, 5\}$ is a clique. The edge set $\{16, 25, 34\}$ forms a perfect matching contained in G. We obtain a proper 3-coloring $\gamma : [6] \rightarrow [3]$ of G by defining $\gamma^{-1}(1) = \{1, 2, 4\}$, $\gamma^{-1}(2) = \{3, 6\}$, and $\gamma^{-1}(3) = \{5\}$.

For $v \in V$, the *neighborhood* of v is the set $N_G(v) = \{w \in V \setminus \{v\} : vw \in E\}$. The *degree* of v is $\deg_G(v) = |N_G(v)|$. For $W \subseteq V$, define the *induced subgraph* $G(W)$ of G on the vertex set W as the pair $(W, E \cap \binom{W}{2})$.

A *matching* on a vertex set V is a graph $G = (V, E)$ such that each vertex $v \in V$ is adjacent to at most one other vertex in G. A matching is *perfect* if each vertex is adjacent to exactly one other vertex.

A vertex set U in G is *stable* if no edge in G is a subset of U; no two vertices in U are adjacent. Some authors refer to stable sets as *independent*. A vertex set W is a *clique* in G if $\binom{W}{2} \subseteq E$; every two vertices in W are adjacent. The *complement* of a graph $G = (V, E)$ is the graph $\bar{G} = (V, \binom{V}{2} \setminus E)$. Note that U is a clique in G if and only if U is an stable set in $\bar{G}$.

A *t-coloring* of a graph $G = (V, e)$ is a function $\gamma : V \rightarrow [t]$. A coloring γ is *proper* if $\gamma(v) \neq \gamma(w)$ whenever $vw \in E$. A graph $G = (V, E)$ is *t-colorable* if there is a proper t-coloring of G.

For $n \geq 1$, K_n denotes the *complete graph* on n vertices containing all $\binom{n}{2}$ possible edges. 2^{K_n} is the family of all graphs on n vertices.

Some of the concepts introduced in this section are illustrated in Figure 2.1.

2.1.2 Paths, Components and Cycles

A *path* in a graph $G = (V, E)$ is a sequence $(\rho_1, \ldots, \rho_r)$ of not necessarily distinct vertices from V such that $\rho_i \rho_{i+1} \in E$ for $1 \leq i \leq r - 1$. If $\rho_1, \ldots, \rho_r$

are all distinct, then the path is *simple*. We obtain an equivalence relation on V by letting v and w be equivalent if and only if there is a (simple) path $(\rho_1, \ldots, \rho_r)$ in G with $\rho_1 = v$ and $\rho_r = w$. The equivalence classes under this relation are the *connected components* of G. We will typically identify the connected components $W_1, \ldots, W_k$ with the corresponding induced subgraphs $G(W_1), \ldots, G(W_k)$. A graph G is *disconnected* if G contains at least two connected components; otherwise, G is *connected*. A vertex v is *isolated* in G if the connected component containing v equals $\{v\}$.

A vertex set W in a graph $G = (V, E)$ is a *cut set* if $G(V \setminus W)$ is disconnected. If $W = \{w\}$, then w is a *cut point*. For $1 \leq k \leq |V|$, we say that G is *k-connected* if G does not contain any cut set of size less than k. For example, G being 1-connected means that G is connected.

A path $(\rho_1, \ldots, \rho_r)$ in a graph G is a *cycle* if $\rho_r \rho_1 \in G$. The cycle is *simple* if it is simple as a path. G contains a cycle if and only if G contains a simple cycle. A *forest* is a cycle-free graph. A *tree* is a forest such that all non-isolated vertices belong to the same connected component. A *spanning tree* is a tree with one single connected component.

A simple path containing all vertices in a graph is a *Hamiltonian path*; a simple cycle containing all vertices is a *Hamiltonian cycle*. A graph is *Hamiltonian* if it contains at least one Hamiltonian cycle and *non-Hamiltonian* otherwise.

2.1.3 Bipartite Graphs

A graph G is *bipartite* if G is 2-colorable. Equivalently, the vertex set of G is the disjoint union of two stable vertex sets U and W; we say that (U, W) is a *bipartition* of G and refer to U and W as the *blocks* of G. Note that the blocks are not uniquely determined unless G is connected. For $m, n \geq 1$, $K_{m,n}$ denotes the *complete bipartite graph* on a vertex set $U \cup W$ such that $U \cap W = \emptyset$, $|U| = m$, and $|W| = n$; this graph contains all mn possible edges uw such that $u \in U$ and $w \in W$.

2.1.4 Digraphs

A (simple and loopless) *digraph* $D = (V, A)$ consists of a finite set V of *vertices* and a set A of ordered pairs $vw = (v, w)$ such that $v \neq w$; $A \subseteq V \times V \setminus \{(v, v) : v \in V\}$. the elements in A are called *directed edges*. The edge vw is *directed* from v to w; v is the *tail* and w is the *head*. For $n \geq 1$, $\overrightarrow{K_n}$ denotes the *complete digraph* on n vertices containing all $n(n-1)$ possible edges.

2.1.5 Directed Paths and Cycles

A *directed path* in a digraph D is a sequence $(\rho_1, \ldots, \rho_r)$ of not necessarily distinct vertices in V such that $\rho_i \rho_{i+1} \in A$ for $1 \leq i \leq r - 1$. A directed path

$(\rho_1, \ldots, \rho_r)$ is a *directed cycle* if $\rho_r \rho_1 \in A$. In a *simple* directed path or cycle, we require all vertices to be distinct. A *directed Hamiltonian path* is a simple directed path containing all vertices; *directed Hamiltonian cycles* are defined analogously. A digraph D is *acyclic* if D does not contain any directed cycles. A digraph is *Hamiltonian* if it contains at least one directed Hamiltonian cycle and *non-Hamiltonian* otherwise. A digraph D is *strongly connected* if every pair of vertices in D are contained in a directed cycle; the cycle need not be simple.

D is a *directed forest* if D is acyclic and each vertex is the head of at most one edge.[1] A *directed tree* is a directed forest such that all non-isolated vertices belong to the same connected component. A *spanning directed tree* is a directed tree with one single connected component. In such a tree, there is a unique element – the *root* – that is not the head of any edge.

2.1.6 Hypergraphs

A (simple) *hypergraph* $H = (V, E)$ consists of a finite set V of vertices and a family E of nonempty subsets of V called *edges*. We denote the edge $\{a_1, a_2, \ldots, a_r\}$ as $a_1 a_2 \ldots a_r$. For a set S of positive integers, H is an *S-hypergraph* if $|e| \in S$ for every $e \in E$. If H is an *$\{r\}$-hypergraph* (i.e., all edges have the same size r), then H is *r-uniform*. For example, ordinary graphs are 2-uniform. For $W \subseteq V$, define the *induced subhypergraph* $G(W)$ of G with respect to the vertex set W as the pair $(W, E \cap 2^W)$; only edges contained in W remain.

2.1.7 General Terminology

Let $G = (V, E)$ be a graph, hypergraph, or digraph. G is *empty* if $E = \emptyset$ and *nonempty* otherwise. A vertex is *covered* in G if the vertex is contained in some edge in G and *uncovered* otherwise. For hypergraphs, the terms "uncovered" and "isolated" (see Section 2.1.2) are not equivalent. Specifically, if the only edge in G containing a given vertex v is the singleton edge $\{v\}$, then v is isolated but not uncovered. Whenever the underlying vertex set V is fixed, we identify G with its set of edges; $e \in G$ means that $e \in E$. For an edge e, we will write $G - e = (V, E \setminus \{e\})$ and $G + e = (V, E \cup \{e\})$. We let $|G|$ denote the size of the edge set of G. Whenever we refer to "the family of all graphs on n vertices with a given property P", we mean to first fix a vertex set V of size n and then consider the family of all graphs G on the vertex set V with property P.

---

[1] Some authors prefer to define directed forests in terms of the dual requirement that each vertex is the *tail* of at most one edge.

2.2 Posets and Lattices

A finite *partially ordered set* or *poset* is a pair $P = (X, \leq)$, where X is a finite set and $\leq$ is a binary relation on X satisfying the following conditions for all $x, y, z \in X$:

- $x \leq x$.
- If $x \leq y$ and $y \leq x$, then $x = y$.
- If $x \leq y$ and $y \leq z$, then $x \leq z$.

An element x is an *atom* in P if $y \not\leq x$ whenever $y \neq x$. Two elements x and y form a *covering relation* in P if $x < y$ (i.e., $x \leq y$ and $x \neq y$) and no element z in X satisfies $x < z < y$. The *direct product* of two posets $P = (X, \leq_P)$ and $Q = (Y, \leq_Q)$ is the poset $P \times Q = (X \times Y, \leq_{P \times Q})$, where $(x, y) \leq_{P \times Q} (x', y')$ if and only if $x \leq_P x'$ and $y \leq_Q y'$. An (order-preserving) *poset map* between two posets $P = (X, \leq_P)$ and $Q = (Y, \leq_Q)$ is a function $f : X \to Y$ such that $f(x) \leq_Q f(y)$ whenever $x \leq_P y$. We will often write $f : P \to Q$.

A *chain* is a set $\{x_1, \ldots, x_r\}$ of elements in X such that $x_1 < x_2 < \cdots < x_r$. A poset is *ranked* of *rank* d if every maximal chain has size d. The *rank* of an element x is the size of a largest chain in which x is the maximal element. It is often useful to introduce a minimal element $\hat{0}$ with rank 0 and a maximal element $\hat{1}$ of rank $d + 1$. $\hat{0}$ is smaller and $\hat{1}$ is larger than all elements in X.

A finite *lattice* is a finite poset $L = (X, \leq_L)$ such that the following hold:

- There are elements $\hat{0}, \hat{1} \in X$ such that $\hat{0} \leq_L x$ and $x \leq_L \hat{1}$ for all $x \in X$.
- Any two elements $x, y \in X$ have a unique greatest lower bound. Thus there exists an element $z \leq_L x, y$ such that $w \leq_L z$ whenever $w \leq_L x, y$.

These conditions imply that any two elements have a unique least upper bound. The *proper part* of a lattice L, denoted $\overline{L}$, is the poset obtained by removing the top element $\hat{1}$ and the bottom element $\hat{0}$ from L.

A *partition* of a finite set V is a family $\{U_1, \ldots, U_k\}$ of nonempty sets such that V is the disjoint union of $U_1, \ldots, U_k$. The *partition lattice* Π_V is the poset of partitions of V ordered under refinement; $\{W_1, \ldots, W_m\}$ is a refinement of – and hence smaller than – $\{U_1, \ldots, U_k\}$ if every W_i is a subset of some U_j. The partition lattice is indeed a lattice [133]. We write $\Pi_n = \Pi_{[n]}$.

Unless otherwise specified, whenever a family Δ of subsets of a set X is referred to as a poset, the underlying order $\leq$ is given by set inclusion;

$$A \leq B \Longleftrightarrow A \subseteq B.$$

2.3 Abstract Simplicial Complexes

We introduce set-theoretic concepts and notation related to abstract simplicial complexes. Throughout the section, all sets and families are finite. Whenever appropriate, we extend our definitions to arbitrary families of sets rather than restricting to the special case of simplicial complexes.

2.3.1 Basic Definitions

An (abstract) *simplicial complex* Δ on a finite set X is a family of subsets of X closed under deletion of elements. We refer to the singleton sets $\{x\}$ in Δ as 0-*cells* or *vertices*. We do *not* require that $\{x\} \in \Delta$ for all $x \in X$. For the purposes of this book, we adopt the convention that the *void complex* $\emptyset$ is a simplicial complex. For geometric reasons, many authors refer to the complex $\{\emptyset\}$, which is different from the void complex, as the *empty complex*. To avoid any confusion, we will consistently refer to any empty family $\emptyset$ as "void" rather than "empty". Members of a simplicial complex Δ are called *faces*. For a face σ and an element $x \in X$, we write $\sigma - x = \sigma \setminus \{x\}$ and $\sigma + x = \sigma \cup \{x\}$. For two simplicial complexes Δ_1 and Δ_2, $\Delta_1 \cong \Delta_2$ means that Δ_1 and Δ_2 are *combinatorially equivalent*. Assuming that X and Y are the vertex sets of Δ_1 and Δ_2, respectively, this means that there exists a bijection $\varphi : X \to Y$ such that $\sigma \in \Delta_1$ if and only if $\varphi(\sigma) \in \Delta_2$ for each set $\sigma \subseteq X$. Note that the same symbol $\cong$ also denotes homeomorphism between topological spaces. Whenever we use the symbol, it will be clear from context how to interpret it. The simplicial complex *generated* by a family $\mathcal{M}$ of sets is the complex of all subsets of sets in $\mathcal{M}$, including $\mathcal{M}$ itself.

2.3.2 Dimension

Define the *dimension* of a set σ as $|\sigma| - 1$. One sometimes refers to a set of dimension d as a d-*face* or d-*cell*. The dimension of a nonvoid family Δ is the maximum dimension among faces of Δ. The *(reduced) Euler characteristic* of Δ is defined as the integer

$$\tilde{\chi}(\Delta) = \sum_{\sigma \in \Delta} (-1)^{\dim \sigma}.$$

For $d \geq -1$, the d-*skeleton* of a family is the family of all sets of dimension at most d. A family is *pure* if all maximal faces (with respect to inclusion) have the same dimension. For a set σ, we refer to the family 2^σ as the *full simplex* on σ. Writing $d = \dim \sigma = |\sigma| - 1$, we say that 2^σ is a d-*simplex*. Note that the (-1)-simplex contains the empty set and nothing else. We sometimes refer to the 0-simplex as a *point*. We obtain the *boundary* $\partial 2^\sigma$ of the d-simplex 2^σ by removing the maximal face σ.

2.3.3 Collapses

A simplicial complex Δ is obtained from another simplicial complex Δ' via an *elementary collapse* if $\Delta' \setminus \Delta = \{\sigma, \tau\}$ and $\sigma \subsetneq \tau$. This means that τ is the only face in Δ' properly containing σ. If Δ can be obtained from Δ' via a sequence of elementary collapses, then Δ' can be *collapsed* to Δ. If Δ' is void or can be collapsed to a 0-simplex $\{\emptyset, \{v\}\}$, then Δ' is *collapsible (to a point)*.

2.3.4 Joins, Cones, Suspensions, and Wedges

The *join* of two families Δ and Γ (assumed to be defined on disjoint ground sets) is the family $\Delta * \Gamma = \{\sigma \cup \tau : \sigma \in \Delta, \tau \in \Gamma\}$. Note that $\Delta * \emptyset = \emptyset$ and $\Delta * \{\emptyset\} = \Delta$. Let x be a 0-cell not in Δ. The *cone* $\mathsf{Cone}(\Delta) = \mathsf{Cone}_x(\Delta)$ over Δ with *cone point* x is the join of Δ with the 0-simplex $\{\emptyset, \{x\}\}$. Cones over simplicial complexes are collapsible. Let y be another 0-cell not in Δ. The *suspension* $\mathsf{Susp}(\Delta) = \mathsf{Susp}_{x,y}(\Delta)$ of Δ with respect to the pair $\{x, y\}$ is the join of Δ with $\{\emptyset, \{x\}, \{y\}\}$. Note that $\mathsf{Susp}_{x,y}(\Delta) = \mathsf{Cone}_x(\Delta) \cup \mathsf{Cone}_y(\Delta)$. We obtain the (one-point) *wedge* $\Delta \vee \Gamma$ of two simplicial complexes Δ and Γ with respect to 0-cells $x \in \Delta$, $y \in \Gamma$ by taking the disjoint union of Δ and Γ and then identifying x and y.

2.3.5 Alexander Duals

For a simplicial complex Δ on a set X, the *Alexander dual* of Δ with respect to X is the simplicial complex $\Delta_X^* = \{\sigma \subseteq X : X \setminus \sigma \notin \Delta\}$. If there is no reference to any underlying set X, it is assumed that X is the set of 0-cells in Δ.

2.3.6 Links and Deletions

For a family Δ of sets and a set σ, the *link* $\mathrm{lk}_\Delta(\sigma)$ is the family of all $\tau \in \Delta$ such that $\tau \cap \sigma = \emptyset$ and $\tau \cup \sigma \in \Delta$. The *deletion* $\mathrm{del}_\Delta(\sigma)$ is the family of all $\tau \in \Delta$ such that $\tau \cap \sigma = \emptyset$. We define the *face-deletion* $\mathrm{fdel}_\Delta(\sigma)$ as the family of all $\tau \in \Delta$ such that $\sigma \not\subseteq \tau$. The link, deletion, and face-deletion of a simplicial complex are all simplicial complexes.

2.3.7 Lifted Complexes

For the purposes of this book, a family Σ of sets is a *lifted complex* over a set σ if Σ is of the form $\Delta * \{\sigma\}$, where Δ is a simplicial complex and σ is a finite set disjoint from all sets in Δ. Any simplicial complex is also a lifted complex; σ may be the empty set.

Given a lifted complex Σ and disjoint sets I and E, define

$$\Sigma(I, E) = \{I\} * \mathrm{lk}_{\mathrm{del}_\Sigma(E)}(I) = \{\tau \in \Sigma : I \subseteq \tau, E \cap \tau = \emptyset\}.$$

If Σ is a lifted complex over σ, then $\Sigma(I, E)$ is a lifted complex over $\sigma \cup I$. Note that $\Sigma(\emptyset, E) = \mathrm{del}_\Sigma(E)$.

2.3.8 Order Complexes and Face Posets

The *order complex* $\Delta(P)$ of a poset $P = (X, \leq)$ is the simplicial complex of all chains in P; a set $A \subseteq X$ belongs to $\Delta(P)$ if and only if $a \leq b$ or $b \leq a$ for all

$a, b \in A$. Whenever we say that a poset P has a certain topological property (e.g., a certain homotopy type), we mean that $\Delta(P)$ has the property. The *face poset* $P(\Delta)$ of a simplicial complex Δ is the poset of *nonempty* faces of Δ ordered by inclusion. $\mathrm{sd}(\Delta) = \Delta(P(\Delta))$ is the *(first) barycentric subdivision* of Δ.

2.3.9 Graph, Digraph, and Hypergraph Complexes and Properties

A *graph complex* on a finite vertex set V is a family Σ of simple graphs on the vertex set V such that Σ is closed under deletion of edges; if $G \in \Sigma$ and $e \in G$, then $G - e \in \Sigma$. Identifying $G = (V, E) \in \Sigma$ with the edge set E, we may interpret Σ as a simplicial complex. Analogously, a *digraph complex* on V is a family of simple and loopless digraphs on V closed under deletion of edges, whereas a *hypergraph complex* on V is a family of simple hypergraphs on V, again closed under deletion of edges. The restriction to *simple* graphs, digraphs, and hypergraphs is for the purposes of this book.

For a graph complex Σ on a vertex set V and a graph $G = (V, E)$, define $\Sigma(G)$ as the graph complex consisting of all graphs H in Σ such that H is a subgraph of G. We refer to $\Sigma(G)$ as the *induced (graph) subcomplex* of Σ. We adopt the same terminology for digraph and hypergraph complexes.

We refer to a digraph complex $\hat{\Delta}$ as the *trivial extension* of a graph complex Δ if the following holds:

- A digraph D is a maximal face of $\hat{\Delta}$ if and only if D equals $\{ab, ba : ab \in G\}$ for some maximal face G of Δ.

For example, the property of being a disconnected digraph is the trivial extension of the property of being a disconnected undirected graph.

A *graph property* is a family Σ of simple graphs on a finite vertex set V such that Σ is closed under permutations of the vertex set V; if $\sigma := \{a_1 b_1, \ldots, a_r b_r\} \in \Sigma$ and $\pi \in \mathfrak{S}_V$, then

$$\pi(\sigma) := \{\pi(a_1)\pi(b_1), \ldots, \pi(a_r)\pi(b_r)\} \in \Sigma.$$

We refer to this action as the *natural action* of $\mathfrak{S}_V$ on Δ.

A *digraph property* is a family Σ of simple and loopless digraphs on a finite vertex set V such that Σ is closed under permutations of the vertex set V. Analogously, a *hypergraph property* is a family of hypergraphs, again on a fixed vertex set, that is closed under permutations of the underlying vertex set.

A graph, digraph, or hypergraph property Σ is *monotone* if Σ is closed under deletion of edges. Equivalently, Σ is a simplicial complex.

2.4 Matroids

A finite *matroid* M is a pair (E, F), where E is a finite set and $\mathsf{F} = \mathsf{F}(M) \subseteq 2^E$ is a nonvoid simplicial complex satisfying the following property:

- If $\sigma, \tau \in \mathsf{F}$ and $|\sigma| < |\tau|$, then there is an element $x \in \tau \setminus \sigma$ such that $\sigma + x \in \mathsf{F}$.

$\mathsf{F}(M)$ is the *independence complex* or *matroid complex* of M. The sets in $\mathsf{F}(M)$ are the *independent sets* in M. Note that F is a pure complex; all maximal faces have the same size. Define the *rank* of M as this size. A *basis* is a maximal independent set. A *circuit* is a minimal dependent set, i.e., a minimal nonface of $\mathsf{F}(M)$.

For a subset τ of E, let $M(\tau)$ denote the pair $(\tau, \mathsf{F} \cap 2^\tau)$. This is a matroid, and we refer to it as the *induced submatroid* of M on the set τ. Define the rank $\rho_M(\tau)$ of τ as the rank of the matroid $M(\tau)$. A set τ is a *flat* in M if the rank of $\tau + x$ exceeds the rank of τ for each x in $E \setminus \tau$. If a flat τ has rank $\rho(E) - 1$, then τ is a *cocircuit* in M.

For $e \in E$, $M - e$ is the pair $(E - e, \mathrm{del}_\mathsf{F}(e))$; $M - e$ is the *deletion* of M with respect to e. M/e is the pair $(E - e, \mathrm{lk}_\mathsf{F}(e))$; M/e is the *contraction* of M with respect to e. The rank function of M/e satisfies the identity

$$\rho_{M/e}(\sigma) = \rho_M(\sigma + e) - \rho_M(\{e\}).$$

The *dual* of M is the matroid M^* on the same ground set E with the property that the rank function ρ^* satisfies

$$\rho^*(\sigma) = |\sigma| + \rho(E \setminus \sigma) - \rho(E). \tag{2.1}$$

Equivalently, σ is a basis of M^* if and only if $E \setminus \sigma$ is a basis of M.

We refer the reader to Oxley [105] or Welsh [147] for more information about matroids.

2.4.1 Graphic Matroids

For a graph $G = ([n], E)$, define $M_n(G)$ to be the pair $(E, \mathsf{F}_n(G))$, where $\mathsf{F}_n(G)$ is the complex of forests contained in G. This is well-known to be a matroid, and the rank function is given by $\rho(H) = n - c(H)$, where $c(H)$ is the number of connected components in H. We refer to $M_n(G)$ as the *graphic matroid* on G. Write $M_n = M_n(K_n)$.

Another matroid that we may associate to G is the *(one-step) truncation* of $M_n(G)$ obtained by redefining the rank function as $\rho(H) = \min\{\rho(H), n - 2\} = n - \max\{2, c(H)\}$. The independent sets in this matroid are exactly all disconnected forests in G. One may pursue this construction further, considering the "k-step" truncation with rank function $\rho(H) = n - \max\{k, c(H)\}$, but we will confine ourselves to the one-step construction.

For a digraph D, let $M_n(D)$ be the matroid with the property that a set of edges is independent if and only if there are no multiple edges or cycles in the underlying undirected graph. The former condition means exactly that $\{ij, ji\}$ is *not* independent. We refer to $M_n(D)$ as the *digraphic* matroid on D. Write $M_n^{\rightarrow} = M_n(K_n^{\rightarrow})$.

2.5 Integer Partitions

For a sequence $\lambda = (\lambda_1, \ldots, \lambda_r)$, define $|\lambda| = \sum_{i=1}^{r} \lambda_i$. Say that λ is a *partition* of n if $\lambda_1 \geq \cdots \geq \lambda_r \geq 1$ and $|\lambda| = n$; we write this as $\lambda \vdash n$. By convention, we set λ_i equal to 0 whenever $i > r$. One may interpret λ as the set $\{(i,j) : 1 \leq j \leq \lambda_i\}$ of lattice points, where (i,j) is the lattice point in the ith row and j^{th} column. Write $D_\lambda = \{(i,i) : \lambda_i \geq i\}$; this is the *diagonal* of λ. Points (i,j) such that $i < j$ are *above* the diagonal, whereas points (i,j) such that $i > j$ are *below* the diagonal.

Given two partitions λ and μ of n, we say that λ *dominates* μ if

$$\sum_{i=1}^{k} \lambda_i \geq \sum_{i=1}^{k} \mu_i$$

for all $i \geq 1$. The *conjugate* λ^T of a partition $\lambda = (\lambda_1, \ldots, \lambda_r)$ is the sequence $(\mu_1, \ldots, \mu_{\lambda_1})$ with the property that μ_j is the largest m such that $\lambda_m \geq j$. Equivalently, the length of the jth row in λ^T equals the length of the jth column in λ for each j. λ is *self-conjugate* if $\lambda = \lambda^T$.

3

Simplicial Topology

We present a brief overview of the theory of homology and homotopy for simplicial complexes and quotients of simplicial complexes. We also list some of the most important classes of simplicial complexes such as contractible and shellable complexes.

In Section 3.1, we consider simplicial homology theory, stating the main definitions and presenting the important Mayer-Vietoris exact sequence. In Section 3.2, we proceed with relative homology and present the long exact sequence for pairs of simplicial complexes. We also state the main result about Alexander duality. Section 3.3 provides the basic definitions from simplicial homotopy theory. In Section 3.4, we discuss acyclic, contractible, collapsible, and nonevasive complexes. We will need some results about quotient complexes, most notably the Contractible Subcomplex Lemma; we present these results in Section 3.5. Section 3.6 is devoted to Cohen-Macaulay, constructible, shellable, and vertex-decomposable complexes. We proceed with balls and spheres in Section 3.7 and conclude the chapter in Section 3.8 with a few comments about the well-known Stanley-Reisner correspondence between simplicial complexes and monomial rings and ideals.

3.1 Simplicial Homology

We review the basic concepts of simplicial homology. Simplicial homology is well-known to coincide with the restriction of singular or cellular homology to simplicial complexes; see Munkres [101, §34, §39].

Throughout this section, let $\mathbb{F}$ be a field or $\mathbb{Z}$, the ring of integers.

Chain Groups

Let Δ be a simplicial complex. For $d \geq -1$, let $\tilde{C}_d(\Delta; \mathbb{F})$ be the free $\mathbb{F}$-module with one basis element, denoted as $[s_1] \wedge \cdots \wedge [s_{d+1}]$, for each d-dimensional face $\{s_1, \ldots, s_{d+1}\}$ of Δ. This means that the rank of $\tilde{C}_d(\Delta; \mathbb{F})$ equals the number of faces of Δ of dimension d. By convention, we set $\tilde{C}_d(\Delta; \mathbb{F})$ equal to 0 for $d < -1$ and for $d > \dim \Delta$. For any permutation $\pi \in \mathfrak{S}_{[d+1]}$ and any face $\sigma = \{s_1, \ldots, s_{d+1}\}$, we define

$$[s_{\pi(1)}] \wedge [s_{\pi(2)}] \wedge \cdots \wedge [s_{\pi(d+1)}] = \operatorname{sgn}(\pi) \cdot [s_1] \wedge [s_2] \wedge \cdots \wedge [s_{d+1}]. \quad (3.1)$$

We will find it convenient to write $[\sigma] = [s_1] \wedge [s_2] \wedge \ldots \wedge [s_{d+1}]$, implicitly assuming that we have a fixed linear order on the 0-cells in Δ. Whenever σ and τ are disjoint faces such that $\sigma \cup \tau \in \Delta$, we define $[\sigma] \wedge [\tau]$ in the natural manner. Note that $[\emptyset] \wedge z = z$ for all z.

Boundary Map

The *boundary map* $\partial_d : \tilde{C}_d(\Delta; \mathbb{F}) \to \tilde{C}_{d-1}(\Delta; \mathbb{F})$ is the homomorphism defined by

$$\partial_d([s_1] \wedge \ldots \wedge [s_{d+1}]) = \sum_{i=1}^{d+1} (-1)^{i-1} [s_1] \wedge \ldots \wedge [s_{i-1}] \wedge [s_{i+1}] \wedge \ldots \wedge [s_{d+1}].$$

One easily checks that this definition is consistent with (3.1). Combining all ∂_d, we obtain an operator ∂ on the direct sum $\tilde{C}(\Delta; \mathbb{F})$ of all $\tilde{C}_d(\Delta; \mathbb{F})$. It is well-known and easy to see that $\partial^2 = 0$. This means that the pair $(\tilde{C}(\Delta; \mathbb{F}), \partial)$ forms a (graded) *chain complex*.

Let Δ_1 and Δ_2 be complexes on disjoint sets of 0-cells. Given any elements $c_1 \in \tilde{C}_{d_1}(\Delta_1; \mathbb{F})$ and $c_2 \in \tilde{C}_{d_2}(\Delta_2; \mathbb{F})$, the element $c_1 \wedge c_2 \in \tilde{C}_{d_1+d_2+1}(\Delta_1 * \Delta_2; \mathbb{F})$ satisfies the following identity:

$$\partial(c_1 \wedge c_2) = \partial(c_1) \wedge c_2 + (-1)^{d_1+1} c_1 \wedge \partial(c_2). \quad (3.2)$$

Homology

For the chain complex $(\tilde{C}(\Delta; \mathbb{F}), \partial)$ on the simplicial complex Δ, we refer to elements in $\partial^{-1}(\{0\})$ as *cycles* and elements in $\partial(\tilde{C}(\Delta; \mathbb{F}))$ as *boundaries*. Define the dth *reduced homology group* of Δ with coefficients in $\mathbb{F}$ as the quotient $\mathbb{F}$-module

$$\tilde{H}_d(\Delta; \mathbb{F}) := \partial_d^{-1}(\{0\}) / \partial_{d+1}(\tilde{C}_{d+1}(\Delta; \mathbb{F})) = \ker \partial_d / \operatorname{im} \partial_{d+1}.$$

Defining $\tilde{C}_{-1}(\Delta; \mathbb{F})$ to be zero, we obtain *unreduced* homology groups, denoted $H_d(\Delta; \mathbb{F})$ ("H" instead of "$\tilde{H}$"). We will be mainly concerned with reduced homology.

Just to give a simple example, we note that $\tilde{H}_d(\Delta; \mathbb{F}) = 0$ for all d whenever $\Delta = \operatorname{Cone}_x(\Sigma)$ for some Σ. Namely, we may write any element c in $\tilde{C}(\Delta; \mathbb{F})$ as $c = [x] \wedge c_1 + c_2$, where c_1 and c_2 are elements in $\tilde{C}(\Sigma; \mathbb{F})$. If c is a cycle, then $\partial(c_2) = -c_1$, which implies that $\partial([x] \wedge c_2) = c$; hence every cycle is a boundary.

Theorem 3.1 (see Munkres [101, Th. 25.1]). *For any pair of simplicial complexes Δ and Γ, we have the Mayer-Vietoris long exact sequence*

$$\cdots \longrightarrow \tilde{H}_{d+1}(\Delta;\mathbb{F}) \oplus \tilde{H}_{d+1}(\Gamma;\mathbb{F}) \longrightarrow \tilde{H}_{d+1}(\Delta \cup \Gamma;\mathbb{F})$$

$$\longrightarrow \tilde{H}_d(\Delta \cap \Gamma;\mathbb{F}) \longrightarrow \tilde{H}_d(\Delta;\mathbb{F}) \oplus \tilde{H}_d(\Gamma;\mathbb{F}) \longrightarrow \tilde{H}_d(\Delta \cup \Gamma;\mathbb{F})$$

$$\longrightarrow \tilde{H}_{d-1}(\Delta \cap \Gamma;\mathbb{F}) \longrightarrow \tilde{H}_{d-1}(\Delta;\mathbb{F}) \oplus \tilde{H}_{d-1}(\Gamma;\mathbb{F}) \longrightarrow \cdots$$

□

Corollary 3.2. *Let Δ and Γ be simplicial complexes. Then the wedge $\Delta \vee \Gamma$ with respect to any identified 0-cells $x \in \Delta$ and $y \in \Gamma$ satisfies*

$$\tilde{H}_d(\Delta \vee \Gamma;\mathbb{F}) \cong \tilde{H}_d(\Delta;\mathbb{F}) \oplus \tilde{H}_d(\Gamma;\mathbb{F}).$$

for all $d \geq -1$.

Proof. We have that $\Delta \cap \Gamma = \{\emptyset, x\}$, which implies that $\tilde{H}_d(\Delta \cap \Gamma;\mathbb{F}) = 0$ for all d. By the Mayer-Vietoris sequence (Theorem 3.1), we are done. □

Remark. Throughout this book, whenever we discuss the homology of a simplicial complex, we are referring to the reduced $\mathbb{Z}$-homology unless otherwise specified.

3.2 Relative Homology

Let $\Delta \subset \Gamma$ be two simplicial complexes. We refer to the family $\Gamma \setminus \Delta$ as a *quotient complex* and denote it as Γ/Δ. We define the *relative* chain complex of Γ/Δ in the following manner: Define the dth chain group $\tilde{C}_d(\Gamma/\Delta;\mathbb{F})$ as the quotient group $\tilde{C}_d(\Gamma;\mathbb{F})/\tilde{C}_d(\Delta;\mathbb{F})$. This means that $\tilde{C}_d(\Gamma/\Delta;\mathbb{F})$ is a free $\mathbb{F}$-module with one generator $[\sigma]$ for each face $\sigma \in \Gamma \setminus \Delta$ of dimension d. Since the boundary map on $\tilde{C}_d(\Gamma;\mathbb{F})$ maps elements in $\tilde{C}_d(\Delta;\mathbb{F})$ to elements in $\tilde{C}_{d-1}(\Delta;\mathbb{F})$, this boundary map induces a boundary map $\partial_d : \tilde{C}_d(\Gamma/\Delta;\mathbb{F}) \to \tilde{C}_{d-1}(\Gamma/\Delta;\mathbb{F})$. If Δ is the void complex, then we obtain the ordinary chain complex of Γ.

Define the dth *relative homology group* of Δ with coefficients in $\mathbb{F}$ as the quotient $\mathbb{F}$-module

$$\tilde{H}_d(\Gamma/\Delta;\mathbb{F}) := \partial_d^{-1}(\{0\})/\partial_{d+1}(\tilde{C}_d(\Delta/\Gamma;\mathbb{F})) = \ker \partial_d/\operatorname{im} \partial_{d+1}.$$

It is clear that this definition depends only on $\Gamma \setminus \Delta$. Specifically, we may replace Γ and Δ with any Γ' and Δ' such that $\Gamma' \setminus \Delta' = \Gamma \setminus \Delta$ without affecting the chain complex structure.

Note that the traditional notation is $\tilde{H}_d(\Gamma, \Delta;\mathbb{F})$ rather than the more streamlined $\tilde{H}_d(\Gamma/\Delta;\mathbb{F})$ that we have chosen.

Theorem 3.3 (see Munkres [101, Th. 23.3]). *For any pair of simplicial complexes* $\Delta \subset \Gamma$, *we have the following long exact sequence for the pair* (Γ, Δ):

$$\cdots \longrightarrow \tilde{H}_{d+1}(\Gamma; \mathbb{F}) \longrightarrow \tilde{H}_{d+1}(\Gamma/\Delta; \mathbb{F})$$

$$\xrightarrow{f} \tilde{H}_d(\Delta; \mathbb{F}) \longrightarrow \tilde{H}_d(\Gamma; \mathbb{F}) \longrightarrow \tilde{H}_d(\Gamma/\Delta; \mathbb{F}) \qquad (3.3)$$

$$\xrightarrow{f} \tilde{H}_{d-1}(\Delta; \mathbb{F}) \longrightarrow \tilde{H}_{d-1}(\Gamma; \mathbb{F}) \longrightarrow \qquad \cdots$$

The map f *is induced by the boundary operator* ∂ *in the chain complex of* Γ. *The other maps are defined in the natural manner.* $\square$

A simple observation is that the relative homology of the pair (Γ, Δ) coincides with the simplicial homology of $\Gamma \cup \mathsf{Cone}(\Delta)$; consider the long exact sequence for the pair $(\Gamma \cup \mathsf{Cone}(\Delta), \mathsf{Cone}(\Delta))$ and observe that $\mathsf{Cone}(\Delta)$ has vanishing reduced homology in all dimensions.

Let $\sigma \in \Gamma$ and write $\Delta = \mathrm{fdel}_\Gamma(\sigma)$. It is immediate from the definition that

$$\tilde{H}_d(\Gamma/\Delta; \mathbb{F}) \cong \tilde{H}_{d-|\sigma|}(\mathrm{lk}_\Gamma(\sigma); \mathbb{F}).$$

By Theorem 3.3, we thus have a long exact sequence relating Γ and the link and face deletion of Γ with respect to σ. We will use this fact in Section 5.2.1 when we examine semi-nonevasive and semi-collapsible complexes.

In situations where there is no torsion, the homology of the Alexander dual of a complex is easy to compute via relative homology:

Theorem 3.4. *Let* $\mathbb{F}$ *be a field or* $\mathbb{Z}$ *and let* Δ *be a simplicial complex on a nonempty set* X *with* $\mathbb{F}$-*free homology. Then*

$$\tilde{H}_d(\Delta; \mathbb{F}) \cong \tilde{H}_{|X|-d-3}(\Delta_X^*; \mathbb{F}). \qquad (3.4)$$

Proof. By Theorem 3.3, $\tilde{H}_d(\Delta; \mathbb{F}) \cong \tilde{H}_{d+1}(2^X/\Delta; \mathbb{F})$ for all d. Almost by definition, we have that $\tilde{H}_{d+1}(2^X/\Delta; \mathbb{F}) \cong \tilde{H}^{|X|-d-3}(\Delta_X^*; \mathbb{F})$, where $\tilde{H}^i(\Delta_X^*; \mathbb{F})$ denotes the ith *cohomology group*; see Munkres [101]. Applying duality between homology and cohomology for complexes with free homology (see Munkres [101, Th. 45.8]), we obtain the desired result. $\square$

We cannot drop the condition that the homology be free; see Munkres [101].

3.3 Homotopy Theory

A *pointed space* is a topological space X together with a *base point* $x_0 \in X$. Let X and Y be pointed spaces with base points $x_0 \in X$ and $y_0 \in Y$. A *(pointed) map* from X to Y is a continuous function $f : X \to Y$ such that

$f(x_0) = y_0$. Let I be the interval $[0,1] = \{x \in \mathbb{R} : 0 \le x \le 1\}$. For maps $f, g : X \to Y$, a *homotopy* from f to g is a continuous function $F : I \times X \to Y$ such that $F_t(x_0) := F(t, x_0) = y_0$ for all $t \in I$ and such that $F_0(x) = f(x)$ and $F_1(x) = g(x)$ for all $x \in X$. We say that f and g are *homotopic* if such a homotopy exists.

X and Y are *homotopy equivalent*, denoted $X \simeq Y$, if there exist maps $f : X \to Y$ and $h : Y \to X$ such that $h \circ f : X \to X$ is homotopic to the identity map on X and $f \circ h : Y \to Y$ is homotopic to the identity map on Y. We will sometimes express this as saying that X has the *homotopy type* of Y. The choice of base point makes a difference only if the space is not path-connected. As almost all our spaces turn out to be path-connected, we will suppress the notion of base point from now on.

Lemma 3.5. *Let Y be a topological space and let X be a subspace. Suppose that there is a homotopy $F : I \times Y \to Y$ such that F_0 is the identity, the restriction of F_1 to X is the identity, and $F_1(Y) = X$. Then X and Y are homotopy equivalent.*

Proof. Define $f : Y \to X$ by $f(y) = F_1(y)$ and $g : X \to Y$ by $g(x) = F_0(x) = x$. We obtain that $f \circ g$ is the identity on X and that $g \circ f = F_1$. Since F_1 is homotopic to the identity F_0 on Y, we are done. $\square$

Let Δ be a nonvoid abstract simplicial complex on a set X, say $X = [n]$. By some abuse of notation, we define the *topological realization* of Δ as any topological space homeomorphic to the following space $\|\Delta\|$: Let $\mathbf{e}_1, \dots, \mathbf{e}_n$ be an orthonormal basis for Euclidean space $\mathbb{R}^n$. For a face σ, let $\|\sigma\|$ denote the set

$$\left\{ \sum_{x \in \sigma} \lambda_x \mathbf{e}_x : \sum_{x \in \sigma} \lambda_x = 1, \lambda_x > 0 \text{ for all } x \in \sigma \right\}. \tag{3.5}$$

Define $\|\Delta\|$ as the union $\bigcup_{\sigma \in \Delta} \|\sigma\|$; this is a disjoint union. Note that $\|2^\sigma\| = \bigcup_{\tau \subseteq \sigma} \|\tau\|$; this is the convex hull of the set $\{\mathbf{e}_x : x \in \sigma\}$. Also note that $\|\{x\}\| = \{\mathbf{e}_x\}$. We refer to $\|\Delta\|$ as the *canonical realization* of Δ

Let Δ and Γ be defined on two disjoint vertex sets X and Y. One easily checks that the canonical realization of the join $\Delta * \Gamma$ is the set

$$\{\lambda x + (1 - \lambda)y : x \in \|\Delta\|, y \in \|\Gamma\|, \lambda \in [0,1]\}.$$

The join operation preserves homeomorphisms and homotopies:

Lemma 3.6. *If $\|\Delta_1\| \cong \|\Delta_2\|$ and $\|\Gamma_1\| \cong \|\Gamma_2\|$, then $\|\Delta_1 * \Gamma_1\| \cong \|\Delta_2 * \Gamma_2\|$. If $\|\Delta_1\| \simeq \|\Delta_2\|$ and $\|\Gamma_1\| \simeq \|\Gamma_2\|$, then $\|\Delta_1 * \Gamma_1\| \simeq \|\Delta_2 * \Gamma_2\|$.*

Proof. Given homeomorphisms $f : \|\Delta_1\| \to \|\Delta_2\|$ and $g : \|\Gamma_1\| \to \|\Gamma_2\|$, a homeomorphism $h : \|\Delta_1 * \Gamma_1\| \to \|\Delta_2 * \Gamma_2\|$ is given by $h(\lambda x + (1 - \lambda)y) = \lambda f(x) + (1 - \lambda)g(y)$ for each $x \in \|\Delta_1\|$, $y \in \|\Gamma_1\|$, and $\lambda \in [0,1]$. This is well-defined, because we may extract λx from $\lambda x + (1 - \lambda)y$ by restricting to the coordinates corresponding to the elements in X, and we may extract λ from λx by summing the coordinates of λx.

In the same manner, one easily establishes the statement about homotopy equivalence. □

We say that an abstract simplicial complex Δ is homotopy equivalent to a pointed space X if the topological realization of Δ is homotopy equivalent to X. More generally, whenever we discuss topological properties of an abstract simplicial complex Δ, we are referring to its topological realization.

The void complex $\emptyset$ is by convention homotopy equivalent to a point (i.e., a 0-simplex).

We will frequently use the following well-known facts without reference; see Munkres [101] for details.

- Two simplicial complexes with the same homotopy type have the same homology (the converse is not true in general).
- The homotopy type of a wedge of two simplicial complexes Δ and Γ with respect to given identified 0-cells $x \in \Delta$ and $y \in \Gamma$ does not depend on the choice of x and y as long as each of Δ and Γ is connected.
- Any simplicial complex is homeomorphic to its first barycentric subdivision.

Occasionally, we will need to consider cell complexes. For a vector $\mathbf{x} = (x_1, \ldots, x_n)$, write $\|\mathbf{x}\| = \sqrt{x_1^2 + \ldots + x_n^2}$. The unit n-ball B^n is the set $\{\mathbf{x} = (x_1, \ldots, x_n) : \|\mathbf{x}\| \leq 1\}$ in $\mathbb{R}^n$. The unit $(n-1)$-sphere S^{n-1} is the boundary $\{\mathbf{x} = (x_1, \ldots, x_n) : \|\mathbf{x}\| = 1\}$ of B^n. By convention, B^0 is a point and S^{-1} is the empty set. Int $B^n = B^n \setminus S^{n-1}$ is the unit *open n-ball*. A topological space D is an *open n-cell* if D is homeomorphic to an open n-ball.

A Hausdorff topological space X is a finite *cell complex* if the following conditions are satisfied [101, §38]:

- X is the disjoint union of a finite number of open cells $\{D_i : i \in I\}$.
- For each open cell D_i, there is a continuous map

$$\varphi_i : B^{n_i} \to X$$

  ($n_i = \dim D_i$) such that the restriction of φ_i to Int B^{n_i} defines a homeomorphism to D_i and such that $\varphi_i(S^{n_i-1})$ is contained in the $(n_i - 1)$-skeleton of X (the union of all open cells D_j of dimension at most $n_i - 1$).
- A set C is closed in X if and only if $C \cap \overline{D}_i$ is closed in $\overline{D}_i$ for each cell D_i, where $\overline{D}_i = \varphi_i(B^{n_i})$.

The topological realization of a nonvoid simplicial complex Δ is a cell complex; for every face σ of Δ of dimension $d \geq 0$, the set $\|\sigma\|$ is homeomorphic to an open d-cell and the boundary of $\|\sigma\|$ is contained in the $(d-1)$-skeleton of $\|\Delta\|$. A simplicial complex is a *regular* cell complex, meaning that each map φ_i defines a homeomorphism to its image and $\varphi_i(S^{n_i-1})$ is equal to a union of smaller cells. We refer to Hatcher [59] or Munkres [101] for a more detailed exposition on cell complexes.

Some results in this book about simplicial complexes generalize to larger classes of cell complexes, but we will not state these generalizations unless we really need them.

We obtain a *wedge* of topological spaces $Y_1, \ldots, Y_r$ by taking the disjoint union of the spaces, choosing points $y_i \in Y_i$, and identifying the points $y_1, \ldots, y_r$. We may interpret a wedge X of spheres as a cell complex; the identified point y is a 0-cell and the space $X \setminus \{y\}$ is a disconnected space in which each component is a cell in X. Many simplicial complexes in this book are homotopy equivalent to such wedges of spheres.

3.4 Contractible Complexes and Their Relatives

We define the classes of acyclic, contractible, collapsible, and nonevasive complexes. In this book, we are particularly interested in the latter two classes, which we will generalize in Chapter 5.

3.4.1 Acyclic and k-acyclic Complexes

Let $\mathbb{F}$ be a field or $\mathbb{Z}$. A simplicial complex Δ is *acyclic* over $\mathbb{F}$ or $\mathbb{F}$-*acyclic* if Δ has no reduced homology over $\mathbb{F}$. By the universal coefficient theorem [59, Th. 3A.3], a complex Δ is $\mathbb{Z}$-acyclic if and only if Δ is $\mathbb{F}$-acyclic for each field $\mathbb{F}$. However, for any field $\mathbb{F}$, there exist $\mathbb{F}$-acyclic complexes that are not $\mathbb{Z}$-acyclic. For example, any triangulation of the real projective plane (e.g., the one in Figure 5.3 in Section 5.2.1) is $\mathbb{F}$-acyclic whenever $\mathbb{F}$ is a field of odd or zero characteristic but not $\mathbb{Z}_2$-acyclic or $\mathbb{Z}$-acyclic.

A complex Δ is k-*acyclic* over $\mathbb{F}$ if the homology group $\tilde{H}_d(\Delta; \mathbb{F})$ vanishes for $d \leq k$. If a complex Δ is k-acyclic over $\mathbb{Z}$, then Δ is k-acyclic over $\mathbb{F}$ for every field, but the converse is again false for $k \geq 1$.

Proposition 3.7. *Let* $d_1, d_2 \geq 0$. *If* Δ *is* $(d_1 - 1)$-*acyclic over* $\mathbb{F}$ *and* Γ *is* $(d_2 - 1)$-*acyclic over* $\mathbb{F}$, *then* $\Delta * \Gamma$ *is* $(d_1 + d_2)$-*acyclic over* $\mathbb{F}$.

Proof. Throughout this proof, c_i and $\hat{c}_i$ denote elements in $\tilde{C}_i(\Delta; \mathbb{F})$ and c'_j denotes an element in $\tilde{C}_j(\Gamma; \mathbb{F})$. Let $a \leq d_1 + d_2$ and let z be a nonzero cycle in $\tilde{C}_a(\Delta * \Gamma; \mathbb{F})$. We can write

$$z = \sum_{i=r}^{s} c_i \wedge c'_{a-i-1} \tag{3.6}$$

for some $r \leq s$, where the first term and the last term are both nonzero. It is clear that c_r and c'_{a-s-1} are cycles. Since $a \leq d_1 + d_2$ and $s \geq r$, we cannot simultaneously have that $r \geq d_1$ and $a - s - 1 \geq d_2$. By symmetry, we may assume that $r \leq d_1 - 1$; hence there is an element $\hat{c}_{r+1}$ such that $\partial(\hat{c}_{r+1}) = c_r$. Consider the element $\hat{z} = \partial(\hat{c}_{r+1} \wedge c'_{a-r-1}) = c_r \wedge c'_{a-r-1} \pm \hat{c}_{r+1} \wedge \partial(c'_{a-r-1})$. If $r = s$, then $\hat{z} = z$; hence z is a boundary. Otherwise, $z - \hat{z}$ is a sum as in (3.6) but from $r + 1$ to s. By induction on $s - r$, $z - \hat{z}$ is a boundary, which concludes the proof. $\square$

3.4.2 Contractible and k-connected Complexes

A simplicial complex Δ is *contractible* if Δ is homotopy equivalent to a single point. A contractible complex Δ is acyclic over $\mathbb{Z}$, but the converse is not necessarily true unless Δ is simply connected; the famous Poincaré homology 3-sphere [106] is one example. For $k \geq 0$, a topological space X is k-*connected* if the following holds for all $d \in [0, k]$:

- Every continuous map $f : S^d \to X$ has a continuous extension $g : B^{d+1} \to X$.

By convention, X is (-1)-connected if and only if X is nonempty. Note that X is 0-connected if and only if X is path-connected. One typically refers to 1-connected complexes as *simply connected*. The *connectivity degree* of X is the largest integer k such that X is k-connected ($+\infty$ if X is k-connected for all k). Increasing the connectivity degree by one, we obtain the *shifted* connectivity degree; this value is the smallest integer k such that X is not k-connected. In many situations, the shifted connectivity degree coincides with the smallest integer d such that the homology in dimension d is nonvanishing:

Theorem 3.8 (see Hatcher [59, Th. 4.32]). *For $k \geq 1$, a simplicial complex Δ is k-connected if and only if Δ is k-acyclic over $\mathbb{Z}$ and simply connected. Δ is contractible if and only if Δ is acyclic over $\mathbb{Z}$ and simply connected. For $k \in \{-1, 0\}$, a complex Δ is k-connected if and only if Δ is k-acyclic.* $\square$

Corollary 3.9. *For $k \geq 0$, if Δ_1 and Δ_2 are k-connected and $\Delta_1 \cap \Delta_2$ is $(k-1)$-connected, then $\Delta_1 \cup \Delta_2$ is k-connected.*

Proof. The corollary is clear if $k = 0$. Assume that $k \geq 1$. By the Mayer-Vietoris exact sequence (Theorem 3.1), $\Delta_1 \cup \Delta_2$ has no homology below dimension k. Now, Δ_1 and Δ_2 are simply connected, whereas $\Delta_1 \cap \Delta_2$ is path-connected. As a consequence, $\Delta_1 \cup \Delta_2$ is simply connected by the van Kampen theorem (see Hatcher [59, Th. 1.20]). Thus we are done by Theorem 3.8. $\square$

Corollary 3.10. *If Δ is a k-connected subcomplex of Γ and the dimension of each face of $\Gamma \setminus \Delta$ is at least $k + 1$, then Γ is k-connected.*

Proof. We are done if $\Delta = \Gamma$. Otherwise, let σ be a maximal face of $\Gamma \setminus \Delta$; by assumption, $\dim \sigma > k$. By induction, $\Gamma \setminus \{\sigma\}$ is k-connected. Now, 2^σ is k-connected, whereas $\partial 2^\sigma$ is $(k-1)$-connected. Since $\Gamma = (\Gamma \setminus \{\sigma\}) \cup 2^\sigma$ and $\partial 2^\sigma = (\Gamma \setminus \{\sigma\}) \cap 2^\sigma$, Corollary 3.9 yields that Γ is k-connected. $\square$

Theorem 3.11. *If Δ is connected and $\dim \Gamma \geq 0$ (i.e., Γ is (-1)-connected), then $\Delta * \Gamma$ is simply connected.*

Proof. If $\Gamma = \{\emptyset, \{x\}\}$, then $\Delta * \Gamma$ is a cone and hence simply connected. Otherwise, let x be a 0-cell in Γ and write $\Gamma_1 = \text{del}_\Gamma(x)$ and $\Gamma_2 = \text{Cone}_x(\text{lk}_\Gamma(x))$. It is clear that $\Gamma = \Gamma_1 \cup \Gamma_2$ and that $\Delta * (\Gamma_1 \cap \Gamma_2)$ is connected. By induction, $\Delta * \Gamma_1$ and $\Delta * \Gamma_2$ are simply connected; each of Γ_1 and Γ_2 is (-1)-connected. By Corollary 3.9, it follows that $\Delta * \Gamma$ is simply connected. $\square$

Corollary 3.12. *Let $d_1, d_2 \geq 0$. If Δ is $(d_1 - 1)$-acyclic over $\mathbb{Z}$ and Γ is $(d_2 - 1)$-acyclic over $\mathbb{Z}$, then $\Delta * \Gamma$ is $(d_1 + d_2)$-connected.*

Proof. The corollary is clearly true for $d_1 = d_2 = 0$. Assume that $d_1 + d_2 \geq 1$. By Proposition 3.7, $\Delta * \Gamma$ is $(d_1 + d_2)$-acyclic. Theorem 3.11 yields that $\Delta * \Gamma$ is simply connected; hence we are done by Theorem 3.8. $\square$

Theorem 3.13. *Let $d \geq 0$. If Δ is $(d - 1)$-connected and $\dim \Delta \leq d$, then Δ is homotopy equivalent to a wedge of spheres of dimension d.*

Proof. The theorem is trivial for $d = 0$. If $d = 1$, then Δ is a connected graph, which is homotopy equivalent to a wedge of circles. Otherwise, Δ is simply connected and $(d-1)$-acyclic by Theorem 3.8. As a consequence, all homology of Δ is concentrated in dimension d. Since $\dim \Delta \leq d$, this homology must be torsion-free and hence of the form $\mathbb{Z}^r$ for some $r \geq 0$. By the homology version of Whitehead's theorem (see Hatcher [59, Prop. 4C.1]), this implies that Δ is homotopy equivalent to a cell complex consisting of r cells of dimension d and one 0-cell, hence a wedge of r spheres of dimension d. $\square$

3.4.3 Collapsible Complexes

Recall that a complex is collapsible if the complex is void or can be collapsed to a point $\{\emptyset, \{v\}\}$. Collapsible complexes are contractible, but not all contractible complexes are collapsible; the dunce hat [150] is one example. One may characterize collapsible complexes in the following manner:

Definition 3.14. We define the class of *collapsible* simplicial complexes recursively as follows:

(i) The void complex $\emptyset$ and any 0-simplex $\{\emptyset, \{v\}\}$ are collapsible.
(ii) If Δ contains a nonempty face σ such that the face-deletion $\text{fdel}_\Delta(\sigma)$ and the link $\text{lk}_\Delta(\sigma)$ are collapsible, then Δ is collapsible.

We discuss further properties of collapsible complexes in Section 5.4.

3.4.4 Nonevasive Complexes

To obtain the class of nonevasive complexes, we use Definition 3.14 with the restriction that the face σ in (ii) must be a 0-cell:

Definition 3.15. We define the class of *nonevasive* simplicial complexes recursively as follows:

(i) The void complex $\emptyset$ and any 0-simplex $\{\emptyset, \{v\}\}$ are nonevasive.
(ii) If Δ contains a 0-cell x such that $\mathrm{del}_\Delta(x)$ and $\mathrm{lk}_\Delta(x)$ are nonevasive, then Δ is nonevasive.

For example, cones are nonevasive. A complex is *evasive* if it is not nonevasive. We explain this terminology in Chapter 5. As Kahn, Saks, and Sturtevant [78] observed, nonevasive complexes are collapsible. The converse is not true in general; in Proposition 5.13, we present a counterexample due to Björner. We discuss further properties of nonevasive complexes in Section 5.4.

3.5 Quotient Complexes

Let $X \subseteq Y$ be two topological spaces such that X is nonempty. Let p be an isolated point not in Y. One defines the *quotient space* Y/X as the set $(Y \setminus X) \cup \{p\}$ equipped with the topology induced by the map $\alpha : Y \to (Y \setminus X) \cup \{p\}$ defined by

$$\alpha(x) = \begin{cases} x \text{ if } x \in Y \setminus X; \\ p \text{ if } x \in X. \end{cases}$$

That is, M is open in Y/X if and only if $\alpha^{-1}(M)$ is open in Y. By convention, we set $Y/\emptyset$ equal to the union of Y and a discrete point $\{p\}$ not in Y.

Let $\Delta \subseteq \Gamma$ be simplicial complexes such that Δ is nonvoid. We define the *topological realization* of the quotient complex Γ/Δ to be any space homeomorphic to $\|\Gamma\|/\|\Delta\|$. One easily checks directly from the definition that $\|\Gamma\|/\|\Delta\|$ is homeomorphic to $\|\Gamma'\|/\|\Delta'\|$ whenever $\Gamma \setminus \Delta = \Gamma' \setminus \Delta'$. Note that $\|\Gamma\|/\|\{\emptyset\}\| = \|\Gamma\| \cup \{p\}$, because $\|\{\emptyset\}\| = \emptyset$.

One may interpret the space $\|\Gamma\|/\|\Delta\|$ as a cell complex. Specifically, we have one cell $\|\sigma\|$ for each face $\sigma \in \Gamma \setminus \Delta$ plus one additional 0-cell $\{p\}$ corresponding to Δ. The boundary of $\|2^\sigma\|$ is the same as in $\|\Gamma\|$ except that we identify all points in $\|\partial 2^\sigma\| \cap \|\Delta\|$ with p.

Whenever we talk about the topology of Γ/Δ, we are referring to the space $\|\Gamma\|/\|\Delta\|$. The following lemma is known as the *Contractible Subcomplex Lemma*.

Lemma 3.16 (see Hatcher [59, Prop. 0.17]). *Let Γ and Δ be simplicial complexes such that Δ is a contractible subcomplex of Γ. Then Γ/Δ and Γ are homotopy equivalent.*

Proof. Let E be the set of 0-cells in Γ. It is well-known and easy to prove that there is a homeomorphism from $\|\Gamma\|$ to $\|\mathrm{sd}(\Gamma)\|$ such that restriction to $\|\Delta\|$ is a homeomorphism to $\|\mathrm{sd}(\Delta)\|$. In particular, $\|\Gamma\|/\|\Delta\|$ and $\|\mathrm{sd}(\Gamma)\|/\|\mathrm{sd}(\Delta)\|$ are homeomorphic. As a consequence, we may assume without loss of generality that Δ coincides with the induced subcomplex of Γ on some set $E_0 \subset E$ of 0-cells; thus $\Delta = \Gamma \cap 2^{E_0}$. Let $\Delta^\perp$ be the induced subcomplex on the set $E \setminus E_0$.

Let $F : I \times \|\Delta\| \to \|\Delta\|$ be a homotopy from the identity to a constant function; $F_0(x) = x$ and $F_1(x) = y$ for some $y \in \|\Delta\|$. Each element x in $\|\Gamma\|$ has a unique representation $x = \lambda q + (1 - \lambda)r$, where $q \in \|\Delta\|$, $r \in \|\Delta^\perp\|$, and $\lambda \in I$. Define $G : I \times \|\Gamma\| \to \|\Gamma\|$ to be the homotopy given by

$$G_t(\lambda q + (1 - \lambda)r) = \begin{cases} (1+t)\lambda q + (1 - (1+t)\lambda)r & \text{if } \lambda \le 1/(1+t); \\ F_{(t+1)\lambda - 1}(q) & \text{if } \lambda \ge 1/(1+t) \end{cases}$$

for all relevant $q \in \|\Delta\|$ and $r \in \|\Delta^\perp\|$. This is indeed a homotopy, because $\lambda = 1/(t+1)$ yields the same result q in both formulas.

We have that G_t induces a homotopy $\tilde{G}_t : \|\Gamma\|/\|\Delta\| \to \|\Gamma\|/\|\Delta\|$. Moreover, G_1 induces a continuous map $\hat{G}_1 : \|\Gamma\|/\|\Delta\| \to \|\Gamma\|$; G_1 maps the entirety of $\|\Delta\|$ to $F_1(\|\Delta\|) = \{y\}$. Define $\alpha : \|\Gamma\| \to \|\Gamma\|/\|\Delta\|$ to be the projection map. Now, $\hat{G}_1 \circ \alpha = G_1$, which is homotopic to the identity G_0. Moreover, $\alpha \circ \hat{G}_1 = \tilde{G}_1$, which is homotopic to the identity $\tilde{G}_0$; hence we are done. $\square$

Corollary 3.17. *Let Γ be a simplicial complex and let Δ be a subcomplex of Γ. Let Σ be a complex on a 0-cell set disjoint from the 0-cell set of Γ such that $\Sigma * \Delta$ is contractible. Then Γ/Δ is homotopy equivalent to $\Gamma \cup (\Sigma * \Delta)$.*

Proof. Since $\Gamma/\Delta = (\Gamma \cup (\Sigma * \Delta))/(\Sigma * \Delta)$, the Contractible Subcomplex Lemma 3.16 implies the desired result. $\square$

Lemma 3.18. *Let Γ be a contractible simplicial complex and let Δ be a subcomplex of Γ. Then Γ/Δ is homotopy equivalent to the suspension of Δ. Moreover, $\tilde{H}_{i+1}(\Gamma/\Delta; \mathbb{F}) = \tilde{H}_i(\Delta; \mathbb{F})$ for $i \ge -1$.*

Proof. Let x and y be two 0-cells not in Γ. By Corollary 3.17, Γ/Δ is homotopy equivalent to $\Gamma \cup \mathrm{Cone}_x(\Delta)$. Since Γ is contractible, the Contractible Subcomplex Lemma 3.16 implies that $\Gamma \cup \mathrm{Cone}_x(\Delta)$ is homotopy equivalent to $(\Gamma \cup \mathrm{Cone}_x(\Delta))/\Gamma$ and hence to $\mathrm{Cone}_x(\Delta)/\Delta$. Another application of Corollary 3.17 yields that $\mathrm{Cone}_x(\Delta)/\Delta$ is homotopy equivalent to $\mathrm{Cone}_x(\Delta) \cup \mathrm{Cone}_y(\Delta) = \mathrm{Susp}_{x,y}(\Delta)$, which concludes the proof. For the last claim, use the long exact sequence in Theorem 3.3. $\square$

In this context, it might be worth stating the following fact about suspensions.

Lemma 3.19 (Björner and Welker [16, Lemma 2.5]). *If $\Delta \simeq \bigvee_{i \in I} S^{d_i}$, then $\mathrm{Susp}(\Delta) \simeq \bigvee_{i \in I} S^{d_i + 1}$. $\square$*

The converse is not true. For example, the suspension of a d-dimensional complex with homology only in top dimension $d \geq 1$ is simply connected by Theorem 3.11 and hence homotopy equivalent to a wedge of spheres by Theorem 3.13.

The following lemma is a special case of a much more general result about homotopy type being preserved under join.

Lemma 3.20. *Let Δ be a simplicial complex and let Γ and Γ' be quotient complexes. If $\Gamma \simeq \Gamma'$, then $\Delta * \Gamma \simeq \Delta * \Gamma'$.*

Proof. Write $\Gamma = \Gamma_1 / \Gamma_0$, where Γ_1 and Γ_0 are simplicial complexes. By Corollary 3.17,

$$\Delta * \Gamma = \frac{\Delta * \Gamma_1}{\Delta * \Gamma_0} \simeq (\Delta * \Gamma_1) \cup \mathsf{Cone}_x(\Delta * \Gamma_0) = \Delta * (\Gamma_1 \cup \mathsf{Cone}_x(\Gamma_0)).$$

By Corollary 3.17 and Lemma 3.6, we obtain that the homotopy type of $\Delta * \Gamma$ is uniquely determined by the homotopy type of each of Δ and Γ, which concludes the proof. $\square$

3.6 Shellable Complexes and Their Relatives

We define the classes of Cohen-Macaulay, constructible, shellable, and vertex-decomposable complexes along with nonpure versions. In Section 3.6.5, we present some basic topological results about these complexes. For our purposes, the class of vertex-decomposable complexes is by far the most important. See Section 6.3 for some specific results related to this class.

3.6.1 Cohen-Macaulay Complexes

Definition 3.21. Let Δ be a pure simplicial complex. Δ is *homotopically Cohen-Macaulay* (CM) if $\mathrm{lk}_\Delta(\sigma)$ is $(\dim \mathrm{lk}_\Delta(\sigma) - 1)$-connected for each σ in Δ. Let $\mathbb{F}$ be a field or $\mathbb{Z}$. Δ is *Cohen-Macaulay over $\mathbb{F}$* (denoted as $CM/\mathbb{F}$) if $\mathrm{lk}_\Delta(\sigma)$ is $(\dim \mathrm{lk}_\Delta(\sigma) - 1)$-acyclic for each σ in Δ.

By Theorem 3.13, $\mathrm{lk}_\Delta(\sigma)$ is $(\dim \mathrm{lk}_\Delta(\sigma) - 1)$-connected if and only if $\mathrm{lk}_\Delta(\sigma)$ is homotopy equivalent to a wedge of spheres of dimension $\dim \mathrm{lk}_\Delta(\sigma)$. See Section 3.8 for the ring-theoretic motivation of Definition 3.21.

Define the *homotopical depth* of a complex Δ as the largest integer k such that the k-skeleton of Δ is homotopically CM. Define the *depth over $\mathbb{F}$* of Δ as the largest integer k such that the k-skeleton of Δ is $CM/\mathbb{F}$. Equivalently, the depth over $\mathbb{F}$ equals

$$\min\{m : \tilde{H}_{m-|\sigma|}(\mathrm{lk}_\Delta(\sigma), \mathbb{F}) \neq 0 \text{ for some } \sigma \in \Delta\}.$$

This is closely related to the ring-theoretic concept of depth; see Section 3.8.

Define the *pure d-skeleton* $\Delta^{[d]}$ of Δ as the subcomplex of Δ generated by all d-dimensional faces of Δ. Stanley [132] extended the concept of Cohen-Macaulayness to nonpure complexes:

Definition 3.22. A simplicial complex Δ is *sequentially homotopy-CM* if the pure d-skeleton $\Delta^{[d]}$ is homotopically CM for every $d \geq 0$. Let $\mathbb{F}$ be a field or $\mathbb{Z}$. Δ is *sequentially CM/$\mathbb{F}$* if the pure d-skeleton $\Delta^{[d]}$ is $CM/\mathbb{F}$ for every $d \geq 0$.

3.6.2 Constructible Complexes

Definition 3.23. We define the class of *constructible* simplicial complexes recursively as follows:

(i) Every simplex (including $\emptyset$ and $\{\emptyset\}$) is constructible.
(ii) If Δ_1 and Δ_2 are constructible complexes of dimension d and $\Delta_1 \cap \Delta_2$ is a constructible complex of dimension $d-1$, then $\Delta_1 \cup \Delta_2$ is constructible.

Hochster [63] introduced constructible complexes.

Let us extend the concept of constructibility to nonpure complexes. For a simplicial complex Δ, define $\mathcal{F}(\Delta)$ to be the family of maximal faces of Δ.

Definition 3.24. We define the class of *semipure constructible* simplicial complexes recursively as follows:

(i) Every simplex (including $\emptyset$ and $\{\emptyset\}$) is semipure constructible.
(ii) Suppose that Δ_1, Δ_2, and $\Gamma = \Delta_1 \cap \Delta_2$ are semipure constructible complexes such that the following conditions are satisfied:
  (a) $\mathcal{F}(\Delta_1 \cup \Delta_2)$ is the disjoint union of $\mathcal{F}(\Delta_1)$ and $\mathcal{F}(\Delta_2)$.
  (b) Every member of $\mathcal{F}(\Gamma)$ is a maximal face of either $\Delta_1 \setminus \mathcal{F}(\Delta_1)$ or $\Delta_2 \setminus \mathcal{F}(\Delta_2)$ (possibly of both).
  Then $\Delta_1 \cup \Delta_2$ is semipure constructible.

Expressed in terms of pure skeletons, condition (b) is equivalent to the identity

$$\Delta_1{}^{[d]} \cap \Delta_2{}^{[d]} = \Gamma^{[d-1]} \cup \Gamma^{[d]}$$

for each d.

One may refer to semipure constructible complexes that are not pure as *nonpure constructible*.

3.6.3 Shellable Complexes

The class of shellable complexes is arguably the most well-studied class of Cohen-Macaulay complexes. Indeed, proving shellability is in many situations the most efficient way of establishing Cohen-Macaulayness; see Björner and Wachs [12] for just one of many examples. In this respect, this book constitutes an exception, as our proofs of Cohen-Macaulayness typically go via vertex-decomposability (see Section 3.6.4). Therefore, we confine ourselves to presenting basic definitions and refer the interested reader to Björner [9] for more information and further references.

Definition 3.25. We define the class of *shellable* simplicial complexes recursively as follows:

(i) Every simplex (including $\emptyset$ and $\{\emptyset\}$) is shellable.
(ii) If Δ is pure and contains a nonempty face σ – a *shedding face* – such that $\mathrm{fdel}_\Delta(\sigma)$ and $\mathrm{lk}_\Delta(\sigma)$ are shellable, then Δ is also shellable.

This way of defining shellability is easily seen to be equivalent to more conventional approaches; see Provan and Billera [108].

We say that a lifted complex $\Sigma = \Delta * \{\rho\}$ (see Section 2.3.7) is shellable if the underlying simplicial complex Δ is shellable. A sequence $(\Sigma_1, \ldots, \Sigma_r = \Sigma)$ is a *shelling* of Σ if each Σ_i is a pure lifted complex over ρ of dimension $\dim \Sigma$ such that $\Sigma_i \setminus \Sigma_{i-1}$ has a unique maximal face τ_i and a unique minimal face σ_i for each $i \in [1, r]$; $\Sigma_0 = \emptyset$. The ith *shelling pair* is the pair (σ_i, τ_i). Note that $\sigma_1 = \rho$.

Let Δ be a lifted complex over ρ. The recursive procedure in (ii) of Definition 3.25 gives rise to a shelling of Δ. Specifically, assume inductively that we have shellings $(\Delta_1, \ldots, \Delta_q)$ of $\mathrm{fdel}_\Delta(\sigma)$ and $(\Delta_{q+1}, \ldots, \Delta_r)$ of $\Delta(\sigma, \emptyset)$ (we lift the link $\mathrm{lk}_\Delta(\sigma)$). If $\Delta = \mathrm{lk}_\Delta(\sigma) * 2^\sigma$, then $(\Delta_{q+1} * 2^\sigma, \ldots, \Delta_r * 2^\sigma)$ is a shelling of Δ. Otherwise, $(\Delta_1, \ldots, \Delta_q, \Delta_{q+1}, \ldots, \Delta_r)$ is a shelling of Δ; the unique minimal element in $\Delta_{q+1} \setminus \Delta_q$ is σ.

Conversely, it is easy to prove that Δ admits a shelling if and only if Δ is shellable in terms of Definition 3.25; use the last minimal face σ_r as the first shedding face.

Björner and Wachs [13] extended shellability to complexes that are not necessarily pure:

Definition 3.26. We define the class of *semipure shellable* simplicial complexes recursively as follows:

(i) Every simplex (including $\emptyset$ and $\{\emptyset\}$) is semipure shellable.
(ii) If Δ contains a nonempty face σ – a *shedding face* – such that $\mathrm{fdel}_\Delta(\sigma)$ and $\mathrm{lk}_\Delta(\sigma)$ are semipure shellable and such that every maximal face of $\mathrm{fdel}_\Delta(\sigma)$ is a maximal face of Δ, then Δ is also semipure shellable.

To see that Definition 3.26 is equivalent to the original definition [13, Def. 2.1], adapt the proof of Björner and Wachs [14, Th. 11.3]. One may refer to semipure shellable complexes that are not pure as *nonpure shellable*.

3.6.4 Vertex-Decomposable Complexes

Definition 3.27. We define the class of *vertex-decomposable* (VD) simplicial complexes recursively as follows:

(i) Every simplex (including $\emptyset$ and $\{\emptyset\}$) is VD.
(ii) If Δ is pure and contains a 0-cell x – a *shedding vertex* – such that $\mathrm{del}_\Delta(x)$ and $\mathrm{lk}_\Delta(x)$ are VD, then Δ is also VD.

Vertex-decomposable complexes were introduced by Provan and Billera [108].

As for shellability, one readily extends vertex-decomposability to lifted complexes. An alternative approach to vertex-decomposability including lifted complexes is as follows:

Definition 3.28. We define the class of VD lifted complexes recursively as follows.

(i) Every simplex (including $\emptyset$ and $\{\emptyset\}$) is VD.
(ii) If Δ contains a 0-cell v such that $\Delta(v, \emptyset)$ and $\Delta(\emptyset, v)$ are VD of the same dimension, then Δ is also VD.
(iii) If Δ is a cone over a VD complex Δ', then Δ is also VD.
(iv) If $\Delta = \Sigma * \{\sigma\}$ and Σ is VD, then Δ is also VD.

The restriction of this definition to simplicial complexes is easily seen to be equivalent to the original Definition 3.27.

Just as for shellability, Björner and Wachs [13] extended the concept of vertex-decomposability to nonpure complexes:

Definition 3.29. We define the class of *semipure VD* simplicial complexes recursively as follows:

(i) Every simplex (including $\emptyset$ and $\{\emptyset\}$) is semipure VD.
(ii) If Δ contains a 0-cell x – a *shedding vertex* – such that $\mathrm{del}_\Delta(x)$ and $\mathrm{lk}_\Delta(x)$ are semipure VD and such that every maximal face of $\mathrm{del}_\Delta(x)$ is a maximal face of Δ, then Δ is also semipure VD.

One may refer to semipure VD complexes that are not pure as *nonpure VD*.

3.6.5 Topological Properties and Relations Between Different Classes

Theorem 3.30. *The properties of being CM, sequentially CM, constructible, semipure constructible, shellable, semipure shellable, VD, and semipure VD are all closed under taking link and join.*

Proof. The properties being closed under taking link is straightforward to prove in all cases. The $CM/\mathbb{F}$ and sequentially $CM/\mathbb{F}$ properties are closed under taking join, because the join of a $(d_1 - 1)$-acyclic complex and a $(d_2 - 1)$-acyclic complex is $(d_1 + d_2)$-acyclic by Proposition 3.7. By Corollary 3.12, the homotopically CM and sequentially homotopy-CM properties are also closed under taking join.

For the remaining properties, use a simple induction argument, decomposing with respect to the first complex in the join and keeping the other complex fixed. In each case, the base case is a join of two simplices, which is again a simplex.

For example, suppose that $\Delta = \Delta_1 \cup \Delta_2$ and Δ' are semipure constructible and that Δ_1 and Δ_2 satisfy the properties in (ii) in Definition 3.24. By induction, $\Delta_1 * \Delta'$, $\Delta_2 * \Delta'$, and $(\Delta_1 \cap \Delta_2) * \Delta'$ are all semipure constructible. Moreover, one easily checks that conditions (a) and (b) in Definition 3.24 hold for $\Delta_1 * \Delta'$ and $\Delta_2 * \Delta'$. It hence follows that $(\Delta_1 \cup \Delta_2) * \Delta'$ is semipure constructible as desired.

The treatment of the other properties is equally straightforward. $\square$

Proposition 3.31. *The following properties hold for any pure simplicial complex Δ:*

(i) *Δ is sequentially CM if and only if Δ is CM in the sense of Definition 3.21.*

(ii) *Δ is semipure constructible if and only if Δ is constructible in the sense of Definition 3.23.*

(iii) *Δ is semipure shellable if and only if Δ is shellable in the sense of Definition 3.25.*

(iv) *Δ is semipure VD if and only if Δ is VD in the sense of Definition 3.27.*

Proof. (i) This is obvious.

(ii) Constructible complexes are easily seen to be semipure constructible. The other direction is obvious if Δ satisfies (i) in Definition 3.24. Suppose that $\Delta = \Delta_1 \cup \Delta_2$ and that the conditions in (ii) are satisfied. Condition (a) yields that Δ_1 and Δ_2 are pure, whereas condition (b) yields that their intersection is pure of dimension one less than Δ_1 and Δ_2. By induction, all these three complexes are constructible, which implies that the same is true for Δ.

(iii) It is clear that Δ is shellable if Δ is semipure shellable. The other direction is immediate, except that we need to check the case that we have a shedding face σ in Definition 3.25 such that the dimension of $\mathrm{fdel}_\Delta(\sigma)$ is strictly smaller than that of Δ. This implies that $\Delta = 2^\sigma * \mathrm{lk}_\Delta(\sigma)$. Namely, Δ is generated by the maximal faces of the lifted complex $\Delta(\sigma, \emptyset)$. By Theorem 3.30, semipure shellability is preserved under join, which implies that Δ is semipure shellable as desired.

(iv) This is proved in exactly the same manner as (iii). $\square$

Lemma 3.32. *Let Δ_1 and Δ_2 be homotopically CM of dimension d such that the $(d-1)$-skeleton of $\Delta_1 \cap \Delta_2$ is homotopically CM. Then $\Delta_1 \cup \Delta_2$ is homotopically CM.*

Proof. Note that $\mathrm{lk}_{\Delta_1 \cup \Delta_2}(\sigma) = \mathrm{lk}_{\Delta_1}(\sigma) \cup \mathrm{lk}_{\Delta_2}(\sigma)$ and analogously for the intersection. As a consequence, the lemma follows immediately from Corollary 3.9. $\square$

Theorem 3.33. *We have the following implications:*

$$VD \implies Shellable \implies Constructible \implies Homotopically\ CM.$$

The analogous implications hold for the semipure variants.

Proof. By Proposition 3.31, it suffices to prove that the implications hold for the semipure variants.

Semipure VD $\Longrightarrow$ *Semipure shellable.* Trivial.

Semipure shellable $\Longrightarrow$ *Semipure constructible.* The theorem is obvious if Δ is a simplex. Otherwise, let σ be a shedding face as in (ii) in Definition 3.26. Write $\Delta_1 = \text{fdel}_\Delta(\sigma)$ and $\Delta_2 = 2^\sigma * \text{lk}_\Delta(\sigma)$; it is clear that $\Delta = \Delta_1 \cup \Delta_2$. Now, Δ_1 is semipure shellable by assumption. Moreover, by Theorem 3.30, $\Delta_2 = 2^\sigma * \text{lk}_\Delta(\sigma)$ is semipure shellable. Finally, the intersection $\Delta_1 \cap \Delta_2$ equals $\partial 2^\sigma * \text{lk}_\Delta(\sigma)$. The boundary of a simplex is well-known to be shellable and hence semipure shellable; hence $\Delta_1 \cap \Delta_2$ is semipure shellable. Induction yields that all these complexes are semipure constructible.

It remains to prove that conditions (a) and (b) in Definition 3.24 are satisfied. Condition (a) follows immediately from Definition 3.26. To prove condition (b), consider a maximal face $(\sigma - x) \cup \tau$ of $\partial 2^\sigma * \text{lk}_\Delta(\sigma)$; $\tau \in \mathcal{F}(\text{lk}_\Delta(\sigma))$. One easily checks that the only maximal face of $\Delta_2 = 2^\sigma * \text{lk}_\Delta(\sigma)$ containing this face is $\sigma \cup \tau$; thus we are done.

Semipure constructible $\Longrightarrow$ *Sequentially homotopy-CM.* This is obvious if Δ satisfies (i) in Definition 3.24. Suppose that $\Delta = \Delta_1 \cup \Delta_2$ and that the conditions in (ii) are satisfied. We need to prove that the pure d-skeleton $\Delta^{[d]}$ is *CM* for every $d \geq 0$.

By induction, each of $\Delta_1{}^{[d]}$ and $\Delta_2{}^{[d]}$ is *CM*. Moreover, by construction, $\Delta_1{}^{[d]} \cap \Delta_2{}^{[d]} = \Gamma^{[d-1]} \cup \Gamma^{[d]}$, where $\Gamma = \Delta_1 \cup \Delta_2$. By induction, each of $\Gamma^{[d-1]}$ and $\Gamma^{[d]}$ is *CM*. Their intersection equals the $(d-1)$-skeleton of $\Gamma^{[d]}$ and is hence *CM*. As a consequence, Lemma 3.32 yields that the $(d-1)$-skeleton of $\Delta_1{}^{[d]} \cap \Delta_2{}^{[d]}$ is *CM*. Another application of the same lemma yields that $\Delta^{[d]} = \Delta_1{}^{[d]} \cup \Delta_2{}^{[d]}$ is *CM*, which concludes the proof. $\square$

All implications in Theorem 3.33 turn out to be strict; see Proposition 5.13, Proposition 5.14, and Björner [9, §11.10]. We also have the following implications for any field $\mathbb{F}$:

$$\text{homotopically } CM \Longrightarrow CM/\mathbb{Z} \Longrightarrow CM/\mathbb{F}.$$

These implications are valid also for sequentially *CM* complexes. Again, all implications are strict.

Corollary 3.34. *Let Δ be a pure complex of dimension d. If Δ is VD, shellable, constructible, or homotopically CM, then the homotopical depth of Δ is equal to d.* $\square$

By the following result due to Björner, Wachs, and Welker, sequentially *CM* complexes have a nice topological structure.

Theorem 3.35 (Björner et al. [15]). *If Δ is sequentially homotopy-CM, then Δ is homotopy equivalent to a wedge of spheres. Moreover, there is no sphere of dimension d in this wedge unless there are maximal faces of Δ of dimension d.* $\square$

Björner and Wachs [13, Th. 4.1] earlier proved Theorem 3.35 in the special case that Δ is semipure shellable.

Let us prove the homology version of Theorem 3.35.

Proposition 3.36 (Wachs [144]). *Assume that Δ is sequentially $CM/\mathbb{Z}$. Then the homology of Δ is torsion-free. Moreover, there is no homology in dimension d unless there are maximal faces of Δ of dimension d. Indeed, the homomorphism $\iota^* : \tilde{H}_d(\Delta \setminus \mathcal{F}(\Delta); \mathbb{Z}) \to \tilde{H}_d(\Delta; \mathbb{Z})$ induced by the inclusion map is zero for all d.*

Proof. We can write ι^* as a composition

$$\tilde{H}_d(\Delta \setminus \mathcal{F}(\Delta); \mathbb{Z}) \to \tilde{H}_d(\Delta^{[d+1]}; \mathbb{Z}) \to \tilde{H}_d(\Delta; \mathbb{Z})$$

of maps induced by inclusion maps; every d-dimensional face of $\Delta \setminus \mathcal{F}(\Delta)$ is contained in a $(d+1)$-dimensional face of Δ. Since $\Delta^{[d+1]}$ is $CM/\mathbb{Z}$, we have that $\tilde{H}_d(\Delta^{[d+1]}; \mathbb{Z}) = 0$ and hence that $\iota^* = 0$ as desired. By the long exact sequence for the pair $(\Delta, \Delta \setminus \mathcal{F}(\Delta))$, it follows that the homology of Δ is torsion-free. $\square$

3.7 Balls and Spheres

We summarize some well-known properties of balls and spheres. Such objects do not play a central part in this book, but they are of some interest in the analysis of the homology of certain complexes. Specifically, in some situations, one may interpret the homology in terms of fundamental cycles of spheres; see Chapters 19 and 20 for the most notable examples.

A simplicial complex Δ is a *d-ball* if there is a homeomorphism $\|\Delta\| \to B^d$. Δ is a *d-sphere* if there is a homeomorphism $\|\Delta\| \to S^d$. For example, the full d-simplex is a d-ball, whereas the boundary of a d-simplex is a $(d-1)$-sphere. We define the *boundary* $\partial\Delta$ of a d-ball Δ as the pure $(d-1)$-dimensional complex with the property that σ is a maximal face of $\partial\Delta$ if and only if σ is contained in exactly one maximal face of Δ.

In general, balls and spheres are not as nice as one may suspect. For example, they are not necessarily homotopically CM; see Björner [9, §11.10]. However, the balls and spheres to be considered in this book are indeed nice.

If a simplicial complex Δ is homeomorphic to a d-dimensional sphere, then $\tilde{H}_d(\Delta; \mathbb{Z})$ is generated by a cycle z that is unique up to sign. We refer to z as the *fundamental cycle* of Δ.

A set $P \subset \mathbb{R}^n$ is a *convex polytope* if P is the convex hull of a finite set P_0 of points in convex position; no point p in P_0 is in the convex hull of $P_0 \setminus \{p\}$. P is homeomorphic to a d-ball for some $d \leq n$; hence P has a well-defined boundary ∂P, which is homeomorphic to a $(d-1)$-sphere. A simplicial complex Δ with 0-cell set Δ_0 is the *boundary complex* of P if there is a bijection $\varphi : \Delta_0 \to P_0$ such that a point $\sum_{i \in \Delta_0} \lambda_i \varphi(i)$ ($\sum_i \lambda_i = 1$, $\lambda_i \geq 0$)

belongs to ∂P if and only if $\{i : \lambda_i > 0\}$ is a face of Δ. We refer to such a complex Δ as *polytopal*. If Δ is the boundary complex of a polytope, then Δ is a shellable sphere; see Bruggesser and Mani [24].

3.8 Stanley-Reisner Rings

We conclude this chapter with a few words about Stanley-Reisner rings. We will only occasionally discuss such rings and include this section merely for completeness.

Let Δ be a simplicial complex on the set Y and let $\mathbb{F}$ be a field. Let R denote the commutative polynomial ring $\mathbb{F}[x_i : i \in Y]$. For each set $\sigma \subseteq Y$, identify σ with the monomial $x_\sigma = \prod_{i \in \sigma} x_i$. Define $I(\Delta)$ to be the monomial ideal in R generated by the (minimal) nonfaces of Δ. This means that a monomial m belongs to $I(\Delta)$ if and only if x_σ divides m for some $\sigma \notin \Delta$. The *Stanley-Reisner ring* or *face ring* of Δ is $R(\Delta) = R/I(\Delta)$.

Let Δ be a $(d-1)$-dimensional simplicial complex with depth $p-1$ over $\mathbb{F}$. Some well-known properties of the Stanley-Reisner ring $R(\Delta)$ are as follows; see a textbook on commutative algebra [26, 43] for ring-theoretic definitions.

- The Krull dimension of $R(\Delta)$ is equal to d.
- The depth of $R(\Delta)$ is equal to p.
- $R(\Delta)$ is a Cohen-Macaulay ring if and only if Δ is $CM/\mathbb{F}$.
- The multiplicity of $R(\Delta)$ is equal to the number of d-dimensional faces of Δ.

We refer the reader to Section 6.2, Reisner [113], and Stanley [132] for details and further references.

Note that this correspondence provides some motivation for examining the topology of simplicial complexes; the problems of counting faces of maximal dimension and examining Cohen-Macaulay properties of a simplicial complex indeed have very natural ring-theoretic counterparts. Moreover, under favorable circumstances, it is possible to combine the theory of Stanley and Reisner with Gröbner basis theory to obtain important information about rings that are not necessarily Stanley-Reisner rings. One example is the work on determinantal ideals by Herzog and Trung [62]; see Section 1.1.6 for more information.

Part II

Tools

4

Discrete Morse Theory

[1] Robin Forman's discrete Morse theory [49] is instrumental in the analysis of many of the complexes in this book. Though ostensibly simple, this theory has proven to be a powerful tool for analyzing the topology of a wide range of different complexes [4, 32, 36, 60, 94, 95, 118]. For an interesting application of discrete Morse theory to geometry, see Crowley [34].

In this book, we confine ourselves to discrete Morse theory for simplicial complexes. For the more general theory, see Forman's original paper [49] and the combinatorial interpretations due to Chari [32] and Shareshian [118].

To facilitate analysis of the homology of certain complexes, we develop a very elementary algebraic version of discrete Morse theory; see Sköldberg [126] and Jöllenbeck and Welker [66] for a more thorough treatment. A discrete Morse matching on a simplicial complex gives rise to a discrete gradient vector; see Forman [49]. One may use this gradient vector to determine a basis for the resulting Morse chain complex in terms of the canonical basis for the original chain complex. Our main object is to provide shortcuts for deriving this basis (or parts thereof) without having to examine the explicit gradient vector. Our methods are mainly useful in situations where it is possible to guess the basis.

In Section 4.2, we provide the combinatorial background. The main topological results appear in Section 4.3, whereas Section 4.4 contains our simplified algebraic Morse theory.

4.1 Informal Discussion

Before proceeding, let us give an informal overview of the basic ideas of discrete Morse theory.

Let Δ be a simplicial complex. One may view discrete Morse theory as a generalization of the theory of simplicial collapses (see Section 2.3.3). Let

---

[1] This chapter is a revised and extended version of Sections 3 and 6 in a paper [67] published in *Journal of Combinatorial Theory, Series A*.

$\{\sigma, \tau\}$ be a pair of faces of Δ such that $\sigma \subset \tau$ and $\dim \sigma = \dim \tau - 1$. For this pair to induce an ordinary elementary collapse, recall that we require τ to be a maximal face and the only face of Δ containing σ; we refer to this as saying that $\{\sigma, \tau\}$ is *free* in Δ. Dropping this condition, we obtain what we refer to as a *generalized* elementary collapse; this terminology is only for the purposes of this section.

Geometrically, we obtain a generalized elementary collapse with respect to σ and τ by first removing the open set $\|\tau\|$ from $\|\Delta\|$ and then identifying $\|2^\sigma\|$ with $\|\partial 2^\tau \setminus \sigma\|$ such that the common boundary of the two balls $\|2^\sigma\|$ and $\|\partial 2^\tau \setminus \sigma\|$ remains the same. Specifically, assuming that τ is a regular simplex (all edge lengths are the same), we identify a point in $\|\partial 2^\tau \setminus \sigma\|$ with its orthogonal projection on $\|2^\sigma\|$. See Figure 4.1 for an example. This identification corresponds to "contracting" the whole of $\|2^\tau\|$ onto $\|\partial 2^\tau \setminus \sigma\|$. In particular, a generalized elementary collapse does not affect the homotopy type. Note that the resulting complex is not necessarily a simplicial complex but rather a more general cell complex.

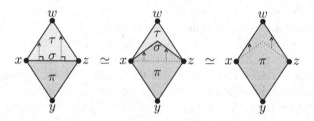

Fig. 4.1. Generalized elementary collapse with respect to σ and τ in the simplicial complex with maximal faces $\tau = wxz$ and $\pi = xyz$. Note that the resulting complex is not simplicial; the new face π has four boundary edges.

Just as we may combine many elementary collapses to form a larger collapse without affecting the homotopy type, we may combine many generalized elementary collapses to form a larger generalized collapse, again without affecting the homotopy type. This is indeed the main principle of discrete Morse theory.

We thus have a number of pairs $\{\sigma_1, \tau_1\}, \ldots, \{\sigma_r, \tau_r\}$ to be collapsed, in this order. One may view the set of all such pairs as a matching on Δ. Accordingly, we refer to faces contained in some pair as *matched* and other faces as *unmatched*.

Let Δ_{i-1} be the resulting cell complex after the first $i - 1$ generalized elementary collapses. Recall that the rule for the pairs to form a sequence of ordinary elementary collapses is that each new pair $\{\sigma_i, \tau_i\}$ should be free in Δ_{i-1}. For generalized elementary collapses, we adopt the same rule, except that we restrict our attention to the family of *matched* faces. More precisely, we do not require $\{\sigma_i, \tau_i\}$ to be free in Δ_{i-1}, but τ_i must be the only *matched* face of $\varDelta_{i-1}$ containing σ_i. Equivalently, for each i, we should have that σ_i is not contained in $\tau_{i+1}, \ldots, \tau_r$. We refer to a matching on $\varDelta$ admitting an ordering with this property as *acyclic*. This terminology is for reasons that we explain in more detail in Section 4.2.

The main theorem of discrete Morse theory states that an acyclic matching induces a homotopy equivalence between $\varDelta$ and the cell complex $\varDelta_r$ resulting from the corresponding generalized collapse. As an immediate corollary, $\varDelta$ is homotopy equivalent to a cell complex with just as many cells of dimension d as there are unmatched faces of $\varDelta$ of dimension d. In particular, we obtain upper bounds on the ranks of the homology groups of $\varDelta$.

If we want more exact results about the homotopy type and homology of $\varDelta$, we typically have to examine the acyclic matching – and the corresponding generalized elementary collapses – in much greater detail. However, in certain special cases, information about the number of cells in each dimension is sufficient to unambiguously determine $\varDelta_r$ and hence the homotopy type of $\varDelta$. For example, this is the case if all unmatched cells are of the same dimension d; in this case, $\varDelta_r$ is a wedge of d-dimensional spheres. In this book, this situation is not at all uncommon.

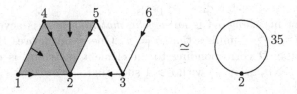

Fig. 4.2. $\varDelta$ is homotopy equivalent to a circle. Arrows indicate faces to be matched.

Example. Consider the simplicial complex $\varDelta$ on the set $\{1, 2, 3, 4, 5, 6\}$ consisting of all subsets of $124, 245, 23, 35$, and 36; 124 denotes the set $\{1, 2, 4\}$ and so on. In Figure 4.2, a geometric realization of $\varDelta$ is illustrated. The figure illustrates an acyclic matching on $\varDelta$ with the property that 35 is the only critical face; an arrow from the face σ to the face τ means that σ and τ are matched. We may also match 2 and the empty set. However, since the empty set has no obvious geometric interpretation, it is convenient to consider 2 as a critical point in the geometric realization. Note that $\varDelta$ is homotopy equivalent to a cell complex consisting of a 1-cell corresponding to 35 and a 0-cell corresponding to 2.

4.2 Acyclic Matchings

We start our exposition by examining acyclic matchings on families of sets. This section is purely combinatorial and does not contain any topology.

Let X be a set and let Δ be a finite family of finite subsets of X. A *matching* on Δ is a family $\mathcal{M}$ of pairs $\{\sigma, \tau\}$ with $\sigma, \tau \in \Delta$ such that no set is contained in more than one pair in $\mathcal{M}$. A set σ in Δ is *critical* or *unmatched* with respect to $\mathcal{M}$ if σ is not contained in any pair in $\mathcal{M}$.

We say that a matching $\mathcal{M}$ on Δ is an *element matching* if every pair in $\mathcal{M}$ is of the form $\{\sigma - x, \sigma + x\}$ for some $x \in X$ and $\sigma \subseteq X$. All matchings considered in this chapter are element matchings.

Consider an element matching $\mathcal{M}$ on a family Δ. Let $D = D(\Delta, \mathcal{M})$ be the digraph with vertex set Δ and with a directed edge from σ to τ if and only if either of the following holds:

1. $\{\sigma, \tau\} \in \mathcal{M}$ and $\tau = \sigma + x$ for some $x \notin \sigma$.
2. $\{\sigma, \tau\} \notin \mathcal{M}$ and $\sigma = \tau + x$ for some $x \notin \tau$.

Thus every edge in D corresponds to an edge in the Hasse diagram of Δ ordered by set inclusion; edges corresponding to pairs of matched sets are directed from the smaller set to the larger set, whereas the other edges are directed the other way around. We write $\sigma \longrightarrow \tau$ if there is a directed path from σ to τ in D. For families $\mathcal{V}$ and $\mathcal{W}$, we write $\mathcal{V} \longrightarrow \mathcal{W}$ if there are $V \in \mathcal{V}$ and $W \in \mathcal{W}$ such that $V \longrightarrow W$. The symbol $\nrightarrow$ is used to denote the non-existence of such a directed path.

An element matching $\mathcal{M}$ is an *acyclic matching* if D is acyclic, that is, $\sigma \longrightarrow \tau$ and $\tau \longrightarrow \sigma$ implies that $\sigma = \tau$. One easily proves that any cycle in a digraph D corresponding to an element matching is of the form $(\sigma_0, \tau_0, \sigma_1, \tau_1, \ldots, \sigma_{r-1}, \tau_{r-1})$ with $r > 1$ such that

$$\sigma_i, \sigma_{(i+1) \bmod r} \subset \tau_i \text{ and } \{\sigma_i, \tau_i\} \in \mathcal{M}; \tag{4.1}$$

for details, see Shareshian [118]. See Figure 4.3 for an illustration. The following two lemmas provide simple but useful methods for dividing a family of sets into smaller subfamilies such that any set of acyclic matchings on the separate subfamilies can be combined to form one single acyclic matching on the original family.

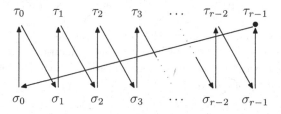

Fig. 4.3. Cycle in a digraph corresponding to a non-acyclic matching.

Lemma 4.1. *Let $\Delta \subseteq 2^X$ and $x \in X$. Define*

$$M_x = \{\{\sigma - x, \sigma + x\} : \sigma - x, \sigma + x \in \Delta\};$$
$$\Delta_x = \{\sigma : \sigma - x, \sigma + x \in \Delta\}.$$

Let $\mathcal{M}'$ be an acyclic matching on $\Delta' := \Delta \setminus \Delta_x$. Then $\mathcal{M} := \mathcal{M}_x \cup \mathcal{M}'$ is an acyclic matching on Δ.

Proof. Assume that $(\sigma_0, \tau_0, \ldots, \sigma_{r-1}, \tau_{r-1})$ is a cycle in $D(\Delta, \mathcal{M})$ satisfying (4.1). Since $\mathcal{M}'$ is an acyclic matching on Δ', there must be some pair $\{\sigma_i, \tau_i\}$ that is included in $\mathcal{M}_x$ rather than in $\mathcal{M}'$; by construction, we then have that $\tau_i = \sigma_i + x$. For simplicity, assume that $i = 0$. Since τ_{r-1} is not matched with σ_0, we must have that $x \notin \tau_{r-1}$. This means that there is some $j \in [1, r-1]$ such that $x \in \tau_{j-1}$ and $x \notin \tau_j$. However, this implies that $\tau_{j-1} = \sigma_j + x$, which is a contradiction, because we would then have that $(\sigma_j, \tau_{j-1}) \in \mathcal{M}_x$ by construction. $\square$

Lemma 4.2. *(Cluster Lemma) Let $\Delta \subseteq 2^X$ and let $f : \Delta \to Q$ be a poset map, where Q is an arbitrary poset. For $q \in Q$, let $\mathcal{M}_q$ be an acyclic matching on $f^{-1}(q)$. Let*

$$\mathcal{M} = \bigcup_{q \in Q} \mathcal{M}_q.$$

Then $\mathcal{M}$ is an acyclic matching on Δ.

Remark. Hersh [60] discovered Lemma 4.2 independently of our work. Björner (personal communication) suggested the formulation in terms of poset maps.

Proof. Let $(\sigma_0, \tau_0, \ldots, \sigma_{r-1}, \tau_{r-1})$ be a cycle in $D(\Delta, \mathcal{M})$ satisfying (4.1). Let $q_0, \ldots, q_{r-1}$ be such that $\sigma_k, \tau_k \in f^{-1}(q_k)$ for $0 \le k \le r-1$. Since $\sigma_{(k+1) \bmod r} \subset \tau_k$, it is clear that $q_{(k+1) \bmod r} = f(\sigma_{(k+1) \bmod r}) \le f(\sigma_k) = q_k$. Via a simple induction argument, this implies that $q_{k'} \le q_k$ for any pair k, k'. Swapping k and k', we obtain $q_k \le q_{k'}$, which implies that $q_k = q_{k'}$; Q is a poset. Hence all sets in the cycle are contained in one single family $f^{-1}(q)$, which is a contradiction. $\square$

The following result is an almost trivial special case of Lemma 4.2.

Lemma 4.3. *Let Δ_0 and Δ_1 be disjoint families of subsets of a finite set such that $\tau \not\subset \sigma$ if $\sigma \in \Delta_0$ and $\tau \in \Delta_1$. If $\mathcal{M}_i$ is an acyclic matching on Δ_i for $i = 0, 1$, then $\mathcal{M}_0 \cup \mathcal{M}_1$ is an acyclic matching on $\Delta_0 \cup \Delta_1$.* $\square$

4.3 Simplicial Morse Theory

Throughout this section, Δ is a simplicial complex containing at least one 0-cell and $\mathcal{M}$ is an acyclic matching on $\mathcal{M}$. We may assume without loss of generality that the empty set is contained in some pair in $\mathcal{M}$. Namely, if all

0-faces are matched with larger faces, then there is a cycle in the digraph $D(\Delta, \mathcal{M})$.

Forman's original discrete Morse theory [49] applies to a much larger class of cell complexes than just the class of simplicial complexes. For this reason, we refer to the theory in the present section as *simplicial Morse theory*.

We begin with some special cases and postpone the most general result until the end of the section.

Theorem 4.4 (Forman [49]). *Suppose that Δ_0 is a subcomplex of Δ such that $\Delta_0 \not\longmapsto \Delta \setminus \Delta_0$ and such that all critical faces belong to Δ_0. Then it is possible to collapse Δ to Δ_0. In particular, Δ and Δ_0 are homotopy equivalent. Hence Δ has no homology in dimensions strictly greater than $\dim \Delta_0$.*

Proof. By assumption, the restriction of the acyclic matching to $\Delta \setminus \Delta_0$ is a perfect matching. Namely, if $\tau \in \Delta \setminus \Delta_0$ is matched with $\sigma \in \Delta_0$, then $\sigma \subset \tau$, which implies that $\sigma \longrightarrow \tau$, a contradiction. We use induction over $|\Delta \setminus \Delta_0|$ to prove the lemma. If $\Delta = \Delta_0$, then we are done. Otherwise, let σ be a face of $\Delta \setminus \Delta_0$ such that no edge in the digraph D corresponding to the matching ends in σ; such a face exists by acyclicity of D and by the assumption that $\Delta_0 \not\longmapsto \Delta \setminus \Delta_0$. It is clear that σ is matched with a larger face τ and that σ is not contained in any other face. In particular, we can collapse Δ to the subcomplex $\Delta \setminus \{\sigma, \tau\}$. By induction, we may collapse $\Delta \setminus \{\sigma, \tau\}$ to Δ_0, which concludes the proof. $\square$

Let us give a very simple example to illustrate the technique.

Proposition 4.5. *Let Δ be a simplicial complex and let x be a 0-cell in Δ. Let y be a new 0-cell and define Δ' to be the complex obtained from Δ by adding $\sigma + y$ and $\sigma + x + y$ whenever $\sigma + x \in \Delta$. Then Δ and Δ' are homotopy equivalent. In particular, if Δ is a graph complex and $\hat{\Delta}$ is the trivial extension of Δ (see Section 2.3.9), then Δ and $\hat{\Delta}$ are homotopy equivalent.*

Proof. Note that Δ' is the disjoint union of $\Delta(\emptyset, y) = \Delta$ and $\Delta(y, \emptyset) = \{\{y\}\} * \Delta(x, \emptyset)$. By Lemma 4.3, any acyclic matchings on these two families yield an acyclic matching on Δ'. Now, define a matching on $\Delta(y, \emptyset)$ by pairing $\sigma - x$ with $\sigma + x$. This is clearly a perfect acyclic matching, which yields an acyclic matching on Δ' such that a face is critical if and only if the face belongs to Δ. As a consequence, Theorem 4.4 yields a collapse from Δ' to Δ. By a simple induction argument, the last claim in the proposition follows. $\square$

Corollary 4.6. *If Δ does not contain any critical faces, then Δ is collapsible and hence contractible to a point.* $\square$

Theorem 4.7 (Forman [49]). *If all critical faces of Δ have dimension at least d, then Δ is $(d-1)$-connected.*

Proof. Let Δ_0 be the subcomplex of Δ consisting of all faces of dimension less than d plus all faces of dimension d that are matched with smaller faces. Since the acyclic matching restricts to a perfect matching on Δ_0, Δ_0 is contractible and hence $(d-1)$-connected. Since Δ_0 contains the entire $(d-1)$-skeleton of Δ, we are done by Corollary 3.10. $\square$

Theorem 4.8 (Forman [49]). *If all critical faces of Δ are of the same dimension d, then Δ is homotopy equivalent to a wedge of spheres of dimension d.*

Proof. Let $\mathcal{C}$ be the family of critical faces. Let Δ_0 be as in the proof of Theorem 4.7; Δ_0 contains all faces of dimension less than d plus all faces of dimension d that are matched with smaller faces. Note that the given acyclic matching restricts to a perfect matching on Δ_0 and also on $\Delta \setminus (\Delta_0 \cup \mathcal{C})$. By Theorem 4.4, we can collapse Δ to $\Delta_0 \cup \mathcal{C}$. Moreover, Corollary 4.6 and the Contractible Subcomplex Lemma 3.16 imply that $\Delta_0 \cup \mathcal{C}$ is homotopy equivalent to $(\Delta_0 \cup \mathcal{C})/\Delta_0 = \mathcal{C}$ and hence to a wedge of $|\mathcal{C}|$ spheres of dimension d. $\square$

Before proceeding, let us apply discrete Morse theory to a situation where we already know the homotopy type.

Proposition 4.9. *Let X be a nonempty finite set. Then $\mathrm{sd}(\partial 2^X)$ admits an acyclic matching with one unmatched face of dimension $|X| - 2$. As a consequence, $\mathrm{sd}(\partial 2^X)$ is homotopy equivalent to a sphere of dimension $|X| - 2$.*

Remark. One may view this proposition as a special case of the more general Theorem 5.31. The proposition also follows from the fact that $\mathrm{sd}(\partial 2^X)$ is shellable [12]; apply Proposition 5.11.

Proof. Pick an element $x \in X$. For each $S \subseteq X \setminus \{x\}$, let $\mathcal{G}(S)$ be the family of faces τ of $\mathrm{sd}(\partial 2^X)$ with the property that S is maximal in τ among sets not containing x. Let Q be the poset of all subsets of $X \setminus \{x\}$ ordered by set inclusion. Define $f : P(\mathrm{sd}(\partial 2^X)) \to Q$ by $f^{-1}(S) = \mathcal{G}(S)$. This is clearly a poset map, which implies that the Cluster Lemma 4.2 applies. For $S \neq X \setminus \{x\}$, we obtain a perfect matching on $\mathcal{G}(S)$ by pairing $\sigma - (S \cup \{x\})$ and $\sigma + (S \cup \{x\})$. The remaining family is $\mathcal{G}(X \setminus \{x\})$, which clearly equals $\{X \setminus \{x\}\} * \mathrm{sd}(\partial 2^{X \setminus \{x\}})$. By an induction argument, $\mathrm{sd}(\partial 2^{X \setminus \{x\}})$ admits an acyclic matching with one critical face of dimension $|X| - 3$. Applying the Cluster Lemma 4.2, we obtain an acyclic matching on $\mathrm{sd}(\partial 2^X)$ with desired properties. The last statement is a consequence of Theorem 4.8. $\square$

For an acyclic matching $\mathcal{M}$ on a simplicial complex Δ, let $\mathcal{U}(\Delta, \mathcal{M})$ be the family of critical faces of Δ with respect to $\mathcal{M}$. For a (possibly void) family $\mathcal{V} \subseteq \mathcal{U}(\Delta, \mathcal{M})$, let

$$\Delta_{\mathcal{V}} = \{\sigma \in \Delta : \mathcal{V} \longrightarrow \sigma\} \cup \{\emptyset, \{x\}\}, \tag{4.2}$$

where $\{x\}$ is the 0-face matched with the empty set in $\mathcal{M}$. If $\mathcal{V}$ is nonvoid, then $\Delta_{\mathcal{V}} = \{\sigma \in \Delta : \mathcal{V} \longrightarrow \sigma\}$.

Lemma 4.10. $\Delta_{\mathcal{V}}$ *is a simplicial complex. That is, if* $\{\sigma, \tau\} \in M$ *with* $\sigma \subset \tau$ *and* $\tau \in \Delta_{\mathcal{V}}$, *then* $\sigma \in \Delta_{\mathcal{V}}$. *In particular,*

$$\mathcal{U}(\Delta_{\mathcal{V}}, \mathcal{M}_{\mathcal{V}}) = \Delta_{\mathcal{V}} \cap \mathcal{U}(\Delta, \mathcal{M}),$$

where $\mathcal{M}_{\mathcal{V}}$ *is the restriction of* $\mathcal{M}$ *to* $\Delta_{\mathcal{V}}$.

Proof. Assume the opposite and let σ be a largest face such that $\sigma \notin \Delta_{\mathcal{V}}$ and such that there is an element $y \in X$ with the property that $\sigma + y \in \Delta_{\mathcal{V}}$. Since there is a $V \in \mathcal{V}$ such that $V \longrightarrow \sigma + y$, we have that $\{\sigma, \sigma + y\} \in M$; otherwise $(\sigma + y, \sigma)$ would be an edge in D. In particular, $\sigma + y \notin \mathcal{U}(\Delta, \mathcal{M})$. This implies that there must be an edge $(\tau, \sigma + y)$ in D such that $\tau \in \Delta_{\mathcal{V}}$. Clearly $\sigma + y \subset \tau$; thus there is a $z \neq y$ such that $\tau = \sigma \cup \{y, z\}$. Since σ is maximal among sets right below $\Delta_{\mathcal{V}}$, we must have that $\sigma + z \in \Delta_{\mathcal{V}}$. However, $(\sigma + z, \sigma)$ is an edge in D, which gives a contradiction. $\square$

We now state and prove a simple result that is indispensable for several proofs in this book. In words, it says the following: Suppose that the family of critical faces with respect to an acyclic matching on a simplicial complex can be divided into two subfamilies such that there are no directed paths between the two subfamilies in the underlying digraph. Then the complex is homotopy equivalent to a wedge of two separate complexes generated as in (4.2) from the two subfamilies.

Theorem 4.11. *Suppose that* $\mathcal{V} \subseteq \mathcal{U} = \mathcal{U}(\Delta, \mathcal{M})$ *has the property that* $\mathcal{U} \setminus \mathcal{V} \nrightarrow \mathcal{V}$ *and* $\mathcal{V} \nrightarrow \mathcal{U} \setminus \mathcal{V}$. *Then* Δ *is homotopy equivalent to* $\Delta_{\mathcal{V}} \vee \Delta_{\mathcal{U} \setminus \mathcal{V}}$. *In particular,* Δ *is homotopy equivalent to* $\Delta_{\mathcal{U}}$.

Proof. By Theorem 4.4 and Lemma 4.10, Δ is homotopy equivalent to $\Delta_{\mathcal{U}}$; thus we may assume that $\Delta = \Delta_{\mathcal{U}} = \Delta_{\mathcal{V}} \cup \Delta_{\mathcal{U} \setminus \mathcal{V}}$. Let $\Sigma = \Delta_{\mathcal{V}} \cap \Delta_{\mathcal{U} \setminus \mathcal{V}}$. By assumption, Σ contains no critical faces and is nonvoid $(\emptyset, \{x\} \in \Sigma)$. By Lemma 4.10 applied to each of $\Delta_{\mathcal{V}}$ and $\Delta_{\mathcal{U} \setminus \mathcal{V}}$, the restriction of $\mathcal{M}$ to Σ is a perfect matching. By Corollary 4.6, this implies that Σ is contractible to a point. By the Contractible Subcomplex Lemma 3.16, Δ is homotopy equivalent to the quotient complex Δ/Σ. By the same lemma, $\Delta_{\mathcal{V}} \vee \Delta_{\mathcal{U} \setminus \mathcal{V}}$ is homotopy equivalent to $(\Delta_{\mathcal{V}}/\Sigma) \vee (\Delta_{\mathcal{U} \setminus \mathcal{V}}/\Sigma)$. Since clearly

$$\Delta/\Sigma \cong (\Delta_{\mathcal{V}}/\Sigma) \vee (\Delta_{\mathcal{U} \setminus \mathcal{V}}/\Sigma),$$

the proof is finished. $\square$

Via a simple induction argument, Theorem 4.11 yields the following result:

Corollary 4.12. *If* $\mathcal{U}$ *is the disjoint union of families* $\mathcal{V}_1, \ldots, \mathcal{V}_r$ *with the property that* $\mathcal{V}_i \nrightarrow \mathcal{V}_j$ *if* $i \neq j$, *then* Δ *is homotopy equivalent to* $\bigvee_{i=1}^{r} \Delta_{\mathcal{V}_i}$.
$\square$

The following is an important special case; use Theorem 4.8.

Corollary 4.13. *Let $\mathcal{V} \subseteq \mathcal{U} = \mathcal{U}(\Delta, \mathcal{M})$ be such that $\mathcal{U} \setminus \{V\} \not\longmapsto V$ and $V \not\longmapsto \mathcal{U} \setminus \{V\}$ for every $V \in \mathcal{V}$. Then Δ is homotopy equivalent to*

$$\left(\bigvee_{V \in \mathcal{V}} S^{|V|-1} \right) \vee \Delta_{\mathcal{U} \setminus \mathcal{V}} . \quad \square$$

Many results of this section are special cases of the following general theorem, which one may refer to as the "Fundamental Theorem of Simplicial Morse Theory":

Theorem 4.14 (Forman [49]). *Let Δ be a simplicial complex and let $\mathcal{M}$ be an acyclic matching on Δ such that the empty set is not critical. Then Δ is homotopy equivalent to a cell complex with one cell of dimension $p \geq 0$ for each critical face of Δ of dimension p plus one additional 0-cell.* $\square$

For a proof sketch of Theorem 4.14, see the informal discussion in Section 4.1, where we outline the transformation of Δ into a cell complex with desired properties. Forman [49] provides a much more detailed description of this transformation. The resulting cell complex is the *discrete Morse complex* of Δ with respect to $\mathcal{M}$.

Finally, we present the "weak Morse inequalities"; they are an immediate consequence of Theorem 4.14 and the existence of a natural isomorphism between simplicial and cellular homology [101, §39]. One may also deduce the inequalities from the theory developed in Section 4.4; see Theorem 4.16.

Theorem 4.15 (Forman [49]). *Let $\mathbb{F}$ be a field, let Δ be a simplicial complex, and let $\mathcal{M}$ be an acyclic matching on Δ. Then the number of critical faces of dimension d is at least $\dim \tilde{H}_d(\Delta; \mathbb{F})$ for each $d \geq -1$.* $\square$

4.4 Discrete Morse Theory on Complexes of Groups

We give an algebraic generalization of Forman's discrete Morse complex [49]. We develop the theory in preparation for Sections 17.2 and 20.2, where we determine linearly independent elements in the homology of the quotient complexes of Hamiltonian graphs and 3-connected graphs, respectively. Sköldberg [126] and Jöllenbeck and Welker [66] developed similar but more general and powerful algebraic interpretations of discrete Morse theory.

Let

$$\mathsf{C} : \cdots \xrightarrow{\partial_{n+2}} C_{n+1} \xrightarrow{\partial_{n+1}} C_n \xrightarrow{\partial_n} C_{n-1} \xrightarrow{\partial_{n-1}} \cdots \qquad (4.3)$$

be a complex of abelian groups; $\partial_{n-1} \circ \partial_n = 0$. Let

$$C = \bigoplus_n C_n \text{ and } \partial = \bigoplus_n \partial_n.$$

Suppose that there are groups $A = \bigoplus_n A_n$, $B = \bigoplus_n B_n$, and $U = \bigoplus_n U_n$ such that

$$C_n \cong A_n \oplus B_n \oplus U_n$$

and such that the function $f : B \to A$ defined as

$$f = \alpha \circ \partial$$

is an isomorphism, where $\alpha(a + b + u) = a$ for $a \in A, b \in B, u \in U$. We say that the pair (A, B) is *removable*.

As an example, consider an element matching on a simplicial complex. Let A be the free group generated by faces matched with larger faces, let B be the free group generated by faces matched with smaller faces, and let U be the free group generated by unmatched faces. We claim that if the matching is acyclic, then (A, B) is a removable pair. Namely, since the digraph D corresponding to the acyclic matching is acyclic, we may assume that the matched pairs $\{\sigma_1, \tau_1\}, \ldots, \{\sigma_r, \tau_r\}$ with $\sigma_k \subset \tau_k$ for $1 \le k \le r$ have the property that the boundary of τ_k does not contain any of the faces $\sigma_1, \ldots, \sigma_{k-1}$ for $2 \le k \le r$. This means that we can write

$$f(\tau_i) = \sum_j \mu_{ij} \sigma_j, \tag{4.4}$$

where $\mu_{ij} = 0$ if $j < i$ and $\mu_{ii} = \pm 1$.

Return to the general case. We want to define a complex U corresponding to Forman's discrete Morse complex [49]. Let $\beta : C \to B$ be defined as

$$\beta = f^{-1} \circ \alpha \circ \partial.$$

Moreover, let

$$\hat{A} = \partial(B) \text{ and } \hat{U} = (\text{Id} - \beta)(U). \tag{4.5}$$

Theorem 4.16. *With notation as above, the sequence*

$$\mathsf{U} : \cdots \xrightarrow{\partial_{n+2}} \hat{U}_{n+1} \xrightarrow{\partial_{n+1}} \hat{U}_n \xrightarrow{\partial_n} \hat{U}_{n-1} \xrightarrow{\partial_{n-1}} \cdots \tag{4.6}$$

is a complex with the same homology as the original complex C *in* (4.3); *the boundary operators are the restrictions of the original boundary operators.*

Proof. Our first claim is that

$$C \cong \hat{A} \oplus B \oplus \hat{U}. \tag{4.7}$$

To prove (4.7), we first show that

$$\hat{A} \oplus B \oplus \hat{U} \cong A \oplus B \oplus \hat{U}. \tag{4.8}$$

Now, (4.8) is an immediate consequence of the fact that $f = \alpha \circ \partial : B \to A$ is an isomorphism. Namely, this implies that $\alpha : \partial(B) \to A$ is an isomorphism, which in turn implies that $\alpha^* : \hat{A} \oplus B \oplus \hat{U} \to A \oplus B \oplus \hat{U}$ defined by $\alpha^*(\hat{a}, b, \hat{u}) = (\alpha(\hat{a}), b, \hat{u})$ is an isomorphism. Next, we show that

$$A \oplus B \oplus \hat{U} \cong A \oplus B \oplus U. \qquad (4.9)$$

This is done by observing that $u \mapsto u - \beta(u)$ is an isomorphism $U \to \hat{U}$; the inverse is the canonical projection function $A \oplus B \oplus U \to U$ restricted to $\hat{U}$. Combining (4.8) and (4.9), we obtain (4.7).

We proceed with a proof of the claim that

$$\partial(\hat{U}) \subset \hat{U}. \qquad (4.10)$$

Let $\hat{u} = u - \beta(u) \in \hat{U}$ and write $\partial(\hat{u}) = a + b + \hat{v}$, where $a \in A$, $b \in B$, and $\hat{v} = v - \beta(v) \in \hat{U}$. Since

$$\alpha \circ \partial(\hat{u}) = \alpha \circ \partial(u) - \alpha \circ \partial \circ \beta(u)$$
$$= \alpha \circ \partial(u) - f \circ f^{-1} \circ \alpha \circ \partial(u) = 0,$$

we must have $a = 0$. Moreover,

$$0 = f^{-1} \circ \alpha \circ \partial^2(\hat{u}) = \beta(b + \hat{v}) = \beta(b) + \beta(v) - \beta^2(v) = b.$$

Hence $\partial(\hat{u}) = \hat{v} \in \hat{U}$, and the claim (4.10) follows.

Now, by (4.7) and (4.10), we have that $\partial(C) \cong \hat{A} \oplus \partial(\hat{U})$ and $\ker \partial \cong \hat{A} \oplus (\ker \partial \cap \hat{U})$, which implies that

$$\ker \partial_n / \partial_{n+1}(C_n) \cong (\ker \partial_n \cap \hat{U}_n)/\partial(\hat{U}_{n+1});$$

hence we are done. $\square$

4.4.1 Independent Sets in the Homology of a Complex

We now turn our attention to the more specific problem of finding a full or partial basis for the resulting complex U in Theorem 4.16. For the remainder of this chapter, $\mathbb{F}$ is an integral domain. While it would be possible to generalize many of our results to larger classes of commutative rings with unity, we restrict our attention to rings without zero divisors; this is to keep the complexity of proofs at a minimum.

Let A_n, B_n, and U_n be $\mathbb{F}$-modules. In Section 20.2, we use the following result to determine a basis for the homology of the complex of 3-connected graphs.

Corollary 4.17. *For every* $u \in U$, *there is a unique* $b \in B$ *such that* $u - b \in \hat{U}$. *As a consequence, if* U_n *is a free* $\mathbb{F}$-*module and* $\{u_1, \ldots, u_k\}$ *is an* $\mathbb{F}$-*basis for* U_n, *then there are unique elements* $b_1, \ldots, b_k \in B_n$ *such that* $\{u_1 - b_1, \ldots, u_k - b_k\}$ *forms an* $\mathbb{F}$-*basis for* $\hat{U}_n$. *If* $\partial_{n+1}(\hat{U}_{n+1}) = \partial_n(\hat{U}_n) = 0$, *then this basis forms a basis for* $H_n(\mathsf{C})$ *as well, and* b_i *is unique in* B_n *such that* $\partial_n(u_i - b_i) = 0$.

Proof. The claims are consequences of the discussion in the previous section; $\beta(u)$ is the unique element b such that $u - b \in \hat{U}$. □

In Section 17.2, we show that a certain basis for the homology of the complex of 2-connected graphs remains an independent set in the homology of the complex of Hamiltonian graphs. This requires a stronger version of Corollary 4.17; the situation is as in Corollary 4.13 with $\mathcal{U} \setminus \mathcal{V}$ nonvoid.

We restrict our attention to the special case that C, A, B, and U are *free* $\mathbb{F}$-modules. We assume that A_n and B_n are finitely generated for each n, but we put no restrictions on U_n. Moreover, A and B may well be of infinite rank if C_n is nonzero for infinitely many n.

Let $\mathcal{C}$ be a graded basis for C such that

$$\mathcal{C} = \mathcal{A} \cup \mathcal{B} \cup \mathcal{U},$$

where $\mathcal{A}, \mathcal{B}$, and $\mathcal{U}$ are bases for A, B, and U, respectively. For any $c_1, c_2 \in \mathcal{C}$, define $\langle c_1, c_2 \rangle$ to be 1 if $c_1 = c_2$ and 0 otherwise. Extend $\langle \cdot, \cdot \rangle$ to an inner product $C \times C \to \mathbb{F}$.

For a basis element $c \in \mathcal{C}$ and an arbitrary element $x \in C$, say that $c \prec x$ if $\langle c, \partial(x) \rangle \neq 0$; the relation $\prec$ does not depend on whether $\langle c, \partial(x) \rangle$ is a unit or not. Let $\mathcal{M}$ be a perfect matching between $\mathcal{A}$ and $\mathcal{B}$ such that $a \prec b$ whenever $a \in \mathcal{A}$ and $b \in \mathcal{B}$ are matched. Such a perfect matching exists, because the determinant associated with f is nonzero. However, the matching need not be unique in general unless f is upper triangular as in (4.4). Let D be the digraph with vertex set $\mathcal{C}$ such that (c_1, c_2) is an edge in D if and only if

$$(\{c_1, c_2\} \in \mathcal{M}, c_1 \in \mathcal{A}, \text{ and } c_2 \in \mathcal{B}) \text{ or } (\{c_1, c_2\} \notin \mathcal{M} \text{ and } c_2 \prec c_1).$$

Note that D is not necessarily acyclic. As in Section 4.3, we let $c_1 \longrightarrow c_2$ mean that there is a directed path from c_1 to c_2 in D.

One may view the following lemma as an algebraic version of Lemma 4.10.

Lemma 4.18. *Let v be an element in $\mathcal{U}$. Let $\mathcal{A}^+$ be the set of all $a \in \mathcal{A}$ such that $v \longrightarrow a$, and let $\mathcal{B}^+$ be the set of all $b \in \mathcal{B}$ such that $v \longrightarrow b$. Then $b \in \mathcal{B}^+$ if and only if the element a matched with b is in $\mathcal{A}^+$.*

Proof. Assume the opposite and let n be maximal with the property that $\mathcal{A} \setminus \mathcal{A}^+$ contains an element a from A_n such that the element b matched with a is contained in $\mathcal{B}^+$. Since $v \longrightarrow b$, there must be an x such that $v \longrightarrow x$ and $b \prec x$. Since $\partial^2(x) = 0$ and $a \prec b$, there is a $c \neq b$ such that $a \prec c \prec x$; here we use the fact that $\mathbb{F}$ is an integral domain. If $\{c, x\} \in \mathcal{M}$, then $x \in \mathcal{B}^+$; the maximality of a implies that $c \in \mathcal{A}^+$. However, this is a contradiction, since $v \longrightarrow c \longrightarrow a$. If $\{c, x\} \notin \mathcal{M}$, then $v \longrightarrow x \longrightarrow c \longrightarrow a$, and another contradiction is obtained. □

Theorem 4.19. *Let $\mathcal{V}$ be a subset of $\mathcal{U}$. Suppose that $\mathcal{U} \setminus \mathcal{V} \nrightarrow \mathcal{V}$ and $\mathcal{V} \nrightarrow \mathcal{U} \setminus \mathcal{V}$. Let $V = \bigoplus_n V_n$ be the submodule generated by $\mathcal{V}$, and let $W = \bigoplus_n W_n$*

be the submodule generated by $\mathcal{W} = \mathcal{U} \setminus \mathcal{V}$. Let $\hat{V}_n = (\mathrm{Id} - \beta)(V_n)$ and $\hat{W}_n = (\mathrm{Id} - \beta)(W_n)$. Then the complex U in (4.6) splits into two complexes

$$\mathsf{V} : \cdots \xrightarrow{\partial_{n+2}} \hat{V}_{n+1} \xrightarrow{\partial_{n+1}} \hat{V}_n \xrightarrow{\partial_n} \hat{V}_{n-1} \xrightarrow{\partial_{n-1}} \cdots ;$$

$$\mathsf{W} : \cdots \xrightarrow{\partial_{n+2}} \hat{W}_{n+1} \xrightarrow{\partial_{n+1}} \hat{W}_n \xrightarrow{\partial_n} \hat{W}_{n-1} \xrightarrow{\partial_{n-1}} \cdots .$$

This implies that
$$H_*(\mathsf{U}) = H_*(\mathsf{V}) \oplus H_*(\mathsf{W}).$$

In particular, if $\mathcal{V} \subseteq U_n$ for some n, then $\{v - \beta(v) : v \in \mathcal{V}\}$ is an $\mathbb{F}$-independent set in $H_n(\mathsf{U})$.

Proof. To prove that $\partial(\hat{V}) \subset \hat{V}$, it suffices to show that if $v \in \mathcal{V}$ and $w \in \mathcal{W}$, then $w \not\prec \hat{v}$, where $\hat{v} = v - \beta(v)$. In fact, since $w \not\prec v$ by assumption, we need only prove that $w \not\prec \beta(v)$.

Consider a basis element $v \in \mathcal{V}$; let k be such that $v \in V_k$. Let $\mathcal{A}^+$ be the set of all $a \in \mathcal{A}$ such that $v \longrightarrow a$, and let $\mathcal{B}^+$ be the set of all $b \in \mathcal{B}$ such that $v \longrightarrow b$. Note that $b \in B_m$ for some $m \leq k$. In particular, the set $\{m : B_m \cap \mathcal{B}^+ \neq \emptyset\}$ has an upper bound. Let $\mathcal{A}^- = \mathcal{A} \setminus \mathcal{A}^+$ and $\mathcal{B}^- = \mathcal{B} \setminus \mathcal{B}^+$.

Define
$$\mu_{ab} = \langle a, \partial(b) \rangle = \langle a, f(b) \rangle$$

for $a \in \mathcal{A}$ and $b \in \mathcal{B}$. This means that

$$f(b) = \sum_{a \in \mathcal{A}} \mu_{ab} \cdot a.$$

Note that $\mu_{ab} = 0$ if $b \in \mathcal{B}^+$ and $a \in \mathcal{A}^-$. Namely, Lemma 4.18 implies that $\{a, b\} \notin M$. Since $f : B_n \to A_{n-1}$ is an isomorphism, we therefore obtain that the matrix

$$(\mu_{ab})_{a \in A_{n-1} \cap \mathcal{A}^-, b \in B_n \cap \mathcal{B}^-} \tag{4.11}$$

is invertible; the matrix is a square matrix by Lemma 4.18. In particular, for any nontrivial linear combination y of elements from $B_n \cap \mathcal{B}^-$, there is some $a \in A_{n-1} \cap \mathcal{A}^-$ such that the coefficient of a in $\partial(y)$ is nonzero. Since $\alpha \circ \partial(\hat{v}) = 0$ and $a \not\prec v$ if $a \in \mathcal{A}^-$, the element $\beta(v)$ is a linear combination of elements in $\mathcal{B}^+$. Hence if $w \prec \beta(v)$, then $v \longrightarrow w$, which is not true.

By symmetry, we obtain that $\partial(\hat{W}) \subset \hat{W}$, which concludes the proof. $\square$

Remark. We emphasize that the conclusion in Theorem 4.19 might be false without the requirement that A_n and B_n be finitely generated. The problem is that we might have a nontrivial linear combination x of elements in $\mathcal{B}^-$ such that $\alpha \circ \partial(x)$ is a linear combination of elements in $\mathcal{A}^+$. Namely, the matrix (4.11) does not have to be invertible if its size is infinite. In particular, it might be the case that $w \prec \beta(v)$ even if $v \not\longrightarrow w$ and $w \not\longrightarrow v$.

To illustrate the problem, suppose that $C_1 = B \oplus (v \cdot \mathbb{F})$ and $C_0 = A \oplus (w \cdot \mathbb{F})$ with

B generated by $\{b_i : i \in \mathbb{Z}\}$ and A generated by $\{a_i : i \in \mathbb{Z}\}$. Let $C_i = 0$ for $i \neq 0, 1$. Define ∂ by

$$\partial(b_{2k}) = a_{2k-1} + a_{2k} + a_{2k+1} + \delta_{k0} \cdot w;$$
$$\partial(b_{2k-1}) = a_{2k-1} + a_{2k};$$
$$\partial(v) = a_1$$

($\delta_{ij} = 1$ if $i = j$ and 0 otherwise). $f : B \to A$ is easily seen to be bijective. Moreover, with b_i matched with a_i for all i, we have that $v \not\mapsto w$ and $w \not\mapsto v$. However, $w \prec \beta(v)$; namely, $\beta(v) = b_0 - b_{-1}$ and $\partial(b_0 - b_{-1}) = a_1 + w$. In fact, $\partial : C_1 \to C_0$ is a bijection, which implies that the homology vanishes. If the statement in Theorem 4.19 were true for this particular case, then we would have had $H_1(\mathsf{C}) = H_0(\mathsf{C}) = \mathbb{F}$.

4.4.2 Simple Applications

We conclude this chapter with two applications of algebraic Morse theory. First, we examine the tensor product of two chain complexes. As a byproduct, we derive well-known results about the join of two simplicial complexes. Second, we present the well-known correspondence between the homology of a simplicial complex and that of its barycentric subdivision. We stress that the main purpose of the section is merely to illustrate the technique.

Let $\mathbb{F}$ be a principal ideal domain. With notation as before, consider a chain complex C with corresponding $\mathbb{F}$-modules A, B, and U. Let C' be another chain complex of $\mathbb{F}$-modules. Throughout this section, $\otimes$ denotes tensor product with respect to $\mathbb{F}$; $\lambda(c_1 \otimes c_2) = c_1 \otimes (\lambda c_2) = (\lambda c_1) \otimes c_2$ whenever $\lambda \in \mathbb{F}$.

For a given constant integer κ, consider the chain complex $(\mathsf{C}^\otimes, \partial^\otimes)$ defined by

$$C^\otimes_{n+\kappa} = \bigoplus_{r+s=n} C_r \otimes C'_s; \tag{4.12}$$
$$\partial^\otimes(c \otimes c') = \partial(c) \otimes c' + (-1)^{r+1}(c \otimes \partial'(c')) \tag{4.13}$$

for $c \in C_r$ and $c' \in C'_s$. Write

$$B^\otimes_{n+\kappa} = \bigoplus_{r+s=n} B_r \otimes C'_s,$$

and define $A^\otimes_{n+\kappa}$ and $U^\otimes_{n+\kappa}$ analogously. It is clear that $C^\otimes_n = A^\otimes_n \oplus B^\otimes_n \oplus U^\otimes_n$. Also, $f^\otimes = \alpha^\otimes \circ \partial^\otimes$ is an isomorphism $B^\otimes \to A^\otimes$, where $\alpha^\otimes(a + b + u) = a$ for $a \in A^\otimes, b \in B^\otimes, u \in U^\otimes$. Namely, for $b \in B$ and $c' \in C'$, we have that

$$f^\otimes(b \otimes c') = \alpha^\otimes \circ \partial^\otimes(b \otimes c') = \alpha^\otimes(\partial(b) \otimes c' \pm b \otimes \partial'(c'))$$
$$= \alpha^\otimes(\partial(b) \otimes c') = (\alpha \circ \partial(b)) \otimes c' = f(b) \otimes c'.$$

Since $\beta^\otimes(u \otimes c') = \beta(u) \otimes c'$, it follows that Theorem 4.16 applies to the complex with chain groups $\hat{U}^\otimes_{n+\kappa} = \bigoplus_{r+s=n} \hat{U}_r \otimes C'_s$. Specifically, we have the following result.

Theorem 4.20. *With notation as above, if $U_r \cong H_r(\mathsf{C})$ for all r, then*

$$H_{n+\kappa}(\mathsf{C}^{\otimes}) \cong \bigoplus_{r+s=n} H_r(\mathsf{C}) \otimes H_s(\mathsf{C}').$$

Proof. With $\hat{u} \in \hat{U}_r$ and $c' \in C'$, we have that $\partial^{\otimes}(\hat{u} \otimes c') = (-1)^{r+1}(\hat{u} \otimes \partial'(c'))$. As a consequence, the homology splits;

$$H_{n+\kappa}(\mathsf{C}^{\otimes}) \cong \bigoplus_{r+s=n} (\hat{U}_r \otimes \ker \partial'_s)/(\hat{U}_r \otimes \partial'_{s+1}(C'_{s+1}))$$

$$\cong \bigoplus_{r+s=n} \hat{U}_r \otimes (\ker \partial'_s/\partial'_{s+1}(C'_{s+1})) \cong \hat{U}_r \otimes H_s(\mathsf{C}'). \square$$

Proposition 4.21. *Assume that $\mathbb{F}$ is a principal ideal domain. If all homology of C is free, then we may write $\mathsf{C} = A \oplus B \oplus U$ such that (A, B) is a removable pair and such that $H_d(\mathsf{C}) \cong U_d$ for all d.*

Proof. Since $\mathbb{F}$ is a principal ideal domain, torsion-free modules and submodules of free modules are free; see Isaacs [65, Th. 16.28]. Write $Z_d = \partial_d^{-1}(\{0\})$; this is a free $\mathbb{F}$-module. Let $B_d \subseteq C_d$ be such that $C_d = Z_d \oplus B_d$; such a B_d exists, because C_d/Z_d is torsion-free and hence free. Define $A_d = \partial(B_{d+1})$; this is again a free $\mathbb{F}$-module. Let $U_d \subseteq Z_d$ be such that $Z_d = A_d \oplus U_d$; such a U_d exists, because $H_d(\mathsf{C}) = Z_d/\partial(C_{d+1}) = Z_d/A_d$ is free by assumption. The desired result follows. $\square$

Corollary 4.22. *With notation as above, if C is a chain complex of free $\mathbb{F}$-modules with $\mathbb{F}$-free homology, then*

$$H_{n+\kappa}(\mathsf{C}^{\otimes}) \cong \bigoplus_{r+s=n} H_r(\mathsf{C}) \otimes H_s(\mathsf{C}').$$

Proof. By Proposition 4.21, we may write $\mathsf{C} = A \oplus B \oplus U$ such that (A, B) is a removable pair and such that $U_r \cong H_r(\mathsf{C})$ for all r. Hence we have the situation in Theorem 4.20. $\square$

For simplicial (or quotient) complexes Δ and Γ, the reduced chain complex of the join $\Delta * \Gamma$ is clearly of the form (4.12) with $\kappa = 1$ and with the boundary operator given by (4.13); compare to (3.2). The same holds for the *unreduced* chain complex of the cell complex $\|\Delta\| \times \|\Gamma\|$ but with $\kappa = 0$; see Munkres [101, Th. 57.1].

Corollary 4.23. *Let Δ and Γ be simplicial complexes and let $\mathbb{F}$ be a field or $\mathbb{Z}$. If $\tilde{H}_n(\Delta; \mathbb{F})$ is free for all n (this is of course always true for fields), then*

$$\tilde{H}_{n+1}(\Delta * \Gamma; \mathbb{F}) \cong \bigoplus_{r=-1}^{n+1} \tilde{H}_r(\Delta; \mathbb{F}) \otimes \tilde{H}_{n-r}(\Gamma; \mathbb{F});$$

$$H_n(\|\Delta\| \times \|\Gamma\|; \mathbb{F}) \cong \bigoplus_{r=0}^{n} H_r(\Delta; \mathbb{F}) \otimes H_{n-r}(\Gamma; \mathbb{F}). \ \square$$

The formula for join remains true if Δ or Γ are quotient complexes. The situation is more complicated if $\mathbb{F} = \mathbb{Z}$ and there is torsion in the homology of both complexes; see Munkres [101, Sec. 59].

Proposition 4.24. *Let Δ be a simplicial complex. For each $\sigma \in \Delta$, let $z(\sigma)$ be the fundamental cycle of $\mathrm{sd}(\partial 2^\sigma)$, appropriately signed. Then the map $\sigma \mapsto [\{\sigma\}] \wedge z(\sigma)$ induces an isomorphism from $\tilde{H}_d(\Delta; \mathbb{F})$ to $\tilde{H}_d(\mathrm{sd}(\Delta); \mathbb{F})$.*

Proof. For $\sigma \in \Delta$, let $\mathcal{F}(\sigma)$ be the family of chains in $P(\Delta)$ with maximal element σ. It is clear that the families $\mathcal{F}(\sigma)$ satisfy the Cluster Lemma 4.2. Now, each $\mathcal{F}(\sigma)$ is of the form $\{\sigma\} * \mathrm{sd}(\partial 2^\sigma)$. By Proposition 4.9, $\mathrm{sd}(\partial 2^\sigma)$ admits an acyclic matching with one unmatched face of dimension $\dim \sigma - 1$. This acyclic matching must have the property that the resulting chain complex in Theorem 4.16 is generated by the fundamental cycle $z(\sigma)$. Namely, this is up to a constant the only cycle of maximum dimension $\dim \sigma - 1$ in the chain complex of $\mathrm{sd}(\partial 2^\sigma)$.

Combining the acyclic matchings on the families $\mathcal{F}(\sigma)$, we obtain an acyclic matching on $\mathrm{sd}(\Delta)$ with exactly one chain of length $|\sigma|$ with top element σ for each face $\sigma \in \Delta$. By the above discussion, the corresponding element $\hat{u}_\sigma$ in the chain complex U in Theorem 4.16 coincides (up to sign) with $[\{\sigma\}] \wedge z(\sigma)$.

Let us examine $z(\sigma)$ in greater detail. Take any total order on the set of 0-cells of Δ and arrange the elements in $\sigma \in \Delta$ in increasing order as $\sigma = \{a_1, \ldots, a_r\}$. For a permutation $\pi \in \mathfrak{S}_\sigma$, define

$$[\pi] = [\pi(\{a_1\})] \wedge [\pi(\{a_1, a_2\})] \wedge \cdots \wedge [\pi(\{a_1, a_2, \ldots, a_{r-1}\})].$$

It is a straightforward exercise to check that, up to sign, $z(\sigma) = \sum_{\pi \in \mathfrak{S}_\sigma} \mathrm{sgn}(\pi) \cdot [\pi]$. With this choice of orientation, one easily checks that

$$z(\sigma) = \sum_{i=1}^{|\sigma|} (-1)^{i-1} \cdot [\{\sigma - a_i\}] \wedge z(\sigma - a_i) = \sum_{i=1}^{|\sigma|} (-1)^{i-1} \cdot \hat{u}_{\sigma - a_i}. \quad (4.14)$$

Since $\partial(\hat{u}_\sigma) = \partial([\{\sigma\}] \wedge z(\sigma)) = z(\sigma)$, it follows that the operator ∂ is isomorphic to the ordinary boundary operator for simplicial complexes. In particular, the complex U is isomorphic to the chain complex of Δ, which concludes the proof. $\square$

5

Decision Trees

[1] We examine topological properties of decision trees on simplicial complexes, the emphasis being on how one may apply decision trees to problems in topological combinatorics. Our work is to a great extent based on Forman's seminal papers [49, 50].

Let Δ be a simplicial complex on the set E. One may view a decision tree on the pair (Δ, E) as a deterministic algorithm A that on input a secret set $\sigma \subseteq E$ asks repeated questions of the form "Is the element x contained in σ?" until all questions but one have been asked. A is allowed to be adaptive, meaning that each question may depend on responses to earlier questions. Let x_σ be the one element that A never queries. σ is *nonevasive* (and A successful) if $\sigma - x_\sigma$ and $\sigma + x_\sigma$ are either both in Δ or both outside Δ. Otherwise, σ is *evasive*.

In this book, we adopt an "intrinsic" approach, meaning that we restrict our attention to the faces of Δ; whether or not a given subset of E outside Δ is evasive is of no interest to us. We may thus interpret A as an algorithm that takes as input a secret face $\sigma \in \Delta$ and tries to save a query x_σ with the property that $\sigma - x_\sigma$ and $\sigma + x_\sigma$ are both in Δ. Clearly, a face σ is evasive if and only if $\sigma + x_\sigma \notin \Delta$. Aligning with this intrinsic approach, we will always assume that the underlying set E is exactly the set of 0-cells in Δ.

Given a simplicial complex Δ, a natural goal is to find a decision tree with as few evasive faces as possible. In general, there is no decision tree such that all faces are nonevasive. Specifically, if Δ is not contractible, then such a decision tree cannot exist; Kahn, Saks, and Sturtevant [78] were the first to observe this. More generally, Forman [50] has demonstrated that a decision tree on Δ gives rise to an acyclic matching on Δ (see Chapter 4) such that a face is unmatched if and only if the face is evasive. One defines the matching by pairing $\sigma - x_\sigma$ with $\sigma + x_\sigma$ for each nonevasive face σ, where x_σ is the element not queried for σ. As a consequence of discrete Morse theory, there

---

[1] This chapter is a revised and extended version of a paper [70] published in *The Electronic Journal of Combinatorics*.

are at least $\dim \tilde{H}_i(\Delta; \mathbb{F})$ evasive faces of Δ of dimension i for any given field $\mathbb{F}$.

The goal of this chapter is two-fold:

- The first goal is to develop some elementary theory about "optimal" decision trees. For a given field $\mathbb{F}$, a decision tree on a complex Δ is $\mathbb{F}$-*optimal* if the number of evasive faces of dimension i is equal to the Betti number $\dim \tilde{H}_i(\Delta; \mathbb{F})$ for each i. We give a recursive definition of the class of *semi-nonevasive* simplicial complexes that admit an $\mathbb{F}$-optimal decision tree. We also generalize the concept of decision trees to allow questions of the form "Is the set τ a subset of σ?" This turns out to yield an alternative characterization of simplicial Morse theory. As a consequence, we may characterize $\mathbb{F}$-optimal acyclic matchings – defined in the natural manner – in terms of generalized decision trees. We will refer to complexes admitting $\mathbb{F}$-optimal acyclic matchings as *semi-collapsible* complexes, aligning with the fact that collapsible complexes are those admitting a perfect acyclic matching. Vertex-decomposable and shellable complexes constitute important examples of semi-nonevasive and semi-collapsible complexes, respectively.
- The second goal is to investigate under what conditions the properties of being semi-nonevasive and semi-collapsible are preserved under standard operations such as taking the join of two complexes or forming the barycentric subdivision or Alexander dual of a complex. The results and proofs are similar in nature to those Welker [146] provided for nonevasive and collapsible complexes.

Optimal decision trees appear in the work of Forman [50] and Soll [129]; Charalambous [30] considered related techniques. Recently, Hersh [60] developed powerful techniques for optimizing acyclic matchings; see Hersh and Welker [61] for an application. The complexity-theoretic aspect of optimization appears in the work of Lewiner, Lopes, and Tavares [93, 91, 92]. For more information about the connection between evasiveness and topology, there are several papers [114, 115, 82, 78, 28] and surveys [9, 19] to consult.

All topological and homological concepts and results in this chapter are defined and stated in terms of simplicial complexes. There are potential generalizations of these concepts and results, either in a topological direction – allowing for a more general class of CW complexes – or in a homological direction – allowing for a more general class of chain complexes. For simplicity and clarity, and in alignment with the main goals of this book, we restrict our attention to simplicial complexes.

For basic definitions and results about decision trees, see Section 5.1. Basic results about optimal decision trees appear in Section 5.2; see Section 5.4 for some operations that preserve optimality. In Section 5.3, we present some useful constructions that we will use throughout the book. We round up the chapter in Section 5.5 with a potential generalization of the concept of semi-collapsibility.

5.1 Basic Properties of Decision Trees

We discuss elementary properties of decision trees and introduce the generalized concept of set-decision trees, the generalization being that arbitrary sets rather than single elements are queried. To distinguish between the two notions, we will refer to ordinary decision trees as "element-decision trees".

5.1.1 Element-Decision Trees

First, we give a recursive definition, suitable for our purposes, of element-decision trees. We are mainly interested in trees on simplicial complexes, but it is convenient to have the concept defined for arbitrary families of sets. Below, the terms "elements" and "sets" always refer to elements and finite subsets of some fixed ground set such as the set of integers.

Definition 5.1. The class of *element-decision trees*, each associated to a finite family of finite sets, is defined recursively as follows:

(i) $T = \mathsf{Win}$ is an element-decision tree on $\emptyset$ and on any 0-simplex $\{\emptyset, \{v\}\}$.
(ii) $T = \mathsf{Lose}$ is an element-decision tree on $\{\emptyset\}$ and on any singleton set $\{\{v\}\}$.
(iii) If Δ is a family of sets, if x is an element, if T_0 is an element-decision tree on $\mathrm{del}_\Delta(x)$, and if T_1 is an element-decision tree on $\mathrm{lk}_\Delta(x)$, then the triple (x, T_0, T_1) is an element-decision tree on Δ.

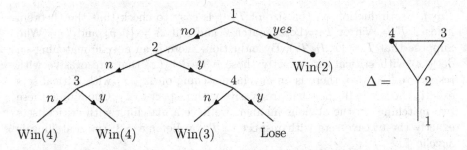

Fig. 5.1. The tree $(1, (2, (3, \mathsf{Win}, \mathsf{Win}), (4, \mathsf{Win}, \mathsf{Lose})), \mathsf{Win})$ on the complex Δ. "$\mathsf{Win}(v)$" means that the complex corresponding to the given leaf is $\{\emptyset, \{v\}\}$; "Lose" means that the complex is $\{\emptyset\}$.

Return to the discussion in the introduction. One may interpret the triple (x, T_0, T_1) as follows for a given set σ to be examined: The element being queried is x. If $x \notin \sigma$, then proceed with $\text{del}_\Delta(x)$, the family of sets not containing x. Otherwise, proceed with $\text{lk}_\Delta(x)$, the family with one set $\tau - x$ for each set τ containing x. Proceeding recursively, we finally arrive at a leaf, either Win or Lose. The underlying family being a 0-simplex $\{\emptyset, \{v\}\}$ means that $\sigma + v \in \Delta$ and $\sigma - v \in \Delta$; we win, as v remains to be queried. The family being $\{\emptyset\}$ or $\{\{v\}\}$ means that we cannot tell whether $\sigma \in \Delta$ without querying all elements; we lose.

Note that we allow for the "stupid" decision tree $(v, \text{Lose}, \text{Lose})$ on $\{\emptyset, \{v\}\}$; this tree queries the element v while it should not. Also, we allow the element x in (iii) to have the property that no set in Δ contains x, which means that $\text{lk}_\Delta(x) = \emptyset$, or that all sets in Δ contain x, which means that $\text{del}_\Delta(x) = \emptyset$.

A set $\tau \in \Delta$ is *nonevasive* with respect to an element-decision tree T on Δ if either of the following holds:

1. $T = \text{Win}$.
2. $T = (x, T_0, T_1)$ for some x not in τ and τ is nonevasive with respect to T_0.
3. $T = (x, T_0, T_1)$ for some x in τ and $\tau - x$ is nonevasive with respect to T_1.

This means that T – viewed as an algorithm – ends up on a Win leaf on input τ; use induction. If a set $\tau \in \Delta$ is not nonevasive, then τ is *evasive*. For example, the edge 24 is the only evasive face with respect to the element-decision tree in Figure 5.1. The following simple but powerful theorem is a generalization by Forman [50] of an observation by Kahn, Saks, and Sturtevant [78].

Theorem 5.2 (Forman [50]). *Let Δ be a finite family of finite sets and let T be an element-decision tree on Δ. Then there is an acyclic matching on Δ such that the critical sets are precisely the evasive sets in Δ with respect to T. In particular, if Δ is a simplicial complex, then Δ is homotopy equivalent to a CW complex with exactly one cell of dimension p for each evasive set in Δ of dimension p and one addition 0-cell.*

Proof. Use induction on the size of T. It is easy to check that the theorem holds if $T = \text{Win}$ or $T = \text{Lose}$; match $\emptyset$ and v if $\Delta = \{\emptyset, v\}$ and $T = \text{Win}$. Suppose that $T = (x, T_0, T_1)$. By induction, there is an acyclic matching on $\text{del}_\Delta(x)$ with critical sets exactly those σ in $\text{del}_\Delta(x)$ that are evasive with respect to T_0. Also, there is an acyclic matching on $\text{lk}_\Delta(x)$ with critical sets exactly those τ in $\text{lk}_\Delta(x)$ that are evasive with respect to T_1. Combining these two matchings in the obvious manner, we have a matching with critical sets exactly the evasive sets with respect to T; by Lemma 4.3, the matching is acyclic. $\square$

5.1.2 Set-Decision Trees

We provide a natural generalization of the concept of element-decision trees.

Definition 5.3. The class of *set-decision trees*, each associated to a finite family of finite sets, is defined recursively as follows:

(i) $T = \mathsf{Win}$ is a set-decision tree on $\emptyset$ and on any 0-simplex $\{\emptyset, \{v\}\}$.

(ii) $T = \mathsf{Lose}$ is a set-decision tree on $\{\emptyset\}$ and on any singleton set $\{\{v\}\}$.

(iii) If Δ is a family of sets, if σ is a nonempty set, if T_0 is a set-decision tree on $\mathrm{fdel}_\Delta(\sigma)$, and if T_1 is a set-decision tree on $\mathrm{lk}_\Delta(\sigma)$, then the triple (σ, T_0, T_1) is a set-decision tree on Δ.

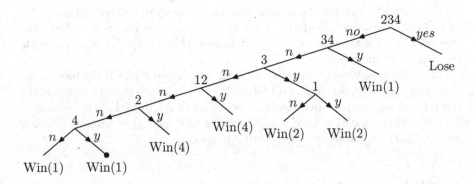

Fig. 5.2. A set-decision tree on the simplicial complex with maximal faces $123, 124, 134, 234$.

A simple example is provided in Figure 5.2. A set $\tau \in \Delta$ is *nonevasive* with respect to a set-decision tree T on Δ if either of the following holds:

1. $T = \mathsf{Win}$.
2. $T = (\sigma, T_0, T_1)$ for some $\sigma \not\subseteq \tau$ and τ is nonevasive with respect to T_0.
3. $T = (\sigma, T_0, T_1)$ for some $\sigma \subseteq \tau$ and $\tau \setminus \sigma$ is nonevasive with respect to T_1.

If a set $\tau \in \Delta$ is not nonevasive, then τ is *evasive*.

Theorem 5.4. *Let Δ be a finite family of finite sets and let T be a set-decision tree on Δ. Then there is an acyclic matching on Δ such that the critical sets are precisely the evasive sets in Δ with respect to T. Conversely, given an acyclic matching $\mathcal{M}$ on Δ, there is a set-decision tree T on Δ such that the evasive sets are precisely the critical sets with respect to $\mathcal{M}$.*

Proof. For the first part, the proof is identical to the proof of Theorem 5.2. For the second part, first consider the case that Δ is a complex as in (i) or (ii) in Definition 5.3. If $\Delta = \emptyset$, then $T = \mathsf{Win}$ is a set-decision tree with the desired properties, whereas $T = \mathsf{Lose}$ is the desired tree if $\Delta = \{\emptyset\}$ or $\Delta = \{\{v\}\}$. For $\Delta = \{\emptyset, \{v\}\}$, $T = \mathsf{Win}$ does the trick if $\emptyset$ and $\{v\}$ are matched, whereas $T = (v, \mathsf{Lose}, \mathsf{Lose})$ is the tree we are looking for if $\emptyset$ and $\{v\}$ are not matched.

Now, assume that Δ is some other family. Pick an arbitrary set $\rho \in \Delta$ of maximum size and go backwards in the digraph D of the matching $\mathcal{M}$ until a source σ in D is found; σ being a source means that there are no edges directed to σ. Such a σ exists, as D is acyclic. It is obvious that $|\rho| - 1 \leq |\sigma| \leq |\rho|$; in any directed path in D, a step up is always followed by and preceded by a step down (unless the step is the first or the last in the path). In particular, σ is adjacent in D to any set τ containing σ. Since σ is matched with at most one such τ and since σ is a source in D, there is at most one set containing σ.

First, suppose that σ is contained in a set τ and hence matched with τ in $\mathcal{M}$. By induction, there is a set-decision tree T_0 on $\mathrm{fdel}_\Delta(\sigma) = \Delta \setminus \{\sigma, \tau\}$ with evasive sets exactly the critical sets with respect to the restriction of $\mathcal{M}$ to $\mathrm{fdel}_\Delta(\sigma)$. Moreover, $\mathrm{lk}_\Delta(\sigma) = \{\emptyset, \tau \setminus \sigma\}$. Since $T_1 = \mathsf{Win}$ is a set-decision tree on $\mathrm{lk}_\Delta(\sigma)$ with no evasive sets, it follows that (σ, T_0, T_1) is a tree with the desired properties.

Next, suppose that σ is maximal in Δ and hence critical. By induction, there is a set-decision tree T_0 on $\mathrm{fdel}_\Delta(\sigma) = \Delta \setminus \{\sigma\}$ with evasive sets exactly the critical sets with respect to the restriction of $\mathcal{M}$ to $\mathrm{fdel}_\Delta(\sigma)$. Moreover, $\mathrm{lk}_\Delta(\sigma) = \{\emptyset\}$; since $T_1 = \mathsf{Lose}$ is a set-decision tree on $\mathrm{lk}_\Delta(\sigma)$ with one evasive set, (σ, T_0, T_1) is a tree with the desired properties. $\square$

5.2 Hierarchy of Almost Nonevasive Complexes

The purpose of this section is to introduce two families of complexes related to the concept of decision trees:

- *Semi-nonevasive complexes* admit an element-decision tree with evasive faces enumerated by the reduced Betti numbers over a given field.
- *Semi-collapsible complexes* admit a set-decision tree with evasive faces enumerated by the reduced Betti numbers over a given field. Equivalently, such complexes admit an acyclic matching with critical faces enumerated by reduced Betti numbers.

One may view these families as generalizations of the well-known families of nonevasive and collapsible complexes defined in Section 3.4:

- *Nonevasive complexes* admit an element-decision tree with no evasive faces.
- *Collapsible complexes* admit a set-decision tree with no evasive faces. Equivalently, such complexes admit a perfect acyclic matching.

In Section 5.2.2, we discuss how all these classes relate to well-known properties such as being shellable and vertex-decomposable. The main conclusion is that the families of semi-nonevasive and semi-collapsible complexes contain the families of vertex-decomposable and shellable complexes, respectively.

Remark. One may characterize semi-collapsible complexes as follows. Given an acyclic matching on a simplicial complex Δ, we may order the critical faces as $\sigma_1, \ldots, \sigma_n$ and form a sequence $\emptyset = \Delta_0 \subset \Delta_1 \subset \cdots \subset \Delta_{n-1} \subset \Delta_n \subseteq \Delta$ of simplicial complexes such that the following is achieved: Δ is collapsible to

Δ_n, σ_i is a maximal face of Δ_i, and $\Delta_i \setminus \{\sigma_i\}$ is collapsible to Δ_{i-1} for $i \in [n]$; compare to the induction proof of Theorem 5.4 (see also Forman [49, Th. 3.3-3.4]). A matching being optimal means that σ_i is contained in a nonvanishing cycle in the homology of Δ_i for each $i \in [n]$; otherwise the removal of σ_i would introduce new homology, rather than kill existing homology. With an "elementary semi-collapse" defined either as an ordinary elementary collapse or as the removal of a maximal face contained in a cycle, semi-collapsible complexes are exactly those complexes that can be transformed into the void complex via a sequence of elementary semi-collapses.

5.2.1 Semi-nonevasive and Semi-collapsible Complexes

Let $\mathbb{F}$ be a field or $\mathbb{Z}$. A set-decision tree (equivalently, an acyclic matching) on a simplicial complex Δ is $\mathbb{F}$-*optimal* if, for each integer i, $\dim \tilde{H}_i(\Delta; \mathbb{F})$ is the number of evasive (critical) faces of dimension i; $\dim \tilde{H}_i(\Delta; \mathbb{Z})$ is the rank of the torsion-free part of $\tilde{H}_i(\Delta; \mathbb{Z})$. We define $\mathbb{F}$-optimal element-decision trees analogously. In this section, we define the classes of simplicial complexes that admit $\mathbb{F}$-optimal element-decision or set-decision trees. See Forman [50] and Soll [129] for more discussion on optimal decision trees.

Definition 5.5. We define the class of *semi-nonevasive* simplicial complexes over $\mathbb{F}$ recursively as follows:

(i) The void complex $\emptyset$, the (-1)-simplex $\{\emptyset\}$, and any 0-simplex $\{\emptyset, \{v\}\}$ are semi-nonevasive over $\mathbb{F}$.
(ii) Suppose Δ contains a 0-cell x – a *shedding vertex* – such that $\mathrm{del}_\Delta(x)$ and $\mathrm{lk}_\Delta(x)$ are semi-nonevasive over $\mathbb{F}$ and such that

$$\tilde{H}_d(\Delta; \mathbb{F}) \cong \tilde{H}_d(\mathrm{del}_\Delta(x); \mathbb{F}) \oplus \tilde{H}_{d-1}(\mathrm{lk}_\Delta(x); \mathbb{F}) \qquad (5.1)$$

for each d. Then Δ is semi-nonevasive over $\mathbb{F}$.

Definition 5.6. We define the class of *semi-collapsible* simplicial complexes over $\mathbb{F}$ recursively as follows:

(i) The void complex $\emptyset$, the (-1)-simplex $\{\emptyset\}$, and any 0-simplex $\{\emptyset, \{v\}\}$ are semi-collapsible over $\mathbb{F}$.
(ii) Suppose that Δ contains a nonempty face σ – a *shedding face* – such that $\mathrm{fdel}_\Delta(\sigma)$ and $\mathrm{lk}_\Delta(\sigma)$ are semi-collapsible over $\mathbb{F}$ and such that

$$\tilde{H}_d(\Delta; \mathbb{F}) \cong \tilde{H}_d(\mathrm{fdel}_\Delta(\sigma); \mathbb{F}) \oplus \tilde{H}_{d-|\sigma|}(\mathrm{lk}_\Delta(\sigma); \mathbb{F}) \qquad (5.2)$$

for each d. Then Δ is semi-collapsible over $\mathbb{F}$.

Clearly, a semi-nonevasive complex over $\mathbb{F}$ is also semi-collapsible over $\mathbb{F}$.

Remark. Let us discuss the identity (5.2); the discussion also applies to the special case (5.1). Let $\Delta_0 = \mathrm{fdel}_\Delta(\sigma)$. Note that the homology group $\tilde{H}_d(\Delta/\Delta_0) = \tilde{H}_d(\Delta/\Delta_0; \mathbb{F})$ is isomorphic to $\tilde{H}_{d-|\sigma|}(\mathrm{lk}_\Delta(\sigma))$ for each d. Assume that $\mathbb{F}$ is a field. By the long exact sequence

$$\cdots \longrightarrow \tilde{H}_d(\Delta_0) \longrightarrow \tilde{H}_d(\Delta) \longrightarrow \tilde{H}_d(\Delta/\Delta_0) \longrightarrow \tilde{H}_{d-1}(\Delta_0) \longrightarrow \cdots \qquad (5.3)$$

for the pair (Δ, Δ_0) (use Theorem 3.3), (5.2) is equivalent to the induced map $\partial_d^* : \tilde{H}_d(\Delta/\Delta_0) \longrightarrow \tilde{H}_{d-1}(\Delta_0)$ being zero for each d, where $\partial_d(z)$ is computed in $\tilde{C}(\Delta)$. This is the case if and only if, for every cycle $z \in \tilde{C}(\Delta/\Delta_0)$, there is a $c \in \tilde{C}(\Delta_0)$ with the same boundary as z in $\tilde{C}(\Delta)$. As an important special case, we have the following observation:

Proposition 5.7. *If* $\tilde{H}_d(\mathrm{fdel}_\Delta(\sigma); \mathbb{F}) = 0$ *whenever* $\tilde{H}_{d-|\sigma|+1}(\mathrm{lk}_\Delta(\sigma); \mathbb{F}) \neq 0$, *then* (5.2) *holds. Hence if* $\tilde{H}_d(\mathrm{del}_\Delta(x); \mathbb{F}) = 0$ *whenever* $\tilde{H}_d(\mathrm{lk}_\Delta(x); \mathbb{F}) \neq 0$, *then* (5.1) *holds.* $\square$

The main result of this section is as follows; we postpone the case $\mathbb{F} = \mathbb{Z}$ until the end of the section.

Theorem 5.8. *Let $\mathbb{F}$ be a field. A complex Δ is semi-collapsible over $\mathbb{F}$ if and only if Δ admits an $\mathbb{F}$-optimal set-decision tree (equivalently, an $\mathbb{F}$-optimal acyclic matching). Δ is semi-nonevasive over $\mathbb{F}$ if and only if Δ admits an $\mathbb{F}$-optimal element-decision tree.*

Proof. First, we show that every semi-collapsible complex Δ over $\mathbb{F}$ admits an $\mathbb{F}$-optimal set-decision tree. This is clear if Δ is as in (i) in Definition 5.6. Use induction and consider a complex derived as in (ii) in Definition 5.6. By induction, $\mathrm{fdel}_\Delta(\sigma)$ and $\mathrm{lk}_\Delta(\sigma)$ admit $\mathbb{F}$-optimal set-decision trees T_0 and T_1, respectively. Combining these two trees, we obtain a set-decision tree $T = (\sigma, T_0, T_1)$ on Δ. (5.2) immediately yields that the evasive faces of Δ are enumerated by the Betti numbers of Δ, and we are done.

Next, suppose that we have an $\mathbb{F}$-optimal set-decision tree $T = (\sigma, T_0, T_1)$; T_0 is a tree on $\mathrm{fdel}_\Delta(\sigma)$, whereas T_1 is a tree on $\mathrm{lk}_\Delta(\sigma)$. We have that $\dim \tilde{H}_d(\Delta) = e_d$, where e_d is the number of evasive faces of dimension d with respect to T. Let a_d and b_d be the number of evasive faces of dimension d with respect to the set-decision trees T_0 and T_1, respectively; clearly, $e_d = a_d + b_{d-|\sigma|}$. By Theorem 4.15, we must have $a_d \geq \dim \tilde{H}_d(\mathrm{fdel}_\Delta(\sigma))$ and $b_{d-|\sigma|} \geq \dim \tilde{H}_{d-|\sigma|}(\mathrm{lk}_\Delta(\sigma))$. We want to prove that equality holds for both a_d and $b_{d-|\sigma|}$. Namely, this will imply (5.2) and yield that T_0 and T_1 are $\mathbb{F}$-optimal set-decision trees; by induction, we will obtain that each of $\mathrm{fdel}_\Delta(\sigma)$ and $\mathrm{lk}_\Delta(\sigma)$ is semi-collapsible and hence that Δ is semi-collapsible. Now, the long exact sequence (5.3) immediately yields that

$$e_d = \dim \tilde{H}_d(\Delta) \leq \dim \tilde{H}_d(\mathrm{fdel}_\Delta(\sigma)) + \dim \tilde{H}_{d-|\sigma|}(\mathrm{lk}_\Delta(\sigma)).$$

Since the right-hand side is bounded by $a_d + b_{d-|\sigma|} = e_d$, the inequality must be an equality; thus (5.2) holds, and we are done.

The last statement in the theorem is proved in the same manner. $\square$

Proposition 5.9. *If a simplicial complex Δ is semi-collapsible over $\mathbb{Q}$, then the $\mathbb{Z}$-homology of Δ is torsion-free; hence $\tilde{H}_d(\Delta; \mathbb{Z}) = \mathbb{Z}^{\beta_d}$, where $\beta_d = \dim \tilde{H}_d(\Delta; \mathbb{Q})$. It follows that semi-nonevasive complexes over $\mathbb{Q}$ have torsion-free $\mathbb{Z}$-homology.*

Proof. This is obvious if (i) in Definition 5.6 holds. Suppose (ii) holds. By induction, the proposition is true for $\mathrm{fdel}_\Delta(\sigma)$ and $\mathrm{lk}_\Delta(\sigma)$. By the remark after Definition 5.6, for every cycle $z \in \tilde{C}(\Delta/\mathrm{fdel}_\Delta(\sigma); \mathbb{Q})$, there is a $c \in \tilde{C}(\mathrm{fdel}_\Delta(\sigma); \mathbb{Q})$ with the same boundary as z in $\tilde{C}(\Delta; \mathbb{Q})$. As a consequence, for every cycle $z \in \tilde{C}(\Delta/\mathrm{fdel}_\Delta(\sigma); \mathbb{Z})$, there is a $c \in \tilde{C}(\mathrm{fdel}_\Delta(\sigma); \mathbb{Z})$ and an integer λ such that $\partial(c) = \lambda\partial(z)$ (computed in $\tilde{C}(\Delta; \mathbb{Z})$). However, since $\tilde{H}(\mathrm{fdel}_\Delta(\sigma); \mathbb{Z})$ is torsion-free, $\lambda\partial(z)$ is a boundary in $\tilde{C}(\mathrm{fdel}_\Delta(\sigma); \mathbb{Z})$ if and only if $\partial(z)$ is a boundary, which implies that there exists a $c' \in \tilde{C}(\mathrm{fdel}_\Delta(\sigma); \mathbb{Z})$ such that $\partial(c') = \partial(z)$. As a consequence, $\partial_d^* : \tilde{H}_d(\Delta/\mathrm{fdel}_\Delta(\sigma); \mathbb{Z}) \longrightarrow \tilde{H}_{d-1}(\mathrm{fdel}_\Delta(\sigma); \mathbb{Z})$ is the zero map. Hence (5.2) holds for $\mathbb{F} = \mathbb{Z}$, and we are done. $\square$

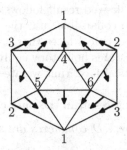

Fig. 5.3. An acyclic matching on a triangulated projective plane with critical faces 23 and 456; 1 is matched with $\emptyset$. This matching is $\mathbb{Z}_2$-optimal but not $\mathbb{Q}$-optimal.

Corollary 5.10. *A complex Δ is semi-collapsible (semi-nonevasive) over $\mathbb{Q}$ if and only if Δ is semi-collapsible (semi-nonevasive) over $\mathbb{Z}$. If this is the case, then Δ is semi-collapsible (semi-nonevasive) over every field.* $\square$

Remark. While the universal coefficient theorem implies that Proposition 5.9 is true for any field of characteristic 0, the proposition does not remain true for coefficient fields of nonzero characteristic. For example, the triangulated projective plane $\mathbb{RP}^2$ in Figure 5.3 is not semi-collapsible over $\mathbb{Q}$, as the homology has torsion. However, the given acyclic matching is $\mathbb{Z}_2$-optimal; $\tilde{H}_1(\mathbb{RP}^2; \mathbb{Z}_2) = \tilde{H}_2(\mathbb{RP}^2; \mathbb{Z}_2) = \mathbb{Z}_2$. In fact, the acyclic matching corresponds to a $\mathbb{Z}_2$-optimal element-decision tree in which we first use 4, 5, and 6 as shedding vertices; thus the complex is semi-nonevasive over $\mathbb{Z}_2$. A semi-nonevasive complex over $\mathbb{Z}_3$ with 3-torsion is provided in Theorem 11.27.

5.2.2 Relations Between Some Important Classes of Complexes

We show how semi-collapsible and semi-nonevasive complexes over $\mathbb{Z}$ relate to vertex-decomposable (VD), shellable, and constructible complexes; see Section 3.6 for definitions. Throughout this section, whenever we refer to a complex as semi-nonevasive or semi-collapsible, we mean over $\mathbb{Z}$ unless otherwise stated.

Chari [32] proved that shellable complexes are semi-collapsible. Let us extend his result to semipure shellable complexes.

Proposition 5.11. *Let Δ be a semipure shellable complex. Then Δ admits an acyclic matching in which all unmatched faces are maximal faces of Δ. In particular, any semipure shellable complex is semi-collapsible.*

Proof. The proposition is clearly true if (i) in Definition 3.26 is satisfied. Suppose (ii) is satisfied. By induction, $\mathrm{fdel}_\Delta(\sigma)$ and $\mathrm{lk}_\Delta(\sigma)$ admit acyclic matchings such that all unmatched faces are maximal faces. Combining these matchings, we obtain an acyclic matching on Δ. Since maximal faces of $\mathrm{fdel}_\Delta(\sigma)$ are maximal faces of Δ, the desired result follows.

To prove that Δ is semi-collapsible, use the fact that we cannot have a directed path between two critical faces that are both maximal. Applying Corollary 4.13, we obtain that Δ is homotopy equivalent to a wedge of spheres with one sphere for each critical face; thus we are done. $\square$

Soll [129] proved the following result in the pure case.

Proposition 5.12. *Semipure VD complexes are semi-nonevasive.*

Proof. Use exactly the same approach as in the proof of Proposition 5.11. $\square$

Proposition 5.13. *Not all shellable complexes are semi-nonevasive.*

Proof. The complex with maximal faces 012, 023, 034, 045, 051, 123, 234, 345, 451, and 512 is well-known to be shellable and collapsible but not nonevasive or VD. This complex is originally due to Björner (personal communication); see Moriyama and Takeuchi [100, Ex. V6F10-6] and Soll [129, Ex. 5.5.5]. $\square$

Proposition 5.14. *Not all constructible complexes are semi-collapsible. Yet, there exist constructible complexes that are nonevasive but not shellable.*

Proof. For the first statement, Hachimori [56] has found a two-dimensional contractible and constructible complex without boundary; a complex with no boundary cannot be collapsible. For the second statement, a cone over a constructible complex is constructible and nonevasive but not shellable unless the original complex is shellable. $\square$

Let us introduce two classes of complexes closely related to the class of constructible complexes. For the purposes of this section, we refer to them as "buildable" and "semi-buildable", but there might be better terms.

Definition 5.15. We define the class of *buildable* simplicial complexes recursively as follows:

 (i) The void complex $\emptyset$ and any d-simplex such that $d \geq 0$ are buildable.
 (ii) Suppose that Δ_1, Δ_2, and $\Gamma = \Delta_1 \cap \Delta_2$ are buildable complexes. Then $\Delta_1 \cup \Delta_2$ is buildable.

Definition 5.16. We define the class of *semi-buildable* simplicial complexes over $\mathbb{F}$ recursively as follows:

 (i) Any simplex (including $\emptyset$ and $\{\emptyset\}$) is semi-buildable over $\mathbb{F}$.
 (ii) Suppose that Δ_1, Δ_2, and $\Gamma = \Delta_1 \cap \Delta_2$ are semi-buildable complexes over $\mathbb{F}$ and that

$$\tilde{H}_d(\Delta_1 \cup \Delta_2; \mathbb{F}) \cong \tilde{H}_d(\Delta_1; \mathbb{F}) \oplus \tilde{H}_d(\Delta_2; \mathbb{F}) \oplus \tilde{H}_{d-1}(\Delta_1 \cap \Delta_2; \mathbb{F}) \quad (5.4)$$

for each d.

Then $\Delta_1 \cup \Delta_2$ is semi-buildable.

Proposition 5.17. *Collapsible complexes are buildable and buildable complexes are contractible.*

Proof. Suppose that Δ is a collapsible complex. If Δ is a simplex, then we are done. Otherwise, let σ be a face such that $\mathrm{fdel}_\Delta(\sigma)$ and $\mathrm{lk}_\Delta(\sigma)$ are collapsible. Write $\Delta_1 = \mathrm{fdel}_\Delta(\sigma)$ and $\Delta_2 = 2^\sigma * \mathrm{lk}_\Delta(\sigma)$. We have that $\Delta_1 \cap \Delta_2 = \partial 2^\sigma * \mathrm{lk}_\Delta(\sigma)$ and $\Delta_1 \cup \Delta_2 = \Delta$. By induction, we obtain that Δ_1, Δ_2, and $\Delta_1 \cap \Delta_2$ are buildable, which implies by definition that $\Delta_1 \cup \Delta_2$ is buildable.

Next, suppose that Δ is a buildable complex. Δ is clearly contractible if Δ is a simplex. Otherwise, suppose that $\Delta = \Delta_1 \cup \Delta_2$, where Δ_1, Δ_2, and $\Delta_1 \cap \Delta_2$ are buildable and hence contractible by induction. By Corollary 3.9, we obtain that Δ is k-connected for every k; hence Theorem 3.8 implies that Δ is contractible as desired. $\square$

Proposition 5.18. *Semi-collapsible complexes over $\mathbb{F}$ are semi-buildable over $\mathbb{F}$. Moreover, semi-buildable complexes over $\mathbb{Z}$ have torsion-free homology.*

Proof. Let Δ be a semi-collapsible complex. If Δ is a simplex, then we are done. Otherwise, suppose that σ is a shedding face of Δ. Write $\Delta_1 = \mathrm{fdel}_\Delta(\sigma)$ and $\Delta_2 = 2^\sigma * \mathrm{lk}_\Delta(\sigma)$. Since $\Delta_1 \cap \Delta_2 = \partial 2^\sigma * \mathrm{lk}_\Delta(\sigma)$, (5.2) is equivalent to (5.4); Δ_2 is contractible and $\tilde{H}_{d-1}(\Delta_1 \cap \Delta_2; \mathbb{F}) = \tilde{H}_{d-|\sigma|}(\mathrm{lk}_\Delta(\sigma); \mathbb{F})$ by Corollary 4.23.

The second statement is immediate from (5.4). $\square$

Proposition 5.19. *Semipure constructible complexes are semi-buildable over $\mathbb{Z}$.*

Proof. Let Δ be semipure constructible. If Δ is a simplex, then we are done. Otherwise, suppose that $\Delta = \Delta_1 \cup \Delta_2$, where Δ_1, Δ_2, and $\Delta_1 \cap \Delta_2$ are semipure constructible complexes satisfying the conditions in (ii) in Definition 3.24. We need to prove that (5.4) holds.

By Theorem 3.33 and Proposition 3.36, we know that the homology of any semipure constructible complex is torsion-free. Hence it suffices to prove that the homomorphism

$$\iota_i^* : \tilde{H}_d(\Delta_1 \cap \Delta_2; \mathbb{Z}) \to \tilde{H}_d(\Delta_i; \mathbb{Z})$$

induced by the inclusion map is zero for $i \in \{1, 2\}$; apply the Mayer-Vietoris sequence (Theorem 3.1). Now, by construction, $\Delta_1 \cap \Delta_2 \subseteq \Delta_i \setminus \mathcal{F}(\Delta_i)$, where $\mathcal{F}(\Delta_i)$ is the family of maximal faces of Δ_i. Hence we can write ι_i^* as a composition

$$\tilde{H}_d(\Delta_1 \cap \Delta_2; \mathbb{Z}) \to \tilde{H}_d(\Delta_i \setminus \mathcal{F}(\Delta_i); \mathbb{Z}) \to \tilde{H}_d(\Delta_i; \mathbb{Z})$$

of maps induced by inclusion maps. As a consequence, Theorem 3.33 and Proposition 3.36 yield that ι_i^* is indeed zero. $\square$

Semi-buildable complexes are well-behaved in the following sense:

Proposition 5.20. *Let Δ be a semi-buildable complex over $\mathbb{Z}$. Then Δ is k-acyclic over $\mathbb{Z}$ if and only if Δ is k-connected.*

Proof. By Theorem 3.8, it suffices to prove that Δ is simply connected whenever Δ is 1-acyclic. This is clear if Δ is a simplex. Otherwise, we have that $\Delta = \Delta_1 \cup \Delta_2$ and that Δ_1, Δ_2, and $\Delta_1 \cap \Delta_2$ are semi-buildable complexes satisfying (5.4). Since Δ is 1-acyclic, (5.4) implies that Δ_1 and Δ_2 are 1-acyclic and that $\Delta_1 \cap \Delta_2$ is 0-acyclic and hence 0-connected. Induction yields that Δ_1 and Δ_2 are simply connected. As a consequence, $\Delta_1 \cup \Delta_2$ is simply connected by Corollary 3.9. $\square$

The results in this section combined with earlier results (see Section 3.6.5 and Björner [9]) yield the diagram of implications in Figure 5.4; "torsion-free" refers to the $\mathbb{Z}$-homology. We conjecture that all implications are strict; this is known to be true in all but two cases:

Problem 5.21. Are there contractible complexes that are not buildable? Are there homotopically Cohen-Macaulay complexes that are not semi-buildable?

We conjecture that any triangulation of the dunce hat [150] is non-buildable; this complex is known to be contractible, non-collapsible, Cohen-Macaulay, and non-constructible.

Ignoring buildable and semi-buildable complexes, two properties are incomparable in the diagram if and only if neither of the properties implies the other. We list the nontrivial cases:

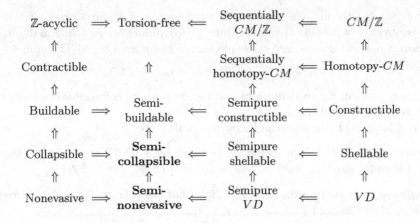

Fig. 5.4. Implications between different classes of simplicial complexes.

- *Collapsible or shellable complexes are not always semi-nonevasive. This is Proposition 5.13.*
- *Contractible or constructible complexes are not always semi-collapsible. This is Proposition 5.14.*

5.3 Some Useful Constructions

Before proceeding, let us introduce some simple but useful constructions that will be used frequently in later sections. For a family Δ of sets, write $\Delta \sim \sum_{i \geq -1} a_i t^i$ if there is an element-decision tree on Δ with exactly a_i evasive sets of dimension i for each $i \geq -1$. This notation has the following basic properties; recall from Section 2.3 that $\Delta(I, E) = \{I\} * \mathrm{lk}_{\mathrm{del}_\Delta(E)}(I)$.

Lemma 5.22. *Let Δ be a finite family of finite sets. Then the following hold:*

(1) *Δ is nonevasive if and only if $\Delta \sim 0$.*
(2) *Assume that Δ is a simplicial complex and let $\mathbb{F}$ be a field. Then Δ is semi-nonevasive over $\mathbb{F}$ if and only if $\Delta \sim \sum_{i \geq -1} \dim \tilde{H}_i(\Delta; \mathbb{F}) t^i$. Moreover, Δ is semi-nonevasive over $\mathbb{Z}$ if and only if $\Delta \sim \sum_{i \geq -1} \dim \tilde{H}_i(\Delta; \mathbb{Q}) t^i$.*
(3) *Let v be a 0-cell. If $\mathrm{del}_\Delta(v) \sim f_\emptyset(t)$ and $\mathrm{lk}_\Delta(v) \sim f_v(t)$, then $\Delta \sim f_\emptyset(t) + f_v(t)t$.*
(4) *Let B be a set of 0-cells. If $\Delta(A, B \setminus A) \sim f_A(t)$ (hence $\mathrm{lk}_{\mathrm{del}_\Delta(B \setminus A)}(A) \sim f_A(t)/t^{|A|}$) for each $A \subseteq B$, then $\Delta \sim \sum_{A \subseteq B} f_A(t)$.*
(5) *Assume that Δ is a simplicial complex such that $\Delta \sim ct^d$. Then Δ is semi-nonevasive and homotopy equivalent to a wedge of c spheres of dimension d.*
(6) *Assume that Δ is a simplicial complex such that $\Delta \sim f(t)t^d$ for some polynomial $f(t)$. Then Δ is $(d-1)$-connected.*

Proof. (1) is obvious. To prove (2), use Theorem 5.8 and Corollary 5.10. (3) is obvious, whereas (4) follows from (3) by induction on $|B|$. Finally, by Theorem 5.4, (5) and (6) are consequences of Theorem 4.8 and Theorem 4.7, respectively. $\square$

One may give analogous definitions and results for semi-collapsible complexes, but we will not need them.

In Chapter 14, we need the following result.

Lemma 5.23. *Let Δ be a simplicial complex on a set E. If $\Delta \sim f(t)$, then the Alexander dual Δ_E^* with respect to E satisfies $\Delta_E^* \sim t^{|E|-3} f(1/t)$.*

Proof. Use induction on the size of E; note that $\mathrm{del}_{\Delta_E^*}(x) = (\mathrm{lk}_\Delta(x))_{E-x}^*$ and $\mathrm{lk}_{\Delta^*(E)}(x) = (\mathrm{del}_\Delta(x))_{E-x}^*$. $\square$

Definition 5.24. *Let Δ be a finite family of finite sets. Let $W = (w_1, \ldots, w_m)$ be a sequence of distinct elements. The first-hit decomposition of Δ with respect to W is the sequence consisting of the families $\Delta(w_j, \{w_1, \ldots, w_{j-1}\})$ for $j \in [m]$ and the family $\Delta(\emptyset, \{w_1, \ldots, w_m\})$.*

The term "first-hit" refers to the natural interpretation of the concept in terms of decision trees; for a given set to be checked, query elements in the sequence until some element from the set is found (a first hit).

Lemma 5.25. *Let Δ be a finite family of finite sets and consider the first-hit decomposition of Δ with respect to a given sequence $(w_1, \ldots, w_m)$ of elements. Suppose that*

$$\Delta(w_j, \{w_1, \ldots, w_{j-1}\}) \sim f_j(t) \quad (j \in [m]);$$
$$\Delta(\emptyset, \{w_1, \ldots, w_m\}) \sim g(t).$$

Then $\Delta \sim g(t) + \sum_{j=1}^m f_j(t)$.

Remark. One may view Lemma 5.25 as a decision-tree version of a result about vertex-decomposability due to Athanasiadis [2, Lemma 2.2].

Proof. We claim that $\Delta(\emptyset, \{w_1, \ldots, w_i\}) \sim g(t) + \sum_{j=i+1}^m f_j(t)$ for $0 \leq i \leq m$; for $i = 0$, we obtain the lemma. The claim is obvious for $i = m$. For $i < m$, we may assume by induction that $\Delta(\emptyset, \{w_1, \ldots, w_{i+1}\}) \sim g(t) + \sum_{j=i+2}^m f_j(t)$. Since $\Delta(w_{i+1}, \{w_1, \ldots, w_i\}) \sim f_{i+1}(t)$, the claim follows by Lemma 5.22. $\square$

We conclude this section with a very simple example, just as an illustration.

Proposition 5.26. *Let $G = (V, E)$ be a simple connected graph with e edges and n vertices. Then $G \sim (e - n + 1)t$.*

Proof. G is clearly nonevasive if G has one vertex. Suppose that G has at least two vertices. Let v be a vertex such that the induced subgraph $G' = G(V \setminus \{v\}) = \mathrm{del}_G(v)$ obtained by removing v is connected (let v be a leaf in a spanning tree). By induction, we obtain that $\mathrm{del}_G(v) \sim (e - |N_v| - (n-1) + 1)t$. Moreover, $\mathrm{lk}_G(v)$ consists of the empty set and the vertices in $N_v = \{w : vw \in G\}$; clearly, $\mathrm{lk}_G(v) \sim (|N_v| - 1)$. By Lemma 5.22, $G \sim (|N_v| - 1)t + (e - |N_v| - n + 2)t = (e - n + 1)t$ as desired. $\square$

5.4 Further Properties of Almost Nonevasive Complexes

We examine to what extent semi-nonevasiveness and semi-collapsibility are preserved under join, barycentric subdivision, direct product, and Alexander duality. The results are either generalizations of results due to Welker [146] or generalizations of weaker results. Open problems are listed at the end of the section.

Theorem 5.27 (Welker [146]). *If at least one of Δ and Γ is collapsible (nonevasive), then the join $\Delta * \Gamma$ is collapsible (nonevasive). If $\Delta * \Gamma$ is nonevasive, then at least one of Δ and Γ is nonevasive.* $\square$

Theorem 5.28. *If Δ and Γ are both semi-collapsible (semi-nonevasive) over $\mathbb{F}$, then the join $\Delta * \Gamma$ is semi-collapsible (semi-nonevasive) over $\mathbb{F}$. If $\Delta * \Gamma$ is semi-nonevasive over $\mathbb{F}$ and evasive, then each of Δ and Γ is semi-nonevasive over $\mathbb{F}$ and evasive.*

Proof. First, consider semi-collapsibility. If Δ satisfies (i) in Definition 5.6, then $\Delta * \Gamma$ is either $\emptyset$, Γ, or a cone over Γ. Each of these complexes is semi-collapsible by assumption. Suppose Δ satisfies (ii) in Definition 5.6 with shedding face σ. By assumption, $\mathrm{fdel}_\Delta(\sigma)$ and $\mathrm{lk}_\Delta(\sigma)$ are both semi-collapsible, which implies by induction that $\mathrm{fdel}_{\Delta * \Gamma}(\sigma) = \mathrm{fdel}_\Delta(\sigma) * \Gamma$ and $\mathrm{lk}_{\Delta * \Gamma}(\sigma) = \mathrm{lk}_\Delta(\sigma) * \Gamma$ are semi-collapsible. For any complex Σ, let $\tilde{\beta}_\Sigma(t) = \sum_{i \geq -1} \dim \tilde{H}_i(\Sigma, \mathbb{F}) t^i$. By Corollary 4.23, we have that

$$\tilde{\beta}_{\Delta * \Gamma}(t)/t = \tilde{\beta}_\Delta(t)\tilde{\beta}_\Gamma(t) = (\tilde{\beta}_{\mathrm{fdel}_\Delta(\sigma)}(t) + t^{|\sigma|}\tilde{\beta}_{\mathrm{lk}_\Delta(\sigma)}(t))\tilde{\beta}_\Gamma(t)$$

$$= (\tilde{\beta}_{\mathrm{fdel}_\Delta(\sigma) * \Gamma}(t) + t^{|\sigma|}\tilde{\beta}_{\mathrm{lk}_\Delta(\sigma) * \Gamma}(t))/t,$$

where the second identity follows from the fact that (5.2) holds for Δ and σ. Thus (5.2) holds for $\Delta * \Gamma$ and σ, and we are done with the first statement. Join preserving semi-nonevasiveness is proved in exactly the same manner.

For the second statement, suppose that $\Delta * \Gamma$ is semi-nonevasive and evasive. If $\Delta * \Gamma = \{\emptyset\}$, then we are done. Otherwise, let x be the first shedding vertex; we may assume that $\{x\} \in \Delta$. Since $\Delta * \Gamma$ is evasive, either the link or the deletion (or both) with respect to x is evasive. By induction, if $\mathrm{del}_\Delta(x) * \Gamma$ is semi-nonevasive and evasive, then the same holds for both $\mathrm{del}_\Delta(x)$ and Γ.

If instead $\mathrm{del}_\Delta(x) * \Gamma$ is nonevasive, then $\mathrm{del}_\Delta(x)$ must be nonevasive by Theorem 5.27; Γ is evasive by assumption. Hence $\mathrm{del}_\Delta(x)$ is semi-nonevasive. A similar argument yields that $\mathrm{lk}_\Delta(x)$ is also semi-nonevasive. Since $\Delta * \Gamma$ and x satisfy (5.1), we obtain that

$$t\tilde{\beta}_\Delta(t)\tilde{\beta}_\Gamma(t) = \tilde{\beta}_{\Delta*\Gamma}(t) = \tilde{\beta}_{\mathrm{del}_\Delta(x)*\Gamma}(t) + t\tilde{\beta}_{\mathrm{lk}_\Delta(x)*\Gamma}(t)$$
$$= t(\tilde{\beta}_{\mathrm{del}_\Delta(x)}(t) + t\tilde{\beta}_{\mathrm{lk}_\Delta(x)}(t))\tilde{\beta}_\Gamma(t).$$

Γ being semi-nonevasive and evasive implies that $\tilde{\beta}_\Gamma(t)$ is nonzero and hence cancels out in this equation. As a consequence, $\tilde{\beta}_\Delta(t) = \tilde{\beta}_{\mathrm{del}_\Delta(x)}(t) + t\tilde{\beta}_{\mathrm{lk}_\Delta(x)}(t)$, which means exactly that (5.1) holds for Δ and x. We are thus done by induction. $\square$

Using exactly the same technique as in the proof of Theorem 5.28, one obtains the following more general result.

Theorem 5.29. *Let Δ and Γ be any families of sets on disjoint ground sets. With notation as in Section 5.3, if $\Delta \sim f(t)$ and $\Gamma \sim g(t)$, then $\Delta * \Gamma \sim tg(t)f(t)$. Moreover, suppose that T_Δ and T_Γ are decision trees on Δ and Γ, respectively. Then there is a decision tree $T_{\Delta*\Gamma}$ on $\Delta * \Gamma$ such that $\sigma \in \Delta$ and $\tau \in \Gamma$ are evasive with respect to T_Δ and T_Γ, respectively, if and only $\sigma \cup \tau$ is evasive with respect to $T_{\Delta*\Gamma}$. The analogous property holds for set-decision trees (i.e., acyclic matchings).* $\square$

Theorem 5.30 (Welker [146]). *If Δ is a collapsible simplicial complex, then the barycentric subdivision $\mathrm{sd}(\Delta)$ of Δ is nonevasive.* $\square$

Theorem 5.31. *If Δ is semi-collapsible over $\mathbb{F}$, then the barycentric subdivision $\mathrm{sd}(\Delta)$ of Δ is semi-nonevasive over $\mathbb{F}$.*

Remark. Theorems 5.30 and 5.31 are closely related to a theorem of Provan and Billera [108, Cor. 3.3.2] stating that $\mathrm{sd}(\Delta)$ is vertex-decomposable whenever Δ is shellable.

Proof. Throughout this proof, we will freely use the fact that homology is preserved under barycentric subdivision; this is Proposition 4.24. Write $\Sigma = \mathrm{sd}(\Delta)$. If Δ satisfies (i) in Definition 5.6, then Σ satisfies (i) in Definition 5.5. Suppose that Δ satisfies (ii) in Definition 5.6 with σ as the shedding face. Note that

$$\mathrm{lk}_\Sigma(\sigma) \cong \mathrm{sd}(\partial 2^\sigma) * \mathrm{sd}(\mathrm{lk}_\Delta(\sigma)).$$

Namely, each chain in $\mathrm{lk}_\Sigma(\sigma)$ consists of nonempty faces that are either proper subsets of σ (i.e., contained in $\partial 2^\sigma \setminus \{\emptyset\}$) or proper supersets of σ (i.e., of the form $\sigma \cup \tau$ for some $\tau \in \mathrm{lk}_\Delta(\sigma) \setminus \{\emptyset\}$). Since $\partial 2^\sigma$ and $\mathrm{lk}_\Delta(\sigma)$ are both semi-collapsible, the corresponding barycentric subdivisions are semi-nonevasive by induction on the size of Δ. By Theorem 5.28, this implies that $\mathrm{lk}_\Sigma(\sigma)$ is semi-nonevasive. By Corollary 4.23, we have that

$$\tilde{H}_i(\mathrm{lk}_\Sigma(\sigma)) \cong \bigoplus_{a+b=i-1} \tilde{H}_a(\partial 2^\sigma) \otimes \tilde{H}_b(\mathrm{lk}_\Delta(\sigma)) \cong \tilde{H}_{i+1-|\sigma|}(\mathrm{lk}_\Delta(\sigma)). \quad (5.5)$$

For the deletion $\mathrm{del}_\Sigma(\sigma)$, let $\tau_1, \ldots, \tau_r$ be the faces of Δ that properly contain σ, arranged in increasing order ($|\tau_i| < |\tau_j| \Rightarrow i < j$). Consider the first-hit decomposition of $\mathrm{del}_\Sigma(\sigma)$ with respect to $(\tau_1, \ldots, \tau_r)$; see Definition 5.24.

We have that

$$\Sigma(\tau_i, \{\sigma, \tau_1, \ldots, \tau_{i-1}\}) \cong \mathrm{sd}(\mathrm{fdel}_{2^{\tau_i}}(\sigma)) * \{\tau_i\} * \mathrm{sd}(\mathrm{lk}_\Delta(\tau_i)).$$

Namely, all faces ρ such that $\sigma \subset \rho \subset \tau_i$ are among the faces $\tau_1, \ldots, \tau_{i-1}$ and hence deleted, whereas all faces ρ such that $\tau_i \subset \rho$ are among the faces $\tau_{i+1}, \ldots, \tau_r$ and hence not yet deleted. It is clear that any element in $\tau_i \setminus \sigma$ is a cone point in $\mathrm{fdel}_{2^{\tau_i}}(\sigma)$, which implies by induction that the corresponding barycentric subdivision is nonevasive. By Theorem 5.27, it follows that $\Sigma(\tau_i, \{\sigma, \tau_1, \ldots, \tau_{i-1}\})$ is nonevasive.

Finally, $\Sigma(\emptyset, \{\sigma, \tau_1, \ldots, \tau_r\}) = \mathrm{sd}(\mathrm{fdel}_\Delta(\sigma))$, which is semi-nonevasive by induction. By Lemma 5.25 (and Proposition 5.7), $\mathrm{del}_\Sigma(\sigma)$ is semi-nonevasive with the same homology as $\mathrm{fdel}_\Delta(\sigma)$. By assumption, (5.2) holds for Δ and σ, which implies by (5.5) that (5.1) holds for Σ and σ, and we are done. $\square$

Before proceeding with direct products, we prove a lemma that may also be of some use in other situations. Let Δ and Γ be families of sets. Say that a map $\varphi : \Gamma \to \Delta$ is *order-preserving* if $\gamma_1 \subseteq \gamma_2$ implies that $\varphi(\gamma_1) \subseteq \varphi(\gamma_2)$. For $\sigma \in \Delta$, let $\Gamma_\sigma = \varphi^{-1}(\sigma)$.

Lemma 5.32. *For nonvoid finite families Δ and Γ of finite sets, let $\mathcal{M}_\Delta$ be an acyclic matching on Δ and let $\varphi : \Gamma \to \Delta$ be an order-preserving map. For each critical set ρ with respect to $\mathcal{M}_\Delta$, let $\mathcal{M}_\rho$ be an acyclic matching on Γ_ρ. For each matched pair $\{\sigma, \tau\}$ with respect to $\mathcal{M}_\Delta$, let $\mathcal{M}_{\sigma,\tau}$ be an acyclic matching on $\Gamma_\sigma \cup \Gamma_\tau$. Then the union $\mathcal{M}_\Gamma$ of all matchings $\mathcal{M}_\rho$ and $\mathcal{M}_{\sigma,\tau}$ is an acyclic matching on Γ.*

Remark. When $\mathcal{M}_\Delta$ is empty, Lemma 5.32 reduces to the Cluster Lemma 4.2.

Proof. Consider a set-decision tree T corresponding to $\mathcal{M}_\Delta$; use Theorem 5.4. If $\Delta = \{\emptyset\}$ or $\Delta = \{\emptyset, \{v\}\}$ with $\emptyset$ and $\{v\}$ matched, then the lemma is trivial since we consider the union of one single matching. Otherwise, suppose that $T = (\sigma, T_0, T_1)$. Let $\Gamma_D = \bigcup_{\tau \in \mathrm{fdel}_\Delta(\sigma)} \Gamma_\tau$ and $\Gamma_L = \bigcup_{\tau \in \mathrm{lk}_\Delta(\sigma)} \Gamma_{\sigma \cup \tau}$. By induction, the union of all matchings $\mathcal{M}_\rho$ and $\mathcal{M}_{\sigma,\tau}$ for $\rho, \sigma, \tau \in \mathrm{fdel}_\Delta(\sigma)$ is an acyclic matching on Γ_D; the analogous property also holds for Γ_L. Now, there are no edges directed from Γ_D to Γ_L in the digraph of $\mathcal{M}_\Gamma$. Namely, that would imply either that some $\gamma_0 \in \Gamma_D$ is matched with some $\gamma_1 \in \Gamma_L$ (which is impossible) or that some $\gamma_0 \in \Gamma_D$ contains some $\gamma_1 \in \Gamma_L$ (which contradicts the fact that φ is order-preserving). As a consequence, $\mathcal{M}_\Gamma$ is acyclic. $\square$

Theorem 5.33 (Welker [146]). *If P and Q are posets such that $\Delta(P)$ and $\Delta(Q)$ are both collapsible (nonevasive), then $\Delta(P \times Q)$ is collapsible (nonevasive). The converse is false for collapsible complexes.* $\square$

Remark. One easily adapts Welker's proof of Theorem 5.33 to a proof that $\Delta(P \times Q)$ is semi-nonevasive whenever $\Delta(P)$ is nonevasive and $\Delta(Q)$ is semi-nonevasive.

Theorem 5.34. *If P and Q are posets such that $\Delta(P)$ and $\Delta(Q)$ are both semi-collapsible over $\mathbb{F}$, then $\Delta(P \times Q)$ is semi-collapsible over $\mathbb{F}$. The converse is false.*

Proof. Our goal is to construct an optimal acyclic matching on $\Gamma = \Delta(P \times Q)$ given optimal acyclic matchings $\mathcal{M}_P$ and $\mathcal{M}_Q$ on $\Delta(P)$ and $\Delta(Q)$, respectively. For technical reasons, we leave the empty set unmatched in both matchings (hence the matchings are only almost optimal). For any complex Σ, let $\beta_\Sigma(t) = \sum_{i \geq 0} \dim H_i(\Sigma, \mathbb{F}) t^i$ (*unreduced* homology). Since $\Gamma \simeq \|\Delta(P)\| \times \|\Delta(Q)\|$ (see Björner [9, (9.6)]), Corollary 4.23 implies that $\beta_\Gamma(t) = \beta_{\Delta(P)}(t)\beta_{\Delta(Q)}(t)$. In particular, we want to find an acyclic matching with one critical face of size $i + j - 1$ for each pair of nonempty critical faces $\sigma \in \Delta(P)$ and $\tau \in \Delta(Q)$ of size i and j, respectively.

Let $\Pi_P : \Delta(P \times Q) \to \Delta(P)$ be the projection map; $\Pi_P(\{(x_i, y_i) : i \in I\}) = \{x_i : i \in I\}$. For $\sigma \in \Delta(P)$, let $\Gamma_\sigma = \Pi_P^{-1}(\sigma)$. It is clear that Π_P is order-preserving. As a consequence, given an acyclic matching on $\Gamma_{\sigma_1} \cup \Gamma_{\sigma_2}$ for each pair $\{\sigma_1, \sigma_2\} \in \mathcal{M}_P$ and an acyclic matching on Γ_ρ for each critical face ρ with respect to $\mathcal{M}_P$, Lemma 5.32 yields that the union of all these matchings is an acyclic matching on Γ.

First, let us use a construction from Welker's proof [146] of Theorem 5.33 to obtain a perfect matching on $\Gamma_{\sigma_1} \cup \Gamma_{\sigma_2}$ for each $\{\sigma_1, \sigma_2\} \in \mathcal{M}_P$; $\sigma_2 = \sigma_1 + x$. Since σ_2 contains at least two elements, x is either not maximal or not minimal in σ_2; by symmetry, we may assume that x is not maximal. Let x' be the smallest element in σ_2 that is larger than x. For a given element γ in $\Gamma_{\sigma_1} \cup \Gamma_{\sigma_2}$ let b_γ be minimal such that $(x', b_\gamma) \in \Gamma$. We obtain a perfect matching by matching $\gamma - (x, b_\gamma)$ with $\gamma + (x, b_\gamma)$. Namely, adding or removing (x, b_γ) does not affect b_γ, and adding (x, b_γ) leads to a new chain due to the minimality of b_γ. The matching is acyclic, as it corresponds to an element-decision tree in which we first query all elements (a, b) such that $a \neq x$ and then query all remaining elements except (x, b_γ) (which only depends on elements queried in the first round).

Next, we want to find a matching on Γ_ρ for each critical face ρ of $\Delta(P)$. Consider the order-preserving projection map $\Pi_Q : \Gamma_\rho \to \Delta(Q)$ and let $\Gamma_{\rho,\tau} = \Pi_Q^{-1}(\tau)$. By Lemma 5.32, given acyclic matchings on $\Gamma_{\rho,\tau_1} \cup \Gamma_{\rho,\tau_2}$ for $\{\tau_1, \tau_2\} \in \mathcal{M}_Q$ and acyclic matchings on $\Gamma_{\rho,\tau}$ for τ critical, the union of all matchings is an acyclic matching on Γ_ρ. We easily obtain a perfect acyclic matching on $\Gamma_{\rho,\tau_1} \cup \Gamma_{\rho,\tau_2}$ in exactly the same manner as we obtained the matching on $\Gamma_{\sigma_1} \cup \Gamma_{\sigma_2}$ above. What remains is the family $\Gamma_{\rho,\tau}$ for each pair of nonempty critical faces $\rho \in \Delta(P)$ and $\tau \in \Delta(Q)$. Write $\rho = x_1 x_2 \cdots x_k$ and $\tau = y_1 y_2 \cdots y_r$; $x_i < x_{i+1}$ and $y_j < y_{j+1}$. It is clear that every face of $\Gamma_{\rho,\tau}$ contains (x_1, y_1). We use induction on $k = |\rho|$ to show that there is an element-decision tree on $\Gamma_{\rho,\tau}$ with exactly one critical face of size $|\rho| + |\tau| - 1$; this will yield the theorem.

For $|\rho| = 1$, $\Gamma_{\rho,\tau}$ consists of one single face of size $|\tau| = |\rho| + |\tau| - 1$. For $|\rho| > 1$, note that the deletion $\Gamma_{\rho,\tau}(\emptyset, (x_1, y_1))$ is void; (x_1, y_1) is present in every face of $\Gamma_{\rho,\tau}$. Write $\Lambda = \Gamma_{\rho,\tau}((x_1, y_1), \emptyset)$ and proceed with the first-hit decomposition of Λ with respect to $((x_2, y_1), (x_2, y_2), \ldots, (x_2, y_k))$; see Definition 5.24. We have that

$$\Lambda((x_2, y_1), \emptyset) = \{\{(x_1, y_1)\}\} * \Gamma_{\rho - x_1, \tau}((x_2, y_1), \emptyset).$$

By induction, $\Gamma_{\rho - x_1, \tau}((x_2, y_1), \emptyset)$ admits an element-decision tree with one critical face of size $|\rho| - 1 + |\tau| - 1$. Adding (x_1, y_1) yields a face of the desired size $|\rho| + |\tau| - 1$. In $\Lambda_i = \Lambda((x_2, y_i), \{(x_2, y_j) : j < i\})$, (x_1, y_i) is a cone point. Namely, we may add the element without destroying the chain structure, and we may delete it, because x_1 and y_i are already contained in (x_1, y_1) and (x_2, y_i), respectively. Thus Λ_i is nonevasive, and we are done by Lemma 5.25.

The final statement is an immediate consequence of Theorem 5.33. □

Proposition 5.35 (Welker [146]). *A simplicial complex Δ on a set X is nonevasive if and only if the Alexander dual Δ_X^* is nonevasive. However, the Alexander dual of a collapsible complex is not necessarily collapsible.* □

Proposition 5.36. *A simplicial complex Δ on a set X is semi-nonevasive over $\mathbb{F}$ if and only if the Alexander dual Δ_X^* is semi-nonevasive over $\mathbb{F}$. However, the Alexander dual of a semi-collapsible complex is not necessarily semi-collapsible.*

Proof. Use induction on the size of X; $\mathrm{del}_{\Delta_X^*}(x) = (\mathrm{lk}_\Delta(x))_{X-x}^*$ and $\mathrm{lk}_{\Delta_X^*}(x) = (\mathrm{del}_\Delta(x))_{X-x}^*$. By (3.4), (5.1) holds for Δ_X^* if and only if it holds for Δ. In the base case, we have the Alexander dual of $\emptyset$, $\{\emptyset\}$, or $\{\emptyset, \{v\}\}$; all three duals are easily seen to be semi-nonevasive over any field. For the final statement, a contractible complex is collapsible if and only if the complex is semi-collapsible. This implies by Proposition 5.35 that the Alexander dual of a semi-collapsible complex is not necessarily semi-collapsible. □

Finally, we present a few important open problems; some of them are due to Welker [146].

Problem 5.37. If $\Delta * \Gamma$ is collapsible, is it true that at least one of Δ and Γ is collapsible?

Problem 5.38. If $\Delta * \Gamma$ is semi-collapsible but not collapsible, is it true that each of Δ and Γ is semi-collapsible?

Problem 5.39. If the barycentric subdivision of Δ is nonevasive, is it true that Δ is collapsible? More generally, if the barycentric subdivision of Δ is semi-nonevasive, is it true that Δ is semi-collapsible?

Problem 5.40. If the barycentric subdivision of Δ is collapsible, is it true that this subdivision is in fact nonevasive? More generally, if the barycentric subdivision of Δ is semi-collapsible, is it true that this subdivision is in fact semi-nonevasive?

Problem 5.41. If $\Delta(P \times Q)$ is nonevasive, is it true that $\Delta(P)$ and $\Delta(Q)$ are both nonevasive?

Problem 5.42. If $\Delta(P)$ and $\Delta(Q)$ are semi-nonevasive and evasive, is it true that $\Delta(P \times Q)$ is semi-nonevasive?

Many of these problems are likely to be very difficult.

5.5 A Potential Generalization

The following is a potential generalization of the concept of semi-collapsibility:

Definition 5.43. Let $\mathcal{C}$ be a family of simplicial complexes. We define the class of $\mathcal{C}$-*collapsible* simplicial complexes over the field $\mathbb{F}$ recursively as follows:

(i) The void complex $\emptyset$ and any complex isomorphic to a complex in $\mathcal{C}$ are $\mathcal{C}$-collapsible over $\mathbb{F}$.
(ii) If Δ contains a nonempty face σ such that $\mathrm{lk}_\Delta(\sigma)$ and $\mathrm{fdel}_\Delta(\sigma)$ are both $\mathcal{C}$-collapsible over $\mathbb{F}$ and such that

$$\tilde{H}_d(\Delta; \mathbb{F}) \cong \tilde{H}_d(\mathrm{fdel}_\Delta(\sigma); \mathbb{F}) \oplus \tilde{H}_{d-|\sigma|}(\mathrm{lk}_\Delta(\sigma); \mathbb{F})$$

for each d, then Δ is $\mathcal{C}$-collapsible over $\mathbb{F}$.

$\mathcal{C}$-nonevasive complexes are defined analogously. Note that if $\mathcal{C}$ consists of $\{\emptyset, \{v\}\}$, then we obtain the collapsible complexes, whereas the family consisting of $\{\emptyset\}$ and $\{\emptyset, \{v\}\}$ yields the semi-collapsible complexes. We do not know whether the given generalization leads to anything useful.

6

Miscellaneous Results

We present some results, mainly from the literature, that will be of some use in later sections. Section 6.1 is devoted to the topology of posets. Section 6.2 contains a discussion about the concept of depth, whereas Section 6.3 deals with the related concept of vertex-decomposability. In Section 6.4 at the end of the chapter, we present a few simple enumerative results.

6.1 Posets

Let P and Q be posets. A *poset map* $f : P \to Q$ is a function with the property that $f(x) \leq f(y)$ whenever $x \leq y$. A poset map $f : P \to P$ is a *closure operator* if $f(x) \geq x$ and $f(f(x)) = f(x)$.

Lemma 6.1 (Closure Lemma; see Björner [9]). *Let P be a poset and let $f : P \to P$ be a closure operator on P. Then $\Delta(P)$ and $\Delta(f(P))$ are homotopy equivalent, and f induces an isomorphism between the homology of $\Delta(P)$ and $\Delta(f(P))$.*

Proof. Write $\Sigma_1 = \Delta(P)$ and $\Sigma_0 = \Delta(f(P))$. For a face σ of $\Sigma_1 \setminus \Sigma_0$, let $x = x(\sigma)$ be maximal in σ such that $x \notin f(P)$. For each $x \in P \setminus f(P)$, define $\mathcal{G}(x)$ to be the family of faces σ of $\Sigma_1 \setminus \Sigma_0$ such that $x(\sigma) = x$. One readily verifies that the families $\mathcal{G}(x)$ satisfy the Cluster Lemma 4.2. Now, if $\sigma \in \mathcal{G}(x)$, then $\sigma + f(x) \in \mathcal{G}(x)$. Namely, consider an element $y \in \sigma$. If $y \leq x$, then clearly $y \leq f(x)$, because $x < f(x)$. If $x \leq y$, then $f(x) \leq y$, because $y = f(y)$ by the maximality of x. As a consequence, $\sigma + f(x)$ remains a chain and is therefore an element in $\mathcal{G}(x)$. In particular, we obtain a perfect acyclic matching on $\mathcal{G}(x)$ by pairing $\sigma - f(x)$ and $\sigma + f(x)$. Hence Σ_1 is collapsible to Σ_0 by Theorem 4.4.

For the final statement, with ι being the natural inclusion map $\Delta(f(P)) \to \Delta(P)$, we have that $f \circ \iota$ is the identity on $\Delta(f(P))$ and hence induces the identity map on the homology of $\Delta(f(P))$. By the long exact sequence for the pair $(\Delta(P), \Delta(f(P)))$ (see Theorem 3.3), ι induces an isomorphism between the homology of $\Delta(f(P))$ and the homology of $\Delta(P)$. As a consequence, f also induces an isomorphism. $\square$

At a few occasions, we will need the following special case of the Nerve Theorem (see Björner [9]):

Theorem 6.2. *For a given simplicial complex Γ, let the* nerve $\mathsf{N}(\Gamma)$ *of Γ be the simplicial complex with one 0-cell for each maximal face of Γ and with $\{\sigma_i : i \in I\}$ a face of $\mathsf{N}(\Gamma)$ if and only if the intersection $\bigcap_{i \in I} \sigma_i$ is nonempty. Then Γ and $\mathsf{N}(\Gamma)$ are homotopy equivalent.*

Proof. We obtain a closure operator f on $P(\Gamma)$ by defining $f(\sigma)$ as the face σ' obtained from σ by adding all elements x that are cone points in $\mathrm{lk}_\Gamma(\sigma)$. The resulting poset Q has the property that σ belongs to Q if and only if $\mathrm{lk}_\Gamma(\sigma)$ contains no cone points. By Lemma 6.1, Γ and $\Delta(Q)$ are homotopy equivalent.

For a face $\mathcal{X}$ of $\mathsf{N}(\Gamma)$, write $\cap\mathcal{X} = \bigcap_{\tau \in \mathcal{X}} \tau$. We obtain a closure operator g on $P(\mathsf{N}(\Gamma))$ by defining $g(\mathcal{X})$ as the face $\mathcal{X}'$ obtained from $\mathcal{X}$ by adding all maximal faces of Γ containing $\cap\mathcal{X}$. In the resulting poset Q', we may identify a given element $\mathcal{X}$ with the face $\sigma(\mathcal{X}) = \cap\mathcal{X} \in \Gamma$. Namely, if two elements $\mathcal{X}$ and $\mathcal{X}'$ in Q' yield the same face σ, then $\cap(\mathcal{X} \cup \mathcal{X}') = \sigma$, which implies that we must have $\mathcal{X} = \mathcal{X}'$. Note that $\mathcal{X}$ is smaller than $\mathcal{Y}$ in Q' if and only if $\sigma(\mathcal{Y})$ is contained in $\sigma(\mathcal{X})$. By Lemma 6.1, $\mathsf{N}(\Gamma)$ and $\Delta(Q')$ are homotopy equivalent.

It remains to prove that a face σ is an element in Q if and only if σ is an element in Q'; this will imply that Q and Q' coincide (except that all order relations are reversed in Q'). Now, σ belongs to Q if and only if $\mathrm{lk}_\Gamma(\sigma)$ contains no cone points. Equivalently, for every $x \notin \sigma$, there is some maximal face of Γ containing σ but not x. This means exactly that $\sigma = \cap\mathcal{X}$ for some family $\mathcal{X}$ of maximal faces of Γ. Hence $\sigma \in Q$ if and only if $\sigma \in Q'$, which concludes the proof. $\square$

6.2 Depth

We discuss the concept of depth, a parameter that we will frequently consider throughout this book. The main objective of the section is to show that we may define the depth over a field or $\mathbb{Z}$ of a complex in terms of deletions rather than in terms of links. This is a well-known fact among ring theorists [128], but we include a complete proof for reference. At the end of the section, we discuss the situation for homotopical depth, which is considerably more complicated.

Let $\mathbb{F}$ be a field or $\mathbb{Z}$ and let Δ be a simplicial complex on the 0-cell set E. Recall that the depth over $\mathbb{F}$ of Δ is defined by

$$\mathrm{depth}_\mathbb{F}(\Delta) = \min\{m : \tilde{H}_{m-|\sigma|}(\mathrm{lk}_\Delta(\sigma); \mathbb{F}) \neq 0 \text{ for some } \sigma \subseteq E\}.$$

Define the *deletion-depth* over $\mathbb{F}$ by

$$\mathrm{deldepth}_{\mathbb{F}}(\Delta) = \min\{m : \tilde{H}_{m-|\sigma|}(\mathrm{del}_{\Delta}(\sigma); \mathbb{F}) \neq 0 \text{ for some } \sigma \subseteq E\}.$$

Theorem 6.3 (Auslander-Buchsbaum Theorem). *For any complex Δ on the 0-cell set E, $\mathrm{depth}_{\mathbb{F}}(\Delta) = \mathrm{deldepth}_{\mathbb{F}}(\Delta)$.*

Remark. For ring-theoretic background and rationale for attributing the theorem to Auslander and Buchsbaum, see later in this section; we refer to Smith [128] for details.

Proof. Let m be minimal such that there is a set σ with the property that

$$\tilde{H}_{m-|\sigma|}(\Delta(\emptyset, \sigma)) \neq 0;$$

we suppress $\mathbb{F}$ from notation. We want to show that there is a face $\tau \in \Delta$ and an integer $a \geq 0$ such that $\tilde{H}_{m-a}(\Delta(\tau, \emptyset)) \neq 0$. This will imply that $\mathrm{depth}_{\mathbb{F}}(\Delta)$ is less than or equal to $m - a$ and hence at most equal to $\mathrm{deldepth}_{\mathbb{F}}(\Delta)$.

If $\sigma = \emptyset$, then we are done, as we may choose $\tau = \emptyset$ and $a = 0$. Otherwise, let $x \in \sigma$. We have the exact sequence

$$\tilde{H}_{m+1-|\sigma|}(\Delta(x, \sigma - x)) \longrightarrow \tilde{H}_{m-|\sigma|}(\Delta(\emptyset, \sigma)) \longrightarrow \tilde{H}_{m-|\sigma|}(\Delta(\emptyset, \sigma - x)),$$

where the group in the middle is nonzero by assumption. Since $m - |\sigma| = m - 1 - |\sigma - x|$, the group to the right is zero by minimality of m, which implies that the group to the left is nonzero. Now,

$$\tilde{H}_{m+1-|\sigma|}(\Delta(x, \sigma - x)) = \tilde{H}_{m-1-|\sigma-x|}((\mathrm{lk}_{\Delta}(x))(\emptyset, \sigma - x)).$$

Since this group is nonzero, we have by induction on the size of Δ that there exists a τ' and an integer $a \geq 0$ such that $\tilde{H}_{m-1-a}((\mathrm{lk}_{\Delta}(x))(\tau', \emptyset)) \neq 0$. Since

$$\tilde{H}_{m-1-a}((\mathrm{lk}_{\Delta}(x))(\tau', \emptyset)) = \tilde{H}_{m-a}(\Delta(\tau' + x, \emptyset)),$$

the claim follows.

Next, let m be the depth of Δ over $\mathbb{F}$ and let τ be minimal such that

$$\tilde{H}_m(\Delta(\tau, \emptyset)) \neq 0.$$

This time, we want to show that there is a set σ and an integer $a \geq 0$ such that $\tilde{H}_{m-|\sigma|-a}(\Delta(\emptyset, \sigma)) \neq 0$. This will imply that $\mathrm{deldepth}_{\mathbb{F}}(\Delta)$ is less than or equal to $m - a$ and hence at most equal to $\mathrm{depth}_{\mathbb{F}}(\Delta)$.

If $\tau = \emptyset$, then we are done, as we may choose $\sigma = \emptyset$ and $a = 0$. Otherwise, let $x \in \tau$. We have the exact sequence

$$\tilde{H}_m(\Delta(\tau - x, \emptyset)) \longrightarrow \tilde{H}_m(\Delta(\tau, \emptyset)) \longrightarrow \tilde{H}_{m-1}(\Delta(\tau - x, x)),$$

where the group in the middle is nonzero by assumption. By minimality of τ, the group to the left is zero, which implies that the group to the right is nonzero. Now,

$$\tilde{H}_{m-1}(\Delta(\tau - x, x)) = \tilde{H}_{m-1}((\mathrm{del}_\Delta(x))(\tau - x, \emptyset)).$$

Since this group is nonzero, we have by induction on the size of Δ that there exists a σ' and an integer $a \geq 0$ such that $\tilde{H}_{m-1-|\sigma'|-a}((\mathrm{del}_\Delta(x))(\emptyset, \sigma')) \neq 0$. Since

$$\tilde{H}_{m-1-|\sigma'|-a}((\mathrm{del}_\Delta(x))(\emptyset, \sigma')) = \tilde{H}_{m-|\sigma'+x|-a}(\Delta(\emptyset, \sigma' + x)),$$

the claim follows. This concludes the proof. $\square$

Corollary 6.4 (Hochster [63]). *Let Δ be a simplicial complex. Then Δ is CM/$\mathbb{F}$ if and only if $\tilde{H}_{i-|\sigma|}(\mathrm{del}_\Delta(\sigma); \mathbb{F}) = 0$ whenever $i < \dim \Delta$ for all $\sigma \subseteq E$. More generally, let m be an integer. Then the m-skeleton of Δ is CM/$\mathbb{F}$ if and only if $\tilde{H}_{i-|\sigma|}(\mathrm{del}_\Delta(\sigma); \mathbb{F}) = 0$ whenever $i < m$ for all σ.* $\square$

As promised, we now give some ring-theoretic background, roughly following Smith [128]. Let $R(\Delta)$ be the Stanley-Reisner ring of Δ; see Section 3.8 for definitions and a textbook on commutative algebra [26, 43] for ring-theoretic definitions. Assume that Δ is $(d-1)$-dimensional and that $\mathrm{depth}_\mathbb{F}(\Delta) = p-1$. As mentioned in Section 3.8, this means that the Krull dimension of $R(\Delta)$ is d and the depth is p. Let n be the size of the vertex set E of Δ. By a theorem of Hochster [63], the homological dimension $\mathrm{hd}(R(\Delta))$ of $R(\Delta)$ satisfies

$$\mathrm{hd}(R(\Delta)) = \max\{i : \tilde{H}_{n-|\sigma|-i-1}(\mathrm{del}_\Delta(\sigma)) \neq 0 \text{ for some } \sigma\}.$$

The Auslander-Buchsbaum theorem [43, Th. 19.9] states that

$$\mathrm{depth}(R(\Delta)) + \mathrm{hd}(R(\Delta)) = n,$$

which immediately implies that

$$\mathrm{depth}(R(\Delta)) = \min\{n - i : \tilde{H}_{n-|\sigma|-i-1}(\mathrm{del}_\Delta(\sigma)) \neq 0 \text{ for some } \sigma\}.$$

Replacing $n - i$ with m and using the identity $\mathrm{depth}_\mathbb{F}(\Delta) = \mathrm{depth}(R(\Delta)) - 1$, we obtain Theorem 6.3. In particular, modulo Hochster's theorem, the Auslander-Buchsbaum Theorem [43, Th. 19.9] is the ring-theoretic counterpart to Theorem 6.3.

Let $\mathbb{F}$ be a field and let E be the vertex set of Δ. By Alexander duality, we have that

$$\tilde{H}_i(\mathrm{del}_\Delta(\sigma); \mathbb{F}) = \tilde{H}_{|E|-i-|\sigma|-3}(\mathrm{lk}_{\Delta_E^*}(\sigma); \mathbb{F});$$
$$\tilde{H}_i(\mathrm{lk}_\Delta(\sigma); \mathbb{F}) = \tilde{H}_{|E|-i-|\sigma|-3}(\mathrm{del}_{\Delta_E^*}(\sigma); \mathbb{F}).$$

As a consequence, we have the following result:

Corollary 6.5. *Let Δ be a simplicial complex on the vertex set E; write $n = |E|$. Then*

$$\mathrm{depth}_{\mathbb{F}}(\Delta_E^*) = \min\{i : \tilde{H}_{n-i-3}(\mathrm{del}_\Delta(\sigma);\mathbb{F}) \neq 0 \text{ for some } \sigma\}$$
$$= \min\{i : \tilde{H}_{n-i-3}(\mathrm{lk}_\Delta(\sigma);\mathbb{F}) \neq 0 \text{ for some } \sigma\}.$$

In particular, if $\tilde{H}_j(\mathrm{lk}_\Delta(\sigma);\mathbb{F}) = 0$ whenever $j > m$, then the $(n-m-3)$-skeleton of Δ_E^* is $CM/\mathbb{F}$. $\square$

Define the *dual depth* over $\mathbb{F}$ of Δ by

$$\mathrm{depth}_{\mathbb{F}}^*(\Delta) = \max\{i : \tilde{H}_i(\mathrm{lk}_\Delta(\sigma);\mathbb{F}) \neq 0 \text{ for some } \sigma\}.$$

By Corollary 6.5, $\mathrm{depth}_{\mathbb{F}}^*(\Delta) + \mathrm{depth}_{\mathbb{F}}(\Delta_E^*) = |E| - 3$.

Corollary 6.6. *Let Δ be a simplicial complex. Then*

$$\mathrm{depth}_{\mathbb{F}}^*(\Delta) = \max\{i : \tilde{H}_i(\mathrm{del}_\Delta(\sigma);\mathbb{F}) \neq 0 \text{ for some } \sigma\}. \square$$ '

A simplicial complex with dual depth at most $d-1$ is sometimes referred to as a *d-Leray complex*. We proceed with a minor result that will be of some use in Chapter 19.

Proposition 6.7. *Let Δ be a simplicial complex with dual depth c and let σ be a nonempty set. If $\tilde{H}_c(\mathrm{del}_\Delta(\tau);\mathbb{F}) = 0$ for some $\tau \subseteq \sigma$ and the dual depth of $\mathrm{lk}_\Delta(y)$ is less than c for all $y \in \sigma \setminus \tau$, then $\tilde{H}_c(\mathrm{del}_\Delta(\sigma);\mathbb{F}) = 0$.*

Proof. If $\sigma = \tau$, then we are done. Otherwise, let $y \in \sigma \setminus \tau$ and assume that $\tilde{H}_c(\Delta(\emptyset,\sigma))$ is nonzero. By the exact sequence

$$\tilde{H}_{c+1}(\Delta(y,\sigma - y)) \longrightarrow \tilde{H}_c(\Delta(\emptyset,\sigma)) \longrightarrow \tilde{H}_c(\Delta(\emptyset,\sigma - y)),$$

this implies that $\tilde{H}_{c+1}(\Delta(y,\sigma - y))$ is nonzero; the group to the right is zero by induction on the size of σ. Since $\tilde{H}_{c+1}(\Delta(y,\sigma - y)) = \tilde{H}_c((\mathrm{lk}_\Delta(y))(\emptyset,\sigma - y))$, the dual depth of $\mathrm{lk}_\Delta(y)$ is at least c by Corollary 6.6, a contradiction. $\square$

By the following theorem, the homotopical depth of a complex Δ on the 0-cell set E is not always equal to the homotopical deletion-depth (defined in the natural manner).

Theorem 6.8. *Suppose that Δ is a simplicial complex on the 0-cell set E such that the depth d over $\mathbb{Z}$ of Δ is strictly greater than the homotopical depth of Δ. For a set Y of 0-cells disjoint from E, let Σ_Y be the 0-dimensional complex with 0-cell set Y. If $|Y| \geq d$, then the homotopical depth of $\Gamma = \Delta * \Sigma_Y$ is strictly smaller than $d+1$, whereas the homotopical deletion-depth is equal to $d+1$.*

Proof. For any $y \in Y$, $\mathrm{lk}_\Gamma(y)$ coincides with Δ, for which the homotopical depth is strictly smaller than d. It follows that the homotopical depth of Γ is strictly smaller than $d+1$.

It remains to prove that $\mathrm{del}_\Gamma(\sigma)$ is $(d-|\sigma|)$-connected for every $\sigma \subseteq E \cup Y$. By properties of join (apply Proposition 3.7), the depth over $\mathbb{Z}$ of Γ is $d+1$. By Theorems 3.8 and 6.3, it suffices to prove that $\mathrm{del}_\Gamma(\sigma)$ is simply connected whenever $d - |\sigma| \geq 1$. Write $\sigma = A \cup X$, where $A \subseteq E$ and $X \subseteq Y$. We obtain that

$$\mathrm{del}_\Gamma(\sigma) = \mathrm{del}_\Delta(A) * \mathrm{del}_{\Sigma_Y}(X) = \mathrm{del}_\Delta(A) * \Sigma_{Y \setminus X}.$$

Since $|\sigma| \leq d - 1$, we have that $|A| \leq d - 1$ and $|X| \leq d - 1$. In particular, $\mathrm{del}_\Delta(A)$ is 0-acyclic and hence 0-connected, whereas $\Sigma_{Y \setminus X}$ is (-1)-connected. As a consequence, $\mathrm{del}_\Gamma(\sigma)$ is simply connected by Theorem 3.11. $\square$

6.3 Vertex-Decomposability

The concept of vertex-decomposability (VD) will be an important tool for us in the analysis of the depth of certain simplicial complexes. This section presents some useful auxiliary lemmas related to the concept.

Let us say that a lifted complex is $VD(d)$ if the d-skeleton is VD. Refer to the complex as $VD^+(d)$ if the complex is $VD(d)$ and admits a decision tree with all evasive sets of dimension d. Note that a $VD(d)$ simplicial complex has homotopical depth at least d and that a $VD^+(d)$ complex is homotopy equivalent to a wedge of spheres of dimension d; use Theorem 5.2 and Theorem 4.8. By convention, the void complex is $VD(d)$ and $VD^+(d)$ for every d.

Lemma 6.9. *Let Δ be a lifted complex and let v be a vertex. If $\mathrm{lk}_\Delta(v)$ is $VD(d-1)$ and $\mathrm{del}_\Delta(v)$ is $VD(d)$, then Δ is $VD(d)$. If $\mathrm{lk}_\Delta(v)$ is $VD^+(d-1)$ and $\mathrm{del}_\Delta(v)$ is $VD^+(d)$, then Δ is $VD^+(d)$.*

Proof. The claims are immediate from Definitions 3.27 and 5.1. $\square$

Lemma 6.10. *Let Δ be a lifted complex and let B be a vertex set. If $\Delta(A, B \setminus A)$ is VD of dimension d for each $A \subseteq B$, then Δ is VD of dimension d. If $\Delta(A, B \setminus A)$ is $VD(d)$ for each $A \subseteq B$, then Δ is $VD(d)$. If $\Delta(A, B \setminus A)$ is $VD^+(d)$ for each $A \subseteq B$, then Δ is $VD^+(d)$.*

Proof. This is immediate by induction on the size of B; use Definition 3.28 and Lemma 6.9. $\square$

The following simple lemma is sometimes useful.

Lemma 6.11. *Let $\Delta_1, \ldots, \Delta_k$ be simplicial complexes and let $d_1, \ldots, d_k$ be integers such that Δ_i is $VD(d_i)$ for each i; $d_i \geq -1$. Then the join $\Delta_1 * \ldots * \Delta_k$ is $VD(\sum_i d_i + k - 1)$. The join is $VD^+(\sum_i d_i + k - 1)$ if each Δ_i is $VD^+(d_i)$.*

Proof. Use induction on the size of $\Delta_1 * \cdots * \Delta_k$ and $d = \sum_i d_i + k - 1$. If $d = -1$, then we are done. Otherwise, let i be such that $d_i \geq 0$; say that $i = k$.

Let v be a shedding vertex for the d_k-skeleton Σ_k of Δ_k; $\mathrm{lk}_{\Sigma_k}(v)$ and $\mathrm{del}_{\Sigma_k}(v)$ are VD. Write $\Delta = \Delta_1 * \ldots * \Delta_{k-1}$.

If v is a cone point in Σ_k, then $\Sigma_k = \Delta_k$; hence the $(d-1)$-skeleton of $\mathrm{del}_{\Delta_k}(v) = \mathrm{lk}_{\Delta_k}(v)$ coincides with $\mathrm{del}_{\Sigma_k}(v) = \mathrm{lk}_{\Sigma_k}(v)$. By induction, $\Delta *$ $\mathrm{lk}_{\Delta_k}(v)$ is $VD(d-1)$. $\Delta * \Delta_k$ is the cone over this complex and hence $VD(d)$.

If v is not a cone point in Σ_k, then $\mathrm{del}_{\Delta_k}(v)$ is $VD(d)$ and $\mathrm{lk}_{\Delta_k}(v)$ is $VD(d-1)$. By induction, $\Delta * \mathrm{del}_{\Delta_k}(v)$ is $VD(d)$, whereas $\Delta * \mathrm{lk}_{\Delta_k}(v)$ is $VD(d-1)$. By Lemma 6.9, we obtain that Δ is $VD(d)$, and we are done.

For the last statement in the lemma, use the above proof and apply Theorem 5.28 and Corollary 4.23. $\square$

By the discussion in Section 3.6.3, a decomposition as in Definition 3.27 (or Definition 3.28) induces a shelling of a complex Δ. We refer to such a shelling as a VD-shelling; each shedding face is a vertex.

Lemma 6.12. *Let Δ be a shellable simplicial complex on the set E with shelling pairs $(\sigma_1, \tau_1), \ldots, (\sigma_r, \tau_r)$ and let Σ be a subcomplex of Δ. Let d be a fixed integer.*

  (i) *If $\Sigma(\sigma_i, E \setminus \tau_i)$ is shellable of dimension d for each shelling pair (σ_i, τ_i), then so is Σ.*

 (ii) *Assume that Δ is VD. If $\Sigma(\sigma_i, E \setminus \tau_i)$ is VD of dimension d for each VD-shelling pair (σ_i, τ_i), then so is Σ. If $\Sigma(\sigma_i, E \setminus \tau_i)$ is $VD(d)$ for each shelling pair (σ_i, τ_i), then so is Σ.*

(iii) *If $\Sigma(\sigma_i, E \setminus \tau_i)$ admits an acyclic matching with all critical faces of dimension d for each shelling pair (σ_i, τ_i), then so does Σ.*

(iv) *Assume that Δ is VD. If $\Sigma(\sigma_i, E \setminus \tau_i)$ admits a decision tree with all evasive faces of dimension d for each VD-shelling pair (σ_i, τ_i), then so does Σ.*

Proof. (i) If there is only one shelling pair, then we are done. Otherwise, consider the first shedding face σ of Δ in Definition 3.25 (ii). Decomposing with respect to σ, we obtain a partition of the family of shelling pairs into two subfamilies. The first subfamily constitutes a shelling of $\mathrm{fdel}_\Delta(\sigma)$ and the other subfamily constitutes a shelling of $\Delta(\sigma, \emptyset)$. By an induction argument, we obtain a shelling of each of $\mathrm{fdel}_\Sigma(\sigma)$ and $\Sigma(\sigma, \emptyset)$, which concludes the proof.

The proofs of (ii)-(iv) are almost identical to that of (i); decompose with respect to the first shedding face or shedding vertex and use induction. $\square$

6.4 Enumeration

We will apply the following simple polynomial formula when examining complexes of bipartite graphs and graphs admitting a p-cover.

Proposition 6.13 (Folklore). *Let $d \geq 0$ and let f be a polynomial of degree at most d. Then, for any integer s and complex number x,*

$$f(x) = \sum_{k=s}^{d+s} (-1)^{d+s-k} \binom{x-s}{k-s} \binom{x-1-k}{d+s-k} f(k);$$

the binomial coefficients are interpreted as polynomials in the natural manner.

Proof. One easily checks that the left-hand and right-hand sides coincide for $x = s, 1+s, \ldots, d+s$. Since the values on any $d+1$ points uniquely determine a polynomial of degree at most d, the proposition follows. $\square$

Proposition 6.14 (see Stanley [133]). *Let $\mathbb{P}$ be the set of positive integers. For a function $f : \mathbb{P} \to \mathbb{Z}$ and a partition $\mathsf{U} = \{U_1, \ldots, U_k\}$ of a finite subset of $\mathbb{P}$, define*

$$f(\mathsf{U}) = f(|U_1|) \cdots f(|U_k|).$$

Let t be a real number. For $n \geq 1$, define

$$h_t(n) = \sum_{\mathsf{U} \in \Pi_n} f(\mathsf{U}) t^{|\mathsf{U}|}, \tag{6.1}$$

where Π_n is the partition lattice on the set $[n]$ and $|\mathsf{U}|$ is the number of sets in the partition U. Then the exponential generating functions $F(x) = \sum_{n \geq 1} f(n) x^n / n!$ and $H_t(x) = \sum_{n \geq 1} h_t(n) x^n / n!$ satisfy

$$H_t(x) = e^{tF(x)} - 1.$$

Proof. For a subset T of $[n-1]$, let $\Pi_n(T)$ be the family of partitions of $[n]$ containing the set $T + n$. It is clear that

$$\frac{h_t(n)}{(n-1)!} - \frac{tf(n)}{(n-1)!} = \sum_{T \subsetneq [n-1]} \sum_{\mathsf{U} \in \Pi_n(T+n)} \frac{f(\mathsf{U}) t^{|\mathsf{U}|}}{(n-1)!}$$

$$= \sum_{T \subsetneq [n-1]} \frac{tf(|T|+1) h_t(n-1-|T|)}{(n-1)!}$$

$$= \sum_{i=1}^{n-1} \binom{n-1}{i-1} \frac{tf(i) h_t(n-i)}{(n-1)!} = \sum_{i=1}^{n-1} \frac{tf(i)}{(i-1)!} \frac{h_t(n-i)}{(n-i)!}.$$

Thus $H_t'(x) - tF'(x) = tF'(x) H_t(x)$. Clearly, $H_t(x) = e^{tF(x)} - 1$ is the unique solution to this equation that satisfies $H_t(0) = 0$. $\square$

Let $|G|$ be the number of edges in the graph G. We interpret Proposition 6.14 in terms of graph properties in the following way.

Corollary 6.15. *For each $n \geq 1$, let Δ_n be a graph property on n vertices such that all graphs in Δ_n are connected (thus Δ_n is not monotone unless $n = 1$). Let Σ_n be the family of graphs G on n vertices with the property that $G(U)$ is isomorphic to a graph in $\Delta_{|U|}$ for each connected component U in G. Let t be a real number. Then the exponential generating functions $F(x) = \sum_{n \geq 1} \tilde{\chi}(\Delta_n) x^n / n!$ and $H_t(x) = \sum_{n \geq 1} \sum_{G \in \Sigma_n} (-1)^{|G|-1} t^{c(G)} x^n / n!$ satisfy*

$$H_t(x) = 1 - e^{-tF(x)}.$$

In particular,

$$H_t(x) = 1 - (1 - H_1(x))^t;$$

note that $H_1(x) = \sum_{n \geq 1} \tilde{\chi}(\Sigma_n) x^n / n!$. The analogous result holds for hypergraph properties and digraph properties.

Proof. Define $f(n) = -\tilde{\chi}(\Delta_n)$ and $h_t(n) = -\sum_{G \in \Sigma_n} (-1)^{|G|-1} t^{c(G)}$. It is easy to see that f and h_t satisfy the equation (6.1) in Proposition 6.14, which implies that $-H_t(x) = e^{-tF(x)} - 1$ (note the change of sign compared to the proposition). $\square$

Corollary 6.16. *Let $s \geq 1$. With notation and assumptions as in Corollary 6.15, let Σ_n^s be the subfamily of Σ_n consisting of all graphs with at least s connected components. Then*

$$\sum_{n \geq 1} \sum_{G \in \Sigma_n^s} (-1)^{|G|-1} t^{c(G)} x^n / n! = \sum_{r=0}^{s-1} \frac{(-tF(x))^r}{r!} - e^{-tF(x)}.$$

Proof. This is an immediate consequence of Corollary 6.15. $\square$

Part III

Overview of Graph Complexes

7

Graph Properties

Recall from Chapter 1 that a monotone graph property is a simplicial complex of graphs – on a fixed vertex set V – that is invariant under the natural action of the symmetric group on V. Roughly speaking, the monotone graph properties to be examined in this book are of four kinds:

- *Properties defined in terms of vertex degree.* The most important example is the property of being a matching, meaning that the degree of each vertex is at most one. The degree being at most two means that every connected component is either a path or a cycle.
- *Properties defined in terms of forbidden cycles.* For example, the property of being a forest means avoiding all cycles, whereas the property of being bipartite means avoiding all cycles of odd length. Restricting avoidance to cycles of a fixed length, we obtain other interesting properties such as the property of being non-Hamiltonian.
- *Properties defined in terms of connectivity.* One example is the property of being disconnected. This generalizes to the two properties of being not k-connected and being not k-edge-connected. The first property means that it is possible to disconnect the given graph by removing $k-1$ vertices. The second property is defined analogously in terms of edges.
- *Properties defined in terms of cliques and stable sets.* This class includes the property of containing a large stable set, the property of being colorable with a certain number of colors, and the property of not containing a matching of a certain size. The second property is about partitioning the vertex set into a small number of stable subsets, whereas the third property is about avoiding partitions of the vertex set into small cliques.

There are certainly many other interesting monotone graph properties that do not fit into any of these four categories. Some examples are the properties of being planar, not containing a certain graph as a minor, or being t-edge-colorable. The reason for not considering these properties is simply that we do not have anything of interest to present about their topology.

In Section 7.1, we present basic definitions. We provide illustrations of the various monotone graph properties in Section 7.2.

Table 7.1. Some monotone graph properties. $f_p(n)$ and $g_k(n)$ are polynomials in n. See Table 10.2 for a formula for $\tilde{\chi}(\mathsf{B}_n)$.

§	Name	Description	Homotopy type	Sec.
§1	M_n	Matching	$\mathbb{Q}$-homology known	11.2
§2	BD_n^d	No vertex of degree $> d$	Not known in general	12.1
§3	F_n	Forest	$\bigvee S^{n-2}$	13.1
§4	B_n	Bipartite	$\bigvee S^{n-2}$	14.1
§5	$\mathsf{B}_{n,p}$	Bipartite, balance number $\leq p$	$\bigvee_{f_p(n)} S^{2p-1}$	14.3
§6	NHam_n	Non-Hamiltonian	Only partially known	17
§7	NC_n	Disconnected	$\bigvee_{(n-1)!} S^{n-3}$	18.1
§8	$\mathsf{NLC}_{n,k}$	$\leq k$ vertices in each component	Wedge of spheres for $n \leq 2k+2$ and $n = 3k+2$	18.2
§9	$\mathsf{SSC}_n^{k,s}$	s components of size $\leq k$	$\bigvee S^{n-s-2}$	18.3
§10	$\mathsf{NC}_{n,p}$	Some p-indivisible component	$\bigvee S^{n-3}$	18.4
§11	NC_n^2	Not 2-connected	$\bigvee_{(n-2)!} S^{2n-5}$	19
§12	NC_n^3	Not 3-connected	$\bigvee_{(n-3)\frac{(n-2)!}{2}} S^{2n-4}$	20
§13	NEC_n^2	Not 2-edge-connected	Only partially known	23
§14	NFC_{2k-1}	Not factor-critical	$\bigvee_{(2k-3)!!} S^{3k-5}$	23.3
§15	$\mathsf{NM}_{n,k}$	No k-matching	$\bigvee_{g_k(n)} S^{3k-4}$	24
§16	Col_n^t	t-colorable	Known for $t = 2, t \geq n-3$	25
§17	$\mathsf{Cov}_{n,p}$	p-coverable	Homology known for $p \leq 3$	26
§18	$\not{\triangleright}_n$	Triangle-free	Only partially known	26.7

7.1 List of Complexes

Below, we present the main monotone graph properties to be examined in this book. See Section 7.2 for illustrations and Table 7.1 for a summary of the main results. Many of these results are due to other authors; see the relevant sections for details.

Properties Defined in Terms of Vertex Degree

§1 *Matchings.* Define M_n to be the complex of matchings on n vertices. In Section 11.2, we discuss the matching complex.

§2 *Bounded-degree graphs.* Let $n \geq 1$ and $d \geq 1$. Define BD_n^d as the complex of graphs G on n vertices such that the degree of each vertex in G is at most d. We devote Section 12.1 to BD_n^d. In Section 12.2, we discuss a variant of BD_n^d in which we allow loops. A loop is an edge with both endpoints in the same vertex, which means that the addition of a loop to a vertex increases the degree of the vertex by two. This variant has certain applications to algebra (see Sections 1.1.2 and 1.1.3) and is for this reason much more well-studied than BD_n^d.

Properties Defined in Terms of Forbidden Cycles

§3 *Forests.* Let F_n denote the simplicial complex of forests on n vertices; the maximal faces of F_n are spanning trees. Being the independence complex of a matroid, F_n has a very attractive topological structure; see Section 13.1 for details.

§4 *Bipartite graphs.* Define B_n as the simplicial complex of bipartite graphs on n vertices. In Section 14.1, we examine B_n.

$\beta = 2$ $\beta = 3$ $\beta = 1$ $\beta = 2$

Fig. 7.1. The balance number of some bipartite graphs on six vertices. In each case, a smallest possible set in a bipartition is indicated with circles.

§5 *Unbalanced bipartite graphs.* Define the *balance number* $\beta(G)$ of a bipartite graph on n vertices as the smallest integer r such that there is a bipartition (U, W) of G with the property that one of U and W has size r. This means that G is a subgraph of a copy of the complete bipartite graph $K_{r,n-r}$. See Figure 7.1 for some examples. To justify this terminology, recall that a bipartition (U, W) is typically referred to as *balanced* if $|U| = |W|$. Let $B_{n,p}$ be the subcomplex of B_n consisting of all bipartite graphs on n vertices with balance number at most p. We analyze $B_{n,p}$ in Section 14.3.

§6 *Non-Hamiltonian graphs.* Let $NHam_n$ be the complex of graphs on n vertices without Hamiltonian cycles. We examine $NHam_n$ in Chapter 17.

Properties Defined in Terms of Connectivity

§7 *Disconnected graphs.* Let NC_n be the complex of disconnected graphs on n vertices. We discuss NC_n in Section 18.1.

§8 *Graphs with no large components.* Let $NLC_{n,k}$ be the complex of graphs on n vertices with all connected components of size at most k; "NLC" stands for "No Large Components." For example, $NLC_{n,2}$ is the matching complex M_n, whereas $NLC_{n,n-1}$ equals NC_n. We examine $NLC_{n,k}$ in Section 18.2.

§9 *Graphs with some small components.* Instead of putting restrictions on all components, one may require just a few of them to be small. Let $n \geq 2$ and $k, s \geq 1$. Define $SSC_n^{k,s}$ ("Some Small Components") to be the simplicial complex of graphs G on n vertices such that at least s connected components in G contain k or fewer vertices. Note that $SSC_n^{n-1,1}$ is the complex NC_n of disconnected graphs; see §7. Moreover, $SSC_n^{1,s}$ is the complex of graphs with at least s isolated vertices. We discuss the complex $SSC_n^{k,s}$ in Section 18.3.

§10 *Graphs with some p-indivisible component.* For $p \in [n]$ such that p divides n, let $NC_{n,p}$ be the complex of graphs on n vertices such that some connected component has a vertex set of size *not* divisible by p. See Sections 13.4.1 and 18.4 for a treatment of this complex.

For $1 \leq k \leq n - 1$, let NC_n^k be the complex of not k-connected graphs on n vertices. Note that $NC_n^1 = NC_n$, which we considered in §7.

§11 *Not 2-connected graphs.* One of the most important objects in this book is the complex NC_n^2 of not 2-connected graphs on n vertices; see Chapter 19 for an overview.

§12 *Not k-connected graphs for $k \geq 3$.* In Chapter 20, we examine the complex NC_n^3 of not 3-connected graphs on n vertices and also NC_n^k for larger k.

§13 *Not k-edge-connected graphs.* For a graph $G = (V, E)$, define the *edge connectivity* of G as the size of a smallest subset S of E such that $G - S = (V, E \setminus S)$ is disconnected. G is *k-edge-connected* if G has edge connectivity at least k. Let NEC_n^k be the simplicial complex of not k-edge-connected graphs on n vertices. For $k = 1$, we obtain the complex NC_n of disconnected graphs. In Chapter 23, we discuss NEC_n^k, concentrating almost entirely on NEC_n^2. ·

§14 *Not factor-critical graphs.* For n odd, a graph G on n vertices is *factor-critical* if $G([n] \setminus \{v\})$ contains a perfect matching for each $v \in [n]$. For $k \geq 1$, let NFC_{2k-1} be the simplicial complex of not factor-critical graphs on $2k - 1$ vertices. NFC_{2k-1} is closely related to the complex $NM_{2k,k}$ of graphs on $2k$ vertices not admitting a perfect matching; see §15. However, the complex appears already in the analysis of NEC_n^2 in Section 23.3.

Properties Defined in Terms of Cliques and Stable Sets

§15 *Graphs with bounded matching size.* Let $\mathsf{NM}_{n,k}$ be the complex of graphs on n vertices that do not contain a k-matching (i.e., k pairwise disjoint edges). We summarize known results about this complex in Chapter 24.

§16 *t-colorable graphs.* For $1 \le t \le n-1$, let Col_n^t be the complex of t-colorable graphs on n vertices. For $t = 2$, we obtain the complex B_n of bipartite graphs; see §4. A *clique partition* of a graph G is a partition $\{U_1, \ldots, U_t\}$ of the vertex set of G such that $G(U_i)$ is a complete graph for each $i \in [t]$. Let $\mathsf{NQP}_{n,t}$ ("No cliQue Partition") be the complex of graphs on n vertices not admitting any clique partition with t sets. For $t \le 0$, it turns out to be convenient to define $\mathsf{NQP}_{n,t}$ as the full simplex on the set $\binom{[n]}{2}$. Note that $\mathsf{NQP}_{n,t}$ is the Alexander dual of Col_n^t. In Chapter 25, we provide an overview of known results about Col_n^t and $\mathsf{NQP}_{n,t}$.

§17 *p-coverable graphs.* The *covering number* of a graph G is the size of a smallest vertex set W such that each edge in G contains some vertex from W. For $1 \le p \le n - 1$, let $\mathsf{Cov}_{n,p}$ be the simplicial complex of graphs on n vertices with covering number at most p. We devote Chapter 26 to $\mathsf{Cov}_{n,p}$.

§18 *Triangle-free graphs.* Let $\not\triangleright_n$ be the complex of triangle-free graphs on n vertices; graphs in $\not\triangleright_n$ do not contain any cliques of size three. See Section 26.7 for a brief discussion about $\not\triangleright_n$.

Table 7.2. Some interesting Alexander duals of monotone graph properties studied in this book.

§	Property	Alexander dual
§1	M_n	NC_n^{n-2}
§2	BD_n^d	Some vertex of degree $\le n - d - 2$
§4	B_n	$\mathsf{NQP}_{n,2}$
§7	NC_n	No complete bipartite subgraph
§16	Col_n^t	$\mathsf{NQP}_{n,t}$
§17	$\mathsf{Cov}_{n,p}$	No clique of size $n - p$
§18	$\not\triangleright_n$	$\mathsf{Cov}_{n,n-3}$

Remarks

Some of the listed properties have interesting Alexander duals; see Table 7.2.

There is some further duality worth noting between some of the complexes. Specifically, the maximal faces of the complex F_n of forests are exactly the minimal nonfaces of the complex NC_n of disconnected graphs. Moreover, the maximal faces of NC_n coincide with the complements of the maximal faces of the complex B_n of bipartite graphs. Analogously, the maximal faces of the complex $SSC_n^{p,1}$ of graphs with some connected component of size at most p coincide with the complements of the maximal faces of the complex $B_{n,p}$ of graphs with balance number p. These dualities are not of Alexander type and hence do not necessarily provide a topological connection between the complexes.

7.2 Illustrations

§1 The 15 maximal graphs in M_6 (and the minimal graphs not in $NM_{6,3}$) are all isomorphic to the following perfect matching.

§2 Each of the 187 maximal graphs in BD_6^2 is isomorphic to one of the following graphs with vertex degree at most two.

§3 Each of the 1296 maximal graphs in F_6 (and each minimal graph not in NC_6) is isomorphic to one of the following spanning trees.

§4 Each of the 31 maximal graphs in B_6 is isomorphic to one of the following complete bipartite graphs. The two graphs to the left yield the 21 maximal faces of the subcomplex $B_{6,2}$ of graphs with balance number at most two.

§5 For an illustration of $B_{6,2}$, see §4.

§6 Each of the 45 maximal graphs in $NHam_5$ is isomorphic to one of the following non-Hamiltonian graphs. The leftmost graph yields the ten maximal faces of the complex $Cov_{5,2}$ of graphs covered by two vertices, whereas the two leftmost graphs yield the 30 maximal faces of the complex NFC_5 of not factor-critical graphs.

§7 Each of the 31 maximal graphs in NC_6 is isomorphic to one of the following disconnected graphs with two components. The two graphs to the left yield the 21 maximal faces of the subcomplex $SSC_6^{2,1}$ of graphs with at least one component of size at most two. The two graphs to the right yield the 16 maximal faces of the subcomplex $NC_{6,2}$ of graphs with some component of odd size.

§8 Each of the 25 maximal graphs in $NLC_{6,3}$ is isomorphic to one of the following graphs with at most three vertices in each component.

§9 For an illustration of $SSC_6^{2,1}$, see §7.
§10 For an illustration of $NC_{6,2}$, see §7.
§11 Each of the 504 maximal graphs in NC_8^2 is isomorphic to one of the following not 2-connected graphs.

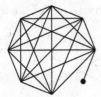  

§12 Each of the 868 maximal graphs in NC_8^3 is isomorphic to one of the following not 3-connected graphs. The bold edges represent sets separating the graphs.

§13 Each of the 1456 maximal graphs in NEC_8^2 is isomorphic to one of the following not 2-edge-connected graphs.

  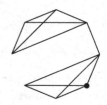

§14 For an illustration of NFC_5, see §6.

§15 Each of the 91 maximal graphs in $\mathsf{NM}_{6,3}$ is isomorphic to one of the following graphs without a perfect matching.

§16 Each of the 75 maximal graphs in Col_6^3 is isomorphic to one of the following 3-colorable graphs; we have colored the vertices with the colors A, B, and C.

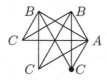 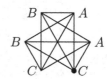

§17 For an illustration of $\mathsf{Cov}_{5,2}$, see §6.

§18 Each of the 27 maximal graphs in $\not\triangleright_5$ is isomorphic to one of the following triangle-free graphs.

8

Dihedral Graph Properties

We discuss monotone dihedral graph properties, which are graph complexes on $[n]$ that are invariant under the natural action of the dihedral group D_n. More precisely, identify the corners of a regular n-gon with the vertices $1, \ldots, n$ arranged in clockwise direction. Representing a given edge ij with the open line segment between the points representing i and j, we obtain the *polygon representation* of a graph. The natural action of D_n on this graph is simply the action induced by the natural action of D_n on the n-gon.

With this geometric representation in mind, we may divide the monotone dihedral graph properties to be examined into two groups:

- *Properties defined in terms of forbidden crossings.* Two edges cross if their representations as open line segments within the n-gon cross. The most important example of a monotone dihedral graph property avoiding crossings is the associahedron, in which each maximal face is a triangulation of the n-gon. We obtain further examples by combining the property of avoiding crossings with a monotone graph property. For example, we consider complexes of noncrossing matchings, forests, and bipartite graphs.
- *Properties defined in terms of connectivity.* In this case, we again look at the polygon representation of a given graph and examine whether it is disconnected or separable as a topological space, the restriction in the latter case being that we only consider cut points that correspond to vertices in the graph. We also consider the property of having a two-separable polygon representation, meaning that we can cut the representation along a line segment into two pieces without crossing any other line segment.

Except for the complex of noncrossing matchings, all dihedral properties in this book have the homotopy type of a wedge of spheres in a fixed dimension.

As it turns out, we may exploit properties of the associahedron in our analysis of other monotone dihedral graph properties. Moreover, as Shareshian and Wachs observed [118], the associahedron plays an important role in the analysis of the monotone graph property of being not 2-connected; see Chapter 19.

8.1 Basic Definitions

A *dihedral graph property* is a family of graphs on the vertex set $[n]$ such that the family is closed under the natural action of the dihedral group on $[n]$; this action has as generators the rotation map $i \mapsto (i+1) \bmod n$ and the reflection map $i \mapsto n+1-i$. If the dihedral graph property Σ is a graph complex (i.e., closed under deletion of edges), then Σ is a *monotone dihedral graph property*.

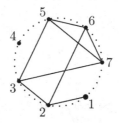

Fig. 8.1. The polygon representation of the graph with vertex set $[7]$ and edge set $\{12, 23, 26, 35, 37, 56, 57, 67\}$ along with a dashed unit circle.

When examining dihedral graph properties, we illustrate a graph in the plane by representing the vertices as points around the unit circle arranged evenly spaced in clockwise direction. More precisely, given that the vertex set is $[n]$, we identify the vertex k with the point $(\cos(2\pi k/n), \sin(-2\pi k/n))$ and the edge ij with the line segment between the points representing the vertices i and j. Note that one may view the points representing the vertices as the corners of a regular n-gon. For this reason, we refer to this representation of a graph as the *polygon representation*; see Figure 8.1 for an example.

We may view the polygon representation of a graph G as a subset of $\mathbb{R}^2$ consisting of the points identifying the vertices in G along with the line segments identifying the edges in G. In particular, we may interpret the polygon representation as a topological space.

Write
$$\begin{cases} \mathrm{Bd}_n = \{12, 23, \ldots, (n-1)n, 1n\}; \\ \mathrm{Int}_n = \binom{[n]}{2} \setminus \mathrm{Bd}_n. \end{cases}$$

We refer to edges in Bd_n as *boundary edges* and edges in Int_n as *interior edges* (some authors would perhaps refer to interior edges as *chords* or *diagonals*).

Two edges ac and bd *cross* if the corresponding *open* line segments intersect. With $a < c$, $a \leq b$, and $b < d$, this means that $a < b < c < d$ with strict inequalities. We refer to a graph without crossing edges as a *noncrossing* graph.

A *closed interval* is a vertex set of the form
$$[a, b] = \begin{cases} \{c : a \leq c \leq b\} & \text{if } 1 \leq a \leq b \leq n; \\ \{c : a \leq c \leq n\} \cup \{c : 1 \leq c \leq b\} & \text{if } 1 \leq b \leq a \leq n. \end{cases}$$

Thus $[a, b]$ consists of those vertices passed – endpoints included – when walk-
ing from a to b in clockwise direction. We define the open interval (a, b) as
the set obtained from $[a, b]$ by removing the endpoints a and b. The half-open
intervals $[a, b)$ and $(a, b]$ are defined in the natural manner.

Table 8.1. List of monotone dihedral graph properties studied in this book. C_n is
the Catalan number $\frac{1}{n+1}\binom{2n}{n}$, whereas F_n is the Fine number recursively defined by
$F_n = (C_{n-1} - F_{n-1})/2$ with $F_1 = 1$.

§	Name	Description	Homotopy type	Sec.
§1	NX_n	Noncrossing	n-fold cone over A_n	16.2
	A_n	As above without cone points	S^{n-4}	
§2	NXM_n	Noncrossing matching	Only partially known	16.3
§3	NXF_n	Noncrossing and cycle-free	$\bigvee S^{n-2}$	16.4
§4	NXB_n	Noncrossing and bipartite	$\bigvee_{F_n} S^{n-2}$	16.5
§5	$NCR_n^{0,0}$	Disconnected polygon representation	$\bigvee_{C_{n-1}} S^{n-3}$	21.2
"	$\cap NX_n$	As above and noncrossing		
§6	$NCR_n^{1,0}$	Separable polygon representation	S^{2n-5}	21.3
"	$\cap NX_n$	As above and noncrossing		
§7	$NCR_n^{1,1}$	Two-separable polygon representation	n-fold cone over $\overline{NCR}_n^{1,1}$	21.4
	$\overline{NCR}_n^{1,1}$	As above without cone points	S^{n-4}	

8.2 List of Complexes

Definitions of the various monotone dihedral graph properties to be examined
are as follows; see Section 8.3 for illustrations. Table 8.1 provides a short summary of the main known results. Table 8.2 lists the dihedral graph properties
along with related monotone graph properties. See the relevant sections for
more details about the results and whom to attribute them to.

Properties Defined in Terms of Forbidden Crossings

§1 *Noncrossing graphs.* We define the *associahedron* A_n as the complex of all
graphs on n vertices with no crossing edges and with no boundary edges
(that is, no edges from Bd_n). Let NX_n be the n-fold cone over A_n with
respect to the boundary edge set Bd_n. We may add any edge in Bd_n to a
graph in A_n without introducing crossings; thus NX_n is the complex of all
noncrossing graphs on $[n]$. We discuss NX_n and A_n in Sections 16.1 and
16.2.

110 110   8 Dihedral Graph Properties

Table 8.2. Comparison between some important monotone dihedral graph properties (MDGPs) and monotone graph properties (MGPs).

§	MDGP	Homotopy type	Related MGP	Description	Homotopy type	Sec.
§2	NXM_n	Only partially known	M_n	Matching	$\mathbb{Q}$-homology known	11.2
§3	NXF_n	$\bigvee S^{n-2}$	F_n	Forest	$\bigvee S^{n-2}$	13.1
§4	NXB_n	$\bigvee_{F_n} S^{n-2}$	B_n	Bipartite	$\bigvee S^{n-2}$	14.1
§5	$NCR_n^{0,0}$	$\bigvee_{C_{n-1}} S^{n-3}$	NC_n	Disconnected	$\bigvee_{(n-1)!} S^{n-3}$	18.1
§6	$NCR_n^{1,0}$	S^{2n-5}	NC_n^2	Not 2-connected	$\bigvee_{(n-2)!} S^{2n-5}$	19
§7	$NCR_n^{1,1}$	"$Cone^n(S^{n-4})$"	NC_n^3	Not 3-connected	$\bigvee_{(n-3)\frac{(n-2)!}{2}} S^{2n-4}$	20

§2 *Noncrossing matchings.* Define NXM_n as the complex of noncrossing matchings on n vertices; $NXM_n = M_n \cap NX_n$. In Section 16.3, we examine NXM_n.

§3 *Noncrossing forests.* Let NXF_n be the complex of noncrossing forests on n vertices; $NXF_n = F_n \cap NX_n$. We discuss NXF_n in Section 16.4.

§4 *Noncrossing bipartite graphs.* We define NXB_n to be the complex of noncrossing bipartite graphs on n vertices; $NXB_n = B_n \cap NX_n$. This is exactly the complex of noncrossing graphs with the property that the boundary of each region in the polygon representation contains an even number of edges. Namely, since the vertices are in convex position, this is equivalent to each cycle containing an even number of edges.

Properties Defined in Terms of Connectivity

For nonnegative integers n, k, l such that $0 \leq k + l \leq n - 2$, let $NCR_n^{k,l}$ be the complex of graphs on n vertices with the following property:

There exist two disjoint half-open intervals $(a - k, a]$ and $(b - l, b]$ of size k and l, respectively, dividing the remaining vertex set into two nonempty pieces $(a, b - l]$ and $(b, a - k]$ with the property that there are no edges between the two pieces $(a, b - l]$ and $(b, a - k]$; $a - k$ and $b - l$ are computed modulo n.

At first sight, this construction may appear as a bit artificial. However, for small values of k and l, there are very natural interpretations:

§5 *Graphs with a disconnected polygon representation.* We may interpret $NCR_n^{0,0}$ as the complex of graphs on n vertices with a disconnected polygon representation. Indeed, NCR is short for "Not Connected Representation".

Namely, the condition for a graph to be in $\mathrm{NCR}_n^{0,0}$ is that there exist a, b such that there are no edges between $(a, b]$ and $(b, a]$ and such that these two intervals are nonempty (thus $a \neq b$). We consider $\mathrm{NCR}_n^{0,0}$ in Section 21.2.

§6 *Graphs with a separable polygon representation.* For $(k, l) = (1, 0)$, we obtain the complex $\mathrm{NCR}_n^{1,0}$ of graphs on n vertices with a *separable* polygon representation. A graph G has this property if there exist a, b such that there is no edge between $(a, b]$ and $(b, a - 1]$ in G and such that these two intervals are nonempty (thus $a \neq b, b + 1$). We refer to a as a *cut point* in G; if we remove the point representing the vertex a from the polygon representation of G, then the result is disconnected. We devote Section 21.3 to $\mathrm{NCR}_n^{1,0}$.

§7 *Graphs with a 2-separable polygon representation.* For $(k, l) = (1, 1)$, the resulting complex $\mathrm{NCR}_n^{1,1}$ consists of all graphs on n vertices with a *two-separable* polygon representation. A graph G has this property if there exist a, b such that there is no edge between $(a, b - 1]$ and $(b, a - 1]$ in G and such that these two intervals are nonempty (thus $a \neq b - 1, b + 1$). We refer to $\{a, b\}$ as a *cut set* in G; if we remove the points representing the vertices a and b from the polygon representation of $G - ab$, then the result is disconnected. All boundary edges turn out to be cone points in $\mathrm{NCR}_n^{1,1}$; we denote by $\overline{\mathrm{NCR}}_n^{1,1}$ the complex obtained by removing all these cone points. Section 21.3 deals with $\mathrm{NCR}_n^{1,1}$ and $\overline{\mathrm{NCR}}_n^{1,1}$.

8.3 Illustrations

§1 We obtain the 42 maximal graphs in NX_7 via rotation and reflection of the following four graphs. We get the maximal graphs in A_7 by removing the seven boundary edges.

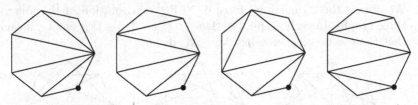

§2 We obtain the 20 maximal faces of NXM_6 via rotation and reflection of the following four noncrossing matchings.

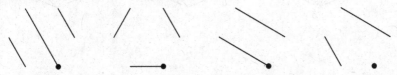

§3 We obtain the 55 maximal faces of NXF_5 via rotation and reflection of the following seven noncrossing spanning trees.

§4 We obtain the 30 maximal faces of NXB$_5$ via rotation and reflection of the following four noncrossing bipartite graphs.

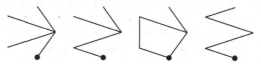

§5 We obtain the 28 maximal faces of NCR$_8^{0,0}$ via rotation of the following four graphs. In each graph, $[a+1,b] \times [b+1,a]$ is empty.

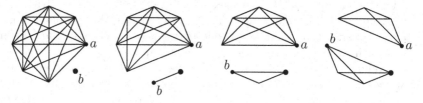

§6 We obtain the 48 maximal faces of NCR$_8^{1,0}$ via rotation and reflection of the following four graphs. In each graph, $[a+1,b] \times [b+1,a-1]$ is empty.

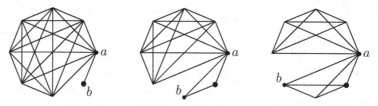

§7 We obtain the 20 maximal faces of NCR$_8^{1,1}$ via rotation of the following three graphs. In each graph, $[a+1,b-1] \times [b+1,a-1]$ is empty, meaning that there is no edge crossing the bold edge ab.

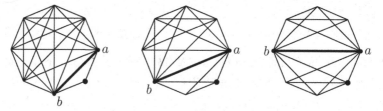

Digraph Properties

We proceed with digraph complexes, which are complexes in which each 0-cell is an ordered – as opposed to unordered – pair. We still denote a pair (i, j) as ij, but now ij and ji are different elements. We may divide the monotone digraph properties to be examined into two groups:

- *Directed variants of forests and bipartite graphs.* We may define directed variants of F_n in several ways. One way is to define the maximal faces as spanning directed trees. Another way is to define the minimal nonfaces as directed cycles. Similarly, there are a variety of ways of defining directed analogues of bipartite graphs, one way being to forbid directed cycles of odd length.
- *Directed variants of disconnected graphs.* Two ways of characterizing connectivity easily adapts to directed variants. The first characterization is that there must be a path between any two vertices; this translates into requiring a directed path from any vertex to any other vertex. The other characterization is that there must be a spanning tree; this translates into requiring the existence of a spanning directed tree.

All digraph properties studied in this book have the homotopy type of a wedge of spheres in a fixed dimension.

9.1 List of Complexes

We define the monotone digraph properties to be examined. Table 9.1 provides a summary of known topological results. Again, only a few of the results are our own; we refer the reader to the relevant section for details.

Directed Variants of Forests and Bipartite Graphs

§1 *Directed forests.* Let DF_n be the complex of directed forests on n vertices. We present the main results about DF_n in Section 15.1.

Table 9.1. Some monotone digraph properties. The Euler characteristic of DNOCy_n equals $-|\tilde{\chi}(\mathsf{B}_n)|$.

§	Name	Description	Homotopy type	Sec.
§1	DF_n	Directed forest	$\bigvee_{(n-1)^{n-1}} S^{n-2}$	15.1
§2	DAcy_n	Acyclic	S^{n-2}	15.2
§3	DB_n	Directed bipartite	S^{n-2}	15.3
	$\mathsf{DAcy}_{n,k}$	Acyclic, no directed path of edge length $k+1$	S^{n-2}	
§4	$\mathsf{DGr}_{n,p}$	Graded modulo p	$\bigvee S^{n-2}$	15.4
§5	DOAC_n	Without non-alternating circuits	$\bigvee S^{n-2}$	15.5
§6	DNOCy_n	Without odd directed cycles	$\bigvee S^{2n-3}$	15.6
§7	DNSC_n	Not strongly connected	$\bigvee_{(n-1)!} S^{2n-4}$	22.1
	$\mathsf{DNSC}_{n,k}$	$P(D)$ has $> k$ elements	$\bigvee S^{2n-k-3}$	
§8	DNSp_n	Non-spanning	$\bigvee_{(n-2)!} S^{2n-5}$	22.3
	$\mathsf{DNSp}_{n,k}$	$P(D)$ has $> k$ atoms	$\bigvee S^{2n-2k-3}$	

§2 *Acyclic digraphs.* Recall that D is acyclic if D contains no directed cycles. Let DAcy_n be the complex of acyclic digraphs on n vertices. In Section 15.2, we list the main results about DAcy_n.

§3 *Bipartite digraphs.* A digraph D is *bipartite* if there is a bipartition (U, W) of the vertex set of D such that the edge set is contained in $U \times W$. Equivalently, for each vertex, either the indegree or the outdegree is zero. Let DB_n be the complex of bipartite digraphs on n vertices. We discuss DB_n in Section 15.3.

Remark. An alternate approach would be to accept edges in both directions between U and W. This would yield the trivial extension of the complex B_n of bipartite graphs and is therefore of limited interest by Proposition 4.5.

§4 *Graded digraphs.* We say that a digraph D with vertex set V is *graded* if there is a function $f : V \to \mathbb{Z}$ such that $f(b) - f(a) = 1$ whenever ab is an edge in D. A graded digraph is necessarily acyclic. For $p \geq 2$, D is *graded modulo p* if there is a function $f : V \to [0, p-1]$ such that $(f(b) - f(a)) \bmod p = 1$ whenever ab is an edge in D. Let DGr_n be the complex of graded digraphs on n vertices and let $\mathsf{DGr}_{n,p}$ be the larger complex of digraphs that are graded modulo p. Note that $\mathsf{DGr}_n = \mathsf{DGr}_{n,p}$ if $p > n$. Clearly, $\mathsf{DGr}_{n,2}$ is the trivial extension of B_n. Yet, for $p \geq 3$, $\mathsf{DGr}_{n,p}$ is not trivial. We discuss DGr_n and $\mathsf{DGr}_{n,p}$ in Section 15.4.

§5 *Digraphs without non-alternating circuits.* For a directed edge ij, define $+ij = (+1)(ij) = ij$ and $-ij = (-1)(ij) = ji$. A circuit

$$\pi = \{s_1(v_1v_2), s_2(v_2v_3), \ldots, s_{r-1}(v_{r-1}v_r), s_r(v_rv_1)\}$$

in the digraphic matroid $M_n^{\rightarrow} = M_n(K_n^{\rightarrow})$ is *alternating* if r is even and the signs s_i form an alternating sequence; $s_{i+1} = -s_i$ for $i \in [r-1]$ and $s_1 = -s_r$. Let DOAC_n be the complex of digraphs on n vertices with no non-alternating circuits ("Only Alternating Circuits"); thus all circuits are alternating. See Section 15.5 for an analysis of DOAC_n.

§6 *Digraphs without odd directed cycles.* We obtain yet another directed variant of B_n by considering digraphs without directed cycles of odd length. Define DNOCy_n to be the complex of such digraphs on n vertices. In this case, there is no underlying bipartition in general. See Section 15.6 for an analysis of DNOCy_n.

Directed Variants of Disconnected Graphs

§7 *Not strongly connected digraphs.* Recall that a digraph D is strongly connected if every pair of vertices in D is contained in a directed cycle. Let DNSC_n be the complex of not strongly connected digraphs on n vertices. See Section 22.1 for a summary of known results about DNSC_n. A digraph D on n vertices is *strongly 2-connected* if D is strongly connected and $D([n] \setminus \{x\})$ is strongly connected for each $x \in [n]$. Hence for every triple of distinct vertices x, y, z, there is a directed path from y to z not using the vertex x. Let DNSC_n^2 be the simplicial complex of not strongly 2-connected digraphs on n vertices. We discuss this complex in Section 22.1.

§8 *Non-spanning digraphs.* Let us say that a digraph D is *spanning* if D contains a spanning directed tree. Let DNSp_n be the complex of non-spanning digraphs on n vertices. Since the minimal nonfaces of DNSp_n are spanning directed trees, this complex is a natural directed analogue of the complex NC_n of disconnected graphs in which undirected spanning trees are minimal nonfaces. Topologically however, DNSp_n has more in common with the complex NC_n^2 of not 2-connected graphs; see Section 22.3.

Interpretations in Terms of Posets

For a digraph D on a vertex set V, define an equivalence relation on V by the rule that v and w are equivalent if and only if there is a directed cycle in D containing both v and w. Equivalently, there is a directed path from v to w and a directed path from w to v. Let $A_1, \ldots, A_r$ be the induced equivalence classes. The poset $P(D)$ *associated to* D is the poset with one element for each equivalence class such that $A_i \leq A_j$ if and only if there is a directed path from some element in A_i to some element in A_j. Equivalently, there is a directed path from *any* element in A_i to *any* element in A_j.

One may interpret some of our digraph properties in terms of associated posets:

§2 $D \in \mathsf{DAcy}_n$ *if and only if* $P(D)$ *has* n *elements.* That is, no two vertices in D are equivalent.

§3 $D \in \mathsf{DB}_n$ *if and only if* $P(D)$ *has* n *elements and no chain in* $P(D)$ *has length three.* Namely, D belongs to DB_n if and only if D does not contain any directed path of edge length two.

§7 $D \in \mathsf{DNSC}_n$ *if and only if* $P(D)$ *has at least two elements.*

§8 $D \in \mathsf{DNSp}_n$ *if and only if* $P(D)$ *has at least two atoms.*

These interpretations suggest three generalizations:

- For $1 \le k \le n-1$, define $\mathsf{DAcy}_{n,k}$ as the complex of acyclic digraphs D on n vertices with no directed path of edge length $k+1$ (i.e., vertex length $k+2$). $k = 1$ yields DB_n, whereas $k = n-1$ yields DAcy_n.
- For $1 \le k \le n-1$, define $\mathsf{DNSC}_{n,k}$ as the complex of digraphs D on n vertices such that $P(D)$ has at least $k+1$ elements. $k = 1$ yields DNSC_n, whereas $k = n-1$ yields DAcy_n.
- For $1 \le k \le n-1$, define $\mathsf{DNSp}_{n,k}$ as the complex of digraphs D on n vertices such that $P(D)$ has at least $k+1$ atoms. This means that every directed forest contained in D has at least $k+1$ connected components. $k = 1$ yields DNSp_n, whereas $k = n-1$ yields the (-1)-simplex $\{\phi\}$.

We discuss the first generalization along with the analysis of DB_n in Section 15.3. We deal with the other two generalizations when we examine DNSC_n and DNSp_n in Chapter 22.

Matrix-Theoretic Remark

We may identify a digraph D on the vertex set $[n]$ with the $n \times n$ matrix M_D defined by

$$\begin{cases} (M_D)_{i,j} = 1 \text{ if } ij \in D; \\ (M_D)_{i,j} = 0 \text{ otherwise.} \end{cases}$$

The way we have defined digraphs, not allowing loops ii, the elements on the diagonal of M_D are always zero. Note that the element on position (i,j) in M_D^k equals the number of directed paths in D from i to j of edge length k.

For a digraph complex Δ, let us write $M_D \in \Delta$ if $D \in \Delta$. We may interpret some of our digraph properties in terms of matrices as follows; all matrix operations are carried out over $\mathbb{Z}$.

§1 $M \in \mathsf{DF}_n$ *if and only if* $M^n = 0$ *and each column in* M *contains at most one nonzero element.*

§2 $M \in \mathsf{DAcy}_n$ *if and only if* $M^n = 0$.

§3 $M \in \mathsf{DB}_n$ *if and only if* $M^2 = 0$.

§4 $M \in \mathsf{DGr}_{n,p}$ *if and only if there is a function* $f : V \to [0, p-1]$ *such that* $(f(b) - f(a)) \bmod p = 1$ *whenever* $M_{a,b} = 1$.

§6 $M \in \mathrm{DNOCy}_n$ *if and only if the diagonal of M^d is zero whenever d is odd.*
§7 $M \in \mathrm{DNSC}_n$ *if and only if there are indices i and j such that $(M^d)_{i,j} = 0$*
    for all $d \geq 1$.
§8 $M \in \mathrm{DNSp}_n$ *if and only if for each i there exists a $j \neq i$ such that $(M^d)_{i,j} =$*
    0 for all $d \geq 1$.

9.2 Illustrations

§1 Each of the 64 maximal digraphs in DF_4 (and the minimal digraphs not
    in DNSp_4) is isomorphic to one of the following spanning directed trees.

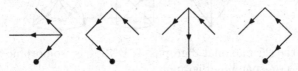

§2 The 24 maximal digraphs in DAcy_4 are all isomorphic to the following
    acyclic digraph.

§3 For an illustration of DB_4, see §4 (ii).
§4 (i) Each of the 74 maximal digraphs in DGr_4 is isomorphic to one of the
    following graded digraphs. Excluding the leftmost digraph, we obtain the
    62 maximal digraphs in the subcomplex DOAC_4 of digraphs with no non-
    alternating cycle.

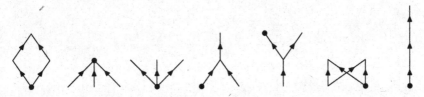

(ii) Each of the 26 maximal digraphs in $\mathrm{DGr}_{4,3}$ is isomorphic to one of the
following digraphs, all graded modulo 3. The three digraphs to the left
yield the 14 maximal faces of the subcomplex DB_4 of bipartite digraphs.

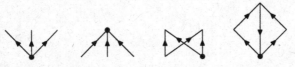

§5 For an illustration of DOAC_4, see §4 (i).

§6 Each of the 49 maximal digraphs in $DNOCy_4$ is isomorphic to one of the following digraphs without directed cycles of odd length.

§7 Each of the 14 maximal digraphs in $DNSC_4$ is isomorphic to one of the following not strongly connected digraphs.

§8 Each of the 25 maximal digraphs in $DNSp_4$ is isomorphic to one of the following non-spanning digraphs.

Main Goals and Proof Techniques

For the complexes introduced in the three preceding chapters, there are five parameters that we are particularly interested in: (1) homology, (2) homotopy type, (3) depth, (4) connectivity degree, and (5) Euler characteristic. In some cases, our analysis leads to more specific information about the complexes such as a large vertex-decomposable skeleton or an optimal decision tree. Still, our focus remains on the five listed parameters.

10.1 Homology

For obvious reasons, homology is a central topic of this book. Our approach to the subject is somewhat simplistic in the sense that we do not take into account group actions such as the natural action of the symmetric group on a given monotone graph property. Instead, we refer the interested reader to the literature [3, 21, 95, 122, 137, 145] for information and further references about this important aspect of the theory.

Almost all homology computations in this book take place in the setting of discrete Morse theory; see Chapter 4. In the vast majority of cases, we use the decision tree variant of the theory discussed in Chapter 5. Indeed, the few exceptions are exactly the cases where we have failed with the decision tree method or where this method would be significantly more cumbersome. Some monotone graph properties for which we apply "raw" discrete Morse theory are NHam_n, NC_n^3, and NEC_n^2.

Many of our decision tree proofs are quite similar to each other; an interesting question is whether one may merge some of the proofs into a more general proof. In at least one case, this is indeed possible. Specifically, the theory developed in Sections 13.2-13.4 applies to F_n, B_n, $\mathsf{DGr}_{n,2}$, DOAC_n, NC_n, and $\mathsf{NC}_{n,p}$. Most of the time however, each individual complex seems to require its own special treatment.

Occasionally, we will spend some time on examining the inner structure of the homology of a complex. The most important example is probably the homology of the quotient complex $2^{K_n}/\mathsf{NC}_n^2$, for which Shareshian [118] was able to present an explicit basis in terms of the fundamental cycle of the associahedron A_n; see Section 19.1. We mimic Shareshian's approach in our analysis of the homology of $2^{K_n}/\mathsf{NHam}_n$ and $2^{K_n}/\mathsf{NC}_n^3$; see Sections 17.2 and 20.2. In Section 21.1, the fundamental cycle of A_n appears once again, this time within the homology of $2^{K_n}/\mathsf{NCR}_n^{1,0}$ and $\mathsf{NCR}_n^{1,1}$. Finally, in Section 23.3, we expose a surprising connection between NEC_n^2 and NFC_n.

One intriguing observation in this book is that the Betti numbers of $\mathsf{B}_{n,p}$ and $\mathsf{Cov}_{n,p}$ are polynomials in n for each fixed p; see Sections 14.3 and 26.4. Add to that the result of Linusson et al. [95] that the same is true for $\mathsf{NM}_{n,k}$ for each fixed k; see Chapter 24. One interesting question is whether this holds for more general classes of monotone graph properties. Unfortunately, we have not been able to prove anything in this direction, but in a separate manuscript [73], we prove some general results about the Euler characteristic being a polynomial under certain conditions; see Chapter 25 for some details.

10.2 Homotopy Type

Being more fine-tuned than homology, homotopy type is often much harder to compute. However, thanks to discrete Morse theory, many homological results in this book are straightforward to translate into the language of homotopy theory. For example, this is the case whenever we can use discrete Morse theory to prove that all reduced homology is concentrated in one dimension; the homotopy type is then that of a wedge of spheres. Via Theorem 4.11, we may also apply discrete Morse theory to more complicated complexes such as NHam_n and $\mathsf{Cov}_{n,p}$, again concluding that the homotopy type is that of a wedge. This time however, some of the components in the wedge are no longer spheres.

10.3 Connectivity Degree

In situations where it is hard to compute the homology and homotopy type of a complex, one may instead head for the presumably easier problem of estimating the connectivity degree. Our main tool for attacking this problem is again discrete Morse theory, the goal being to find an acyclic matching such that the dimension of the smallest unmatched faces is as large as possible. We apply this technique to a number of complexes, including BD_n^d and $\mathsf{NLC}_{n,k}$. See Table 10.1 for a summary of known results.

10.4 Depth

There is an obvious upper bound on the depth of a simplicial complex given by the shifted connectivity degree. Intriguingly, this bound is sharp for many

Table 10.1. Homotopical depth and *shifted* connectivity degree of some graph, dihedral graph, and digraph properties. For the properties to the left, the shifted connectivity degree coincides with the depth, *possible* exceptions being complexes where we only have a lower bound on the depth (and, of course, contractible complexes). For the properties to the right, we typically do not have any nontrivial lower bound on the depth.

§	Complex	Depth	Sec.
§7.1	M_n	$\lceil \frac{n-4}{3} \rceil$	11.2
§7.3	F_n	$n-2\ (VD)$	13.1
§7.4	B_n	$n-2$	14.1
§7.5	$B_{n,p}$	$2p-1$	14.3
§7.7	NC_n	$n-3$	18.1
§7.8	$NLC_{n,k}$	$\geq \frac{(k-1)(n-1)}{k+1} - 1$	18.2
§7.9	$SSC_n^{k,s}$	$n-s-2$	18.3
§7.10	$NC_{n,p}$	$n-3$	18.4
§7.11	NC_n^2	$2n-5$	19
§7.17	$Cov_{n,p}$	$\geq 2p-1$	26
§7.18	$\not{b}_n$	$n-2$	26.7
§8.1	NX_n	$2n-4\ (VD)$	16.2
	A_n	$n-4\ (VD)$	
§8.2	NXM_n	$\lceil \frac{n-4}{3} \rceil$	16.3
§8.3	NXF_n	$n-2\ (VD)$	16.4
§8.4	NXB_n	$n-2$	16.5
§8.5	$NCR_n^{0,0}$	$n-3$	21.2
§8.6	$NCR_n^{1,0}$	$2n-5$	21.3
§9.1	DF_n	$n-2\ (VD)$	15.1
§9.2	$DAcy_n$	$n-2$	15.2
§9.3	DB_n	$n-2$	15.3
§9.4	$DGr_{n,p}$	$n-2$	15.4
§9.5	$DOAC_n$	$n-2$	15.5
§9.7	$DNSC_n$	$2n-4$	22.1

§	Complex	Shifted connectivity degree	Sec.
§7.2	BD_n^2	$\geq \frac{7n-13}{9}$	12.1
	BD_n^3	$\geq \frac{11n-13}{9}$	
	BD_n^4	$\geq \frac{27n-25}{16}$	
§7.6	$NHam_n$	$\geq \frac{3n-4}{2}\ (n \geq 6)$	17
§7.12	NC_n^3	$2n-4$	20
§7.13	NEC_{2k}^2	$\geq 3k-3$	23
	NEC_{2k-1}^2	$3k-5$	
§7.14	NFC_{2k-1}	$3k-5$	23.3
§7.15	$NM_{n,k}$	$3k-4$	24
§7.16	Col_n^t	$\geq \frac{(t-1)(n-1)}{2} - 1$	25
§8.7	$NCR_n^{1,1}$	$n-4$	21.4
§9.6	$DNOCy_n$	$2n-3$	15.6
§9.8	$DNSp_n$	$2n-5$	22.3

graph complexes. Moreover, the relevant skeleton is often not only Cohen-Macaulay but also vertex-decomposable. For example, this holds for M_n, F_n, B_n, $B_{n,p}$, NC_n, and $\not{b}_n$. Our proof techniques are typically very similar to those used to determine homology and homotopy type. Keeping in mind that a proof of vertex-decomposability is basically all about defining a decision tree, this is perhaps not so surprising.

In situations where a vertex decomposition is difficult to find, one may instead adapt the technique that Shareshian used in his analysis of NC_n^2 [117], establishing Cohen-Macaulayness via a separate treatment of each individual link. We take this approach in our analysis of $NLC_{n,k}$ and $DNSC_n$.

Complexes in this book with a known depth – or a known nontrivial lower bound on the depth – are listed to the left in Table 10.1. For the complexes to the right in the same table, we know very little about the depth.

Table 10.2. Reduced Euler characteristic of some graph, dihedral graph, and digraph properties. "$Poly_k(n)$" denotes a polynomial of degree at most k. The functions in the table to the right are exponential generating functions (plus or minus some leading terms), except for dihedral properties, in which case they are ordinary generating functions.

§	Complex	Euler char. (up to sign)	Sec.
§7.5	$B_{n,p}$	$Poly_{2p}(n)$	14.3
§7.7	NC_n	$(n-1)!$	18.1
§7.9	$SSC_n^{1,s}$	$n\binom{n-2}{s-1} - \binom{n-1}{s-1}$	18.3
§7.11	NC_n^2	$(n-2)!$	19
§7.12	NC_n^3	$(n-3)\frac{(n-2)!}{2}$	20
§7.14	NFC_{2k-1}	$((2k-3)!!)^2$	23.3
§7.15	$NM_{n,k}$	$Poly_{3k-3}(n)$	24
§7.16	$(Col_n^{n-r})^*$	$Poly(n)$	25
§7.17	$Cov_{n,p}$	$Poly_{2p}(n)$	26
§8.1	NX_n	0	16.2
	A_n	1	
§8.5	$NCR_n^{0,0}$	$\frac{1}{n}\binom{2n-2}{n-1}$	21.2
§8.6	$NCR_n^{1,0}$	1	21.3
§8.7	$\overline{NCR_n^{1,1}}$	1	21.4
§9.1	DF_n	$(n-1)^{n-1}$	15.1
§9.2	$DAcy_n$	1	15.2
§9.3	DB_n	1	15.3
§9.7	$DNSC_n$	$(n-1)!$	22.1
§9.8	$DNSp_n$	$(n-2)!$	22.3

§	Complex	Gen. function	Sec.
§7.1	M_n	$-e^{-x+x^2/2}$	11.2
§7.2	BD_n^2	$-\frac{\exp(\frac{x}{2+2x}+x)}{\exp(\frac{x^2}{4})\sqrt{1+x}}$	12.3
	$(BD_n^{n-3})^*$	$e^{-x^2/2}(x+e^{-x})$	
§7.3	F_n	See Th. 13.3	13.1
§7.4	B_n	$-\sqrt{2e^x-1}$	14.1
§7.4	$B_n \setminus NC_n$	$-\frac{1}{2}\ln(2e^x-1)$	14.2
§7.6	$NHam_n$	Not known	17
§7.8	$NLC_{n,k}$	See Prop. 18.9	18.2
§7.9	$SSC_n^{k,s}$	See Th. 18.19	18.3
§7.10	$NC_{n,p}$	$(1-(-x)^p)^{1/p}$	18.4
§7.12	$(NC_n^{n-3})^*$	$\frac{EGF(BD_n^2)}{\exp(x^4/8)}$	20.4
§7.13	NEC_n^2	See Th. 23.2	23
§7.18	$\not\vDash_n$	Not known	26.7
§8.2	NXM_n	$\frac{1-x-\sqrt{1-2x+5x^2}}{2x^2}$	16.3
§8.3	NXF_n	See Th. 16.13	16.4
§8.4	NXB_n	$\frac{1-2x-\sqrt{1+4x}}{4-2x}$	16.5
§9.4	DGr_n	$-\frac{1}{2}\sqrt{4e^x-3}$	15.4
	$DGr_{n,3}$	$-(3e^x-2)^{1/3}$	
	$DGr_{n,4}$	$-(2e^{2x}-1)^{1/4}$	
§9.5	$DOAC_n$	Not known	15.5
§9.6	$DNOCy_n$	$-\sqrt{2e^{-x}-1}$	15.6

10.5 Euler Characteristic

For several complexes in this book, the Euler characteristic admits a closed expression, either explicitly or via a generating function. For the matching complex M_n and the complex NC_n of disconnected graphs, the existence of such a nice formula is an immediate consequence of Corollary 6.15, but in general we need to work a bit harder. In some instances, we can deduce the Euler characteristic directly from an explicit description of the critical faces with respect to a given acyclic matching or decision tree. In other cases, the procedure goes via generating functions, and the most frequently used approach in this book is to search for a nice recursive identity of the form $\tilde{\chi}(\Delta_n) = F_n(\tilde{\chi}(\Delta_1), \ldots, \tilde{\chi}(\Delta_{n-1}))$. In favorable instances, we may extract from this identity a closed expression – explicit or implicit – for the (exponential) generating function for $\tilde{\chi}(\Delta_n)$. For a fairly straightforward example, see Theorem 16.12. We present formulas for the Euler characteristic of some complexes in Table 10.2.

10.6 Remarks on Nonevasiveness and Related Properties

In Section 1.1.9, we discussed the evasiveness conjecture. Let us say that a simplicial complex Σ is *nearly nonevasive* if the deletion of Σ with respect to some vertex is nonevasive. For example, vertex-decomposable combinatorial spheres are nearly nonevasive. Clearly, any nonevasive complex, the 0-simplex excluded, is also nearly nonevasive.

For complexes with a vertex-transitive automorphism group, a particular vertex deletion is nonevasive if and only if all vertex deletions are nonevasive. In particular, monotone graph and digraph properties satisfy this condition. An interesting research project would be to characterize the family of nearly nonevasive monotone graph and digraph properties. While a complete characterization is probably very hard to achieve, we believe that any example is likely to have a rich and beautiful structure, thanks to the vertex-transitive structure.

So far, we have discovered the following nearly nonevasive monotone graph and digraph properties:

- The graph property of not being the complete graph on n vertices.
- The digraph property of not being the complete digraph on n vertices.
- The graph property $NM_{4,2}$ of not containing a perfect matching on four vertices. To see that $NM_{4,2}$ is nearly nonevasive, note that the deletion with respect to the edge 12 is a cone with cone point 34. In fact, $NM_{4,2}$ is isomorphic to the octahedron.
- The digraph property of not containing a cycle (ij, ji) of length 2. Namely, the deletion with respect to 12 is a cone with cone point 21.
- The Alexander dual of he digraph property $DAcy_n$ of being acyclic on n vertices; use the proof of Theorem 15.3.

- The digraph property DB_n of being directed bipartite on n vertices; use the proof of Theorem 15.7.
- The graph property NC_n^2 of being not 2-connected on n vertices; use the proof of Theorem 19.9.

Another question related to the evasiveness conjecture is whether there are collapsible monotone graph properties that are not nonevasive. More generally, one may ask whether there are semi-collapsible monotone graph properties that are not semi-nonevasive. Not surprisingly, the answer to the second question is yes:

Let Δ be the complex of all graphs on the vertex set $\{1, 2, 3, 4, 5\}$ that are contained in a copy of $\{12, 34, 35\}$. This complex is collapsible to the matching complex M_5 on five vertices (see Chapter 11); collapse all pairs $(\{cd, ce\}, \{ab, cd, ce\})$. Since M_5 is semi-collapsible by Theorem 11.27, the same is true for Δ. However, Δ is not semi-nonevasive. Namely, the three 1-cells $\{34, 35\}$, $\{34, 45\}$, $\{35, 45\}$ form a cycle in $lk_\Delta(12)$, which implies that $lk_\Delta(12)$ has nonvanishing homology in its top dimension; by symmetry, the same is true for $lk_\Delta(x)$ for any x. Since Δ has no homology in its top dimension, it follows that Δ cannot be semi-nonevasive.

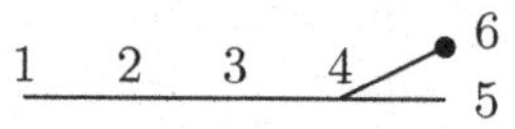

Fig. 10.1. The graph $\{12, 23, 34, 45, 46\}$.

It may also be worth mentioning that there exists a $\mathbb{Q}$-acyclic monotone graph property that is not $\mathbb{Z}$-acyclic: Let Δ be the complex of all bipartite graphs on the vertex set $\{1, 2, 3, 4, 5, 6\}$ that do not contain a subgraph isomorphic to the graph in Figure 10.1. Using the computer program homology [42], one may conclude that the only nonzero homology group is $\tilde{H}_3(\Delta; \mathbb{Z}) \cong \mathbb{Z}_2^{16} \oplus \mathbb{Z}_3^4 \oplus \mathbb{Z}_9$. We have not found a simple proof of this fact.

Part IV

Vertex Degree

11

Matchings

We discuss simplicial complexes of matchings. Recall that a matching is a graph in which each vertex is adjacent to at most one other vertex.

For any graph G on n vertices, let $\mathsf{M}(G) = \mathsf{M}_n(G)$ be the simplicial complex of matchings contained in G. Arguably the most important special case is the full complex $\mathsf{M}_n = \mathsf{M}(K_n)$ of all matchings on the vertex set $[n]$. For an excellent survey of results on M_n, see Wachs [145]. In Section 11.2, we give a summary of some of these results:

- Bouc [21] derived a formula for the rational homology of M_n; see also the work of Karaguezian [80] and Reiner and Roberts [111]. A consequence of the formula is that $\tilde{H}_d(\mathsf{M}_n; \mathbb{Q})$ is nonzero if and only if $\left\lceil \frac{n - \lfloor \sqrt{n} \rfloor - 2}{2} \right\rceil \leq d \leq \left\lfloor \frac{n-3}{2} \right\rfloor$.

- Shareshian and Wachs [122] provided a shelling of the ν_n-skeleton of M_n, where $\nu_n = \left\lceil \frac{n-4}{3} \right\rceil$, thereby giving a new proof of a result of Björner, Lovász, Vrećica, and Živaljević about the connectivity degree of M_n. Athanasiadis [2] proved that the ν_n-skeleton is vertex-decomposable.

- Bouc [21] proved that M_n has nonvanishing homology in dimension ν_n for all $n \neq 2$;[1] hence the shifted connectivity degree of M_n is ν_n. Combining results of Bouc with new ideas, Shareshian and Wachs [122] were able to deduce that the group $\tilde{H}_{\nu_n}(\mathsf{M}_n; \mathbb{Z})$ is a finite group of exponent three if $n \in \{7, 10, 12, 13\}$ or $n \geq 15$.

We present a proof of the important result about the connectivity degree of M_n, basically following the approach of Shareshian and Wachs [122]. Moreover, using the 3-torsion result of Shareshian and Wachs, we show that $\tilde{H}_d(\mathsf{M}_n; \mathbb{Z})$ contains 3-torsion whenever $\nu_n \leq d \leq \frac{n-6}{2}$. In Section 11.1, we generalize Athanasiadis' result to general graphs, proving that $\mathsf{M}_n(G)$ is $VD(\frac{n-t}{2} - 1)$ whenever the vertex set of G admits a partition into t cliques of size at most three.

---

[1] Bouc did not explicitly mention the case $n \bmod 3 = 2$; see Shareshian and Wachs [122].

Another very important and well-studied special case is the *chessboard complex* $M_{m,n} = M(K_{m,n})$, where $K_{m,n}$ is the complete bipartite graph with block sizes m and n. Again, the rational homology is given by a beautiful formula; see Friedman and Hanlon [52]. In Section 11.3, we list some important results due to Björner et al. [11], Ziegler [151], and Shareshian and Wachs [122], the main conclusion being that the depth and the shifted connectivity degree of $M_{m,n}$ are equal to $\min\{m-1, \lceil \frac{m+n-4}{3} \rceil\}$. Using results of Shareshian and Wachs and our own result about 3-torsion in $\tilde{H}_d(M_n; \mathbb{Z})$, we prove that $\tilde{H}_d(M_{m,n}; \mathbb{Z})$ contains 3-torsion whenever $\frac{m+n-4}{3} \leq d \leq m-4$ and whenever $d = m-3$ and $m+1 \leq n \leq 2m-5$.

In Section 11.4, we proceed with some results due to Kozlov [86] about matching complexes on paths and cycles; we will need these results in later sections. There are many other potentially interesting matching complexes, e.g., on rectangular grids, honeycomb graphs, and Kneser graphs, but the analysis of such complexes falls outside the scope of this book. See a separate manuscript [69] for a treatment of grids.

Occasionally, we will say a few words about matching complexes on hypergraphs. The most important example is the matching complex HM_n^k of all k-hypergraphs on $[n]$ with mutually disjoint edges.

The matching complex and its relatives have found applications in several areas of mathematics; see Sections 1.1.1, 1.1.2, 1.1.3, and 1.1.7 for discussion.

11.1 Some General Results

We consider a general graph G and present some lower bounds on the depth of $M(G)$.

Theorem 11.1. *Let G be a graph on the vertex set V. Suppose that there is a partition $\{U_1, \ldots, U_t\}$ of V such that $|U_i| \leq 3$ for each i and such that $G(U_i)$ is isomorphic to either K_1, K_2, K_3, or $\mathsf{Pa}_3 = ([3], \{12, 23\})$. Suppose further that whenever $G(U_i)$ is of the form $(\{a, b, c\}, \{ab, bc\})$ (thus isomorphic to Pa_3), the vertex b is not adjacent in G to any other vertices than a and c. Then $M(G)$ is $VD(\nu)$, where*

$$\nu = \left\lceil \frac{|V| - t}{2} \right\rceil - 1.$$

In particular, this holds whenever $\{U_1, \ldots, U_t\}$ is a clique partition of G such that each U_i has size at most three.

Proof. Let σ be the union of the sets of edges within the induced subgraphs $G(U_i)$ for $i \in [1, k]$. If the edge set of G is σ, then

$$M(G) = M(G(U_1)) * \cdots * M(G(U_t)),$$

which is $VD(\nu)$ by Lemma 6.11; $M(H)$ is $VD(0)$ if $H \in \{K_2, K_3, \mathsf{Pa}_3\}$ and $VD(-1)$ if $H = K_1$. Otherwise, let e be any edge in $G - \sigma$. By induction on the number of edges in G, $\mathrm{del}_{\mathsf{M}(G)}(e) = \mathsf{M}(G - e)$ is $VD(\nu)$. Moreover, $\mathrm{lk}_{\mathsf{M}(G)}(e)$ equals $\mathsf{M}(G(V \setminus e))$, where V is the vertex set of G; we remove all edges containing either of the two vertices in e. Removing the endpoints of e from the appropriate sets U_i in the partition, we obtain a partition of $V \setminus e$ with at most t sets. Moreover, each of the corresponding induced subgraphs is either a complete graph or isomorphic to $\mathsf{Pa_3}$; if one of the endpoints of e lies in an induced subgraph $G(U_i)$ isomorphic to $\mathsf{Pa_3}$, then $G(U_i \setminus e)$ must be isomorphic to K_2 by assumption. By induction, $G(V \setminus e)$ is hence $VD(\nu')$, where

$$\nu' = \left\lceil \frac{|V| - 2 - t}{2} \right\rceil - 1 = \nu - 1.$$

By Lemma 6.10, we are done. $\square$

As an immediate consequence, we obtain the following result:

Corollary 11.2. *Let G be a graph on the vertex set $[n]$. Suppose that there is a clique partition of G into $t = \lceil \frac{n}{3} \rceil$ parts, each of size at most three. Then $\mathsf{M}(G)$ is $VD(\nu_n)$, where $\nu_n = \lceil \frac{n-4}{3} \rceil$. In particular, this is true for $G = K_n$.*

Remark. Athanasiadis was the first to prove that $\mathsf{M}_n = \mathsf{M}(K_n)$ is $VD(\nu_n)$; see Section 11.2 for more information.

Proof. Let k be such that $n = 3k - r$ and $r \in \{0, 1, 2\}$. We obtain that

$$\left\lceil \frac{n - \lceil n/3 \rceil}{2} \right\rceil = \left\lceil \frac{3k - r - k}{2} \right\rceil = \left\lceil k - \frac{r}{2} \right\rceil = \left\lceil k - \frac{r+1}{3} \right\rceil = \left\lceil \frac{n-1}{3} \right\rceil.$$

Thus we are done by Theorem 11.1. $\square$

The following corollary to Theorem 11.1 is less significant but still somewhat interesting.

Corollary 11.3. *Let G be a graph on n vertices admitting a perfect matching. Then $\mathsf{M}(G)$ is $VD(\lceil \frac{n}{4} \rceil - 1)$.* $\square$

The dimension $\lceil \frac{n}{4} \rceil - 1$ in Corollary 11.3 is best possible. Namely, suppose that G is a graph with $n = 4m$ vertices and with m connected components, all isomorphic to $\mathsf{Pa_4} = ([4], \{12, 23, 34\})$. Then G admits a perfect matching of size $2m = \frac{n}{2}$. However, since $\mathsf{M}(\mathsf{Pa_4}) \sim t^0$ (use Proposition 11.42 below), we have that $\mathsf{M}(G) = \mathsf{M}(\mathsf{Pa_4}) * \cdots * \mathsf{M}(\mathsf{Pa_4}) \sim t^{m-1}$ by Theorem 5.29; this implies that the shifted connectivity degree of $\mathsf{M}(G)$ is $m - 1 = \frac{n}{4} - 1$.

Nevertheless, in many special cases, the value in Corollary 11.3 is way below the actual depth. This is true not only for graphs admitting a partition with many triangles, such as the graphs in Corollary 11.2, but also for several triangle-free graphs such as the complete bipartite graph discussed in Section 11.3.

The following powerful result is worth mentioning for its use in the work of Athanasiadis [2]. We will apply it in Section 11.3.

Theorem 11.4 (Athanasiadis [2, Th. 4.1]). *Let G be a graph on the vertex set V, let $d \geq 0$, and let $e = ab$ be an edge in G. Suppose that $\mathsf{M}(G(V \setminus S))$ is $VD(d-1)$ whenever $S = \{a, x\}$ for some x such that $ax \in G$ (including $x = b$) or $S = \{a, b, y\}$ for some y such that $by \in G$. Then G is $VD(d)$.* $\square$

The proof idea is to decompose $\mathsf{M}(G)$ with respect to the edge set $\{ax : ax \in G, x \neq b\}$ and then decompose the resulting deletion with respect to the set $\{yb : yb \in G, y \neq a\}$. The deletion with respect to all these edges is the join of $\{\emptyset, ab\}$ and $\mathsf{M}(G(V \setminus \{a, b\}))$, whereas each link in the decomposition coincides with some of the other induced subcomplexes mentioned in the theorem.

11.2 Complete Graphs

We consider the full matching complex $\mathsf{M}_n = \mathsf{M}(K_n)$.

11.2.1 Rational Homology

The rational homology of M_n is given by a surprisingly beautiful formula. A *standard Young tableau* T on a partition $\lambda \vdash n$ (see Section 2.5) is a bijection from the set λ to $[n]$ such that $T(a, b) \leq T(c, d)$ whenever $a \leq c$ and $b \leq d$. Thus T increases along each row and each column in λ. The *hook* of an element (a, b) in λ is the set

$$H_\lambda(a, b) = \{(a, b') \in \lambda : b' \geq b\} \cup \{(a', b) \in \lambda : a' \geq a\}.$$

The number f_λ of standard Young tableaux on λ is given by the celebrated *hook length formula* [51]:

$$f_\lambda = \frac{|\lambda|!}{\prod_{(a,b) \in \lambda} |H_\lambda(a, b)|}. \tag{11.1}$$

Recall that D_λ is the set $\{(i, i) : \lambda_i \geq i\}$ of diagonal elements in λ.

Theorem 11.5 (Bouc [21]). *Let notation be as in Section 2.5. For $n \geq 1$ and $d \geq 0$,*

$$\dim \tilde{H}_{d-1}(\mathsf{M}_n; \mathbb{Q}) = \sum_\lambda f_\lambda,$$

where the sum is over all self-conjugate partitions $\lambda \vdash n$ such that $|D_\lambda| = n - 2d$ (i.e., $d = \frac{n - |D_\lambda|}{2}$) and f_λ is the number of standard Young tableaux on λ (use equation (11.1)). In particular, M_n has homology over $\mathbb{Q}$ in dimension d if and only if

$$\alpha_n = \left\lceil \frac{n - \lfloor \sqrt{n} \rfloor - 2}{2} \right\rceil \leq d \leq \left\lfloor \frac{n-3}{2} \right\rfloor.$$

As a consequence, the depth of M_n over $\mathbb{Q}$ equals α_n. $\square$

For the last statement in Theorem 11.5, use the fact that $\alpha_{n-2k} \geq \alpha_n - k$ for all n, k such that $n > 2k$. Theorem 11.5 was rediscovered by Karaguezian [80] and by Reiner and Roberts [111]. Dong and Wachs [37] gave an elegant proof in terms of the combinatorial Laplacian. See Wachs' survey [145] for more information about the rational homology of M_n.

11.2.2 Homotopical Depth and Bottom Nonvanishing Homology

Let $\nu_n = \lceil \frac{n-4}{3} \rceil$. Björner, Lovász, Vrećica, and Živaljević proved the first significant result about the homotopical depth of M_n, which turns out to be significantly smaller than the depth over $\mathbb{Q}$:

Theorem 11.6 (Björner et al. [11]). *For $n \geq 1$, the ν_n-skeleton of M_n is homotopically CM. Moreover, M_n has no homology above dimension $\lfloor \frac{n-3}{2} \rfloor$.* □

As we will see, ν_n is indeed equal to the depth and the shifted connectivity degree of M_n.

Shareshian and Wachs [120, 122] strengthened Theorem 11.6, showing that the ν_n-skeleton of M_n is shellable. Athanasiadis [2] extended this result to hypergraphs:

Theorem 11.7 (Athanasiadis [2]). *Let $n \geq k \geq 2$. Then HM_n^k is $VD(\nu_{n,k})$, where $\nu_{n,k} = \lceil \frac{n-2k}{k+1} \rceil$.* □

The special case $k = 2$ is a consequence of Corollary 11.2. Another approach to proving this special case would be to apply Theorem 11.4. Specifically, with notation and assumptions as in Theorem 11.4, one readily verifies that the graph $K_n(V \setminus S)$ is a complete graph on either $n - 3$ or $n - 2$ vertices. A simple induction argument yields the desired result. For $k \geq 3$ and $n \geq 3k+2$, Theorem 11.7 provides a significant and surprising improvement to the bound $\lceil \frac{n-k-2}{2k-1} \rceil$ of Ksontini [88] on the shifted connectivity degree of HM_n^k. One easily checks by hand that HM_n^3 is homotopy equivalent to a nonempty wedge of spheres of dimension $\nu_{n,3}$ for $n \in [4, 9]$. Moreover, a computer calculation yields that $\tilde{H}_1(\mathsf{HM}_{10}^3; \mathbb{Z}) \cong \mathbb{Z}^{42}$. and $\tilde{H}_2(\mathsf{HM}_{10}^3; \mathbb{Z}) \cong \mathbb{Z}^{861}$.

Define

$$P_d^n = \bigoplus_{i=2}^n \tilde{H}_{d-1}(\mathsf{M}_{[2,n]\setminus\{i\}}; \mathbb{Z}) \cdot \mathsf{p}_i;$$

$$Q_d^n = \bigoplus_{i \neq j \in [3,n]} \tilde{H}_{d-2}(\mathsf{M}_{[3,n]\setminus\{i,j\}}; \mathbb{Z}) \cdot \mathsf{q}_{i,j};$$

$$R_d^n = \bigoplus_{a=1}^2 \bigoplus_{i=3}^n \tilde{H}_{d-1}(\mathsf{M}_{[3,n]\setminus\{i\}}; \mathbb{Z}) \cdot \mathsf{r}_{a,i}.$$

Here, p_i, $\mathsf{q}_{i,j}$, and $\mathsf{r}_{a,i}$ are formal variables and M_X is the matching complex on the complete graph with vertex set X. Let Δ_n be the subcomplex of M_n

consisting of all matchings G such that at least one of the vertices 1 and 2 is isolated in $G - 12$. Thus M_n/Δ_n consists of all matchings G such that $1i, 2j \in G$ for some distinct $i, j \in [3, n]$.

Lemma 11.8 (see Bouc [21]). *We have isomorphisms*

$$f : P_d^n \to \tilde{H}_d(\mathsf{M}_n/\mathsf{M}_{[2,n]}; \mathbb{Z});$$
$$g : Q_d^n \to \tilde{H}_d(\mathsf{M}_n/\Delta_n; \mathbb{Z});$$
$$h : R_d^n \to \tilde{H}_d(\Delta_n; \mathbb{Z})$$

given coefficient-wise by $f(z \cdot \mathsf{p}_i) = [1i] \wedge z$, $g(z \cdot \mathsf{q}_{i,j}) = [1i] \wedge [2j] \wedge z$, *and* $h(z \cdot \mathsf{r}_{a,i}) = ([ai] - [12]) \wedge z$.

Proof. First, consider $\mathsf{M}_n/\mathsf{M}_{[2,n]}$. We have that

$$\mathsf{M}_n/\mathsf{M}_{[2,n]} = \bigcup_{i \in [2,n]} \{\{1i\}\} * \mathsf{M}_{[2,n]\setminus\{i\}}.$$

Since the families in the union form an antichain with respect to inclusion, we immediately obtain that f defines an isomorphism.

Next, note that

$$\mathsf{M}_n/\Delta_n = \bigcup_{i \neq j \in [3,n]} \{\{1i, 2j\}\} * \mathsf{M}_{[3,n]\setminus\{i,j\}}.$$

Again, we have an antichain, which implies that g defines an isomorphism.

Finally, consider Δ_n. We have that Δ_n is homotopy equivalent to the quotient complex $\tilde{\Delta}_n = \Delta_n/(\{\emptyset, \{12\}\} * \mathsf{M}_{[3,n]})$ by the Contractible Subcomplex Lemma 3.16. It is clear that

$$\tilde{\Delta}_n = \bigcup_{a=1}^{2} \bigcup_{i \in [3,n]} \{\{ai\}\} * \mathsf{M}_{[3,n]\setminus\{i\}}.$$

As a consequence, we have an isomorphism $\tilde{h}$ from R_d^n to $\tilde{\Delta}_n$ given by $\tilde{h}(z \cdot \mathsf{r}_{a,i}) = [ai] \wedge z$. Now, observe that we obtain $\tilde{\Delta}_n$ from Δ_n via the acyclic matching defined by pairing $\sigma + 12$ with $\sigma - 12$ whenever possible. Applying the theory in Section 4.4 to this matching, one readily verifies that an isomorphism from $\tilde{H}_d(\tilde{\Delta}_n; \mathbb{Z})$ to $\tilde{H}_d(\Delta_n; \mathbb{Z})$ is given by the map $[ai] \wedge z \mapsto ([ai] - [12]) \wedge z$. Composing this map with $\tilde{h}$, we obtain h, which concludes the proof. $\square$

For a cycle z in $\tilde{C}_d(\mathsf{M}_n; \mathbb{Z})$ and a sequence $(e_1, \ldots, e_r)$ of edges, we define $z_{e_1,\ldots,e_r}$ as the unique cycle in $\tilde{C}_d(\mathsf{M}_{[n]\setminus\bigcup_i e_i}; \mathbb{Z})$ such that

$$z - [e_1] \wedge \cdots \wedge [e_r] \wedge z_{e_1,\ldots,e_r} \in \tilde{C}_d(\mathrm{fdel}_{\mathsf{M}_n}(\{e_1, \ldots, e_r\}); \mathbb{Z}).$$

Corollary 11.9. *For each n and d, we have an exact sequence*

$$\tilde{H}_d(\mathsf{M}_{[2,n]};\mathbb{Z}) \longrightarrow \tilde{H}_d(\mathsf{M}_n;\mathbb{Z}) \overset{\omega}{\longrightarrow} P_d^n$$

$$\longrightarrow \tilde{H}_{d-1}(\mathsf{M}_{[2,n]};\mathbb{Z}) \longrightarrow \tilde{H}_{d-1}(\mathsf{M}_n;\mathbb{Z}),$$

where $\omega(z) = \sum z_{1i} \cdot \mathsf{p}_i$. Moreover, we have an exact sequence

$$R_d^n \overset{\varphi}{\longrightarrow} \tilde{H}_d(\mathsf{M}_n;\mathbb{Z}) \overset{\kappa}{\longrightarrow} Q_d^n \overset{\psi}{\longrightarrow} R_{d-1}^n \longrightarrow \tilde{H}_{d-1}(\mathsf{M}_n;\mathbb{Z}),$$

where $\psi(z \cdot \mathsf{q}_{i,j}) = z \cdot \mathsf{r}_{2,j} - z \cdot \mathsf{r}_{1,i}$, $\varphi(z \cdot \mathsf{r}_{a,i}) = ([ai] - [12]) \wedge z$, and $\kappa(z) = \sum_{i,j} z_{1i,2j} \cdot \mathsf{q}_{i,j}$.

Proof. This is an immediate consequence of Lemma 11.8 and the long exact sequences for the pairs $(\mathsf{M}_n, \mathsf{M}_{[2,n]})$ and (M_n, Δ_n); see Theorem 3.3. $\square$

For $N \bmod 3 = 0$, define

$$\gamma_N = ([12] - [23]) \wedge ([45] - [56]) \wedge ([78] - [89]) \qquad (11.2)$$
$$\wedge \cdots \wedge ([(N-2)(N-1)] - [(N-1)N]);$$

this is a cycle in $\tilde{C}_{\nu_N}(\mathsf{M}_N;\mathbb{Z})$.

Lemma 11.10 (Shareshian and Wachs [122]). *Write $N = 3\lfloor\frac{n}{3}\rfloor$. For $n \bmod 3 \in \{0,1\}$, $\tilde{H}_{\nu_n}(\mathsf{M}_n;\mathbb{Z})$ is generated by $\{\pi(\gamma_N) : \pi \in \mathfrak{S}_{[n]}\}$, the action of $\mathfrak{S}_{[n]}$ on $\tilde{H}_{\nu_n}(\mathsf{M}_n;\mathbb{Z})$ being the one induced by the natural action on M_n.*

Proof. One easily checks the statement for $n = 3, 4$. By Corollary 11.9, we have an exact sequence

$$R_{\nu_n}^n \overset{\varphi}{\longrightarrow} \tilde{H}_{\nu_n}(\mathsf{M}_n;\mathbb{Z}) \longrightarrow Q_{\nu_n}^n,$$

where $\varphi(z \cdot \mathsf{r}_{a,i}) = ([ai]-[12]) \wedge z$. Now, since $\nu_{n-4} = \nu_n - 1$ if $n \bmod 3 \in \{0,1\}$, we have that $\tilde{H}_{\nu_n - 2}(\mathsf{M}_{n-4};\mathbb{Z}) = 0$ and hence that $Q_{\nu_n}^n = 0$, which implies that φ is surjective. The desired claim easily follows by induction. $\square$

Theorem 11.11 (Bouc [21]). *For $n \geq 7$ and $n \bmod 3 = 1$, $\tilde{H}_{\nu_n}(\mathsf{M}_n;\mathbb{Z}) \cong \mathbb{Z}_3$.*

Proof. A computer calculation yields the statement for $n = 7$. Assume that $n \geq 10$. By Corollary 11.9, we have an exact sequence

$$Q_{\nu_n+1}^n \overset{\psi}{\longrightarrow} R_{\nu_n}^n \longrightarrow \tilde{H}_{\nu_n}(\mathsf{M}_n;\mathbb{Z}) \longrightarrow 0, \qquad (11.3)$$

where $\psi(z \cdot \mathsf{q}_{i,j}) = z \cdot \mathsf{r}_{2,j} - z \cdot \mathsf{r}_{1,i}$. The rightmost group being zero is a consequence of Lemma 11.8 and the fact that $\nu_n - 2 < \nu_{n-4}$.

Now, induction on n yields that $\tilde{H}_{\nu_n-1}(\mathsf{M}_{[3,n-1]};\mathbb{Z}) \cong \mathbb{Z}_3$; $\nu_n - 1 = \nu_{n-3}$. Consider the cycle

$\gamma_{n-4}^{(2)} = ([34] - [45]) \wedge ([67] - [78]) \wedge \cdots \wedge ([(n-4)(n-3)] - [(n-3)(n-2)]);$

this is an element in $\tilde{H}_{\nu_n - 1}(\mathsf{M}_{[3,n-1]}; \mathbb{Z})$.[2] By Lemma 11.10 and symmetry, we must have that $\pi(\gamma_{n-4}^{(2)})$ is nonzero in $\tilde{H}_{\nu_n - 1}(\mathsf{M}_{[3,n-1]}; \mathbb{Z})$ for every permutation $\pi \in \mathfrak{S}_{[3,n-1]}$. In particular, $\pi(\gamma_{n-4}^{(2)}) = \pm\gamma_{n-4}^{(2)}$. Transposing $\pi(3)$ and $\pi(5)$ in $\pi(\gamma_{n-4}^{(2)})$, we obtain $-\pi(\gamma_{n-4}^{(2)})$, which implies that $\mathfrak{S}_{[3,n-1]}$ acts on $\tilde{H}_{\nu_n - 1}(\mathsf{M}_{[3,n-1]}; \mathbb{Z})$ by $\pi(z) = \text{sgn}(\pi) \cdot z$.

Let ϵ_i be the permutation $(i, i+1, \ldots, n-1, n)$ in $\mathfrak{S}_{[3,n]}$; r is mapped to $r+1$ for $r \in [i, n-1]$ and n is mapped to i. By symmetry, it is immediate that the map $\hat{\epsilon}_i : \tilde{H}_{\nu_n - 1}(\mathsf{M}_{[3,n-1]}; \mathbb{Z}) \to \tilde{H}_{\nu_n - 1}(\mathsf{M}_{[3,n]\setminus\{i\}}; \mathbb{Z})$ defined by $\hat{\epsilon}_i(z) = (-1)^{n-i}\epsilon_i(z)$ is an isomorphism.

Now, let $i, j \in [3, n]$ and consider an element $z \cdot \mathsf{q}_{i,j}$ in $Q_{\nu_n + 1}^n$; any element in $Q_{\nu_n + 1}^n$ is a linear combination of such elements. We may write $z = \pi(\hat{z})$, where $\hat{z}$ is an element in $\tilde{H}_{\nu_n - 3}(\mathsf{M}_{[3,n-2]}; \mathbb{Z})$ and π is a permutation in $\mathfrak{S}_{[3,n]}$ such that $\pi(n-1) = i$ and $\pi(n) = j$. Note that we may view $\epsilon_i \circ \pi^{-1} \circ (i,j)$ as a permutation in $\mathfrak{S}_{[3,n]\setminus\{i\}}$ and $\epsilon_j \circ \pi^{-1}$ as a permutation in $\mathfrak{S}_{[3,n]\setminus\{j\}}$. Moreover, $(i,j) \circ \pi(\hat{z}) = \pi(\hat{z}) = z$. As consequence, we obtain that

$$\begin{aligned}
\psi(z \cdot \mathsf{q}_{i,j}) &= z \cdot \mathsf{r}_{2,j} - z \cdot \mathsf{r}_{1,i} = \pi(\hat{z}) \cdot \mathsf{r}_{2,j} - (i,j) \circ \pi(\hat{z}) \cdot \mathsf{r}_{1,i} \\
&= \text{sgn}(\epsilon_j \circ \pi^{-1})\epsilon_j(\hat{z}) \cdot \mathsf{r}_{2,j} - \text{sgn}(\epsilon_i \circ \pi^{-1} \circ (i,j))\epsilon_i(\hat{z}) \cdot \mathsf{r}_{1,i} \\
&= \text{sgn}(\pi)\left((-1)^{n-j}\epsilon_j(\hat{z}) \cdot \mathsf{r}_{2,j} - (-1)^{n-i-1}\epsilon_i(\hat{z}) \cdot \mathsf{r}_{1,i}\right) \\
&= \text{sgn}(\pi)\left(\hat{\epsilon}_j(\hat{z}) \cdot \mathsf{r}_{2,j} + \hat{\epsilon}_i(\hat{z}) \cdot \mathsf{r}_{1,i}\right).
\end{aligned}$$

Note that $\gamma_{n-4}^{(2)}$ is a generator of $\tilde{H}_{\nu_n - 1}(\mathsf{M}_{[3,n-1]}; \mathbb{Z})$. We define $e_{a,i} = \hat{\epsilon}_i(\gamma_{n-4}^{(2)}) \cdot \mathsf{r}_{a,i}$. Observe that the image under ψ is generated by $\{e_{1,i} + e_{2,j} : i, j \in [3, n], i \neq j\}$. It remains to prove that this image has codimension one when viewed as a vector space over $\mathbb{Z}_3$; by the exactness of the sequence in (11.3), this will imply that $\tilde{H}_{\nu_n}(\mathsf{M}_n; \mathbb{Z}) \cong \mathbb{Z}_3$. Now,

$$e_{1,3} + e_{2,3} = (e_{1,3} + e_{2,5}) + (e_{1,4} + e_{2,3}) - (e_{1,4} + e_{2,5})$$
$$e_{1,3} - e_{1,j} = (e_{1,3} + e_{2,3}) - (e_{1,j} + e_{2,3})$$

for any $j \in [4, n]$, which yields that the image has codimension at most one. Moreover, define $\eta : R_{\nu_n}^n \to \tilde{H}_{\nu_n - 1}(\mathsf{M}_{[3,n-1]}; \mathbb{Z})$ by $\eta(e_{a,i}) = (-1)^a \gamma_{n-4}^{(2)}$. Clearly, η is nonzero, whereas $\eta \circ \psi = 0$, which implies that ψ is not onto. It follows that the codimension is at least one and hence exactly one. $\square$

Theorem 11.12 (Bouc [21], Shareshian and Wachs [122]). *For $n = 1$ and $n \geq 3$, the homology group $\tilde{H}_{\nu_n}(\mathsf{M}_n; \mathbb{Z})$ is nonzero.*

Proof. One easily checks $n \leq 4$ by hand; thus assume that $n \geq 5$. We established the case $n \bmod 3 = 1$ in Theorem 11.11. For the case $n \bmod 3 = 0$, consider the exact sequence

---

[2] The number 2 in the exponent of $\gamma_{n-4}^{(2)}$ indicates a two-step "shift"; we replace $[ij]$ with $[(i+2)(j+2)]$. Compare to Theorem 11.20.

$$\tilde{H}_{\nu_n}(\mathsf{M}_{[2,n+1]};\mathbb{Z}) \longrightarrow \tilde{H}_{\nu_n}(\mathsf{M}_{n+1};\mathbb{Z}) \longrightarrow P_{\nu_n}^{n+1};$$

apply Corollary 11.9. The group $P_{\nu_n}^{n+1}$ is a direct sum of groups isomorphic to the group $\tilde{H}_{\nu_n-1}(\mathsf{M}_{n-1};\mathbb{Z})$. This group is zero, because $\nu_n - 1 = \nu_{n-1} - 1$. As a consequence, the map $\tilde{H}_{\nu_n}(\mathsf{M}_{[2,n+1]};\mathbb{Z}) \to \tilde{H}_{\nu_n}(\mathsf{M}_{n+1};\mathbb{Z})$ is onto. Since the image is nonzero by Theorem 11.11, it follows that $\tilde{H}_{\nu_n}(\mathsf{M}_{[2,n+1]};\mathbb{Z})$ is nonzero.

For the case $n \bmod 3 = 2$, consider the exact sequence

$$\tilde{H}_{\nu_n}(\mathsf{M}_n;\mathbb{Z}) \longrightarrow Q_{\nu_n}^n \longrightarrow R_{\nu_n-1}^n.$$

Now, $Q_{\nu_n}^n$ is a nonempty direct sum of groups isomorphic to $\tilde{H}_{\nu_n-2}(\mathsf{M}_{n-4};\mathbb{Z})$ and hence nonzero since $\nu_n - 2 = \nu_{n-4}$. Moreover, $R_{\nu_n-1}^n$ is a direct sum of groups isomorphic to $\tilde{H}_{\nu_n-2}(\mathsf{M}_{n-3};\mathbb{Z})$ and hence zero. It follows that $\tilde{H}_{\nu_n}(\mathsf{M}_n;\mathbb{Z})$ is nonzero. $\square$

Corollary 11.13. *For $n \geq 3$, the shifted connectivity degree and the homotopical depth of M_n are both equal to ν_n.* $\square$

Corollary 11.14 (Shareshian and Wachs [122]). *For $n \bmod 3 \in \{0,1\}$, the cycle γ_N in (11.3) is a nonzero element in $\tilde{H}_{\nu_n}(\mathsf{M}_n;\mathbb{Z})$; $N = 3\lfloor\frac{n}{3}\rfloor$.*

Proof. By symmetry, $\tilde{H}_{\nu_n}(\mathsf{M}_n;\mathbb{Z})$ is nonzero if and only if every generator in Lemma 11.10 is nonzero. By Theorem 11.12, we are done. $\square$

Corollary 11.15. *Let G be a graph on a vertex set V of size n such that $n \bmod 3 \in \{0,1\}$. Suppose that there is a partition $\{U_1,\ldots,U_t\}$ of V such that $|U_i| = 3$ for each $i < t$ and $|U_t| = n \bmod 3$ and such that $G(U_i)$ is isomorphic to either K_3 or $\mathsf{Pa}_3 = ([3],\{12,23\})$ for each $i < t$. Then $\mathsf{M}(G)$ has nonvanishing homology in dimension ν_n.*

Proof. We may assume that $U_i = \{3i - 2, 3i - 1, 3i\}$ and that $(3i - 2)(3i - 1)$ and $(3i - 1)(3i)$ belong to G for $i < t$. This means that the cycle γ_N defined in (11.3) is a cycle in the chain complex of $\mathsf{M}(G)$. Since γ_N is not a boundary in M_n, the same holds in $\mathsf{M}(G)$; hence we are done. $\square$

Shareshian and Wachs [122] have an even more precise description of the bottom nonvanishing homology group of M_n:

Theorem 11.16 (Shareshian & Wachs [122]). *For $n \in \{7,10,12,13\}$ and for $n \geq 15$, $\tilde{H}_{\nu_n}(\mathsf{M}_n;\mathbb{Z})$ is of the form $(\mathbb{Z}_3)^{e_n}$ for some $e_n \geq 1$. For $n = 14$, $\tilde{H}_{\nu_n}(\mathsf{M}_n;\mathbb{Z})$ is a finite group with nonvanishing 3-torsion.* $\square$

Somewhat surprisingly, it turns out that $\tilde{H}_4(\mathsf{M}_{14};\mathbb{Z})$ contains 5-torsion:

Theorem 11.17 (Jonsson [75]). *$\tilde{H}_4(\mathsf{M}_{14};\mathbb{Z})$ is a finite group of exponent a multiple of 15.*

Let us give a brief outline of the proof [75] of Theorem 11.17. Recall that $\overline{BD}_n^2$ is the complex of graphs on n vertices (loops allowed) such that the degree of each vertex is at most two. The main ingredient in the proof is a result due to Andersen [1] stating that $\tilde{H}_4(BD_7^2;\mathbb{Z}) \cong \mathbb{Z}_5$; see Theorem 12.19 in Section 12.2. One may relate Andersen's result to the homology of M_{14} via a map π from $\tilde{H}_4(M_{14};\mathbb{Z})$ to $\tilde{H}_4(BD_7^2;\mathbb{Z})$; this map is induced by the natural action on M_{14} by the Young group $(\mathfrak{S}_2)^7$. Using a standard representation-theoretic argument, one may construct an "inverse" φ of π with the property that $\pi \circ \varphi(z) = |(\mathfrak{S}_2)^7| \cdot z$ for all $z \in \tilde{H}_4(BD_7^2;\mathbb{Z})$. To conclude the proof, one observes that $\varphi(z)$ is nonzero unless the order of z divides the order of $(\mathfrak{S}_2)^7$. Since the latter order is 128, the image under φ of any nonzero element of order five is again a nonzero element of order five.

Table 11.1. The homology of M_n for $n \leq 14$. T_1 and T_2 are nontrivial finite groups of exponent a multiple of 3 and 15, respectively; see Proposition 11.22 and Theorem 11.17.

$\tilde{H}_i(M_n;\mathbb{Z})$	$i=0$	1	2	3	4	5
$n=3$	$\mathbb{Z}^2$	-	-	-	-	-
4	$\mathbb{Z}^2$	-	-	-	-	-
5	-	$\mathbb{Z}^6$	-	-	-	-
6	-	$\mathbb{Z}^{16}$	-	-	-	-
7	-	$\mathbb{Z}_3$	$\mathbb{Z}^{20}$	-	-	-
8	-	-	$\mathbb{Z}^{132}$	-	-	-
9	-	-	$\mathbb{Z}_3^8 \oplus \mathbb{Z}^{42}$	$\mathbb{Z}^{70}$ -	-	-
10	-	-	$\mathbb{Z}_3$	$\mathbb{Z}^{1216}$ -	-	-
11	-	-	-	$\mathbb{Z}_3^{45} \oplus \mathbb{Z}^{1188}$	$\mathbb{Z}^{252}$	-
12	-	-	-	$\mathbb{Z}_3^{56}$	$\mathbb{Z}^{12440}$	-
13	-	-	-	$\mathbb{Z}_3$	$T_1 \oplus \mathbb{Z}^{24596}$	$\mathbb{Z}^{924}$
14	-	-	-	-	T_2	$\mathbb{Z}^{138048}$

By Theorem 11.11, $e_n = 1$ whenever $n \bmod 3 = 1$ and $n \geq 7$. Table 11.1 lists the homology of M_n for $n \leq 14$; see Wachs [145] for more information.

11.2.3 Torsion in Higher-Degree Homology Groups

We apply the theory in the preceding section to detect 3-torsion in higher-degree homology groups.

First, let us state an elementary but useful result; the proof is straightforward.

Lemma 11.18. *Let $k \geq 1$ and let G be a graph on $2k$ vertices. Then $\mathsf{M}(G)$ admits a collapse to a complex of dimension at most $k - 2$.* □

Let $k_0 \geq 0$ and let $\mathcal{G} = \{G_k : k \geq k_0\}$ be a family of graphs such that the following conditions hold:

- For each $k \geq k_0$, the vertex set of G_k is $[2k + 1]$.
- For each $k > k_0$ and for each vertex s such that $1s$ is an edge in G_k, the induced subgraph $G_k([2k + 1] \setminus \{1, s\})$ is isomorphic to G_{k-1}.

We say that such a family is *compatible*.

Proposition 11.19. *In each of the following three cases, $\mathcal{G} = \{G_k : k \geq k_0\}$ is a compatible family:*

(1) $G_k = K_{2k+1}$ for all k.
(2) $G_k = K_{k+1,k}$ for all k, where $K_{k+1,k}$ is the complete bipartite graph with blocks $[k + 1]$ and $[k + 2, 2k + 1]$.
(3) $G_k = K_{2k+1} \setminus \{23, 45, 67, \ldots, 2k(2k + 1)\}$ for all k.

Proof. It suffices to prove that $G_k([2k + 1] \setminus \{1, s\})$ is isomorphic to G_{k-1} whenever $1s$ is an edge in G_k and $k > k_0$. This is immediate in all three cases. □

Now, fix $k_0, n, d \geq 0$. Let $\mathcal{G} = \{G_k : k \geq k_0\}$ be a family of compatible graphs and let γ be an element in $\tilde{H}_{d-1}(\mathsf{M}_n; \mathbb{Z})$. For each $k \geq k_0$, define a map

$$\begin{cases} \theta_k : \tilde{H}_{k-1}(\mathsf{M}(G_k); \mathbb{Z}) \to \tilde{H}_{k-1+d}(\mathsf{M}_{2k+1+n}; \mathbb{Z}) \\ \theta_k(z) = z \wedge \gamma^{(2k+1)}, \end{cases}$$

where we obtain $\gamma^{(2k+1)}$ from γ by replacing each occurrence of the vertex i with $i + 2k + 1$ for every $i \in [n]$. Note that $\tilde{H}_{k-1}(\mathsf{M}(G_k); \mathbb{Z})$ is the top homology group of $\mathsf{M}(G_k)$ (provided G_k contains matchings of size k). For any prime p, we have that θ_k induces a homomorphism

$$\theta_k \otimes_{\mathbb{Z}} \iota_p : \tilde{H}_{k-1}(\mathsf{M}(G_k); \mathbb{Z}) \otimes_{\mathbb{Z}} \mathbb{Z}_p \to \tilde{H}_{k-1+d}(\mathsf{M}_{2k+1+n}; \mathbb{Z}) \otimes_{\mathbb{Z}} \mathbb{Z}_p,$$

where $\iota_p : \mathbb{Z}_p \to \mathbb{Z}_p$ is the identity.

Theorem 11.20. *With notation and assumptions as above, if $\theta_{k_0} \otimes_{\mathbb{Z}} \iota_p$ is a monomorphism, then $\theta_k \otimes_{\mathbb{Z}} \iota_p$ is a monomorphism for each $k \geq k_0$. If, in addition, the exponent of γ in $\tilde{H}_{d-1}(\mathsf{M}_n; \mathbb{Z})$ is p, then we have a monomorphism*

$$\begin{cases} \hat{\theta}_k : \tilde{H}_{k-1}(\mathsf{M}(G_k); \mathbb{Z}) \otimes_{\mathbb{Z}} \mathbb{Z}_p \to \tilde{H}_{k-1+d}(\mathsf{M}_{2k+1+n}; \mathbb{Z}) \\ \hat{\theta}_k(z \otimes_{\mathbb{Z}} \lambda) = \theta_k(\lambda z) = \lambda z \wedge \gamma^{(2k+1)} \end{cases}$$

for each $k \geq k_0$. In particular, the group $\tilde{H}_{k-1+d}(\mathsf{M}_{2k+1+n}; \mathbb{Z})$ contains p-torsion of rank at least the rank of $\tilde{H}_{k-1}(\mathsf{M}(G_k); \mathbb{Z})$.

Proof. To prove the first part of the theorem, we use induction on k; the base case $k = k_0$ is true by assumption. Assume that $k > k_0$ and consider the head end of the long exact sequence for the pair $(M(G_k), M(G_k \setminus \{1\}))$, where $G_k \setminus \{1\} = G_k([2k + 1] \setminus \{1\})$:

$$0 \longrightarrow \tilde{H}_{k-1}(M(G_k \setminus \{1\}); \mathbb{Z})$$

$$\longrightarrow \tilde{H}_{k-1}(M(G_k); \mathbb{Z}) \xrightarrow{\hat{\omega}} P_{k-2}(G_k) \longrightarrow \tilde{H}_{k-2}(M(G_k \setminus \{1\}); \mathbb{Z}).$$

Here,

$$P_{k-2}(G_k) = \bigoplus_{s:1s \in G_k} 1s \otimes \tilde{H}_{k-2}(M(G_k \setminus \{1, s\}); \mathbb{Z})$$

and $\hat{\omega}$ is defined in the natural manner.

Now, the group $\tilde{H}_{k-1}(M(G_k \setminus \{1\}); \mathbb{Z})$ is zero by Lemma 11.18. As a consequence, $\hat{\omega}$ is a monomorphism. Moreover, all groups in the second row of the above sequence are torsion-free. Namely, the dimensions of $M(G_k)$ and $M(G_k \setminus \{1, s\})$ are at most $k - 1$ and $k - 2$, respectively, and Lemma 11.18 yields that $M(G_k \setminus \{1\})$ is homotopy equivalent to a complex of dimension at most $k - 2$. It follows that the induced homomorphism

$$\hat{\omega} \otimes \iota_p : \tilde{H}_{k-1}(M(G_k); \mathbb{Z}) \otimes \mathbb{Z}_p \to P_{k-2}(G_k) \otimes \mathbb{Z}_p$$

remains a monomorphism.

Now, consider the following diagram:

$$
\begin{array}{ccc}
\tilde{H}_{k-1}(M(G_k); \mathbb{Z}) \otimes \mathbb{Z}_p & \xrightarrow{\hat{\omega} \otimes \iota_p} & P_{k-2}(G_k) \otimes \mathbb{Z}_p \\
{\scriptstyle \theta_k \otimes \iota_p} \downarrow & & \downarrow {\scriptstyle \theta_{k-1}^{\oplus} \otimes \iota_p} \\
\tilde{H}_{k-1+d}(M_{2k+1+n}; \mathbb{Z}) \otimes \mathbb{Z}_p & \xrightarrow{\omega \otimes \iota_p} & P_{k-2+d}^{2k-1+n} \otimes \mathbb{Z}_p.
\end{array}
$$

Here,

$$P_{k-2+d}^{2k-1+n} = \bigoplus_{s=2}^{2k+1+n} 1s \otimes \tilde{H}_{k-2+d}(M_{[2,2k+1+n] \setminus \{s\}}; \mathbb{Z}),$$

ω is defined as in Corollary 11.9, and $\theta_{k-1}^{\oplus}$ is defined by

$$\theta_{k-1}^{\oplus}(1s \otimes z) = [1s] \otimes z \wedge \gamma^{(2k+1)}.$$

One easily checks that the diagram commutes; going to the right and then down or going down and then to the right both give the same map

$$\left(c_1 + \sum_{s:1s \in G_k} [1s] \wedge z_{1s}\right) \otimes 1 \mapsto \sum_{s:1s \in G_k} \left([1s] \otimes z_{1s} \wedge \gamma^{(2k+1)}\right) \otimes 1,$$

where c_1 is a sum of oriented simplices from $M(G_k \setminus \{1\})$ and each z_{1s} is a sum of oriented simplices from $M(G_k \setminus \{1, s\})$ satisfying $\partial(z_{1s}) = 0$ and

$\partial(c_1) + \sum_s z_{1s} = 0$. Moreover, $\theta_{k-1}^\oplus \otimes \iota_p$ is a monomorphism, because the restriction to each summand is a monomorphism by induction on k. Namely, since $\mathcal{G}$ is compatible, $G_k \setminus \{1, s\}$ is isomorphic to G_{k-1} for each s such that $1s \in G_k$. As a consequence, $(\theta_{k-1}^\oplus \circ \hat{\omega}) \otimes \iota_p$ is a monomorphism, which implies that $\theta_k \otimes \iota_p$ is a monomorphism.

For the very last statement, it suffices to prove that $\hat{\theta}_k$ is a well-defined homomorphism, which is true if and only if $\theta_k(pz) = 0$ for each $z \in \tilde{H}_{k-1}(\mathrm{M}(G_k); \mathbb{Z})$. Now, let $c \in \tilde{C}_d(\mathrm{M}_n; \mathbb{Z})$ be such that $\partial(c) = p\gamma$; such a c exists by assumption. We obtain that

$$\partial(z \wedge c^{(2k+1)}) = \pm z \wedge (p\gamma^{(2k+1)}) = \pm(pz) \wedge \gamma^{(2k+1)};$$

hence $\theta_k(pz) = 0$ as desired. $\square$

We will find the following transformation very useful:

$$\begin{cases} k = 3d - n + 4 \\ r = n - 2d - 3 \end{cases} \Longleftrightarrow \begin{cases} n = 2k + 1 + 3r \\ d = k - 1 + r. \end{cases} \tag{11.4}$$

Theorem 11.21. *For $k \geq 0$ and $r \geq 4$, there is 3-torsion of rank at least $\binom{2k}{k}$ in $\tilde{H}_{k-1+r}(\mathrm{M}_{2k+1+3r}; \mathbb{Z})$. Moreover, for $k \geq 0$, there is 3-torsion of rank at least $\binom{k+1}{\lfloor (k+1)/2 \rfloor}$ in $\tilde{H}_{k+2}(\mathrm{M}_{2k+10}; \mathbb{Z})$. To summarize, $\tilde{H}_{k-1+r}(\mathrm{M}_{2k+1+3r}; \mathbb{Z})$ contains nonvanishing 3-torsion whenever $k \geq 0$ and $r \geq 3$.*

Proof. For the first statement, consider the compatible family $\{K_{2k+1} : k \geq 0\}$ and the cycle $\gamma_{3r} \in \tilde{H}_{r-1}(\mathrm{M}_{3r}; \mathbb{Z})$ defined as in (11.3). By Theorem 11.11 and Lemma 11.10,

$$\theta_0 \otimes \iota_3 : \tilde{H}_{-1}(\mathrm{M}_1; \mathbb{Z}) \otimes_{\mathbb{Z}} \mathbb{Z}_3 \cong \mathbb{Z} \otimes_{\mathbb{Z}} \mathbb{Z}_3 \to \tilde{H}_{r-1}(\mathrm{M}_{3r+1}; \mathbb{Z}) \otimes_{\mathbb{Z}} \mathbb{Z}_3$$

defines an isomorphism, where $\theta_0(\lambda) = \lambda \gamma_{3r}^{(1)}$. By Lemma 11.10 and Theorem 11.16, γ_{3r} has exponent 3 in $\tilde{H}_{r-1}(\mathrm{M}_{3r}; \mathbb{Z})$; hence Theorem 11.20 yields that the group $\tilde{H}_{k-1+r}(\mathrm{M}_{2k+1+3r}; \mathbb{Z})$ contains 3-torsion of rank at least the rank of the group $\tilde{H}_{k-1}(\mathrm{M}_{2k+1}; \mathbb{Z})$. By Theorem 11.5, this rank equals $\binom{2k}{k}$.

For the second statement, consider the compatible family $\{G_k = K_{2k+1} \setminus \{23, 45, 67, \ldots, 2k(2k+1)\} : k \geq 1\}$ and the cycle $\gamma_6 = ([12] - [23]) \wedge ([45] - [56]) \in \tilde{H}_1(\mathrm{M}_7; \mathbb{Z})$. For $k = 1$, we obtain that G_1 is the graph P_3 on three vertices with edge set $\{12, 13\}$; clearly, $\tilde{H}_0(\mathrm{M}(P_3); \mathbb{Z}) \cong \mathbb{Z}$. As a consequence,

$$\theta_1 \otimes \iota_3 : \tilde{H}_0(\mathrm{M}(P_3); \mathbb{Z}) \otimes_{\mathbb{Z}} \mathbb{Z}_3 \to \tilde{H}_2(\mathrm{M}_{10}; \mathbb{Z}) \otimes_{\mathbb{Z}} \mathbb{Z}_3$$

is an isomorphism; apply Theorem 11.11. Proceeding as in the first case and using the fact that γ_6 has exponent 3 in $\tilde{H}_1(\mathrm{M}_7; \mathbb{Z})$, we conclude that $\tilde{H}_{k+1}(\mathrm{M}_{2k+8}; \mathbb{Z})$ contains 3-torsion of rank at least the rank of $\tilde{H}_{k-1}(\mathrm{M}(G_k); \mathbb{Z})$ for each $k \geq 1$.

It remains to show that the rank of $\tilde{H}_{k-1}(\mathrm{M}(G_k); \mathbb{Z})$ is at least $\binom{k}{\lfloor k/2 \rfloor}$. Let A be any subset of the removed edge set

$$E = \{23, 45, \ldots, 2k(2k+1)\}$$

such that $|A| = \lfloor k/2 \rfloor$; write $B = E \setminus A$. Consider the complete bipartite graph G_k^A with one block equal to $\{1\} \cup \bigcup_{e \in A} e$ and the other block equal to $\bigcup_{e \in B} e$. For even k, the size of the "A" block is $k + 1$; for odd k, the size of the "A" block is k. It is clear that G_k^A is a subgraph of G_k.

Label the vertices in $[2, 2k+1]$ as $s_1, t_1, s_2, t_2, \ldots, s_k, t_k$ such that $s_i t_i \in A$ for even i and $s_i t_i \in B$ for odd i. Consider the matching

$$\sigma_A = \{1s_1, t_1 s_2, t_2 s_3, \ldots, t_{k-1} s_k\}.$$

One easily checks that $\sigma_A \in \mathsf{M}(G_k^{A'})$ if and only if $A = A'$. Now, as observed by Shareshian and Wachs [122, (6.2)], $\mathsf{M}(G_k^A)$ is an orientable pseudomanifold. Defining z_A to be the fundamental cycle of $\mathsf{M}(G_k^A)$, we obtain that $\{z_A : A \subset E, |A| = \lfloor k/2 \rfloor\}$ forms an independent set in $\tilde{H}_{k-1}(\mathsf{M}(G_k); \mathbb{Z})$, which concludes the proof. $\square$

Let $G_k = K_{2k+1} \setminus \{23, 45, 67, \ldots, 2k(2k+1)\}$ be the graph in the above proof. Based on computer calculations for $k \leq 5$, we conjecture that the rank r_k of $\tilde{H}_{k-1}(\mathsf{M}(G_k); \mathbb{Z})$ equals the coefficient of x^k in $(1 + x + x^2)^k$; this is sequence A002426 in Sloane's Encyclopedia [127]. Equivalently,

$$\sum_{k \geq 0} r_k x^k = \frac{1}{\sqrt{1 - 2x - 3x^2}}.$$

Proposition 11.22 (Jonsson [75]). *We have that* $\tilde{H}_4(\mathsf{M}_{13}; \mathbb{Z}) \cong T \oplus \mathbb{Z}^{24596}$, *where T is a finite group containing $\mathbb{Z}_3^{10}$ as a subgroup.* $\square$

Corollary 11.23. *For $n \geq 1$, there is nonvanishing 3-torsion in the homology group* $\tilde{H}_d(\mathsf{M}_n; \mathbb{Z}) = \tilde{H}_{k-1+r}(\mathsf{M}_{2k+1+3r}; \mathbb{Z})$ *whenever*

$$\left\lceil \frac{n-4}{3} \right\rceil \leq d \leq \left\lfloor \frac{n-6}{2} \right\rfloor \iff \begin{cases} k \geq 0 \\ r \geq 3 \end{cases}$$

or $r = 2$ and $k \in \{0, 1, 2, 3\}$. Moreover, $\tilde{H}_d(\mathsf{M}_n; \mathbb{Z})$ is nonzero if and only if

$$\left\lceil \frac{n-4}{3} \right\rceil \leq d \leq \left\lfloor \frac{n-3}{2} \right\rfloor \iff \begin{cases} k \geq 0 \\ r \geq 0. \end{cases}$$

Proof. The first statement is a consequence of Theorem 11.21, Proposition 11.22, and Table 11.1. For the second statement, Theorem 11.5 yields that the group $\tilde{H}_{k-1+r}(\mathsf{M}_{2k+1+3r}; \mathbb{Z})$ is infinite if and only if $r \geq 0$ and $k \geq \binom{r}{2}$. In particular, the group is infinite for all $k \geq 0$ and $0 \leq r \leq 2$ except $(k, r) = (0, 2)$. Since $\tilde{H}_{k-1+r}(\mathsf{M}_{2k+1+3r}; \mathbb{Z}) \cong \mathbb{Z}_3$ when $k = 0$ and $r = 2$, we are done by Theorem 11.6 and Lemma 11.18. $\square$

Corollary 11.23 suggests the following conjecture:

Conjecture 11.24. *The group* $\tilde{H}_d(M_n;\mathbb{Z}) = \tilde{H}_{k-1+r}(M_{2k+1+3r};\mathbb{Z})$ *contains 3-torsion if and only if*

$$\left\lceil \frac{n-4}{3} \right\rceil \leq d \leq \left\lfloor \frac{n-5}{2} \right\rfloor \Longleftrightarrow \begin{cases} k \geq 0 \\ r \geq 2. \end{cases}$$

By Corollary 11.23, the conjecture remains unsettled if and only if $r = 2$ and $k \geq 4$; for the cases $r = 0$ and $r = 1$, one easily checks that the homology is free. The conjecture would follow if we were able to solve Problem 11.28 in Section 11.2.4 to the affirmative.

For sufficiently small d, the group $\tilde{H}_d(M_n;\mathbb{Z})$ is a 3-group:

Theorem 11.25 (Jonsson [74]). *If*

$$\left\lceil \frac{n-4}{3} \right\rceil \leq d \leq \left\lfloor \frac{2n-9}{5} \right\rfloor \Longleftrightarrow 0 \leq k \leq r-2,$$

then $\tilde{H}_d(M_n;\mathbb{Z}) = \tilde{H}_{k-1+r}(M_{2k+1+3r};\mathbb{Z})$ *is a nontrivial 3-group.*

Define k and r as in (11.4). Let $\beta_d^n = \dim_{\mathbb{Z}_3} \tilde{H}_d(M_n;\mathbb{Z}_3)$ and write $\hat{\beta}_{k,r} = \beta_d^n$. The following theorem provides polynomial bounds on β_d^n; these bounds are not sharp.

Theorem 11.26 (Jonsson [74]). *For each* $k \geq 0$, *there is a polynomial* $f_k(r)$ *of degree* $3k$ *with dominating term* $\frac{3^k}{k!}r^{3k}$ *such that* $\hat{\beta}_{k,r} \leq f_k(r)$ *for all* $r \geq k+2$. *Equivalently,* $\beta_d^n \leq f_{3d-n+4}(n-2d-3)$ *for all* $n \geq 7$ *and* $\lceil \frac{n-4}{3} \rceil \leq d \leq \lfloor \frac{2n-9}{5} \rfloor$.

For $k \leq 2$, this provides upper bounds on the dimension of the bottom non-vanishing homology group. For $k = 1$ and $r \geq 4$, we have the precise bound

$$\beta_r^{3r+3} \leq \frac{6r^3 + 9r^2 + 5r}{2} - 73$$

[74]. Again, this bound is not sharp.

11.2.4 Further Properties

As promised in Section 5.2.1, we now present a $\mathbb{Z}_3$-optimal decision tree on a complex with 3-torsion in its homology.

Theorem 11.27. M_7 *is semi-nonevasive over* $\mathbb{Z}_3$, *but not over* $\mathbb{Z}$.

Proof. We have that $\tilde{H}_1(M_7;\mathbb{Z}) \cong \mathbb{Z}_3$, $\tilde{H}_2(M_7;\mathbb{Z}) \cong \mathbb{Z}^{20}$, and $\tilde{H}_i(M_7,\mathbb{Z}) = 0$ if $i \neq 1, 2$; see Table 11.1. This means that

$$\tilde{H}_1(M_7;\mathbb{Z}_3) \cong \mathbb{Z}_3;$$
$$\tilde{H}_2(M_7;\mathbb{Z}_3) \cong \mathbb{Z}_3^{21}.$$

We want to find an element-decision tree with 1+21 evasive faces. For this, consider the first-hit decomposition with respect to the sequence

$$(13, 14, 35, 46, 56, 12, 57, 67, 23, 24, 37, 47, 16, 25, 17, 27, 34, 36, 45, 15);$$

all edges but the edge 26 appear in the sequence. Let b_i be the ith edge in the sequence and let $B_k = \{b_i : i \leq k\}$. It is easy to check that $\mathsf{M}_7(b_i, B_{i-1}) \sim c_i t^2$ for $1 \leq i \leq 6$, where $c_1 = c_2 = 6$, $c_3 = 4$, $c_4 = c_5 = 2$, and $c_6 = 1$; the corresponding links are connected graphs, so optimal element-decision trees exist by Proposition 5.26. Moreover, $\mathsf{M}_7(b_{18}, B_{17}) \sim t$, whereas $\mathsf{M}_7(\emptyset, B_{20}) \sim 0$ and $\mathsf{M}_7(b_i, B_{i-1}) \sim 0$ for $i \in \{7, \dots, 17\} \cup \{19, 20\}$. Applying Lemma 5.25, we obtain that $\mathsf{M}_7 \sim t + 21t^2$ as desired. $\square$

Since there is 5-torsion in the homology of M_{14}, not all matching complexes are semi-nonevasive over $\mathbb{Z}_3$.

In this context, the following problem might be worth mentioning:

Problem 11.28. For each n and d, is it true that

$$\tilde{H}_d(\mathsf{M}_n; \mathbb{Z}) \cong \tilde{H}_d(\mathrm{del}_{\mathsf{M}_n}(e); \mathbb{Z}) \oplus \tilde{H}_{d-1}(\mathsf{M}_{n-2}; \mathbb{Z}), \tag{11.5}$$

where e is any edge in K_n?

Note that we consider homology over $\mathbb{Z}$. If (11.5) were true for all n, then $\tilde{H}_d(\mathsf{M}_n; \mathbb{Z})$ would contain p-torsion whenever $\tilde{H}_{d-1}(\mathsf{M}_{n-2}; \mathbb{Z})$ contains p-torsion. In particular, $\tilde{H}_d(\mathsf{M}_{2d+5}; \mathbb{Z})$ would contain 3-torsion for all $d \geq 1$, which would settle Conjecture 11.24. Moreover, $\tilde{H}_d(\mathsf{M}_{2d+6})$ would contain 5-torsion for all $d \geq 4$; use Theorem 11.17. We have verified (11.5) for $n \leq 11$.

For completeness, we mention the following very simple and well-known result about the Euler characteristic of M_n.

Proposition 11.29. *Let $f_{n,i}$ be the number of faces of M_n of dimension $i-1$ and define $f_n(t) = \sum_i f_{n,i} t^i$. Then*

$$\sum_{n \geq 1} f_n(t) \frac{x^n}{n!} = e^{x + tx^2/2} - 1.$$

In particular, the reduced Euler characteristic of M_n satisfies

$$\sum_{n \geq 1} \tilde{\chi}(\mathsf{M}_n) \frac{x^n}{n!} = 1 - e^{x - x^2/2}.$$

Proof. Apply Corollary 6.15. $\square$

11.3 Chessboards

We examine the chessboard complex $\mathsf{M}_{m,n}$, which is the matching complex on the complete bipartite graph $K_{m,n}$. Aligning with the notation of Shareshian and Wachs [122], we identify the two parts of $K_{m,n}$ with the sets $[m] = \{1, 2, \ldots, m\}$ and $[\overline{n}] = \{\overline{1}, \overline{2}, \ldots, \overline{n}\}$; the latter set should be interpreted as a disjoint copy of $[n]$. Hence each edge is of the form $i\overline{j}$, where $i \in [m]$ and $j \in [n]$. Sometimes, it will be useful to view $\mathsf{M}_{m,n}$ as a subcomplex of the matching complex M_{m+n} on the complete graph K_{m+n}. In such situations, we identify the vertex $\overline{j}$ in $K_{m,n}$ with the vertex $m + j$ in K_{m+n} for each $j \in [n]$.

11.3.1 Bottom Nonvanishing Homology

For $1 \le m \le n$, define

$$\nu_{m,n} = \min\{m - 1, \lceil \tfrac{m+n-4}{3} \rceil\} = \begin{cases} \lceil \tfrac{m+n-4}{3} \rceil & \text{if } m \le n \le 2m - 2; \\ m - 1 & \text{if } n > 2m - 2. \end{cases} \quad (11.6)$$

Theorem 11.30 (Ziegler [151]). *Let $1 \le m \le n$. Then $\mathsf{M}_{m,n}$ is $VD(\nu_{m,n})$. In particular, $\mathsf{M}_{m,n}$ is VD whenever $n \ge 2m - 1$.*

Proof. We apply Athanasiadis' Theorem 11.4. The case $m = 1$ is trivially true, as $\nu_{m,n} = 0$. Assume that $m > 1$ and consider the edge $m\overline{n}$.

First, we need to prove that $\mathsf{M}(K_{m,n}(([m] \cup [\overline{n}]) \setminus S))$ is $VD(\nu_{m,n} - 1)$ for each $S = \{x, \overline{n}\}$ such that $x\overline{n} \in K_{m,n}$. By symmetry, it suffices to consider the case $x = m$. Now, $\mathsf{M}(K_{m,n}(([m] \cup [\overline{n}]) \setminus S)) = \mathsf{M}_{m-1,n-1}$. Since

$$\nu_{m-1,n-1} + 1 = \min\{m - 2, \lceil \tfrac{m+n-6}{3} \rceil\} + 1 \ge \nu_{m,n},$$

the claim follows by induction.

Second, we need to prove that $\mathsf{M}(K_{m,n}(([m] \cup [\overline{n}]) \setminus S))$ is $VD(\nu_{m,n} - 1)$ for each $S = \{m, \overline{y}, \overline{n}\}$ such that $m\overline{y} \in K_{m,n}$. By symmetry, it suffices to consider the case $y = n - 1$. Now, $\mathsf{M}(K_{m,n}(([m] \cup [\overline{n}]) \setminus S)) = \mathsf{M}_{m-1,n-2}$. If $n > m$, note that

$$\nu_{m-1,n-2} + 1 = \min\{m - 2, \lceil \tfrac{m+n-7}{3} \rceil\} + 1 = \nu_{m,n}.$$

If $n = m$, note that $\nu_{m-2,m-1} + 1 = \lceil \tfrac{2m-7}{3} \rceil + 1 = \nu_{m,m}$. Again, the claim follows by induction, and we are done. $\square$

Björner, Lovász, Vrećica, and Živaljević [11] earlier proved that the connectivity degree of $\mathsf{M}_{m,n}$ is at least $\nu_{m,n}$.

Ziegler [151] proved a generalization of Theorem 11.30, extending to certain subgraphs of $K_{m,n}$. One example is the subgraph of $K_{n,n}$ obtained by removing the diagonal elements $i\overline{i}$, $i \in [n]$. Note that one may identify the matching complex on this graph with the complex of digraphs such that each vertex has in- and outdegree at most one. See Björner and Welker [17] and Shareshian and Wachs [122] for more information about this complex. Athanasiadis [2] generalized Theorem 11.30 to k-hypergraph matchings on k-dimensional chessboards.

Friedman and Hanlon [52] proved a chessboard analogue of Bouc's Theorem 11.5; see Wachs [145] for an overview. For our purposes, the most important consequence is the following result:

Theorem 11.31 (Friedman and Hanlon [52]). *For $1 \leq m \leq n$, we have that $\tilde{H}_d(\mathsf{M}_{m,n}; \mathbb{Z})$ is infinite if and only if $(m - d - 1)(n - d - 1) \leq d + 1$, $m \geq d + 1$, and $n \geq d + 2$. In particular, $\tilde{H}_{\nu_{m,n}}(\mathsf{M}_{m,n}; \mathbb{Z})$ is finite if and only if $n \leq 2m - 5$ and $(m, n) \notin \{(6, 6), (7, 7), (8, 9)\}$.*

By Theorems 11.30 and 11.31, the shifted connectivity degree of $\mathsf{M}_{m,n}$ is exactly $\nu_{m,n}$ whenever $n \geq 2m - 4$ or $(m, n) \in \{(6, 6), (7, 7), (8, 9)\}$. Shareshian and Wachs [122] extended this to all $(m, n) \neq (1, 1)$:

Theorem 11.32 (Shareshian & Wachs [122]). *If $m \leq n \leq 2m - 5$ and $(m, n) \neq (8, 9)$, then there is nonvanishing 3-torsion in $\tilde{H}_{\nu_{m,n}}(\mathsf{M}_{m,n}; \mathbb{Z})$. If in addition $(m + n) \bmod 3 = 1$, then $\tilde{H}_{\nu_{m,n}}(\mathsf{M}_{m,n}; \mathbb{Z}) \cong \mathbb{Z}_3$.* $\square$

By Theorem 11.37 in Section 11.3.2, there is nonvanishing 3-torsion also in $\tilde{H}_{\nu_{8,9}}(\mathsf{M}_{8,9}; \mathbb{Z})$; in that theorem, choose $(k, a, b) = (2, 1, 2)$.

Conjecture 11.33 (Shareshian & Wachs [122]). *Let $1 \leq m \leq n$. The group $\tilde{H}_{\nu_{m,n}}(\mathsf{M}_{m,n}; \mathbb{Z})$ is torsion-free if and only if $n \geq 2m - 4$.*

The conjecture is known to be true in all cases but $n = 2m - 4$ and $n = 2m - 3$; Shareshian and Wachs [122] settled the case $n = 2m - 2$ and verified the two special cases $(m, n) \in \{(6, 6), (7, 7)\}$ using computer.

Corollary 11.34 (Shareshian & Wachs [122]). *For all (m, n) except $(m, n) = (1, 1)$, $\tilde{H}_{\nu_{m,n}}(\mathsf{M}_{m,n}; \mathbb{Z})$ is nonzero. In particular, the shifted connectivity degree and the homotopical depth of $\mathsf{M}_{m,n}$ are both equal to $\nu_{m,n}$.* $\square$

Assume that $(m + n) \bmod 3 = 0$ and $m \leq n \leq 2m$. Define the cycle $\gamma_{m,n}$ in $\tilde{H}_{\nu_{m,n}}(\mathsf{M}_{m,n}; \mathbb{Z})$ recursively as follows, the base case being $\gamma_{1,2} = [1\bar{1}] - [1\bar{2}]$:

$$\gamma_{m,n} = \begin{cases} \gamma_{m-1,n-2} \wedge ([m(\overline{n-1})] - [m\bar{n}]) & \text{if } m < n; \\ \gamma_{m-2,n-1} \wedge ([(m-1)\bar{n}] - [m\bar{n}]) & \text{if } m = n. \end{cases} \quad (11.7)$$

For $n > m$, we define $\gamma_{n,m}$ by replacing $i\bar{j}$ with $j\bar{i}$ in $\gamma_{m,n}$ for each $i \in [m]$ and $j \in [n]$.

Shareshian and Wachs [122] provided more detailed information about the structure of the group $\tilde{H}_{\nu_{m,n}}(\mathsf{M}_{m,n}; \mathbb{Z})$. For example, the following is true:

Theorem 11.35 (Shareshian & Wachs [122]). *Assume that $(m+n)$ mod $3 \in \{0,1\}$ and $m \leq n \leq 2m+1$. We define $(M,N) = (m, 3\lfloor \frac{n}{3} \rfloor)$ unless $m = n$ and $(m+n)$ mod $3 = 1$, in which case we define $(M,N) = (m-1,m)$. Then $\gamma_{M,N}$ is a nonzero element in $\tilde{H}_{\nu_{m,n}}(\mathsf{M}_{m,n}; \mathbb{Z})$. Moreover, if $\tilde{H}_{\nu_{m,n}}(\mathsf{M}_{m,n}; \mathbb{Z})$ is finite, then the exponent of $\gamma_{M,N}$ in $\tilde{H}_{\nu_{m,n}}(\mathsf{M}_{m,n}; \mathbb{Z})$ is divisible by three.*

Proof. We may view $\gamma_{M,N}$ as an element in $\tilde{H}_{\nu_{m+n}}(\mathsf{M}_{m+n}; \mathbb{Z})$; $\nu_{m+n} = \nu_{m,n}$. Since $\gamma_{M,N}$ is nonzero in this group by Corollary 11.14, the same must be true in $\tilde{H}_{\nu_{m,n}}(\mathsf{M}_{m,n}; \mathbb{Z})$; apply Corollary 11.15. For the second statement, the exponent of $\gamma_{M,N}$ is nonzero by the first statement and finite by assumption. Now, $\gamma_{M,N}$ has exponent three in $\tilde{H}_{\nu_{m+n}}(\mathsf{M}_{m+n}; \mathbb{Z})$. Namely, since $m \leq n \leq 2m-5$ by Theorem 11.31, we have that $m+n \geq 10$; hence the claim is a consequence of Theorem 11.16. Since the exponent of $\gamma_{M,N}$ in $\tilde{H}_{\nu_{m+n}}(\mathsf{M}_{m+n}; \mathbb{Z})$ divides the exponent of $\gamma_{M,N}$ in $\tilde{H}_{\nu_{m,n}}(\mathsf{M}_{m,n}; \mathbb{Z})$, we are done. $\square$

11.3.2 Torsion in Higher-Degree Homology Groups

As already mentioned in the proof of Theorem 11.21 in Section 11.2.3, we have that $\mathsf{M}_{k,k+1}$ is an orientable pseudomanifold of dimension $k-1$; hence $\tilde{H}_{k-1}(\mathsf{M}_{k,k+1}; \mathbb{Z}) \cong \mathbb{Z}$. Shareshian and Wachs [122] observed that this group is generated by the cycle

$$z_{k,k+1} = \sum_{\pi \in \mathfrak{S}_{[k+1]}} \operatorname{sgn}(\pi) \cdot \overline{1\pi(1)} \wedge \cdots \wedge \overline{k\pi(k)}.$$

Note that the sum is over all permutations on $k+1$ elements. Theorem 11.20 implies the following result.

Corollary 11.36. *With notation and assumptions as in Theorem 11.20, defining $G_k = K_{k,k+1}$, if $(z_{k_0,k_0+1} \wedge \gamma^{(2k_0+1)}) \otimes 1$ is nonzero in the group $\tilde{H}_{k_0-1+d}(\mathsf{M}_{2k_0+1+n}; \mathbb{Z}) \otimes \mathbb{Z}_p$, then $(z_{k,k+1} \wedge \gamma^{(2k+1)}) \otimes 1$ is nonzero in the group $\tilde{H}_{k-1+d}(\mathsf{M}_{2k+1+n}; \mathbb{Z}) \otimes \mathbb{Z}_p$ for all $k \geq k_0$.* $\square$

Define

$$\begin{cases} k = -m-n+3d+4 \\ a = -m+n \\ b = m -d-1 \end{cases} \Leftrightarrow \begin{cases} m = k+a+3b-1 \\ n = k+2a+3b-1 \\ d = k+a+2b-2. \end{cases} \qquad (11.8)$$

Recall that $\nu_{m,n} = \frac{m+n-4}{3}$ whenever $m \leq n \leq 2m-2$.

Theorem 11.37. *There is 3-torsion in $\tilde{H}_d(\mathsf{M}_{m,n}; \mathbb{Z})$ whenever*

$$\begin{cases} m+1 \leq n \leq 2m-5 \\ \lceil \frac{m+n-4}{3} \rceil \leq d \leq m-3 \end{cases} \iff \begin{cases} k \geq 0 \\ a \geq 1 \\ b \geq 2, \end{cases}$$

where k, a, and b are defined as in (11.8). Moreover, there is 3-torsion in $\tilde{H}_d(\mathsf{M}_{m,m}; \mathbb{Z})$ whenever

$$\left\lceil \frac{2m-4}{3} \right\rceil \le d \le m-4 \Longleftrightarrow \begin{cases} k \ge 0 \\ a = 0 \\ b \ge 3. \end{cases}$$

Proof. Assume that $k \ge 0$, $a \ge 1$, and $b \ge 2$. Writing $m_0 = a + 3b - 2$ and $n_0 = 2a + 3b - 3$, we have the inequalities

$$a + 3b - 2 \le 2a + 3b - 3 \le 2a + 6b - 9 \Longleftrightarrow m_0 \le n_0 \le 2m_0 - 5. \qquad (11.9)$$

Note that $m_0 + n_0 = 3a + 6b - 5 \equiv 1 \pmod 3$. Define

$$w_{k+1} = z_{k+1,k+2} \wedge \gamma_{m_0,n_0-1}^{(k+1,k+2)},$$

where we obtain $\gamma_{m_0,n_0-1}^{(k+1,k+2)}$ from the cycle γ_{m_0,n_0-1} defined in (11.7) by replacing $i\bar{j}$ with $(i+k+1)(\overline{j+k+2})$. View γ_{m_0,n_0-1} as an element in the homology of M_{m_0,n_0}. Since $z_{k+1,k+2} \in \tilde{H}_k(\mathsf{M}_{k+1,k+2};\mathbb{Z})$ and $\gamma_{m_0,n_0-1} \in \tilde{H}_{a+2b-3}(\mathsf{M}_{a+3b-2,2a+3b-3};\mathbb{Z})$, we obtain that

$$w_{k+1} \in \tilde{H}_{k+a+2b-3+1}(\mathsf{M}_{k+1+a+3b-2,k+2+2a+3b-3};\mathbb{Z}) = \tilde{H}_d(\mathsf{M}_{m,n};\mathbb{Z}).$$

Choosing $k = 0$, we obtain that

$$w_1 = z_{1,2} \wedge \gamma_{m_0,n_0-1}^{(1,2)} \cong \gamma_{m_0+1,n_0+1}.$$

Since $m_0, n_0 \ge 5$, we have that γ_{m_0+1,n_0+1} has exponent three when viewed as an element in $\tilde{H}_{\frac{m_0+n_0-1}{3}}(\mathsf{M}_{m_0+n_0+3};\mathbb{Z}) = \tilde{H}_{a+2b-2}(\mathsf{M}_{3a+6b-2};\mathbb{Z})$; apply Lemma 11.10 and Theorem 11.11 and note that γ_{m_0+1,n_0+1} is isomorphic to the cycle $\gamma_{m_0+n_0+2}$ defined in (11.3).

Applying Corollary 11.36, we conclude that $w_{k+1} \otimes 1$ is a nonzero element in the group $\tilde{H}_{k+a+2b-2}(\mathsf{M}_{2k+3a+6b-2};\mathbb{Z}) \otimes \mathbb{Z}_3 = \tilde{H}_d(\mathsf{M}_{m+n};\mathbb{Z}) \otimes \mathbb{Z}_3$ for every $k \ge 0$. As a consequence, $w_{k+1} \otimes 1$ is nonzero also in

$$\tilde{H}_{k+a+2b-2}(\mathsf{M}_{k+a+3b-1,k+2a+3b-1};\mathbb{Z}) \otimes \mathbb{Z}_3 = \tilde{H}_d(\mathsf{M}_{m,n};\mathbb{Z}) \otimes \mathbb{Z}_3$$

for every $k \ge 1$. Since $\tilde{H}_{a+b-3}(\mathsf{M}_{m_0,n_0};\mathbb{Z})$ is an elementary 3-group by Theorem 11.32 and (11.9), the exponent of γ_{m_0,n_0-1} in $\tilde{H}_r(\mathsf{M}_{m_0,n_0};\mathbb{Z})$ is three. It follows that the exponent of w_{k+1} in $\tilde{H}_d(\mathsf{M}_{m,n};\mathbb{Z})$ is three as well.

The remaining case is $m = n$, in which case the upper bound on d is $m-4$ rather than $m-3$. Since $a = 0$, we get

$$\begin{cases} k = -2m + 3d + 4 \\ b = m - d - 1 \end{cases} \Leftrightarrow \begin{cases} m = k + 3b - 1 \\ d = k + 2b - 2. \end{cases}$$

Clearly, $k \ge 0$ and $b \ge 3$.

Consider the cycle $w_{k+1} = z_{k+1,k+2} \wedge \gamma_{3b-2,3b-4}^{(k+1,k+2)}$. By Corollary 11.36, $w_{k+1} \otimes 1$ is nonzero in $\tilde{H}_{k+2b-2}(\mathsf{M}_{2k+6b-2};\mathbb{Z}) \otimes \mathbb{Z}_3$. Namely, w_1 is isomorphic to γ_{6b-3} in (11.3), which is a nonzero element with exponent three in

$\tilde{H}_{2b-2}(\mathsf{M}_{6b-2};\mathbb{Z})$ by Lemma 11.10 and Theorem 11.11; $b \geq 3$. We conclude that $w_{k+1} \otimes 1$ is a nonzero element in $\tilde{H}_{k+2b-2}(\mathsf{M}_{k+3b-1,k+3b-1};\mathbb{Z}) \otimes \mathbb{Z}_3 = \tilde{H}_d(\mathsf{M}_{m,m};\mathbb{Z}) \otimes \mathbb{Z}_3$. Since $3b - 3 \geq 6$, we have that $\gamma_{3b-2,3b-4}$ must have exponent three in $\tilde{H}_{2b-3}(\mathsf{M}_{3b-2,3b-3};\mathbb{Z})$; apply Theorem 11.32. This implies that the same must be true for w_{k+1} in $\tilde{H}_d(\mathsf{M}_{m,m};\mathbb{Z})$. $\square$

Corollary 11.38. *The group* $\tilde{H}_5(\mathsf{M}_{8,9};\mathbb{Z}) = \tilde{H}_{\nu_{8,9}}(\mathsf{M}_{8,9};\mathbb{Z})$ *contains nonvanishing 3-torsion. As a consequence, there is nonvanishing 3-torsion in* $\tilde{H}_{\nu_{m,n}}(\mathsf{M}_{m,n};\mathbb{Z})$ *whenever* $m \leq n \leq 2m - 5$.

Proof. The first statement is a consequence, of Theorem 11.37; choose $k = 2$, $a = 1$, and $b = 2$. For the second statement, apply Theorem 11.32.

Theorem 11.39. *For* $1 \leq m \leq n$, *the group* $\tilde{H}_d(\mathsf{M}_{m,n};\mathbb{Z})$ *is nonzero if and only if either*

$$\left\lceil \frac{m+n-4}{3} \right\rceil \leq d \leq m - 2 \Longleftrightarrow \begin{cases} k \geq 0 \\ a \geq 0 \\ b \geq 1 \end{cases}$$

or

$$\begin{cases} m \geq 1 \\ n \geq m+1 \\ d = m - 1 \end{cases} \Longleftrightarrow \begin{cases} k \geq 2-a \\ a \geq 1 \\ b = 0, \end{cases}$$

where k, a, *and* b *are defined as in* (11.8).

Proof. For homology to exist, we certainly must have that $b \geq 0$, and we restrict to $a \geq 0$ by assumption. Moreover, $b = 0$ means that $d = m - 1$, in which case there is homology only if $m \leq n-1$, hence $a \geq 1$ and $k+a \geq 2$; for the latter inequality, recall that we restrict our attention to $m \geq 1$. Finally, $k < 0$ reduces to the case $b = 0$, because we then have homology only if $n \geq 2m + 2$ and $d = m - 1$; apply Theorem 11.30.

For the other direction, Theorem 11.37 yields that we only need to consider the following cases:

- $k \geq 0$, $a = 0$, and $b = 2$. By Theorem 11.31, we have infinite homology for $a = 0$ and $b = 2$ if and only if $k \geq (b-1)(a+b-1) = 1$. The remaining case is $(k,a,b) = (0,0,2) \Longleftrightarrow (m,n,d) = (5,5,2)$, in which case we have nonzero homology by Theorem 11.32.
- $k \geq 0$, $a \geq 0$, and $b = 1$. This time, Theorem 11.31 yields infinite homology for $a \geq 0$ and $b = 1$ as soon as $k \geq 0$.
- $k \geq 2-a$, $a \geq 1$, and $b = 0$. By yet another application of Theorem 11.31, we have infinite homology for $b = 0$ whenever $a \geq 1$, $k \geq 1-a$, and $k+a \geq 2$. Since the third inequality implies the second, we are done. $\square$

Conjecture 11.40 (Shareshian & Wachs [122]). *For all* $m, n \geq 1$, *the group* $\tilde{H}_d(\mathsf{M}_{m,n};\mathbb{Z})$ *contains 3-torsion if and only if*

$$\begin{cases} m \le n \le 2m - 5 \\ \lceil \frac{m+n-4}{3} \rceil \le d \le m - 3 \end{cases} \iff \begin{cases} k \ge 0 \\ a \ge 0 \\ b \ge 2; \end{cases}$$

k, a, and b are defined as in (11.8).

Note that Conjecture 11.40 implies Conjecture 11.33. Conjecture 11.40 remains unsettled in the following cases:

- $d = m - 2$: $9 \le m + 2 \le n \le 2m - 3$. Equivalently, $k \ge 1$, $a \ge 2$, and $b = 1$. *Conjecture:* There is no 3-torsion.
- $d = m - 3$: $8 \le m = n$. Equivalently, $k \ge 3$, $a = 0$, and $b = 2$. *Conjecture:* There is 3-torsion.

The conjecture is fully settled for $n = m + 1$ and $n \ge 2m - 2$; see Shareshian and Wachs [122] for the case $n = 2m - 2$ and use Theorem 11.30 for the case $n \ge 2m - 1$. For the case $n = m + 1$, we have that $\tilde{H}_{m-2}(\mathsf{M}_{m,m+1}; \mathbb{Z})$ is torsion-free, because $\mathsf{M}_{m,m+1}$ is an orientable pseudomanifold; see Spanier [130, Ex. 4.E.2].

Let

$$\beta_d^{m,n} = \dim_{\mathbb{Z}_3} \tilde{H}_d(\mathsf{M}_{m,n}; \mathbb{Z}_3)$$

and write $\hat{\beta}_k^{a,b} = \beta_d^{m,n}$, where k, a, and b are defined as in (11.8). The following theorem provides a chessboard analogue of Theorem 11.26.

Theorem 11.41 (Jonsson [76]). *For each $k \ge 0$, there is a polynomial $f_k(a,b)$ of degree $3k$ such that $\hat{\beta}_k^{a,b} \le f_k(a,b)$ whenever $a \ge 0$ and $b \ge k + 2$ and such that*

$$f_k(a,b) = \frac{1}{3^k k!} \left((a + 3b)^3 - 9b^3 \right)^k + \epsilon_k(a,b)$$

for some polynomial $\epsilon_k(a,b)$ of degree at most $3k - 1$. Equivalently,

$$\beta_d^{m,n} \le f_{3d-m-n+4}(n - m, m - d - 1)$$

for $m \le n \le 2m - 5$ and $\frac{m+n-4}{3} \le d \le \frac{2m+n-7}{4}$.

11.4 Paths and Cycles

We consider matching complexes on paths and cycles. These complexes have an extremely simple structure, but they are still worth discussing, as they appear naturally in many situations. For some examples, see Section 16.3 and Kozlov [86].

For $n \ge 0$, define Pa_n to be the graph with edge set $\{i(i+1) : i \in [n-1]\}$; we define Pa_n as the empty graph if $n \in \{0, 1\}$. $\mathsf{M}(\mathsf{Pa}_n)$ is isomorphic to the complex of stable sets in Pa_{n-1}; Kozlov [86, Prop. 4.6] determined the homotopy type of this complex.

Proposition 11.42. *Let $n \geq 0$ and $\nu_n = \lceil \frac{n-4}{3} \rceil$. Then $\mathsf{M}(\mathsf{Pa}_n)$ is $VD(\nu_n)$ and*

$$\mathsf{M}(\mathsf{Pa}_n) \simeq \begin{cases} \text{point} & \text{if } n \bmod 3 = 2; \\ S^{\nu_n} & \text{if } n \bmod 3 \neq 2. \end{cases} \tag{11.10}$$

Proof. One readily verifies the proposition for $n \leq 3$. Assume that $n \geq 4$. To prove vertex-decomposability, decompose with respect to the second edge 23. Clearly, 12 is a cone point in $\mathrm{del}_{\mathsf{M}(\mathsf{Pa}_n)}(23)$, and the underlying complex is isomorphic to $\mathsf{M}(\mathsf{Pa}_{n-2})$;

$$\mathrm{del}_{\mathsf{M}(\mathsf{Pa}_n)}(23) \cong \{\emptyset, \{12\}\} * \mathsf{M}(\mathsf{Pa}_{n-2}).$$

By induction on n, this implies that $\mathrm{del}_{\mathsf{M}(\mathsf{Pa}_n)}(23)$ is $VD(\nu_{n-2}+1)$ and hence $VD(\nu_n)$; $\nu_n \leq \nu_{n-2} + 1$. Moreover, $\mathrm{lk}_{\mathsf{M}(\mathsf{Pa}_n)}(23)$ is isomorphic to $\mathsf{M}(\mathsf{Pa}_{n-3})$, which implies that $\mathrm{lk}_{\mathsf{M}(\mathsf{Pa}_n)}(23)$ is $VD(\nu_n - 1)$; $\nu_n = \nu_{n-3} + 1$. It follows that $\mathsf{M}(\mathsf{Pa}_n)$ is $VD(\nu_n)$. To prove (11.10), note that $\mathsf{M}(\mathsf{Pa}_n)$ is homotopy equivalent to the suspension of $\mathsf{M}(\mathsf{Pa}_{n-3})$ for $n \geq 3$; apply Lemma 3.18 and use the facts that $\mathrm{del}_{\mathsf{M}(\mathsf{Pa}_n)}(23)$ is collapsible and $\mathrm{lk}_{\mathsf{M}(\mathsf{Pa}_n)}(23) \simeq \mathsf{M}(\mathsf{Pa}_{n-3})$. Indeed, the procedure just described is easily extended to a decision tree on $\mathsf{M}(\mathsf{Pa}_n)$ with at most one evasive set (of dimension ν_n if present). $\square$

For $n \geq 3$, define Cy_n as the graph with edge set $\{1n\} \cup \{i(i+1) : i \in [n-1]\}$. $\mathsf{M}(\mathsf{Cy}_n)$ is isomorphic to the complex of stable sets in Cy_n; Kozlov [86, Prop. 5.2] determined the homotopy type of this complex.

Proposition 11.43. *Let $n \geq 3$ and $\nu_n = \lceil \frac{n-4}{3} \rceil$. Then $\mathsf{M}(\mathsf{Cy}_n)$ is $VD(\nu_n)$ and*

$$\mathsf{M}(\mathsf{Cy}_n) \simeq \begin{cases} S^{\nu_n} \vee S^{\nu_n} & \text{if } n \bmod 3 = 0; \\ S^{\nu_n} & \text{if } n \bmod 3 \neq 0. \end{cases} \tag{11.11}$$

Proof. Decompose with respect to the edge $1n$. $\mathrm{lk}_{(\mathsf{Cy}_n)}(1n)$ is easily seen to be isomorphic to $\mathsf{M}(\mathsf{Pa}_{n-2})$. Moreover, $\mathrm{del}_{\mathsf{M}(\mathsf{Cy}_n)}(1n)$ equals $\mathsf{M}(\mathsf{Pa}_n)$. By Proposition 11.42, $\mathsf{M}(\mathsf{Pa}_n)$ is $VD(\nu_n)$ and $\mathsf{M}(\mathsf{Pa}_{n-2})$ is $VD(\nu_{n-2})$, which implies that $\mathsf{M}(\mathsf{Cy}_n)$ is $VD(\nu_n)$; we have that $\nu_{n-2} \geq \nu_n - 1$. To prove formula (11.11), we have optimal decision trees on $\mathsf{M}(\mathsf{Pa}_n)$ and $\mathsf{M}(\mathsf{Pa}_{n-2})$ by the proof of Proposition 11.42; combining these decision trees, we obtain a decision tree on $\mathsf{M}(\mathsf{Cy}_n)$ with one or two evasive sets of dimension ν_n. $\square$

Graphs of Bounded Degree

We consider the complex BD_n^d of graphs G on n vertices such that the degree of each vertex in G is at most d. Note that BD_n^1 coincides with the matching complex M_n. A natural generalization of BD_n^d is as follows. Let $\lambda = (\lambda_1, \ldots, \lambda_n)$ be an arbitrary sequence of integers. Define BD_n^λ as the complex of graphs G on n vertices such that $\deg_G(i) \leq \lambda_i$ for each $i \in [n]$. Clearly, $\mathsf{BD}_n^{(d,\ldots,d)} = \mathsf{BD}_n^d$. We discuss the connectivity degree of BD_n^λ in Section 12.1. The main result is that the shifted connectivity degrees of BD_n^2 and BD_n^3 are at least $\frac{7n-13}{9}$ and $\frac{11n-13}{9}$, respectively. For general BD_n^d, we obtain the lower bound $\frac{(d-1)n}{2} + \frac{3n}{2(d+4)} - \epsilon$, where ϵ is a small term; note that this bound is strictly weaker than those just given for $d \in \{2,3\}$. In addition, we demonstrate that the depth of BD_n^2 is at least $\frac{3n-7}{4}$.

Let us define an *L-graph* to be a pair $G = (V, E)$ such that E is a subset of $\binom{V}{1} \cup \binom{V}{2}$. Thus an L-graph is simply a [2]-hypergraph. We refer to the elements in $\binom{V}{1}$ as *loops* and denote the loop $\{i\}$ as ii. For an L-graph G, let G^- be the simple graph obtained by removing all loops from G. Define the degree of a vertex i in G as follows:

$$\deg_G(i) = \begin{cases} \deg_{G^-}(i) & \text{if } ii \notin G; \\ \deg_{G^-}(i) + 2 & \text{if } ii \in G. \end{cases}$$

Hence a loop increases the degree by two. For a sequence $\lambda = (\lambda_1, \ldots, \lambda_n)$ of nonnegative integers, define $\overline{\mathsf{BD}}_n^\lambda$ as the complex of L-graphs G such that $\deg_G(i) \leq \lambda_i$ for each $i \in [n]$. In Section 12.2, we review the most important known results about the homology and homotopy type of $\overline{\mathsf{BD}}_n^\lambda$:

- Reiner and Roberts [111] computed the rational homology of $\overline{\mathsf{BD}}_n^\lambda$, thereby generalizing Bouc's Theorem 11.5 about the matching complex M_n.
- Dong [36] characterized all partitions λ such that $\overline{\mathsf{BD}}_n^\lambda$ is collapsible. Dong also demonstrated that the Alexander dual of $\overline{\mathsf{BD}}_n^\lambda$ is homotopy equivalent to the $(n-1)$-fold suspension of $\overline{\mathsf{BD}}_n^{n^n-\lambda}$; $n^n - \lambda = (n - \lambda_1, \ldots, n - \lambda_n)$.

In addition, we show that our bounds on the connectivity degree of the smaller complex BD_n^λ remain valid for $\overline{\mathsf{BD}}_n^\lambda$.

In Section 12.3, we derive formulas for the reduced Euler characteristic of BD_n^d for $d \in \{1, 2, n-3, n-2\}$ and of $\overline{\mathsf{BD}}_n^d$ for $d \in \{1, 2, n-2, n-1, n\}$. In the latter case, we apply Dong's Alexander duality result.

For a few applications of the variant $\overline{\mathsf{BD}}_n^\lambda$ admitting loops, see Sections 1.1.2 and 1.1.3.

12.1 Bounded-Degree Graphs Without Loops

For a real number ν, say that a family Δ is $AM(\nu)$ if Δ admits an acyclic matching such that all unmatched sets are of dimension at least $\lceil \nu \rceil$. For a sequence $\lambda = (\lambda_1, \ldots, \lambda_n)$, define $|\lambda| = \sum_{i=1}^n \lambda_i$. For a statement P, let $\chi(P) = 1$ if P is true and $\chi(P) = 0$ otherwise. Our first result is similar in nature to Athanasiadis' Theorem 11.4.

Lemma 12.1. *Let α be an integer and let $(\lambda_1, \ldots, \lambda_n)$ be a sequence of non-negative integers such that $\lambda_1, \lambda_2 > 0$. Suppose that the following hold:*

(i) *For each set $U \subseteq [3, n]$ such that $|U| = \lambda_1$, $\mathsf{BD}_{n-1}^{\lambda'}$ is $AM(\frac{|\lambda'|-\alpha+1}{2} - 1)$, where*
$$\lambda' = (\lambda_2 - 1, \lambda_3 - \chi(3 \in U), \ldots, \lambda_n - \chi(n \in U)).$$

(ii) *For each set $U \subseteq [3, n]$ such that $|U| = \lambda_2$, $\mathsf{BD}_{n-1}^{\lambda'}$ is $AM(\frac{|\lambda'|-\alpha}{2} - 1)$, where*
$$\lambda' = (\lambda_1, \lambda_3 - \chi(3 \in U), \ldots, \lambda_n - \chi(n \in U)).$$

Then BD_n^λ is $AM(\frac{|\lambda|-\alpha}{2} - 1)$.

Proof. Define an acyclic matching on BD_n^λ by pairing $G - 12$ and $G + 12$ whenever both graphs belong to the complex. The remaining family $\mathcal{C}$ consists of all graphs G such that $12 \notin G$ and such that either $\deg_G(1) = \lambda_1$ or $\deg_G(2) = \lambda_2$.

Let $\mathcal{A}$ be the subfamily of $\mathcal{C}$ consisting of all graphs G such that $\deg_G(1) = \lambda_1$ and $\deg_G(2) < \lambda_2$. Let $\mathcal{B} = \mathcal{C} \setminus \mathcal{A}$; the graphs G in $\mathcal{B}$ satisfy $\deg_G(2) = \lambda_2$. It is clear that the Cluster Lemma 4.2 applies to $\mathcal{A}$ and $\mathcal{B}$.

First, consider $\mathcal{A}$. For each set $U \subseteq [3, n]$ such that $|U| = \lambda_1$, let $\mathcal{A}_U$ be the family of graphs G in $\mathcal{A}$ such that $N_G(1) = U$. The families $\mathcal{A}_U$ form an antichain with respect to inclusion; hence the Cluster Lemma 4.2 applies. Define
$$\lambda' = (\lambda_2 - 1, \lambda_3 - \chi(3 \in U), \ldots, \lambda_n - \chi(n \in U)).$$

Note that $\sum_i \chi(i \in U) = \lambda_1$. By assumption, $\mathsf{BD}_{n-1}^{\lambda'}$ is $AM(\frac{|\lambda'|-\alpha+1}{2} - 1)$. One readily verifies that
$$\mathcal{A}_U = \{iu : u \in U \cup \{2\}\} * \mathsf{BD}_{n-1}^{\lambda'};$$

thus it follows that $\mathcal{A}_U$ is $AM(r)$, where

$$r = \lambda_1 + \frac{|\lambda'| - \alpha + 1}{2} - 1 = \frac{|\lambda'| - \alpha + 2\lambda_1 + 1}{2} - 1 = \frac{|\lambda| - \alpha}{2} - 1;$$

the last equality is because $|\lambda| - |\lambda'| = 2\lambda_1 + 1$.

Next, consider $\mathcal{B}$. This time, for each $U \subseteq [3, n]$ such that $|U| = \lambda_2$, let $\mathcal{B}_U$ be the family of graphs G in $\mathcal{B}$ such that $N_G(2) = U$. As in the previous case, the Cluster Lemma 4.2 applies. Define

$$\lambda' = (\lambda_1, \lambda_3 - \chi(3 \in U), \ldots, \lambda_n - \chi(n \in U));$$

note that $\sum_i \chi(i \in U) = \lambda_2$. By assumption, $\mathrm{BD}_{n-1}^{\lambda'}$ is $AM(\frac{|\lambda'| - \alpha}{2} - 1)$. One readily verifies that

$$\mathcal{B}_U = \{iu : u \in U\} * \mathrm{BD}_{n-1}^{\lambda'};$$

thus it follows that $\mathcal{B}_U$ is $AM(r)$, where

$$r = \lambda_2 + \frac{|\lambda'| - \alpha}{2} - 1 = \frac{|\lambda'| - \alpha + 2\lambda_2}{2} - 1 = \frac{|\lambda| - \alpha}{2} - 1;$$

the last equality is because $|\lambda| - |\lambda'| = 2\lambda_2$. Combining all acyclic matchings, we obtain an acyclic matching on $\mathcal{C}$ such that all unmatched faces have dimension at least $\frac{|\lambda| - \alpha}{2} - 1$. $\square$

12.1.1 The Case $d = 2$

Using Lemma 12.1, one may easily compute lower bounds on the connectivity degree of BD_n^λ for small λ. Specifically, let us consider sequences λ of the form $2^a 1^b$, i.e., $\lambda_i = 2$ for $i \in [a]$ and $\lambda_{a+j} = 1$ for $j \in [b]$.

Theorem 12.2. *For $a \geq 0$ and $b \geq 0$ such that $(a, b) \neq (1, 0)$, we have that* $\mathrm{BD}_{a+b}^{2^a 1^b}$ *is $AM(\beta_{a,b})$, where*

$$\beta_{a,b} = \frac{7a + 3b - 12 - \epsilon_{a,b}}{9};$$

$\epsilon_{a,b} = 1$ *if $b = 0$ or if $(a, b) \in \{(4, 1), (1, 2)\}$; $\epsilon_{a,b} = 0$ otherwise. In particular,* BD_n^2 *is $AM(\frac{7n-13}{9})$ for $n \neq 1$.*

Remark. See Table 12.1 for bounds on the shifted connectivity degree for $a \leq 16$ and $b \leq 8$.

Proof. We use induction on $|\lambda|$. The case $a = 0$ follows by Theorem 11.1. One easily checks the statement by hand for $a + b \leq 3$ and also for $a + b = 4$, at least when $b \geq 1$. Moreover, it is a straightforward task to check that $\mathrm{BD}_5^{2^2 1^3}$ is collapsible; apply Lemma 12.1 with $\lambda_1 = \lambda_2 = 2$.

Table 12.1. Lower bound $\beta_{a,b}$ on the shifted connectivity degree of $\mathrm{BD}_{a+b}^{2^a 1^b}$ for $a \leq 16$ and $b \leq 8$. "–" means that the complex under consideration is contractible.

$\beta_{a,b}$	a = 0	1	2	3	4	5	6	7	8	9	10	11	12	13	14	15	16
b = 0	−1	−1	–	–	2	3	4	4	5	6	7	8	8	9	10	11	11
1	−1	–	–	–	2	3	4	5	6	6	7	8	9	10	10	11	12
2	–	0	1	2	3	4	4	5	6	7	8	8	9	10	11	11	12
3	0	1	–	2	3	4	5	6	6	7	8	9	9	10	11	12	13
4	0	1	2	3	4	4	5	6	7	7	8	9	10	11	11	12	13
5	1	2	2	3	4	5	5	6	7	8	9	9	10	11	12	12	13
6	1	2	3	3	4	5	6	7	7	8	9	10	10	11	12	13	14
7	1	2	3	4	5	5	6	7	8	8	9	10	11	12	12	13	14
8	2	3	3	4	5	6	6	7	8	9	10	10	11	12	13	13	14

First, assume that $a \geq 1$ and $b \geq 1$ and that $a + b \geq 5$ and $(a,b) \neq (2,3)$. Arrange the elements in λ such that $\lambda_1 = 2$ and $\lambda_2 = 1$. In (i) in Lemma 12.1, each sequence λ' is of the form $2^{a'} 1^{b'}$, where $(a',b') = (a - r, b - 5 + 2r)$ for some $r \in \{1,2,3\}$. One easily checks that we cannot have $(a',b') = (1,0)$. Thus by induction, $\mathrm{BD}_{a'+b'}^{2^{a'} 1^{b'}}$ is $AM(\mu)$, where

$$\mu = \frac{7a' + 3b' - 12 - \epsilon_{a',b'}}{9} = \frac{7a + 3b - r - 9 - \epsilon_{a',b'}}{9} - 2.$$

If $r < 3$ or $\epsilon_{a',b'} = 0$, then $\mu \geq \beta_{a,b} - 2$. If $r = 3$ and $\epsilon_{a',b'} = 1$, then we must have that $(a',b') \in \{(4,1), (1,2)\}$; hence $(a,b) \in \{(7,0), (4,1)\}$. Since $\epsilon_{a,b} = 1$ for these values, it follows that $\mu \geq \beta_{a,b} - 2$ in all cases.

In Lemma 12.1 (ii), each sequence λ' is of the form $2^{a'} 1^{b'}$, where $(a',b') = (a - r, b - 2 + 2r)$ for some $r \in \{0,1\}$. By induction, $\mathrm{BD}_{a'+b'}^{2^{a'} 1^{b'}}$ is $AM(\mu)$, where

$$\mu = \frac{7a' + 3b' - 12 - \epsilon_{a',b'}}{9} = \frac{7a + 3b - r - 9 - \epsilon_{a',b'}}{9} - 1 \geq \beta_{a,b} - 1.$$

Lemma 12.1 yields that $\mathrm{BD}_{a+b}^{2^a 1^b}$ is $AM(\beta_{a,b})$.

It remains to consider the case $b = 0$ and $a \geq 4$. In Lemma 12.1 (i), each sequence λ' is of the form $2^{a-4} 1^3$. By induction, $\mathrm{BD}_{a-1}^{2^{a-4} 1^3}$ is $AM(\mu)$, where

$$\mu = \frac{7(a-4) + 3 \cdot 3 - 12 - \epsilon_{a-4,3}}{9} = \frac{7a - 13}{9} - 2 = \beta_{a,0} - 2.$$

In Lemma 12.1 (ii), each sequence λ' is of the form $2^{a-3} 1^2$. One easily checks that $\mathrm{BD}_{a-1}^{2^{a-3} 1^2}$ is $AM(\beta_{a,0} - 2)$; hence Lemma 12.1 yields that $\mathrm{BD}_{a+b}^{2^a 1^b}$ is $AM(\beta_{a,b})$. $\square$

Table 12.2. The homology of BD_n^2 for $n \le 8$.

$\tilde{H}_i(BD_n^2, \mathbb{Z})$	$i = 1$	2	3	4	5	6	7
$n = 4$	-	$\mathbb{Z}^3$	-	-	-	-	-
5	-	-	$\mathbb{Z}^9$	-	-	-	-
6	-	-	-	$\mathbb{Z}^{36}$	-	-	-
7	-	-	-	$\mathbb{Z}$	$\mathbb{Z}^{181}$	-	-
8	-	-	-	-	$\mathbb{Z}_2 \oplus \mathbb{Z}^{125}$	$\mathbb{Z}^{890}$	-

We have some hope that the bound in Theorem 12.2 is best possible:

Conjecture 12.3. *The shifted connectivity degree of* BD_n^2 *equals* $\lceil \frac{7n-13}{9} \rceil$.

The conjecture is true for small values of n; see Table 12.2. Note by the way that there is 2-torsion in the homology of BD_8^2.

12.1.2 The General Case

We now consider general λ. Let $\mu = (\mu_1, \ldots, \mu_n)$ be a sequence. Let $\Lambda(n, \mu)$ be the set of all nonnegative sequences $(\lambda_1, \ldots, \lambda_n)$ such that $\lambda_i \le \mu_i$ for all i. For $n \ge 1$ and $\mu = (\mu_1, \ldots, \mu_n)$, define

$$\alpha(n, \mu) = \min\{\alpha : BD_n^\lambda \text{ is } AM(\tfrac{|\lambda| - \alpha}{2} - 1)\}, \tag{12.1}$$

where the minimum is taken over all sequences $\lambda \in \Lambda(n, \mu)$. In particular, BD_n^λ is $(\frac{|\lambda| - \alpha(n, \mu)}{2} - 2)$-connected for each λ such that $\lambda_i \le \mu_i$ for all i. It follows that there is no homology below dimension $\frac{|\lambda| - \alpha(n, \mu)}{2} - 1$.

Write $\Lambda(n, d) = \Lambda(n, (d, \ldots, d))$ and $\alpha(n, d) = \alpha(n, (d, \ldots, d))$. It is immediate that $\alpha(n', d') \le \alpha(n, d)$ whenever $n' \le n$ and $d' \le d$, because if $(\lambda_1, \ldots, \lambda_{n'})$ belongs to $\Lambda(n', d')$, then $(\lambda_1, \ldots, \lambda_{n'}, 0^{n-n'})$ belongs to $\Lambda(n, d)$.

Corollary 12.4. *For $n \ge 1$, we have that $\alpha(n, 1) = \lfloor \frac{n+2}{3} \rfloor$. For $n \ge 3$, we have that $\alpha(n, 2) \le \lfloor \frac{4n+8}{9} \rfloor$.*

Proof. Let $\lambda = 2^a 1^b 0^c$ and $n = a + b + c$. If $a = 1$ and $b = 0$, then BD_n^λ is only $AM(\frac{|\lambda| - 2}{2} - 1)$; hence $\alpha(n, 2) \ge 2$, which explains the bound $n \ge 3$ in the second statement. Otherwise, Theorem 12.2 yields that BD_n^λ is $AM(\beta_{a,b})$, where

$$\beta_{a,b} = \frac{7a + 3b - 12 - \epsilon_{a,b}}{9} = \frac{2a + b}{2} - \frac{1}{2} \cdot \frac{4a + 3b + 6 + 2\epsilon_{a,b}}{9} - 1$$

$$= \frac{|\lambda|}{2} - \frac{1}{2} \cdot \frac{4n - b - 4c + 6 + 2\epsilon_{a,b}}{9} - 1.$$

Now, $\frac{4n-b-4c+6+2\epsilon_{a,b}}{9}$ is at most $\frac{4n+8}{9}$ in the general case and at most $\frac{3n+6}{9}$ if $a = 0$ ($\epsilon_{0,b} = 0$). Observing that the latter bound is sharp by Theorem 11.12, we are done. $\square$

Our main goal is to estimate $\alpha(n,d)$ for small n and use this result to estimate the connectivity degree of general complexes BD_n^{λ}. We stress that our estimates are unlikely to be sharp.

For a sequence $\lambda = (\lambda_1, \ldots, \lambda_n)$ and a set $U \subseteq [n]$, let λ_U be the subsequence of λ consisting of all λ_u such that $u \in U$. The following result generalizes parts of Theorem 11.1.

Theorem 12.5. *Let G be a graph on the vertex set V. Let $\{U_1, \ldots, U_t\}$ be a clique partition of G and let $\lambda = (\lambda_1, \ldots, \lambda_n)$ and $\mu = (\mu_1, \ldots, \mu_n)$ be sequences of nonnegative integers such that $\lambda_i \leq \mu_i$ for all i. Then $BD_n^{\lambda}(G)$ is $AM(\nu)$, where*

$$\nu = \frac{|\lambda|}{2} - \frac{1}{2} \sum_{j=1}^{t} \alpha(|U_j|, \mu_{U_j}) - 1.$$

In particular, $\alpha(n,d) \leq \sum_{j=1}^{t} \alpha(n_j, d)$ for every $d \geq 1$ and every sequence $(n_1, \ldots, n_t)$ of positive integers summing to n.

Proof. Let σ be the union of the sets of edges within the induced subgraphs $G(U_i)$ for $i \in [1,k]$. If the edge set of G is σ, then

$$BD_n^{\lambda}(G) = BD_{|U_1|}^{\lambda_{U_1}}(G(U_1)) * \cdots * BD_{|U_t|}^{\lambda_{U_t}}(G(U_t)),$$

which satisfies the theorem; use Theorem 5.29 and the fact that $BD_{|U_j|}^{\lambda_{U_j}}(G(U_j))$ is $AM(\frac{|\lambda_{U_j}| - \alpha(|U_j|, \mu_{U_j})}{2} - 1)$.

Otherwise, let $e = ab$ be any edge in $G - \sigma$. By induction on the number of edges in G, $\mathrm{del}_{BD_n^{\lambda}(G)} = BD_n^{\lambda}(G - e)$ is $AM(\nu)$. Moreover, $\mathrm{lk}_{BD_n^{\lambda}(G)}(e)$ equals $BD_n^{\lambda'}(G - e)$, where we obtain λ' from λ by subtracting one from each of λ_a and λ_b. By induction on λ, $BD_n^{\lambda'}(G - e)$ is $AM(\nu')$, where

$$\nu' = \frac{|\lambda'|}{2} - \sum_{j=1}^{t} \frac{\alpha(|U_j|, \mu_{U_j})}{2} - 1 = \nu - 1.$$

This implies that $BD_n^{\lambda}(G)$ is $AM(\nu)$, and we are done. $\square$

Lemma 12.6. *Let $d \geq 1$ and $1 \leq n \leq d + 2$. Then $\alpha(n,d) = d$.*

Proof. If $\lambda = (d, 0, \ldots, 0)$, then obviously $BD_n^{\lambda} = \{\phi\}$, which implies that $\alpha(n,d) \geq d$; we must have that $\frac{|\lambda| - \alpha(n,d)}{2} \leq 0$ in this case.

It remains to prove that $\alpha(n,d) \leq d$. We use double induction on d and n, proving that BD_n^{λ} is $AM(\frac{|\lambda| - d}{2} - 1)$ whenever $\lambda \in \Lambda(n,d)$. The case $d = 1$ is clear by Corollary 12.4. The case $n = 1$ is obvious.

Let $d \geq 2$ and let $(\lambda_1, \ldots, \lambda_n)$ be a sequence such that $d \geq \lambda_1 \geq \ldots \geq \lambda_n$. Without loss of generality, we may assume that all elements in the sequence are positive; otherwise, just remove all vertices i such that $\lambda_i = 0$.

We want to apply Lemma 12.1 with $\alpha = d$. First, we verify that (i) in the lemma holds. Consider $BD_{n-1}^{\lambda'}$, where

$$\lambda' = (\lambda_2 - 1, \lambda_3 - \chi(3 \in U), \ldots, \lambda_n - \chi(n \in U)),$$

$U \subseteq [3, n]$, and $|U| = \lambda_1$. We have that $\lambda' \in \Lambda(n - 1, d - 1)$. Namely, if $\lambda_1 = d$, then $\chi_i = 1$ for all $i \in [3, n]$, because $n - 2 \leq d$. As a consequence, $BD_{n-1}^{\lambda'}$ is $AM(\frac{|\lambda'| - d + 1}{2} - 1)$ by induction on d.

Next, we verify that (ii) in the lemma holds. Consider $BD_{n-1}^{\lambda'}$, where

$$\lambda' = (\lambda_1, \lambda_3 - \chi(3 \in U), \ldots, \lambda_n - \chi(n \in U)),$$

$U \subseteq [3, n]$, and $|U| = \lambda_2$. This time, $BD_{n-1}^{\lambda'}$ is $AM(\frac{|\lambda'| - d}{2} - 1)$ by induction on n.

As a consequence, we are done by Lemma 12.1. $\square$

Lemma 12.7. *Let $d \geq 0$ and let $(\lambda_1, \ldots, \lambda_{d+4})$ be a weakly decreasing sequence of integers such that $\lambda_1 \leq d + 1$ and $\lambda_3 \leq d$. Then BD_{d+4}^{λ} is $AM(\frac{|\lambda| - d - 1}{2} - 1)$. In particular, $\alpha(d + 3, d) \leq \alpha(d + 4, d) \leq d + 1$.*

Proof. The lemma is obvious if $d = 0$; thus assume that $d \geq 1$. Moreover, the lemma is obvious if $\lambda_2 = 0$, because $\lambda_1 \leq d + 1$; thus assume that $\lambda_2 \geq 1$.

As in the previous proof, we want to apply Lemma 12.1, this time with $\alpha = d + 1$. First, consider (i) in the lemma. We are interested in $BD_{d+3}^{\lambda'}$, where

$$\lambda' = (\lambda_2 - 1, \lambda_3 - \chi(3 \in U), \ldots, \lambda_{d+4} - \chi(d + 4 \in U)),$$

$U \subseteq [3, d+4]$, and $|U| = \lambda_1$. If we can prove that at most two elements in the sequence λ' are equal to d and that no elements are larger than d, then it will follow by induction that $BD_{d+3}^{\lambda'}$ is $AM(\frac{|\lambda'| - d}{2} - 1)$ as desired. Now, the given condition on λ' trivially holds if $\lambda_1 \leq d - 1$. If $\lambda_1 = d$, then we subtract one from λ_2 and from all but two elements in the sequence $(\lambda_3, \ldots, \lambda_{d+4})$. Since each of these elements is at most d, it follows that $\lambda_i' = d$ for at most two i. If $\lambda_1 = d + 1$, then we subtract one from λ_2 and from all but one element in $(\lambda_3, \ldots, \lambda_{d+4})$. Again, it follows that at most two elements λ_i' equal d.

Next, consider (ii) in the lemma. This time, we are interested in $BD_{d+3}^{\lambda'}$, where

$$\lambda' = (\lambda_1, \lambda_3 - \chi(3 \in U), \ldots, \lambda_{d+4} - \chi(d + 4 \in U)),$$

$U \subseteq [3, d+4]$, and $|U| = \lambda_2$. Now, $BD_{d+3}^{\lambda'} = BD_{d+4}^{\lambda''}$, where we obtain λ'' by adding a zero at the end of λ'. By induction on $|\lambda|$, it follows that $BD_{d+3}^{\lambda'}$ is $AM(\frac{|\lambda| - d - 1}{2} - 1)$. Thus we are done by Lemma 12.1. $\square$

Define $\max \lambda = \max\{\lambda_i : i \in [n]\}$.

Theorem 12.8. *Let* $d \geq 2$ *and* $n \geq d + 1$ *and let* $\lambda = (\lambda_1, \ldots, \lambda_n)$ *be a sequence of nonnegative integers such that* $\max \lambda \leq d$. *Write* $n = (d+4)k + r$, *where* $d + 1 \leq r \leq 2d + 4$. *Then* BD_n^λ *is* $AM(\nu)$, *where*

$$\nu = \frac{|\lambda|}{2} - \frac{(d+1)n}{2(d+4)} - \frac{\epsilon_d(r)}{2} - 1$$

and

$$\epsilon_d(r) = \frac{3r}{d+4} - \begin{cases} 1 & \text{if } r = d + 1; \\ 2 & \text{if } d + 2 \leq r \leq d + 3; \\ 3 & \text{if } d + 4 \leq r \leq 2d + 3; \\ 4 & \text{if } r = 2d + 4. \end{cases}$$

In particular, BD_n^d *is* $AM(\nu_n^d)$, *where*

$$\nu_n^d = \frac{(d^2 + 3d - 1)n}{2(d+4)} - \frac{\epsilon_d(r)}{2} - 1.$$

Remark. One easily checks that the "error term" $\epsilon_d(r)$ satisfies $0 \leq \epsilon_d(r) \leq 3(d-1)/(d+4)$; the maximum is obtained for $r = 2d + 3$. As a consequence,

$$\nu_n^d \geq \frac{(d^2 + 3d - 1)n - 5(d+1)}{2(d+4)}. \tag{12.2}$$

Proof. Divide $[n]$ into k sets of size $d+4$ and one set of size r. By Lemma 12.6, $\alpha(r, d) = d$ for $r \leq d+2$. Moreover, Lemma 12.7 implies that $\alpha(d+3, d), \alpha(d+4, d) \leq d+1$. Using Lemma 12.1, one easily checks that $\alpha(n, d) \leq \alpha(n-1, d)+1$ for $n \geq 2$, which implies that

$$\alpha(r, d) \leq \alpha(d + 4, d) + r - d - 4 \leq r - 3$$

for $r \geq d + 4$. Finally, $\alpha(2d + 4, d) \leq 2\alpha(d + 2, d) = 2d$ by Theorem 12.5. As a consequence,

$$\alpha(r, d) \leq \epsilon_d(r) + \frac{(d+1)r}{d+4}$$

for $d + 1 \leq r \leq 2d + 4$. By Theorem 12.5, it follows that

$$\alpha(n, d) \leq k(d+1) + \epsilon_d(r) + \frac{(d+1)r}{d+4} = \frac{n-r}{d+4}(d+1) + \epsilon_d(r) + \frac{(d+1)r}{d+4}$$

$$= \frac{(d+1)n}{d+4} + \epsilon_d(r),$$

which concludes the proof. $\square$

As already indicated, we do not believe that the derived bound is actually equal to the shifted connectivity degree. Indeed, for $d = 2$, we obtain $\nu_n^2 = \frac{3n-5}{4}$, which is substantially smaller than the bound $\frac{7n-13}{9}$ in Theorem 12.2. We now show how to improve on Theorem 12.8 for $d = 3$.

Lemma 12.9. *We have that $\alpha(9,3) \leq 5$.*

Proof. Throughout this proof, we assume that λ is weakly decreasing. In the proof, we repeatedly apply Lemma 12.1 to prove that BD_n^λ is $AM\left(\frac{|\lambda|-\alpha}{2} - 1\right)$ for a given α whenever n and λ are bounded by certain values. For each such application, it will be obvious by induction on $|\lambda|$ that (ii) in the lemma is satisfied; hence we suppress this step in the below discussion.

First, we note that BD_6^λ is $AM\left(\frac{|\lambda|-2}{2} - 1\right)$ whenever $|\lambda| \leq 7$ and $\lambda_i \leq 2$ for all i; apply Theorem 12.2 or consult Table 12.1.

Second, we prove that BD_7^λ is $AM\left(\frac{|\lambda|-3}{2} - 1\right)$ whenever $|\lambda| \leq 12$, $\lambda_1 \leq 3$, and $\lambda_3 \leq 2$. This is clear if $\lambda_i \leq 1$ for all i. Otherwise, use Lemma 12.1. Each resulting sequence λ' in (i) has the property that $|\lambda'| \leq 7$; $\lambda_1 \geq 2$. Moreover, each element in λ' is at most two; we remove λ_1 and subtract one from λ_2. As a consequence, the claim follows by the first claim.

The third step is to prove that BD_8^λ is $AM\left(\frac{|\lambda|-4}{2} - 1\right)$ whenever $|\lambda| \leq 19$, $\lambda_1 \leq 3$, and $\lambda_5 \leq 2$. Since $\alpha(8,2) \leq 4$ by Corollary 12.4, we may assume that $\lambda_1 = 3$. Each resulting sequence λ' in Lemma 12.1 (i) has the property that $|\lambda'| \leq 12$ and has at most two elements equal to 3. As a consequence, the second result yields the claim.

Finally, we prove that BD_9^λ is $AM\left(\frac{|\lambda|-5}{2} - 1\right)$ whenever $|\lambda| \leq 26$ and $\max \lambda \leq 3$. Since $\alpha(9,2), \alpha(8,3) \leq 5$, we may assume that $\lambda_1 = 3$ and $\lambda_9 \neq 0$. Each resulting sequence λ' in Lemma 12.1 (i) has the property that $|\lambda'| \leq 19$ and has at most four elements equal to 3, because we remove λ_1 and subtract one from four of the remaining λ_i. As a consequence, the third result yields the claim.

It remains to consider BD_9^3. By Theorem 12.8, BD_9^3 is $AM\left(\frac{|\lambda|-6}{2} - 1\right) = AM(9.5)$, which clearly implies that the complex is $AM\left(\frac{|\lambda|-5}{2} - 1\right) = AM(10)$. The lemma follows. $\square$

Theorem 12.10. *Let $n \geq 4$ and let $\lambda = (\lambda_1, \ldots, \lambda_n)$ be a sequence of non-negative integers such that $\max \lambda \leq 3$. Then BD_n^λ is $AM(\nu)$, where*

$$\nu = \frac{9|\lambda| - 5n - 26}{18}.$$

In particular, BD_n^3 is $AM\left(\frac{11n-13}{9}\right)$.

Remark. For $d = 3$, the bound in (12.2) equals $(17n - 20)/14$.

Proof. By previous lemmas, we know that $\alpha(4,3) = \alpha(5,3) = 3$, $\alpha(7,3) \leq 4$, and $\alpha(9,3) \leq 5$. Moreover, Theorem 12.5 implies that $\alpha(10,3) \leq 2\alpha(5,3) = 6$ and $\alpha(12,3) \leq \alpha(5,3) + \alpha(7,3) \leq 7$. To summarize,

$$\alpha(r,3) \leq \frac{5r + 8}{9}$$

for $4 \leq r \leq 12$. Write $n = 9k + r$, where $4 \leq r \leq 12$. By Theorem 12.5, it follows that

$$\alpha(n, 3) \leq k\alpha(9, 3) + \alpha(r, 3) \leq 5k + \frac{5r + 8}{9} = \frac{5n + 8}{9},$$

which immediately implies the desired result. $\square$

Table 12.3. Lower bounds on the shifted connectivity degree of BD_n^d for small d.

d	Best known bound	Reference	d	Best known bound	Reference
1	$\dfrac{n - 4}{3}$	Th. 11.6	4	$\dfrac{27n - 25}{16}$	Th. 12.8
2	$\dfrac{7n - 13}{9}$	Th. 12.2	5	$\dfrac{13n - 10}{6}$	,,
3	$\dfrac{11n - 13}{9}$	Th. 12.10	6	$\dfrac{53n - 35}{20}$	,,

See Table 12.3 for a summary of our results on the shifted connectivity degree of BD_n^d. In many cases, this value is strictly larger than the depth. Specifically, assume that $d \geq 7$. By Theorem 12.8, the shifted connectivity degree of BD_{2d+4}^d is at least $d^2 + d - 1$. Now, let G be a graph on $2d + 4$ vertices consisting of two connected components. The first component is a clique of vertex size four, whereas the second component is a graph on $2d$ vertices in which every vertex has degree d. It is clear that G is maximal in BD_n^d, because all vertices of degree less than d are already adjacent to each other. However, the number of edges in G is

$$\frac{d \cdot (2d)}{2} + 6 = d^2 + 6 < d^2 + d,$$

which implies that the depth of BD_{2d+4}^d is strictly less than $d^2 + d - 1$.

For the special case $d = 2$, we have been able to prove the following about the depth:

Proposition 12.11. *For* $n \geq 3$, BD_n^2 *is* $VD(\lceil \frac{3n-7}{4} \rceil)$. *In particular, the depth of* BD_n^2 *is at least* $\lceil \frac{3n-7}{4} \rceil$.

Proof. Write $n = 4(k - 1) + r$ such that $r \in [1, 4]$. Divide $[n]$ into $k - 1$ sets $U_1, \ldots, U_{k-1}$ of size 4 and one set U_k of size r. Let Y be the set of edges between different sets U_i and U_j. For each $A \subseteq Y$, we have that

$$\mathsf{BD}_n^2(A, Y \setminus A) = \{A\} * \mathsf{BD}_4^{\lambda^1} * \ldots * \mathsf{BD}_4^{\lambda^{k-1}} * \mathsf{BD}_r^{\lambda^k},$$

where the coefficient in λ^i corresponding to a given vertex $u \in U_i$ is two minus the number of edges in A that are adjacent to u. Adding zeros at the end of λ^k, we may identify $\mathsf{BD}_r^{\lambda^k}$ with a complex of the form $\mathsf{BD}_4^{\lambda'}$.

It suffices to prove that $\mathsf{BD}_4^{\lambda^i}$ is $VD(\lceil |\lambda^i|/2 \rceil - 2)$ for each $i < k$ and that the same is true for $\mathsf{BD}_r^{\lambda^k} = \mathsf{BD}_4^{\lambda'}$. Namely, this will imply that $\mathsf{BD}_n^2(A, Y \setminus A)$ is $VD(\lceil \alpha \rceil)$, where

$$\alpha = |A| + \sum_{i=1}^{k}(|\lambda^i|/2 - 1) - 1 = n - k - 1 = \left\lceil \frac{3n - 7}{4} \right\rceil.$$

Thus let $\lambda = (\lambda_1, \ldots, \lambda_4)$ be a sequence such that $\max \lambda \leq 2$ and $\lambda_1 \geq \lambda_2 \geq \lambda_3 \geq \lambda_4$. We want to prove that BD_4^λ is $VD(\lceil |\lambda|/2 \rceil - 2)$. If some λ_i is negative, then $\mathsf{BD}_4^\lambda = \emptyset$. Otherwise, we have the following four cases:

- $|\lambda| \leq 2$. Then we are done, as BD_4^λ is obviously $VD(-1)$.
- $3 \leq |\lambda| \leq 4$. Then λ_1 and λ_2 are both nonzero, which implies that BD_4^λ contains the 0-cell $\{12\}$. As a consequence, BD_4^λ is $VD(0)$.
- $5 \leq |\lambda| \leq 6$. Then $\lambda_1 = 2$. Decompose with respect to the edge set $Z = \{23, 24, 34\}$. Let $B \subseteq Z$. If $|B| \geq 2$, then $\mathsf{BD}_4^\lambda(B, Z \setminus B)$ is clearly $VD(1)$. If $B = \{e\}$ for some e, then some vertex $i > 1$ has the property that i is incident to less than λ_i edges in B; $\lambda_2 + \lambda_3 + \lambda_4 \geq 3$. As a consequence, $\{e, 1i\}$ belongs to $\mathsf{BD}_4^\lambda(\{e\}, Z \setminus \{e\})$, which implies that $\mathsf{BD}_4^\lambda(\{e\}, Z \setminus \{e\})$ is $VD(1)$. Finally, consider $B = \emptyset$. Since $\lambda_2, \lambda_3 \geq 1$, we have that $\mathsf{BD}_4^\lambda(\emptyset, Z)$ is either $2^{\{12,13\}}$ or $2^{\{12,13,14\}}$, which are both $VD(1)$.
- $|\lambda| \geq 7$. In this case, $\lambda_1 = \lambda_2 = \lambda_3 = 2$. Decompose with respect to the edge set $Z = \{14, 24, 34\}$. Let $B \subseteq Z$. If $|B| > \lambda_4$, then $\mathsf{BD}_4^\lambda(B, Z \setminus B)$ is void. Otherwise, $\mathsf{BD}_4^\lambda(B, Z \setminus B)$ is isomorphic to $\{B\} * \mathsf{BD}_3^{\lambda'}$, where $\lambda_i' = 2$ if $i \leq 3 - |B|$ and $\lambda_i' = 1$ if $i > 3 - |B|$. Now, $\mathsf{BD}_3^{(2,2,2)} = 2^{\{12,13,23\}}$ and $\mathsf{BD}_3^{(2,2,1)} = 2^{\{12\}} * \{\emptyset, \{13\}, \{23\}\}$; hence these complexes are $VD(2)$ and $VD(1)$, respectively. Moreover, $\mathsf{BD}_3^{(2,1,1)}$ contains 0-cells and is hence $VD(0)$. It follows that $\mathsf{BD}_4^\lambda(B, Z \setminus B)$ is $VD(2)$ in all cases; as a consequence, BD_4^λ is $VD(2)$ as desired.

The desired result follows. $\square$

Problem 12.12. Determine the connectivity degree and the homotopical depth of BD_n^d for general d.

12.2 Bounded-Degree Graphs with Loops

We proceed with the complex $\overline{\mathsf{BD}}_n^\lambda$ of L-graphs H such that $\deg_H(i) \leq \lambda_i$ for each $i \in [n]$. Reiner and Roberts generalized Bouc's Theorem 11.5 about the homology of M_n to $\overline{\mathsf{BD}}_n^\lambda$. We confine ourselves to presenting an immediate consequence of their result:

Theorem 12.13 (Reiner and Roberts [111]). *Let notation be as in Section 2.5. For $n \geq 1$, $\overline{BD}_n^\lambda$ has homology over $\mathbb{Q}$ in dimension $d-1$ if and only if there is a self-conjugate partition $\mu \vdash |\lambda|$ such that $|D_\mu| = |\lambda| - 2d$ and such that μ dominates λ.* $\square$

Regarding the homotopy type, very little is known. Dong gave a complete characterization of all partitions λ such that $\overline{BD}_n^\lambda$ is collapsible:

Theorem 12.14 (Dong [36]). *Let $n \geq 1$ and let $\lambda_1 \geq \lambda_2 \geq \cdots \geq \lambda_n \geq 1$. The complex $\overline{BD}_n^\lambda$ is collapsible if and only if there is no self-conjugate partition $\mu \vdash |\lambda|$ such that μ dominates λ. As a consequence, $\overline{BD}_n^\lambda$ is collapsible if and only if $\overline{BD}_n^\lambda$ is $\mathbb{Q}$-acyclic.* $\square$

We refer to a partition λ as *diagonal-balanced* if

$$\sum_i \max\{\lambda_i - i, 0\} = \sum_i \min\{i - 1, \lambda_i\}.$$

This means that there are just as many elements above the diagonal as below the diagonal. Dong [36] uses the term "balanced" to denote diagonal-balanced partitions.

Theorem 12.15 (Dong [36]). *Let $n \geq 1$ and let λ be a diagonal-balanced partition. Then $\overline{BD}_n^\lambda$ is semi-collapsible and has the homotopy type of a wedge of spheres of dimension $\frac{|\lambda| + |D_\lambda|}{2} - 1$.* $\square$

There is a simple lower bound on the connectivity degree of a general complex $\overline{BD}_n^\lambda$. In most cases, this bound is far from sharp:

Proposition 12.16. *Let $n \geq 1$ and let G be an L-graph on $[n]$ such that $ii \in G$ for each $i \in [n]$. Let $\lambda = (\lambda_1, \ldots, \lambda_n)$ be such that $\lambda_i \geq 1$. Then $\overline{BD}_n^\lambda(G)$ is $AM(\frac{|\lambda|-n}{2} - 1)$.*

Proof. For an L-graph H, let H^- be the simple graph obtained by removing all loops from H. For each $I \subseteq [n]$, let $\overline{BD}_n^\lambda(G, I)$ be the subfamily of $\overline{BD}_n^\lambda(G)$ consisting of all L-graphs H such that

$$\deg_{H^-}(i) \begin{cases} \leq \lambda_i - 2 \text{ if } i \in I; \\ \geq \lambda_i - 1 \text{ if } i \in [n] \setminus I. \end{cases}$$

This partition of $\overline{BD}_n^\lambda(G)$ clearly satisfies the Cluster Lemma 4.2. For $I \neq \phi$, let $i = \min I$. We obtain a perfect matching on $\overline{BD}_n^\lambda(G, I)$ by pairing $H - ii$ and $H + ii$. Namely, by assumption, the degree of i with the loop ii excluded is at most $\lambda_i - 2$. The remaining family is $BD_n^\lambda(G, \phi)$. In an L-graph in this family, the degree of vertex i is at least $\lambda_i - 1$, which implies that the total number of edges is at least

$$\sum_{i=1}^{n} \frac{\lambda_i - 1}{2} = \frac{|\lambda| - n}{2}. \quad \square$$

Using results from the preceding section, we may obtain better bounds in certain cases:

Theorem 12.17. *Let $\Lambda(n, \mu)$ and $\alpha(n, \mu)$ be defined as in (12.1) in Section 12.1.2. If $\lambda \in \Lambda(n, \mu)$, then $\overline{\mathrm{BD}}_n^{\lambda}$ is $AM\left(\frac{|\lambda| - \alpha(n,\mu)}{2} - 1\right)$. In particular, Theorems 12.2 (for $n \geq 3$; cf. Corollary 12.4), 12.8, and 12.10 all apply to $\overline{\mathrm{BD}}_n^{\lambda}$.*

Proof. Let W be the set of loops; $W = \{ii : i \in [n]\}$. For each $A \subseteq W$, consider the family $\Sigma_A = \overline{\mathrm{BD}}_n^{\lambda}(A, W \setminus A)$. It is clear that Σ_A is isomorphic to $\{A\} * \mathrm{BD}_n^{\lambda^A}$, where

$$\lambda^A = (\lambda_1 - 2 \cdot \chi(1 \in A), \ldots, \lambda_n - 2 \cdot \chi(n \in A)).$$

Now, λ^A either contains negative elements (which yields that $\mathrm{BD}_n^{\lambda^A}$ is void) or belongs to $\Lambda(n, \mu)$. It follows that $\mathrm{BD}_n^{\lambda^A}$ is $AM\left(\frac{|\lambda^A| - \alpha(n,\mu)}{2} - 1\right)$ and hence that Σ_A is $AM\left(\frac{|\lambda| - \alpha(n,\mu)}{2} - 1\right)$; $|\lambda| = |\lambda^A| + 2|A|$. Combining appropriate acyclic matchings on the families Σ_A, we obtain an acyclic matching on $\overline{\mathrm{BD}}_n^{\lambda}$ with all unmatched faces of dimension at least $\frac{|\lambda| - \alpha(n,\mu)}{2} - 1$; this concludes the proof. $\square$

Proposition 12.18. *For $n \geq 3$, $\overline{\mathrm{BD}}_n^2$ is $VD(\lceil \frac{3n-7}{4} \rceil)$. In particular, the depth of $\overline{\mathrm{BD}}_n^2$ is at least $\lceil \frac{3n-7}{4} \rceil$.*

Proof. Apply the proof of Proposition 12.11; the only modification is that we add all loops to the edge set Y. $\square$

Table 12.4. The homology of $\overline{\mathrm{BD}}_n^2$ for $n \leq 7$.

$\tilde{H}_i(\overline{\mathrm{BD}}_n^2, \mathbb{Z})$	$i = 0$	1	2	3	4	5	6
$n = 2$	$\mathbb{Z}$	-	-	-	-	-	-
3	-	$\mathbb{Z}^2$	-	-	-	-	-
4	-	-	$\mathbb{Z}^6$	-	-	-	-
5	-	-	-	$\mathbb{Z}^{28}$	-	-	-
6	-	-	-	-	$\mathbb{Z}^{140}$	-	-
7	-	-	-	-	$\mathbb{Z}_5$	$\mathbb{Z}^{732}$	-

In Table 12.4, we provide the homology groups of $\overline{\mathrm{BD}}_n^2$ for $n \leq 7$. Note that there is 5-torsion in the homology of $\overline{\mathrm{BD}}_7^2$; this result is due to Andersen:

Theorem 12.19 (Andersen [1]). *We have that*

$$\tilde{H}_i(\mathrm{BD}_7^2; \mathbb{Z}) \cong \begin{cases} \mathbb{Z}_5 & \text{if } i = 4; \\ \mathbb{Z}^{732} & \text{if } i = 5; \\ 0 & \text{otherwise.} \end{cases}$$

Table 12.4 and Theorem 12.2 suggest the following conjecture:

Conjecture 12.20. *The shifted connectivity degree of $\overline{\mathrm{BD}}_n^2$ equals $\lceil \frac{7n-13}{9} \rceil$.*

Problem 12.21. Determine the connectivity degree and the homotopical depth of $\overline{\mathrm{BD}}_n^d$ for general d.

For $k \in [n]$, define $\overline{\mathrm{PBD}}_n^{\lambda;k}$ ("Partially Bounded Degree") to be the complex of L-graphs H with the property that there is a vertex set I of size at least k such that $\deg_H v \leq \lambda_v$ for all $v \in I$. We define $\mathrm{PBD}_n^{\lambda;k}$ analogously in terms of simple graphs; thus $\mathrm{PBD}_n^{\lambda;k}$ is the induced subcomplex of $\overline{\mathrm{PBD}}_n^{\lambda;k}$ on the set $\binom{[n]}{2}$. Clearly, $\overline{\mathrm{PBD}}_n^{\lambda;n} = \overline{\mathrm{BD}}_n^\lambda$. Moreover, $\overline{\mathrm{PBD}}_n^{\lambda;1}$ is the Alexander dual of $\overline{\mathrm{BD}}_n^{n^{n-2}-\lambda}$ with respect to the set $\binom{[n]}{1} \cup \binom{[n]}{2}$ and $\mathrm{PBD}_n^{\lambda;1}$ is the Alexander dual of $\mathrm{BD}_n^{(n-2)^n-\lambda}$ with respect to the set $\binom{[n]}{2}$; $k^n - \lambda = (k - \lambda_1, \ldots, k - \lambda_n)$.

Theorem 12.22 (Dong [36]). *Let $n \geq \lambda_1 \geq \lambda_2 \geq \cdots \geq \lambda_n \geq 0$ and $k \in [n]$. For a simplicial complex Σ and an integer p, let $\Sigma^{(p)}$ denote the p-skeleton of Σ. Then $\overline{\mathrm{PBD}}_n^{\lambda;k} \simeq \left(2^{[n]}\right)^{(n-k-1)} * \overline{\mathrm{BD}}_n^\lambda$. In particular,*

$$\tilde{H}_i(\overline{\mathrm{PBD}}_n^{\lambda;k}; \mathbb{Z}) \cong \bigoplus_{\binom{n-1}{k-1}} \tilde{H}_{i+k-n}(\overline{\mathrm{BD}}_n^\lambda; \mathbb{Z})$$

for all i.

Remark. If $k = 1$, then $\left(2^{[n]}\right)^{(n-k-1)}$ is an $(n-2)$-sphere.

Proof. Write $L_n = \binom{[n]}{1}$. First, note that we may collapse $\overline{\mathrm{PBD}}_n^{\lambda;k}$ to $\overline{\mathrm{PBD}}_n^{\lambda;k} \cap (2^{L_n} * \mathrm{BD}_n^\lambda)$; for a given L-graph H in the difference, match with the smallest ii such that $\deg_{H^-}(i) > \lambda_i$. Moreover, the subcomplex $\overline{\mathrm{PBD}}_n^{\lambda;k} \cap \left(2^{L_n} * (\mathrm{BD}_n^\lambda \cap \mathrm{PBD}_n^{\lambda-2^n;1})\right)$ is collapsible; given an L-graph H in the subcomplex, match with the smallest ii such that $\deg_{H^-}(i) \leq \lambda_i - 2$. By the Contractible Subcomplex Lemma 3.16, the conclusion is that $\overline{\mathrm{PBD}}_n^{\lambda;k}$ is homotopy equivalent to the quotient complex

$$\frac{\overline{\mathrm{PBD}}_n^{\lambda;k} \cap (2^{L_n} * \mathrm{BD}_n^{\lambda})}{\overline{\mathrm{PBD}}_n^{\lambda;k} \cap \left(2^{L_n} * (\mathrm{BD}_n^{\lambda} \cap \mathrm{PBD}_n^{\lambda-2^n;1})\right)} = \left(2^{L_n}\right)^{(n-k-1)} * \Gamma_n^{\lambda},$$

where Γ_n^{λ} is the quotient complex of all simple graphs G such that $\deg_G(i) \in \{\lambda_i - 1, \lambda_i\}$ for all $i \in [n]$. Inserting $k = n$, we obtain that $\overline{\mathrm{BD}}_n^{\lambda} \simeq \Gamma_n^{\lambda}$, which concludes the proof of the first statement in the theorem.

The second statement in the theorem is an immediate consequence of Corollary 4.23 and the fact that $\left(2^{[n]}\right)^{(n-k-1)}$ is homotopy equivalent to a wedge of $\binom{n-1}{k-1}$ spheres of dimension $n - k - 1$. $\square$

12.3 Euler Characteristic

We discuss the Euler characteristic of BD_n^d for $d \leq 2$ and $d \geq n - 3$ and the Euler characteristic of $\overline{\mathrm{BD}}_n^d$ for $d \leq 2$ and $d \geq n - 2$. For intermediate values of d, we do not have any results of interest to present.

For $d = 1$, we obtain M_n; in this case, the exponential generating function equals $1 - e^{x - x^2/2}$ by Proposition 11.29. For $d = 2$, the situation is as follows:

Theorem 12.23 (Babson et al. [3]). *We have that*

$$\sum_{n \geq 1} \tilde{\chi}(\mathrm{BD}_n^2) \frac{x^n}{n!} = 1 - \frac{\exp(\frac{x}{2+2x} + x - \frac{x^2}{4})}{\sqrt{1+x}};$$

$$\sum_{n \geq 1} \tilde{\chi}(\overline{\mathrm{BD}}_n^2) \frac{x^n}{n!} = 1 - \frac{\exp(\frac{x}{2+2x} - \frac{x^2}{4})}{\sqrt{1+x}}.$$

Proof. Let Δ_n be the family of connected graphs in BD_n^2. It is clear that $\tilde{\chi}(\Delta_1) = -1$ and $\tilde{\chi}(\Delta_2) = 1$. For $n \geq 3$, Δ_n contains all $(n-1)!/2$ Hamiltonian cycles and all $n!/2$ Hamiltonian paths. As a consequence,

$$\tilde{\chi}(\Delta_n) = (-1)^n \left(n!/2 - (n-1)!/2 \right).$$

We obtain that

$$\sum_{n \geq 1} \chi(\Delta_n) \frac{x^n}{n!} = -x + \frac{x^2}{2} + \sum_{n \geq 3} (-1)^n \left(\frac{1}{2} - \frac{1}{2n} \right) x^n$$

$$= -x + \frac{x^2}{4} + \sum_{n \geq 1} (-x)^n \left(\frac{1}{2} - \frac{1}{2n} \right)$$

$$= -x + \frac{x^2}{4} - \frac{x}{2 + 2x} + \frac{\ln(1+x)}{2}.$$

The desired result for BD_n^2 now follows immediately from Corollary 6.15. For $\overline{\mathsf{BD}}_n^2$, the families Δ_n are the same as for BD_n^2, except that Δ_1 now contains not only the empty graph but also the L-graph with edge set $\{11\}$. Hence $\tilde{\chi}(\Delta_1)$ equals 0 instead of -1, which yields the desired result. $\square$

We now proceed with large values of d, starting with BD_n^d and postponing $\overline{\mathsf{BD}}_n^d$ until later. As it turns out, it is easier to analyze the Alexander dual $\mathsf{PBD}_n^{r;1} = (\mathsf{BD}_n^{n-2-r})^*$. For $r = 0$, we obtain the complex SSC_n^1 of graphs with at least one isolated vertex. We analyze this complex (which is very simple) in Corollary 18.20. For $r = 1$, the complex under consideration is the complex in which at least one vertex has degree at most one.

Theorem 12.24. *We have that*

$$\sum_{n \geq 1} \tilde{\chi}(\mathsf{PBD}_n^{1;1}) \frac{x^n}{n!} = -1 - x + e^{-x^2/2}\left(x + e^{-x}\right).$$

Proof. Let Σ_n be the quotient complex $2^{K_n}/\mathsf{PBD}_n^{1;1}$. Thus Σ_n contains all graphs in which each vertex has degree at least two. Write $a_n = \tilde{\chi}(\Sigma_n)$. It is clear that $a_1 = a_2 = 0$. We want to prove that

$$\sum_{n \geq 1} a_n \frac{x^n}{n!} = 1 - e^{-x^2/2}\left(x + e^{-x}\right).$$

Assume that $n \geq 3$. For any set $S \subseteq [n-1]$, let Σ_n^S be the subfamily of Σ_n consisting of all graphs G such that S is the set of vertices with degree one in the induced subgraph $G([n-1])$; all other vertices have degree at least two. Note that any $G \in \Sigma_n^S$ must have the property that $sn \in G$ for all $s \in S$; the degree of s in G is at least two.

First, consider $S = \emptyset$. For any given graph $H \in \Sigma_{n-1}$, a graph G such that $G([n-1]) = H$ belongs to $\Sigma_n^\emptyset$ if and only if the degree of n in G is at least 2. Thus

$$\tilde{\chi}(\Sigma_n^\emptyset) = (n-2)\tilde{\chi}(\Sigma_{n-1}) = (n-2)a_{n-1}. \tag{12.3}$$

Next, consider the case that $2 \leq |S| \leq n-2$. Since n is adjacent in G to all vertices in S whenever $G \in \Sigma_n^S$, we obtain a perfect matching on Σ_n^S by pairing $G - xn$ and $G + xn$ for any $x \in [n-1] \setminus S$. In particular,

$$\tilde{\chi}(\Sigma_n^S) = 0. \tag{12.4}$$

The third case is that $S = [n-1]$. If $G \in \Sigma_n^{[n-1]}$, then all vertices in $G([n-1])$ have degree one, which means that $G([n-1])$ constitutes a perfect matching. Moreover, all edges sn such that $s \in [n-1]$ belong to G. It follows that

$$\tilde{\chi}(\Sigma_n^{[n-1]}) = \begin{cases} 0 & \text{if } n \text{ is even;} \\ (-1)^{k-1}\frac{(2k)!}{k!2^k} & \text{if } n = 2k+1 \text{ and } k \geq 1. \end{cases} \tag{12.5}$$

The remaining case is that $|S| = 1$. For any vertices $p, q \in [n]$, let $\Gamma_n^{p,q}$ be the family of graphs G such that q is the only neighbor of p and such that the degree of all vertices but p is at least two. Define $b_n = \tilde{\chi}(\Gamma_n^{p,q})$; this definition does not depend on the choice of p and q. Write $\Gamma_n^p = \bigcup_{q \in [n] \setminus \{p\}} \Gamma_n^{p,q}$; by symmetry, $\tilde{\chi}(\Gamma_n^p) = (n-1)b_n$. It is clear that G belongs to $\Sigma_n^{\{p\}}$ if and only if $G([n-1])$ belongs to Γ_{n-1}^p and n is adjacent to p and at least one other vertex. As a consequence,

$$\tilde{\chi}(\Sigma_n^{\{p\}}) = \sum_{q \in [n-1] \setminus \{p\}} \tilde{\chi}(\Gamma_{n-1}^{p,q}) = \tilde{\chi}(\Gamma_{n-1}^p) = (n-2)b_n. \tag{12.6}$$

Summing over all sets S and using (12.3)-(12.6), we obtain that

$$a_n = (n-2)a_{n-1} + \tilde{\chi}(\Sigma_n^{[n-1]}) + (n-1)(n-2)b_{n-1}.$$

Thus

$$A'(x) = xA'(x) - A(x) + 1 - e^{-x^2/2} + B''(x)x^2, \tag{12.7}$$

where $A(x) = \sum_{n \geq 1} a_n x^n / n!$ and $B(x) = \sum_{n \geq 1} b_n x^n / n!$. Note that $e^{x^2/2}$ is the exponential generating function for the number of perfect matchings on n vertices.

Now, we may write $\Gamma_n^{n,q}$ as the disjoint union of $\{nq\} * \Sigma_{n-1}$ and $\{nq\} * \Gamma_{n-1}^q$. As a consequence, $b_n = -a_{n-1} - (n-2)b_{n-1}$, which implies that

$$B''(x) = -A'(x) - B''(x)x \iff B''(x) = \frac{-A'(x)}{1+x}.$$

Substituting for $B''(x)$ in (12.7), we obtain that

$$A'(x) = xA'(x) - A(x) + 1 - e^{-x^2/2} + \frac{-A'(x)x^2}{1+x}$$

$$\iff \frac{A'(x)}{1+x} = -A(x) + 1 - e^{-x^2/2}$$

$$\iff \frac{d}{dx}\left(Ae^{x^2/2+x}\right) = (1+x)(e^{x^2/2+x} - e^x).$$

The desired result now easily follows. $\square$

Finally, we examine $\overline{\mathrm{BD}}_n^d$ for large values of d. Again, we consider the Alexander dual.

Theorem 12.25. *We have that*

$$\overline{\mathrm{PBD}}_n^{0;1} \simeq S^{n-2};$$

$$\sum_{n \geq 1} \tilde{\chi}(\overline{\mathrm{PBD}}_n^{1;1}) \frac{x^n}{n!} = e^{-x^2/2-x} - 1;$$

$$\sum_{n \geq 1} \tilde{\chi}(\overline{\mathrm{PBD}}_n^{2;1}) \frac{x^n}{n!} = \frac{\exp(\frac{x}{2x-2} - \frac{x^2}{4})}{\sqrt{1-x}} - 1.$$

Proof. By Theorem 12.22, the first statement is obvious, because $\overline{BD}_n^0 = \{\emptyset\}$. The same theorem yields the two other statements; use Proposition 11.29 and Theorem 12.23. $\square$

Part V

Cycles and Crossings

13

Forests and Matroids

Let M be a matroid on a finite set E with rank function ρ. The independence complex $\mathsf{F}(M)$ is well-known to have attractive topological properties. More precisely, $\mathsf{F}(M)$ is known to be shellable – in fact even vertex-decomposable – of dimension $\rho(E) - 1$; see Provan and Billera [108, 107] and Björner [8]. We give a summary in Section 13.1.

One of the main goals of this chapter is to extend this result to a larger class of simplicial complexes. Specifically, we define a complex Δ to be a *pseudo-independence* or *PI* complex over M by the property that if τ is a face of Δ and e is an element such that $\rho(\tau + e) > \rho(\tau)$, then $\tau + e$ is also a face of Δ. In Section 13.2, we show that every PI complex over M has a vertex-decomposable skeleton of dimension $\rho(E) - 1$.

What distinguishes an arbitrary PI complex from the independence complex $\mathsf{F}(M)$ is that the latter complex satisfies the following property: If τ is a face and e is an element not in τ such that $\rho(\tau + e) = \rho(\tau)$, then $\tau + e$ is *not* a face. We may relax this condition a bit, requiring that *either $\tau + e \notin \Delta$ or* the element e is a cone point in $\mathrm{lk}_\Delta(\tau)$. We refer to a PI complex with this property as a *strong* PI or *SPI* complex. As we will see in Section 13.3, such a complex has the homotopy type of a wedge of spheres in dimension $\rho(E) - 1$. As a consequence, from a homotopical point of view, SPI complexes are very similar to the independence complex.

An attractive property of the families of PI and SPI complexes is that they are closed under deletion and contraction of an element in the matroid; the latter operation corresponds to taking the link of the complex with respect to the contracted element. Indeed, this is the crucial property that allows us to derive the mentioned topological results.

We also consider PI* and SPI* complexes, which are Alexander duals of PI and SPI complexes, respectively. In terms of the dual M^* (with rank function ρ^*) of the underlying matroid M, a PI* complex has the property that each maximal face τ is a flat; the rank of $\tau + x$ exceeds the rank of τ for each x not in τ. As a fairly immediate consequence, a PI* complex is homotopy equivalent to the order complex of a certain order ideal in the lattice of flats in M^*. In particular, all homology is concentrated in dimensions below $\rho^*(E) - 1$. An SPI* complex Δ satisfies the additional condition that if τ is a nonface of Δ and e is an element such that $\rho^*(\tau + e) > \rho^*(\tau)$, then e is a cone point in the induced subcomplex of Δ on the set $\tau + e$. In Section 13.4, we show that every SPI* complex has a vertex-decomposable $(\rho^*(E) - 2)$-skeleton. Note that this is not an immediate consequence of the corresponding result for SPI complexes.

The independence complex over the graphic matroid $M_n = M_n(K_n)$ coincides with the complex F_n of forests on n vertices. Chari [31] and later Linusson and Shareshian [94] examined the larger complex B_n of bipartite graphs on n vertices. It is easy to see that B_n is an SPI complex, which implies the result of Chari [31] that B_n is homotopy equivalent to a wedge of spheres of dimension $n - 2$; see Section 14.1 for more information about B_n.

A graph is bipartite if and only if the graph does not contain an odd cycle. More generally, for a matroid M on a set E, one may consider the complex B_M of subsets of E that do not contain any circuits of odd cardinality. In Section 13.3.1, we show that B_M is SPI whenever M admits a representation over the field with two elements. We generalize this observation to representations over arbitrary fields, but the resulting complex is no longer equal to B_M in general.

The generic example of an SPI* complex is the complex NC_n of disconnected graphs on n vertices. For each p dividing n, another SPI* complex is the complex of graphs such that the number of vertices in some connected component is not divisible by p; see Section 13.4.1.

13.1 Independence Complexes

We give a brief overview of some basic properties of the independence complex of a matroid. From our perspective, the most important fact is the following result, which implies that independence complexes are shellable.

Theorem 13.1 (Provan and Billera [107, 108]). *Let M be a matroid of rank r. Then $\mathsf{F}(M)$ is VD and hence homotopy equivalent to a wedge of spheres of dimension $r - 1$.* $\square$

For the graphic matroid and its one-step truncation (see Section 2.4.1), we have the following well-known consequences:

Corollary 13.2. *For any graph G on n vertices, the following hold:*

- $\mathsf{F}_n(G)$ *is VD of dimension $n - c(G) - 1$. More generally, for any graph $H \subseteq G$, $\mathrm{lk}_{\mathsf{F}_n(G)}(H)$ is VD of dimension $c(H) - c(G) - 1$.*
- $\mathsf{F}_n(G) \cap \mathsf{NC}_n$ *is VD of dimension $n - \max\{c(G), 2\} - 1$.*

Table 13.1. The reduced Euler characteristic of F_n for small values on n.

n	1	2	3	4	5	6	7	8	9	10
$\tilde{\chi}(\mathsf{F}_n)$	-1	0	-1	6	-51	560	-7575	122052	-2285353	48803904

For each positive integer t, one may easily extend the second statement in the corollary to the complex of forests with at least t connected components; replace $\max\{c(G), 2\}$ with $\max\{c(G), t\}$.

The reduced Euler characteristic of F_n is complicated and is perhaps best expressed in terms of the Hermite polynomial $H_n(t)$, which is defined by the equation

$$\sum_{n \geq 0} H_n(t) \frac{x^n}{n!} = e^{tx + x^2/2}.$$

Note that $H_{n+1}(t) = tH_n(t) + nH_{n-1}(t)$ for $n \geq 1$ and that $H_n(t) = t^n f(1/t^2)$, where $f(t)$ is the polynomial in Proposition 11.29. Equivalently, the coefficient of t^{n-2k} in $H_n(t)$ equals the number of k-matchings on n vertices.

Theorem 13.3 (Novik et al. [103, Th. 5.8]). *For $n \geq 3$, we have that* $\tilde{\chi}(\mathsf{F}_n) = (-1)^n(n-2)H_{n-3}(n-1)$. $\square$

We present $\tilde{\chi}(\mathsf{F}_n)$ for small n in Table 13.1.

The following immediate consequence of Theorem 13.1 and Proposition 5.12 might be worth mentioning.

Corollary 13.4. *Let M be a matroid of rank r. Then $\mathsf{F}(M)$ is semi-nonevasive with all evasive faces of dimension $r - 1$. Moreover, the complex $\mathsf{NC}(M)$ of all sets of rank at most $r - 1$ is semi-nonevasive with all evasive faces of dimension $r - 2$.* $\square$

For the second statement in the corollary, use Proposition 5.36 and the fact that $\mathsf{NC}(M)$ is the Alexander dual of $\mathsf{F}(M^*)$, where M^* is the dual matroid of M. In Section 13.4, we show that the $(r - 2)$-skeleton of $\mathsf{NC}(M)$ is VD.

Remark. For a connected graph G on n vertices, the complex $\mathsf{NC}(M_n(G))$ in Corollary 13.4 coincides with the complex $\mathsf{NC}_n(G)$ of disconnected subgraphs of G. See Section 18.1 for more information about this complex.

13.2 Pseudo-Independence Complexes

We introduce the concept of a pseudo-independence complex over a matroid and show that a certain skeleton of such a complex is VD.

Let $M = (E, \mathsf{F})$ be a matroid with rank function ρ_M. Say that a simplicial complex Δ of subsets of M is a *pseudo-independence complex* over M if the following holds:

- If $\tau \in \Delta$, $x \in E \setminus \tau$, and $\rho_M(\tau + x) > \rho_M(\tau)$, then $\tau + x \in \Delta$.

The rationale for this terminology is that a nonvoid pseudo-independence complex over M contains the independence complex F of M. To simplify notation, we say that a pseudo-independence complex over M is *PI over M*. By convention, we consider the void complex to be a PI complex over any matroid.

For any subset τ of E, let $\Delta(\tau)$ be the induced subcomplex of Δ on the vertex set τ. It is clear that $\Delta(\tau)$ is PI over $M(\tau)$ whenever Δ is PI over M. In particular, $\mathrm{del}_\Delta(e)$ is PI over $M - e$ for any $e \in E$. Also, $\mathrm{lk}_\Delta(e)$ is PI over M/e. Namely, $\rho_{M/e}(\tau + x) > \rho_{M/e}(\tau)$ is equivalent to

$$\rho_M(\tau + e + x) - \rho_M(e) > \rho_M(\tau + e) - \rho_M(e) \iff \rho_M(\tau + e + x) > \rho_M(\tau + e).$$

Theorem 13.5. *Let Δ be a PI simplicial complex over a matroid $M = (E, \mathsf{F})$. Then, for any subset τ of E, $\Delta(\tau)$ is $VD(\rho_M(\tau) - 1)$.*

Proof. It is clear that the $(\rho_M(\tau) - 1)$-skeleton of $\Delta(\tau)$ is pure; $\Delta(\tau)$ being PI implies that all maximal faces of $\Delta(\tau)$ have full rank $\rho_M(\tau)$. We are done if $\rho_M(\tau) = 0$. Suppose $\rho_M(\tau) > 0$, and let $e \in \tau$ be such that $\rho_M(e) = 1$. There are two cases:

- $\rho_M(\tau) = \rho_M(\tau - e) + 1$. As $\Delta(\tau)$ is PI, this implies that e is a cone point in $\Delta(\tau)$. By induction, $\mathrm{lk}_{\Delta(\tau)}(e) = \mathrm{del}_{\Delta(\tau)}(e) = \Delta(\tau - e)$ is $VD(\rho_M(\tau - e) - 1) = VD(\rho_M(\tau) - 2)$, and we are done.
- $\rho_M(\tau) = \rho_M(\tau - e)$. By induction, $\mathrm{del}_{\Delta(\tau)}(e) = \Delta(\tau - e)$ is $VD(\rho_M(\tau) - 1)$ and $\mathrm{lk}_{\Delta(\tau)}(e)$ is $VD(\rho_{M/e}(\tau) - 1) = VD(\rho_M(\tau) - 2)$, the latter complex being PI over $M(\tau)/e$. By Lemma 6.9, this implies that $\Delta(\tau)$ is $VD(\rho_M(\tau) - 1)$, and we are done. $\square$

A set S in a matroid M is *isthmus-free* if no element in the induced submatroid $M(S)$ is an isthmus; $\rho_M(S) = \rho_M(S - e)$ for all $e \in S$.

Theorem 13.6. *Let $M = (E, \mathsf{F})$ be a matroid and let $\Delta \subseteq 2^E$ be a simplicial complex. Then Δ is PI if and only if all minimal nonfaces of Δ are isthmus-free. In particular, such a complex has the property that $\Delta(\tau)$ is $VD(\rho(\tau) - 1)$ for each $\tau \subseteq E$.*

Proof. Suppose that all minimal nonfaces of Δ are isthmus-free. We want to show that Δ is PI. Suppose $\tau \in \Delta$ and $e \in E \setminus \tau$ have the property that $\rho_M(\tau + e) = \rho_M(\tau) + 1$. We have to show that $\tau + e \in \Delta$. Assume the opposite and let σ be a minimal nonface of Δ contained in $\tau + e$. Since e is an isthmus in $\tau + e$, e cannot be contained in σ; σ is isthmus-free by assumption. However, this means that σ is contained in τ, and we have a contradiction to the assumption that $\tau \in \Delta$. Thus Δ is PI.

Next, suppose that some minimal nonface σ is not isthmus-free. Let e be an isthmus in σ; since σ is a minimal nonface, $\sigma - e \in \Delta$. However, $\rho(\sigma) = \rho(\sigma - e) + 1$, which implies that Δ is not PI. $\square$

Proposition 13.7. *Let $M = (E, \mathsf{F})$ be a matroid and assume that Σ and Δ are PI complexes over M. Then the intersection $\Sigma \cap \Delta$ and the union $\Sigma \cup \Delta$ are also PI over M. As a consequence, the set of PI complexes over M ordered by inclusion is a lattice with union and intersection as join and meet operators, respectively.*

Proof. Let σ be a minimal nonface of $\Sigma \cap \Delta$; assume that $\sigma \notin \Delta$. Then σ is a minimal nonface of Δ and hence isthmus-free. By Theorem 13.6, this implies that $\Sigma \cap \Delta$ is a PI complex.

Now, let σ be a minimal nonface of $\Sigma \cup \Delta$. Suppose that σ is not isthmus-free and let e be an isthmus in σ. Since $\sigma - e$ is not a minimal nonface of $\Sigma \cup \Delta$, $\sigma - e$ belongs to at least one of Σ and Δ, say Δ. However, since Δ is PI, this implies that σ belongs to Δ; $\rho(\sigma - e) < \rho(\sigma)$. This a contradiction; thus σ must be isthmus-free and $\Sigma \cup \Delta$ is hence a PI complex by Theorem 13.6. $\square$

13.2.1 PI Graph Complexes

We consider the special case of graphic matroids. A graph (or digraph) G is *isthmus-free* if each connected component in G is 2-edge-connected. Equivalently, $c(G) = c(G - e)$ for each edge $e \in G$. This means exactly that the graphic (or digraphic) matroid on G with rank function $\rho(H) = n - c(H)$ is isthmus-free. The following result is an immediate consequence of Theorem 13.6.

Corollary 13.8. *Let Δ be a complex of graphs on n vertices such that all minimal nonfaces are isthmus-free, let G be a graph on n vertices, and let $H \in \Delta(G)$. Then $\mathrm{lk}_{\Delta(G)}(H)$ is $VD(c(H) - c(G) - 1)$. In particular, $\Delta(G)$ is $VD(n - c(G) - 1)$. The analogous property holds whenever Δ is a complex of digraphs on n vertices such that all minimal nonfaces are isthmus-free.* $\square$

While the depth and the shifted connectivity degree of many PI complexes over M_n are way above the bound $n - 2$ given in Corollary 13.8, the following theorem shows that there are also plenty of complexes for which the bound is tight:

Theorem 13.9. *Let $n \geq 3$ and let Δ be a nonvoid complex of graphs on n vertices. If Δ is PI over M_n and Δ does not contain any triangles $\{ab, ac, bc\}$, then the shifted connectivity degree and homotopical depth of Δ is $n - 2$.*

Proof. Δ being nonvoid and PI implies that Δ contains the complex F_n of forests. Let G_n be the graph with edge set $\{in : i \in [n - 1]\}$. This graph is a maximal face of Δ. Namely, G_n is a spanning tree, and $G_n + ij$ is not in Δ for any $i, j \in [n-1]$; $G_n + ij$ contains the triangle $\{ij, in, jn\}$. Thus it suffices to prove that there is a cycle in the chain group $\tilde{C}_{n-2}(\Delta, \mathbb{Z})$ such that the coefficient of $[G_n]$ is nonzero. Since Δ contains F_n, we may instead consider $\tilde{C}_{n-2}(\mathsf{F}_n, \mathbb{Z})$.

To obtain the desired property, we define an optimal decision tree on F_n such that G_n is evasive. Since all evasive faces have the same dimension, Corollary 4.17 will then yield that $[G_n]$ is indeed contained in a cycle. First, decompose with respect to $1n$. By Corollary 13.2, $\Delta(\emptyset, 1n) \sim ct^{n-2}$, where $c \geq 0$. Decompose $\Delta(1n, \emptyset)$ with respect to the set Y, where $Y = \{in : i \in [2, n-1]\}$. For each $A \subseteq Y$, we have that $\Delta(A + 1n, Y \setminus A) \sim c_A t^{n-2}$, where $c_A \geq 0$, again by Corollary 13.2. This yields a decision tree with desired properties. Namely, G_n is evasive, as there are no other members of the family $\Delta(Y + 1n, \emptyset)$. $\square$

13.3 Strong Pseudo-Independence Complexes

We say that a PI complex Δ over a matroid $M = (E, \mathsf{F})$ is a *strong* PI complex if, for each $\sigma \in \Delta$ and each element $e \in E \setminus \sigma$ such that $\rho(\sigma + e) = \rho(\sigma)$, the element e is either a cone point in $\mathrm{lk}_\Delta(\sigma)$ or not contained at all in $\mathrm{lk}_\Delta(\sigma)$. We use the abbreviation SPI to denote a strong PI complex.

For any subset $\tau \subseteq E$, it is clear that $\Delta(\tau)$ is SPI over $M(\tau)$ whenever Δ is SPI over M. In particular, $\mathrm{del}_\Delta(e)$ is SPI over $M - e$ for any $e \in E$. Similarly, it is clear that $\mathrm{lk}_\Delta(e)$ is SPI over M/e; $\rho_{M/e}(\tau + x) = \rho_{M/e}(\tau)$ if and only if $\rho_M(\tau + e + x) = \rho_M(\tau + e)$.

Theorem 13.10. *Let Δ be an SPI complex over a matroid $M = (E, \mathsf{F})$. Then, for any subset τ of E, $\Delta(\tau)$ is $VD^+(\rho(\tau)-1)$. In particular, $\Delta(\tau)$ is homotopy equivalent to a wedge of spheres of dimension $\rho(\tau) - 1$.*

Proof. By Theorem 13.5, we already know that $\Delta(\tau)$ is $VD(\rho(\tau) - 1)$. Hence it suffices to prove that $\Delta(\tau) \sim ct^{\rho(\tau)-1}$ for some integer c; recall that this means that there is a decision tree on $\Delta(\tau)$ with c evasive sets of dimension $\rho(\tau) - 1$.

For any set $\sigma \subseteq \tau$, we will show that

$$\mathrm{lk}_{\Delta(\tau)}(\sigma) \sim ct^{\rho(\tau)-\rho(\sigma)-1}$$

for some integer c. If $\rho(\tau) = \rho(\sigma)$, then all elements in the link are cone points; Δ is SPI. Hence $\mathrm{lk}_{\Delta(\tau)}(\sigma)$ is either the (-1)-simplex or nonevasive as desired. Suppose $\rho(\tau) > \rho(\sigma)$, and let $e \in \tau \setminus \sigma$ be such that $\rho(\sigma + e) = \rho(\sigma) + 1$. There are two cases:

- $\rho(\tau) = \rho(\tau - e) + 1$. As Δ is PI, this implies that e is a cone point in $\mathrm{lk}_{\Delta(\tau)}(\sigma)$, and we are done.
- $\rho(\tau) = \rho(\tau - e)$. Let $\Sigma = \mathrm{lk}_{\Delta(\tau)}(\sigma)$. By induction, there are integers $c_\emptyset$ and c_e such that

$$\mathrm{del}_\Sigma(e) = \mathrm{lk}_{\Delta(\tau-e)}(\sigma) \sim c_\emptyset t^{\rho(\tau-e)-\rho(\sigma)-1} = c_\emptyset t^{\rho(\tau)-\rho(\sigma)-1};$$

$$\mathrm{lk}_\Sigma(e) = \mathrm{lk}_{\Delta(\tau)}(\sigma + e) \sim c_e t^{\rho(\tau)-\rho(\sigma+e)-1} = c_e t^{\rho(\tau)-\rho(\sigma)-2}.$$

Lemma 5.22 implies that $\Sigma \sim (c_\emptyset + c_e) t^{\rho(\tau)-\rho(\sigma)-1}$ as desired.

The final statement is a consequence of Lemma 5.22. □

We have no general formula for the Euler characteristic of an SPI complex. However, using the following lemma, it is possible to give a simple criterion for when an SPI complex is nonevasive.

Lemma 13.11. *Let Δ be a simplicial complex with at least one 0-cell. If Δ is not a cone, then there is a 0-cell x such that $\mathrm{del}_\Delta(x)$ and $\mathrm{lk}_\Delta(x)$ are not both cones.*

Remark. Note that there is nothing to prove if Δ has only one 0-cell.

Proof. For $x, y \in \Delta$, define $x \sim y$ if, for each maximal face σ, $x \in \sigma$ if and only if $y \in \sigma$. Let $E_1, \ldots, E_r$ be the equivalence classes with respect to $\sim$. Define a partial order on $\{E_1, \ldots, E_r\}$ by $E_i \leq E_j$ if and only if every maximal face containing E_i also contains E_j. Let E_j be maximal with respect to this partial order.

First, assume that E_j contains one single element x. Then $\mathrm{lk}_\Delta(x)$ is not a cone by construction; for each $y \neq x$, there is some maximal face σ of Δ containing x but not y.

Second, assume that E_j contains at least two elements; let x and y be distinct elements in E_j. We claim that $\mathrm{del}_\Delta(x)$ is not a cone. Namely, suppose that z is a cone point in $\mathrm{del}_\Delta(x)$. If $z \in E_j$, then z is a cone point in $\mathrm{lk}_\Delta(x)$ and hence in Δ, a contradiction. Thus assume that $z \notin E_j$. Let σ be a maximal face of Δ. If $x \notin \sigma$, then $z \in \sigma$ by assumption. If $x \in \sigma$, then $(\sigma - x) + z$ is a face of Δ; z is a cone point in $\mathrm{del}_\Delta(x)$. Now, $y \in \sigma$, because y is a cone point in $\mathrm{lk}_\Delta(x)$. Since x is a cone point in $\mathrm{lk}_\Delta(y)$, we deduce that $((\sigma - x) + z) + x = \sigma + z$ is a face of Δ. However, σ is maximal, which implies that $z \in \sigma$. Since σ was arbitrary, z is a cone point in Δ, another contradiction. □

Theorem 13.12. *Let Δ be an SPI complex over a matroid $M = (E, \mathsf{F})$. Then Δ is nonevasive if and only if Δ is a cone. In particular, the homology of Δ is nonzero whenever Δ is not a cone.*

Proof. By Theorem 13.10, we have that $\Delta \sim ct^{\rho(E)-1}$ for some $c \geq 0$. If Δ is a cone, then clearly $c = 0$, which means that Δ is nonevasive. Suppose that Δ is not a cone. By Lemma 13.11, there is an element e such that either $\mathrm{lk}_\Delta(e)$ or $\mathrm{del}_\Delta(e)$ is not a cone. Now, M is isthmus-free, because any isthmus would be a cone point in Δ. In particular, $\rho(E - e) = \rho(E)$. By the proof of Theorem 13.10, there are integers $c_\emptyset$ and c_e such that $c = c_\emptyset + c_e$ and such that $\mathrm{del}_\Delta(e) \sim c_\emptyset t^{\rho(E)-1}$ and $\mathrm{lk}_\Delta(e) \sim c_e t^{\rho(E)-2}$. By induction on the size of Δ, either $\mathrm{lk}_\Delta(e)$ or $\mathrm{del}_\Delta(e)$ is evasive, which implies that $c = c_\emptyset + c_e > 0$; hence Δ is evasive. □

Theorem 13.13. *Let $M = (E, \mathsf{F})$ be a matroid and let Δ and Σ be simplicial complexes on E. Assume that Δ is VD and that each maximal face of Δ has full rank $\rho(E)$. If Σ is PI, then $\Delta \cap \Sigma$ is $VD(\rho(E) - 1)$. Assume in addition* *that Δ admits a VD-shelling (see Section 6.3) in which each minimal face in the shelling belongs to* F. *If Σ is SPI, then $\Delta \cap \Sigma$ is $VD^+(\rho(E) - 1)$.*

Proof. By Lemma 6.12, it suffices to prove that the desired property holds for $(\Delta \cap \Sigma)(\sigma, E \setminus \tau)$ for each shelling pair (σ, τ). Since $\tau \in \Delta$, we have that

$$(\Delta \cap \Sigma)(\sigma, E \setminus \tau) = \Sigma(\sigma, E \setminus \tau).$$

If Σ is PI over M, then $\Sigma(\tau)$ is $VD(\rho(\tau) - 1)$ over the induced submatroid $M(\tau)$; use Theorem 13.5. By assumption, $\rho(\tau) = \rho(E)$, which implies that $\Sigma(\sigma, E \setminus \tau)$ is $VD(\rho(E) - 1)$; use Theorem 3.30.

For the second claim, if Σ is SPI over M, then $\Sigma(\tau)$ is SPI over $M(\tau)$. Since $\rho(\sigma) = |\sigma|$ and $\rho(\tau) = \rho(E)$, Theorem 13.10 (or rather its proof) yields that $\mathrm{lk}_{\Sigma(\tau)}(\sigma) \sim ct^{\rho(E) - |\sigma| - 1}$ for some c. This is equivalent to $\Sigma(\sigma, E \setminus \tau) \sim ct^{\rho(E)-1}$, which concludes the proof. $\square$

For a matroid $M = (E, \mathsf{F})$, let $\mathcal{B}(M)$ be the family of bases (maximal independent sets) in M. Let Δ be a nonvoid SPI complex over M. By construction, each basis σ is contained in a unique maximal face of Δ; each element not in σ is either a cone point or not present in $\mathrm{lk}_\Delta(\sigma)$. Among PI complexes, this property characterizes SPI complexes:

Theorem 13.14. *Let $M = (E, \mathsf{F})$ be a matroid and let Δ be a nonvoid PI complex over M. Then Δ is SPI over M if and only if every basis is contained in a unique maximal face of Δ.*

Proof. By the above discussion, it suffices to prove that Δ is SPI whenever every basis is contained in a unique maximal face. Thus assume that the latter holds. It is clear that every face of Δ of maximal rank $\rho(E)$ is contained in a unique maximal face.

Let σ be a face of Δ and assume that x is an element such that $\rho(\sigma + x) = \rho(\sigma)$ and $\sigma + x \in \Delta$. We need to prove that $\sigma' + x \in \Delta$ whenever $\sigma' \in \Delta$ and $\sigma \subseteq \sigma'$. Use induction in decreasing order on the rank of σ. If $\rho(\sigma) = \rho(E)$, then we are done, because σ is contained in a unique maximal face.

Suppose that $\rho(\sigma) < \rho(E)$. First, assume that $\sigma' \setminus \sigma$ contains an element y such that $\rho(\sigma + y) > \rho(\sigma)$. Since Δ is PI, $(\sigma + x) + y \in \Delta$. Thus by induction, $\sigma' + x$ belongs to Δ. Next, assume that $\rho(\sigma') = \rho(\sigma)$. Pick an element y from E such that $\rho(\sigma + y) > \rho(\sigma)$. Since Δ is PI, $\sigma + y$, $\sigma \cup \{x, y\}$, and $\sigma' + y$ all belong to Δ. Thus by induction, $\sigma' \cup \{x, y\}$ belongs to Δ, and we are done. $\square$

The following simple observation is often useful.

Proposition 13.15. *Let $M = (E, \mathsf{F})$ be a matroid and let Δ be a nonvoid SPI complex over M. Let σ be a circuit of M contained in Δ and let x be an element in σ. Then x is a cone point in $\mathrm{lk}_\Delta(\sigma - x)$. Equivalently, whenever $\tau \subseteq E$ is a set containing σ, we have that $\tau - x$ belongs to Δ if and only if τ belongs to Δ.*

Proof. We have that $\rho(\sigma - x) = \rho(\sigma)$, because σ is a circuit. Since Δ is SPI and since $\sigma \in \Delta$, the desired claim follows. $\square$

In Theorem 13.6, we proved that all minimal nonfaces of a PI complex are isthmus-free. The minimal nonfaces of an SPI complex have an even more specific structure:

Proposition 13.16. *Let $M = (E, \mathsf{F})$ be a matroid and let $\Delta \subseteq 2^E$ be a nonvoid SPI complex. Then every minimal nonface of Δ is a circuit of M.*

Proof. Let τ be a minimal nonface of Δ and let σ be a circuit contained in τ. Pick some $x \in \sigma$. By the minimality assumption on τ, we have that $\tau - x \in \Delta$. As a consequence, $\sigma \notin \Delta$ by Proposition 13.15. Since τ is a minimal nonface, the only possibility is that $\sigma = \tau$, which concludes the proof. $\square$

Remark. A complex is not necessarily SPI just because all minimal nonfaces are circuits. For example, the complex of triangle-free graphs is not SPI; see Theorem 13.24.

We say that two SPI complexes Δ and Γ are *nearly identical*, written as $\Delta \approx \Gamma$, if all faces of $\Delta \setminus \Gamma$ and $\Gamma \setminus \Delta$ are circuits. This is clearly an equivalence relation. Note that each circuit in the difference is necessarily a maximal face of one of the complexes and hence has full rank $\rho(E)$ and size $\rho(E) + 1$. Namely, proper subsets of circuits are not circuits. Identifying nearly identical complexes is a way of simplifying classification problems; see Section 13.3.2.

Lemma 13.17. *Let $M = (E, \mathsf{F})$ be a matroid and let Δ be an SPI complex over M. Assume that there exists a maximal face σ of Δ such that σ is a circuit of M. Then $\Delta \setminus \{\sigma\}$ is also an SPI complex over M.*

Proof. $\Delta \setminus \{\sigma\}$ is obviously PI. To prove that $\Delta \setminus \{\sigma\}$ is SPI, it suffices to prove that each basis is contained in a unique maximal face; use Theorem 13.14. Since Δ is SPI, the only bases we need to check are the ones contained in σ. However, they are themselves maximal faces of $\Delta \setminus \{\sigma\}$, which concludes the proof. $\square$

Let M be a matroid with rank function ρ and let Δ be a nonvoid SPI complex over M. As we have already indicated, for each face $\sigma \in \Delta$, there is a unique face $f(\sigma)$ such that $f(\sigma)$ is maximal among all faces τ of Δ containing σ and satisfying $\rho(\tau) = \rho(\sigma)$. We claim that f defines a poset map and hence a closure operator on the face poset $P(\Delta)$ of Δ. Namely, let τ be a face of Δ containing σ. By construction, every element in $f(\sigma) \setminus \sigma$ is a cone point in $\mathrm{lk}_\Delta(\sigma)$. As a consequence, every element in $f(\sigma) \setminus \tau$ is a cone point in $\mathrm{lk}_\Delta(\tau)$, which implies that $f(\tau)$ contains $f(\sigma)$ as desired.

Theorem 13.18. *Let $M = (E, \mathsf{F})$ be a matroid and let Δ be a nonvoid SPI complex over M. Let f be defined as above. Then $Q = f(P(\Delta))$ coincides* *with the proper part of a lattice with meet operation being set intersection. Moreover, Q is Cohen-Macaulay of dimension $\rho(E) - 1$.*

Proof. Let σ and τ be elements in Q such that $\sigma \cap \tau \neq \emptyset$. We have that $f(\sigma \cap \tau) = \sigma \cap \tau$, because $f(\sigma \cap \tau)$ is contained in each of σ and τ and contains $\sigma \cap \tau$. Hence $\sigma \cap \tau \in Q$, which shows that Q is the proper part of a lattice with meet operation being intersection.

It remains to prove that Q is Cohen-Macaulay. It suffices to prove that the order complex of each interval $Q_{>\sigma,<\tau} = \{\pi \in Q : \sigma \subsetneq \pi \subsetneq \tau\}$ is $(\rho(\tau) - \rho(\sigma) - 3)$-connected and that the order complex of each interval $Q_{>\sigma} = \{\pi \in Q : \sigma \subsetneq \pi\}$ (including $Q = Q_{>\{\emptyset\}}$) is $(\rho(E) - \rho(\sigma) - 2)$-connected. Namely, each link in the order complex of Q is a join of such complexes, and if Γ_i is $(d_i - 1)$-connected for $i \in [s]$, then $\Gamma_1 * \cdots * \Gamma_s$ is $(d_1 + \ldots + d_s + s - 2)$-connected by Corollary 3.12.

First, consider an interval $Q_{>\sigma}$. The order complex of this interval has the same homotopy type as $\mathrm{lk}_\Delta(\sigma)$; the restriction of f to $P(\{\sigma\} * \mathrm{lk}_\Delta(\sigma))$ is a closure operator with image $Q_{>\sigma}$. Hence the order complex is $(\rho(E) - \rho(\sigma) - 2)$-connected by Theorem 13.10.

Next, consider an interval $Q_{>\sigma,<\tau}$; $\sigma \subsetneq \tau$. Let Σ be the subcomplex of $\mathrm{lk}_{\Delta(\tau)}(\sigma)$ consisting of all faces π such that $\rho(\sigma \cup \pi) < \rho(\tau)$. It is clear that Σ contains the $(\rho(\tau) - \rho(\sigma) - 2)$-skeleton of $\mathrm{lk}_{\Delta(\tau)}(\sigma)$, because if π contains σ and has rank $\rho(\tau)$, then $|\pi| - |\sigma| \geq \rho(\tau) - \rho(\sigma)$. Since $\mathrm{lk}_{\Delta(\tau)}(\sigma)$ is $(\rho(\tau) - \rho(\sigma) - 2)$-connected by Theorem 13.10, it follows that Σ is $(\rho(\tau) - \rho(\sigma) - 3)$-connected. Now, the restriction of f to $P(\{\sigma\} * \Sigma)$ is a closure operator with image $Q_{>\sigma,<\tau}$. Hence the order complex of $Q_{>\sigma,<\tau}$ is $(\rho(\tau) - \rho(\sigma) - 3)$-connected as desired. $\square$

Proposition 13.19. *Let $M = (E, \mathsf{F})$ be a matroid and assume that Δ and Γ are SPI complexes over M. Then the intersection $\Delta \cap \Gamma$ is also SPI over M. As a consequence, the set of nonvoid SPI complexes over M ordered by inclusion is a lattice with intersection as the meet operator.*

Proof. By Proposition 13.7 and Theorem 13.14, it suffices to prove that every basis σ is contained in a unique maximal face of $\Delta \cap \Gamma$. However, each of $\mathrm{lk}_\Delta(\sigma)$ and $\mathrm{lk}_\Gamma(\sigma)$ is a simplex, which immediately implies that the same is true for the intersection $\mathrm{lk}_{\Delta \cap \Gamma}(\sigma)$; hence the claim follows. $\square$

Proposition 13.20. *Let $M = (E, \mathsf{F})$ be a matroid. For a given family $\mathcal{C}$ of circuits, define $\Delta_{\mathcal{C}}$ to be the simplicial complex with minimal nonfaces all circuits of M that are not in $\mathcal{C}$. Then Δ is an atom in the lattice of SPI complexes over M if and only if $\Delta = \Delta_{\{\sigma\}}$ for some circuit σ.*

Proof. First, we prove that $\Delta_{\{\sigma\}}$ is an SPI complex whenever σ is a circuit. $\Delta_{\{\sigma\}}$ is clearly a PI complex. Let τ be a basis in M. If $\tau \cap \sigma$ is of the form $\sigma - x$ for some x in σ, then $\tau + x$ is the unique maximal face containing τ. Otherwise, τ is itself a maximal face. By Theorem 13.14, it follows that $\Delta_{\{\sigma\}}$ is SPI.

Next, Proposition 13.16 yields that every SPI complex over M is of the form $\Delta_{\mathcal{C}}$ for some family $\mathcal{C}$ of circuits. Since $\Delta_{\mathcal{C}}$ is a proper subfamily of $\Delta_{\mathcal{C}'}$ whenever $\mathcal{C}$ is a proper subfamily of $\mathcal{C}'$, it follows that the atoms of the lattice are exactly the complexes $\Delta_{\{\sigma\}}$. $\square$

13.3.1 Sets in Matroids Avoiding Odd Cycles

The purpose of this section is to introduce a certain family of complexes defined in terms of linear representations of matroids and show that all complexes in this family are SPI. As a special case, we have the complex of bipartite subgraphs of a graph.

A *linear representation* over a field $\mathbb{F}$ of a matroid $M = (E, \mathsf{F})$ is a map $\varphi : E \to \mathbb{F}^k$ with $k \geq 1$ such that a set τ is independent in M if and only if the set $\varphi(\tau)$ is linearly independent. For example, if $G = ([n], E)$ is a graph, then we obtain a representation over $\mathbb{F}_2$ of $M_n(G)$ by defining $\varphi(ab) = \mathsf{e}_a + \mathsf{e}_b$, where e_i denotes the ith unit vector in $(\mathbb{F}_2)^n$. Note that not every matroid admits a linear representation.

Let $\varphi : E \to \mathbb{F}^k$ be a linear representation of the matroid M. Let $\psi : E \to \mathbb{F}^m$ be an arbitrary function; $m \geq 0$. We say that a set $\tau \subseteq E$ contains an *odd cycle* with respect to the pair (φ, ψ) if there are scalars $\{\lambda_x : x \in \tau\}$ in $\mathbb{F}$ such that

$$\sum_{x \in \tau} \lambda_x \varphi(x) = 0 \quad \text{and} \quad \sum_{x \in \tau} \lambda_x \psi(x) \neq 0.$$

Define $\mathsf{B}_{M,\varphi,\psi}$ as the complex of subsets τ of E such that τ does not contain any odd cycle with respect to (φ, ψ). For example, for the representation of $M_n(G)$ specified above and with $\psi(e) = 1 \in \mathbb{Z}_2$ for all edges e, we obtain the complex $\mathsf{B}_n(G)$ of bipartite subgraphs of G examined in Chapter 14. For another example, see Section 15.5.

Theorem 13.21. *Let $M = (E, \mathsf{F})$ be a matroid, let $\varphi : E \to \mathbb{F}^k$ be a linear representation of M, and let $\psi : E \to \mathbb{F}^m$ be an arbitrary function. Then the complex $\mathsf{B}_{M,\varphi,\psi}$ is SPI. In particular, for any $\tau \subseteq E$, $\mathsf{B}_{M,\varphi,\psi}(\tau)$ is $VD^+(\rho(\tau) - 1)$.*

Proof. First, we prove that $\mathsf{B}_{M,\varphi,\psi}$ is PI. Let τ be a nonface of $\mathsf{B}_{M,\varphi,\psi}$ containing an isthmus e. This means that e is independent of all other elements in τ. In particular, every linear combination $\sum_{x \in \tau} \lambda_x \varphi(x) = 0$ has the property that $\lambda_e = 0$. As a consequence, if τ contains an odd cycle, then so does $\tau - e$, which implies that τ is not a minimal nonface.

Next, we prove that $\mathsf{B}_{M,\varphi,\psi}$ is SPI. Let $\sigma \in \mathsf{B}_{M,\varphi,\psi}$ and let $e \in E \setminus \sigma$ be such that $\sigma + e \in \mathsf{B}_{M,\varphi,\psi}$ and $\rho(\sigma) = \rho(\sigma + e)$. This means that there are scalars λ_x such that

$$\varphi(e) = \sum_{x \in \sigma} \lambda_x \varphi(x) \tag{13.1}$$

and such that $\psi(e) = \sum_{x \in \sigma} \lambda_x \psi(x)$; $\sigma + e$ contains no odd cycle by assumption. We need to prove that e is a cone point in the link of $\mathsf{B}_{M,\varphi,\psi}$ with respect to σ.

Suppose that τ is a face of $\mathsf{B}_{M,\varphi,\psi}$ containing σ such that $\tau + e$ contains an odd cycle. Since τ does not contain any odd cycle, this means that there are scalars μ_y such that

$$\varphi(e) = \sum_{y \in \tau} \mu_y \varphi(y) \qquad (13.2)$$

and such that $\psi(e) \neq \sum_{y \in \tau} \mu_y \psi(y)$. Combining (13.1) and (13.2), we obtain the linear combination

$$\sum_{x \in \sigma} \lambda_x \varphi(x) = \sum_{y \in \tau} \mu_y \varphi(y).$$

Since

$$\sum_{x \in \sigma} \lambda_x \psi(x) = \psi(e) \neq \sum_{y \in \tau} \mu_y \psi(y),$$

this implies that τ contains an odd cycle, which is a contradiction to the assumption that τ belongs to $\mathsf{B}_{M,\varphi,\psi}$. Thus e is a cone point, and we are done. $\square$

To demonstrate that a complex coincides with $\mathsf{B}_{M,\varphi,\psi}$, it suffices to prove that the minimal nonfaces are exactly those circuits that contain odd cycles:

Corollary 13.22. *Let $M = (E, \mathsf{F})$ be a matroid, let $\varphi : E \to \mathbb{F}^k$ be a linear representation of M, and let $\psi : E \to \mathbb{F}^m$ be an arbitrary function. Then the minimal nonfaces of $\mathsf{B}_{M,\varphi,\psi}$ are the circuits of M that contain odd cycles.*

Proof. This is an immediate consequence of Proposition 13.16 and Theorem 13.21. $\square$

13.3.2 SPI Graph Complexes

First, let us state an immediate consequence of Theorem 13.10.

Corollary 13.23. *Let G be a graph on n vertices and let Δ be an SPI complex over the graphic matroid on G. Let $H \in \Delta$. Then $\mathrm{lk}_\Delta(H)$ is $VD^+(c(H) - c(G) - 1)$. In particular, Δ is $VD^+(n - c(G) - 1)$. $\square$*

For the remainder of the section, we concentrate on SPI monotone graph properties, proving that there are only four of them on each fixed vertex set. We stress that we consider only graph properties; there are plenty of SPI graph complexes that are not invariant under permutations of the underlying vertex set.

Theorem 13.24. *Let Δ be a nonvoid monotone graph property on n vertices. Then Δ is SPI over the graphic matroid M_n if and only if Δ is either of the following complexes:*

(1) *The complex* F_n *of forests.*
(2) *The complex* FH_n *of forests and full (Hamiltonian) cycles.*
(3) *The complex* B_n *of bipartite graphs.*
(4) *The full simplex.*

Proof. First, let us show that the four listed complexes are all SPI monotone graph properties. They are clearly PI. To prove strength, we show that every spanning tree is contained in a unique maximal face. This is trivially true for F_n and the full simplex and obvious for FH_n. For B_n, any spanning tree T admits a unique bipartition (U, W), which immediately implies that the complete bipartite graph with blocks U and W is the unique maximal face of B_n containing T. Applying Theorem 13.14, the desired claim follows.

Next, we show that any SPI monotone graph property Δ coincides with one of the four listed complexes. If Δ does not contain any r-cycles such that $r \in [3, n-1]$, then Δ must be either F_n or FH_n. Thus assume that Δ does contain all r-cycles for some $r \in [3, n-1]$. We want to show that Δ is either the complex of all bipartite graphs or the full simplex.

To start with, let us show that Δ contains all 4-cycles. Since Δ contains everything in F_n, Δ contains the Hamiltonian path π with edges $i(i+1)$ for $1 \le i \le n-1$. By Proposition 13.15, each of $\pi + 1r$ and $\pi + 2(r+1)$ belongs to Δ. Since Δ is SPI, this implies that $\pi' = \pi \cup \{1r, 2(r+1)\}$ belongs to Δ. Since π' contains the 4-cycle $\gamma_4 = \{12, 2(r+1), r(r+1), 1r\}$, we obtain that $\gamma_4 \in \Delta$ as desired.

As a consequence, $\pi + 14$ belongs to Δ by Proposition 13.15. We show by induction on k that $\pi \cup \{14, 16, \ldots, 1k\}$ belongs to Δ whenever $k \le n$ and k is even. By induction hypothesis, $\pi \cup \{14, 16, \ldots, 1(k-2)\}$ belongs to Δ. Hence since $\{1(k-2), (k-2)(k-1), (k-1)k, 1k\} \in \Delta$, we immediately obtain the desired claim, again by Proposition 13.15. The conclusion is that Δ contains all cycles of even length. Since Δ is SPI, this is equivalent to saying that Δ contains B_n.

It remains to show that if Δ contains an odd cycle of length $2s + 1$, then Δ contains all cycles and is hence the full simplex. Again, consider the Hamiltonian path π. We obtain that $1(2s)$ and $1(2s + 1)$ are both cone points in $\mathrm{lk}_\Delta(\pi)$, which implies that the triangle $\{1(2s), 1(2s+1), (2s)(2s+1)\}$, and hence any triangle, is contained in Δ. Thus for any $5 \le 2r + 1 \le n$, we have that $1(2r)$ and $(2r-1)(2r+1)$ are cone points in $\mathrm{lk}_\Delta(\pi)$. This implies that the $(2r+1)$-cycle is contained in Δ, which concludes the proof. $\square$

Note that $\mathsf{F}_n \approx \mathsf{FH}_n$ (see Lemma 13.17). In particular, modulo the equivalence relation $\approx$, the only nontrivial SPI monotone graph properties are F_n and B_n.

Remark. In a separate manuscript [68], we classify all SPI monotone digraph properties over the digraphic matroid $M_n^{\rightarrow}$ modulo the equivalence relation $\approx$. It turns out that we have the following properties:

(1) The complex $\mathsf{F}(M_n^{\rightarrow})$ of digraphs without circuits.

(2) The complex DOAC_n of digraphs without non-alternating circuits.
(3) The complex $\mathsf{DGr}_{n,p}$ of digraphs that are graded modulo p for some integer $p \geq 1$.
(4) The trivial extensions of F_n, FH_n, B_n, and the full simplex.

See Sections 15.4 and 15.5 for more information about $\mathsf{DGr}_{n,p}$ and DOAC_n, respectively.

13.4 Alexander Duals of SPI Complexes

Let $M = (E, \mathsf{F})$ be a matroid. We say that a complex Δ is PI* over M if the Alexander dual $\Delta^* = \Delta_E^*$ is PI over the dual matroid M^* with rank function as in (2.1). We say that Δ is SPI* over M if Δ^* is SPI over M^*. For example, the complex $\mathsf{NC}(M)$ of all sets of rank at most $r - 1$ is SPI*, where r is the rank of M. Note that

$$\rho(\tau) < \rho(\tau + x) \Longleftrightarrow \rho^*(E \setminus \tau) = \rho^*(E \setminus (\tau + x)).$$

A complex Δ is PI* if and only if the first of the following two properties holds. Δ is SPI* if and only if both properties hold.

(i) If $\tau \in \Delta$ and $\rho(\tau) = \rho(\tau + x)$, then $\tau + x \in \Delta$.
(ii) If $\tau \notin \Delta$ and $\rho(\tau) < \rho(\tau + x)$, then x is a cone point in the induced subcomplex $\Delta(\tau + x) = \mathrm{del}_\Delta(E \setminus (\tau + x))$.

To see that the second property holds when Δ is SPI*, use the fact that $(\Delta(\tau + x))_{\tau+x}^* = \mathrm{lk}_{\Delta^*}(E \setminus (\tau + x))$.

An important special case is $\mathsf{NC}_n(G)$, the complex of disconnected subgraphs of a graph G. We will examine this complex in Section 18.1.

Theorem 13.25. *Let Δ be an SPI* complex over a matroid $M = (E, \mathsf{F})$. Then, for any subset τ of E, $\Delta(\tau)$ is $VD^+(\rho(\tau) - 2)$. In particular, $\Delta(\tau)$ is homotopy equivalent to a wedge of spheres of dimension $\rho(\tau) - 2$.*

Proof. For any set $\sigma \subset \tau$ such that $\rho(\sigma) < \rho(\tau)$, we will show that $\mathrm{lk}_{\Delta(\tau)}(\sigma)$ is $VD^+(\rho(\tau) - \rho(\sigma) - 2)$. Write $\rho(\sigma, \tau) = \rho(\tau) - \rho(\sigma)$.

If $\rho(\sigma, \tau) = 1$ and $\rho(\sigma, \sigma + x) = 1$ for each $x \in \tau \setminus \sigma$, then $\mathrm{lk}_{\Delta(\tau)}(\sigma)$ is a simplex; if $\sigma \cup A_0$ and $\sigma \cup A_1$ belong to Δ, then so does $\sigma \cup (A_0 \cup A_1)$ by condition (i). As a consequence, $\mathrm{lk}_{\Delta(\tau)}(\sigma)$ is $VD^+(\rho(\sigma, \tau) - 2) = VD^+(-1)$.

If there is an $x \in \tau \setminus \sigma$ such that $\rho(\sigma) = \rho(\sigma + x)$, then this x is a cone point in $\mathrm{lk}_{\Delta(\tau)}(\sigma)$. By induction on $|\tau| - |\sigma|$, $\mathrm{lk}_{\Delta(\tau)}(\sigma + x)$ is $VD^+(\rho(\sigma, \tau) - 2)$; hence the same is true for $\mathrm{lk}_{\Delta(\tau)}(\sigma)$.

Finally, suppose that $\rho(\sigma, \tau) \geq 2$ and $\rho(\sigma, \sigma + x) = 1$ for each $x \in \tau \setminus \sigma$. If there is an $x \in \tau \setminus \sigma$ such that $\rho(\tau - x, \tau) = 0$, then, by induction on $|\tau| - |\sigma|$, $\mathrm{lk}_{\Delta(\tau)}(\sigma + x)$ is $VD^+(\rho(\sigma, \tau) - 3)$ and $\mathrm{lk}_{\Delta(\tau-x)}(\sigma)$ is $VD^+(\rho(\sigma, \tau) - 2)$. This implies that $\mathrm{lk}_{\Delta(\tau)}(\sigma)$ is $VD^+(\rho(\sigma, \tau) - 2)$. If $\rho(\tau - x, \tau) = 1$ for each $x \in \tau \setminus \sigma$, then $\rho(\sigma, \tau) = |\tau| - |\sigma|$. We have two cases:

- If $\tau - x \in \Delta(\tau)$ for each $x \in \tau \setminus \sigma$, then $\mathrm{lk}_{\Delta(\tau)}(\sigma)$ is either a $(|\tau| - |\sigma| - 1)$-simplex or the boundary of such a simplex and is hence $VD^+(\rho(\tau, \sigma) - 2)$.
- If there is an $x \in \tau \setminus \sigma$ such that $\tau - x \notin \Delta(\tau)$, then x is a cone point in $\Delta(\tau)$ by property (ii) and hence a cone point in $\mathrm{lk}_{\Delta(\tau)}(\sigma)$. By induction on $|\tau| - |\sigma|$, $\mathrm{lk}_{\Delta(\tau)}(\sigma + x)$ is $VD^+(\rho(\sigma, \tau) - 3)$, which implies that $\mathrm{lk}_{\Delta(\tau)}(\sigma)$ is $VD^+(\rho(\sigma, \tau) - 2)$. $\square$

Remark. $\mathrm{lk}_{\Delta(\tau)}(\sigma)$ being homotopy equivalent to a wedge of spheres of dimension $\rho(\tau) - \rho(\sigma) - 2$ is an immediate consequence of Alexander duality and Theorem 13.10. Namely, the Alexander dual of $\mathrm{lk}_{\Delta(\tau)}(\sigma)$ with respect to $\tau \setminus \sigma$ equals $(\Delta(\tau))^*_\tau(\tau \setminus \sigma)$, which admits a decision tree with all evasive sets of dimension $\rho^*(\tau \setminus \sigma) - 1$. By Lemma 5.23, this implies that $\mathrm{lk}_{\Delta(\tau)}(\sigma)$ admits a decision tree with all evasive sets of dimension

$$|\tau| - |\sigma| - \rho^*(\tau \setminus \sigma) - 2 = \rho(\tau) - \rho(\sigma) - 2;$$

use (2.1). Also, since this exceeds $\rho(\tau) - |\sigma| - 2$, we have that the $(\rho(\tau) - 2)$-skeleton of $\Delta(\tau)$ is homotopically Cohen-Macaulay.

Corollary 13.26. *For any matroid M of rank r, the complex $\mathsf{NC}(M)$ of all sets of rank at most $r - 1$ is $VD^+(r - 2)$.* $\square$

Let us translate some of the results in Section 13.3 to the language of SPI* complexes.

Theorem 13.27. *Let Δ be an SPI* complex over a matroid $M = (E, \mathsf{F})$. Then Δ is nonevasive if and only if Δ is a cone. In particular, the homology of Δ is nonzero whenever Δ is not a cone.*

Proof. This is an immediate consequence of Theorem 13.12 and Alexander duality. $\square$

Theorem 13.28. *Let $M = (E, \mathsf{F})$ be a matroid and let Δ be a PI* complex over M different from 2^E. Then Δ is SPI* over M if and only if every basis contains a unique minimal nonface of Δ.*

Proof. This is an immediate consequence of Theorem 13.14 and Alexander duality. $\square$

Proposition 13.29. *Let $M = (E, \mathsf{F})$ be a matroid and let $\Delta \subseteq 2^E$ be an SPI* complex different from 2^E. Then every maximal face of Δ is a cocircuit of M.*

Proof. This is an immediate consequence of Proposition 13.16 and Alexander duality; the complement of a circuit in a matroid is a cocircuit in the dual matroid and vice versa. $\square$

Proposition 13.30. *Let $M = (E, \mathsf{F})$ be a matroid and assume that Σ and Δ are SPI* complexes over M. Then the union $\Sigma \cup \Delta$ is also SPI* over M. As a consequence, the set of SPI* complexes over M ordered by inclusion is a lattice with union as the join operator.*

Proof. This is an immediate consequence of Proposition 13.19. □

Similarly to the way we excluded the void complex from the lattice of SPI complexes, we exclude the full simplex 2^E from the lattice of SPI* complexes. Instead, the complex $\mathsf{NC}(M)$ of sets of rank at most $\rho(E) - 1$ is the maximal element in the lattice.

Proposition 13.31. *Let $M = (E, \mathsf{F})$ be a matroid. For a given family $\mathcal{F}$ of cocircuits, define $\Delta_{\mathcal{F}}$ to be the simplicial complex with maximal faces all cocircuits of M that are not in $\mathcal{F}$. Then Δ is a coatom in the lattice of SPI* complexes over M if and only if $\Delta = \Delta_{\{\sigma\}}$ for some cocircuit σ.*

Proof. This is an immediate consequence of Proposition 13.20. □

13.4.1 SPI* Monotone Graph Properties

Let us proceed with the classification of all monotone graph properties that are SPI* over the graphic matroid M_n. The Alexander duals of these complexes are exactly all monotone graph properties that are SPI over the dual of M_n. For $p \in [1, n]$ such that p divides n, recall that $\mathsf{NC}_{n,p}$ is the complex of graphs with some component of size not divisible by p.

Theorem 13.32. *Let Δ be a monotone graph property on n vertices. Then Δ is SPI* over M_n if and only if Δ is the full simplex or equal to $\mathsf{NC}_{n,p}$ for some p dividing n.*

Proof. Let Δ be an SPI* monotone graph property. Say that a graph is *closed* if each connected component is a clique. First, note that any maximal graph in Δ is closed. Namely, by property (i) in Section 13.4, we may add edges not decreasing the number of components in a graph in Δ without ending up outside Δ. Moreover, by the proof of Theorem 13.25, we have that $\mathrm{lk}_{\Delta}(H)$ is $VD^+(c(H) - 3)$ for each graph $H \in \Delta$. In particular, since a maximal graph H in Δ is not $VD(0)$, a maximal graph cannot have more than two connected components.

If there is a connected graph in Δ, then K_n belongs to Δ, which means that Δ is the full simplex. The remaining case is that all maximal graphs in Δ have exactly two connected components. If all such graphs appear in Δ, then we obtain $\mathsf{NC}_n = \mathsf{NC}_{n,n}$, whereas the other extreme with no such graphs means that we obtain the void complex $\mathsf{NC}_{n,1}$. Assume that some but not all graphs with two connected components belong to Δ.

For a multiset $A = \{a_1, \ldots, a_k\}$ of positive integers summing to n, say that a graph G has type A if we may order the vertex sets of the connected components of G as $U_1, \ldots, U_k$ such that $|U_i| = a_i$. For example, a graph with two components of size a and $n - a$ has type $\{a, n - a\}$. Since Δ is a monotone graph property and since every maximal graph in Δ is closed, we have that a given graph G belongs to Δ if and only if every graph with the same type as G belongs to Δ. For simplicity, say that a given type belongs to Δ if every graph of this type belongs to Δ.

Let $A = \{a_1, \ldots, a_k\}$ be a type not belonging to Δ and assume that $a_i < a_j$ for some i, j. Then the multiset

$$A' = \{a_1, \ldots, a_{j-1}, a_j - a_i, a_i, a_{j+1}, \ldots, a_k\}$$

does not belong to Δ. Namely, let G be a closed graph of type A and let e be an edge joining two components U_i and U_j of size a_i and a_j, respectively. By property (ii) in Section 13.4, we have that e is a cone point in the induced subcomplex $\Delta(G + e)$. Construct a graph H of type A' from G by splitting the component U_j into one component U_j' of size $a_j - a_i$ and one component U_j'' of size a_i; make sure the edge e has one endpoint in the component U_j' of size $a_j - a_i$. If H belongs to Δ, then so does $H + e$. However, $H + e$ has the same type A as G; $|U_i| + |U_j'| = a_j$. This is a contradiction; hence $H \notin \Delta$.

For p dividing n, let A_p be the type consisting of n/p copies of the integer p. By induction, we obtain that if $A = \{a_1, \ldots, a_k\}$ does not belong to Δ, then the type A_d does not belong to Δ either, where d is the greatest common divisor of $a_1, \ldots, a_k$. As a consequence, each minimal nonface of Δ is of type A_d for some integer d.

To prove that $\Delta = \mathsf{NC}_{n,p}$ for some p, it suffices to demonstrate that all minimal nonfaces of Δ are of the very same type A_p. Let x be minimal such that A_x does not belong to Δ. Suppose that there is a y such that A_y does not belong to Δ and such that y is not divisible by x. Let y be minimal with this property. Note that the types $\{x, n - x\}$ and $\{y, n - y\}$ do not belong to Δ. We claim that $\{y, x, n - y - x\}$ does not belong to Δ. Namely, let G be a closed graph of type $\{y, n - y\}$ and let H be a subgraph of type $\{y - x, x, n - y\}$. Let e be an edge joining the components of size $y - x$ and $n - y$. Since $H + e$ is of type $\{x, n - x\}$ and since e is a cone point in $\Delta(G + e)$ by property (ii), we obtain that H does not belong to Δ. However, this implies that the type $\{y - x, n - y + x\}$ does not belong to Δ. Since y is not divisible by x, the same is true for $y - x$. Yet, since $y - x < y$, this contradicts the minimality of y.

It remains to prove that $\mathsf{NC}_{n,p}$ is an SPI* complex whenever p divides n. One readily verifies that property (i) in Section 13.4 holds. To prove property (ii), suppose that G is a graph not contained in $\mathsf{NC}_{n,p}$ and that e is an edge joining two connected components in G. Suppose that H is a subgraph of G such that the size of each connected component in $H + e$ is divisible by p. Then H has the same property. Namely, let U and W be the components in G containing the endpoints of e. By assumption, $|U|$ and $|W|$ are both divisible by p. Let $U_1, \ldots, U_r$ be the components in H that are contained in U; let U_1 be the component containing one endpoint of e. By assumption, $|U_i|$ is divisible by p for $i > 1$, but this clearly implies that $|U_1|$ is also divisible by

p; $|U_1| = |U| - \sum_{i>1} |U_i|$. By symmetry, the same holds for all components in H that are contained in W. As a consequence, H is not contained in $NC_{n,p}$, which implies that e is a cone point as desired. This concludes the proof. $\square$

We provide an implicit formula for the Euler characteristic of $NC_{n,p}$ in Section 18.4.

Remark. One may also examine when a monotone digraph property is SPI* over $M_n^{\rightarrow}$. One readily verifies that the classification of such properties reduces to the classification of monotone graph properties that are SPI* over M_n. Specifically, a monotone digraph property Δ is SPI* if and only if Δ is the trivial extension of an SPI* monotone graph property.

Bipartite Graphs

We examine complexes of graphs with the important property of being bipartite. Recall that a graph G is bipartite if G contains no cycles of odd length. Equivalently, G admits a bipartition (U, W), meaning that the vertex set V can be partitioned into two stable subsets U and W.

In Section 14.1, we discuss the complex B_n of all bipartite graphs on n vertices. For any graph G on n vertices, Chari [31] showed that the complex $\mathsf{B}_n(G)$ consisting of all bipartite subgraphs of G is homotopy equivalent to a wedge of spheres of dimension $n - c(G) - 1$, where $c(G)$ is the number of connected components in G. In particular, B_n is homotopy equivalent to a wedge of spheres of dimension $n - 2$; Linusson and Shareshian [94] gave an alternative proof of this result. Using the theory developed in Chapter 13 about strong pseudo-independence complexes over matroids, we give a new proof of Chari's result. We also show that the $(n - c(G) - 1)$-skeleton of $\mathsf{B}_n(G)$ is vertex-decomposable and present a formula due to Stanley [134] for the Euler characteristic of B_n.

In Section 14.2, we obtain similar results for the complex $\mathsf{NCB}_n(G) = \mathsf{B}_n(G) \cap \mathsf{NC}_n$ of disconnected bipartite subgraphs of G. Again, we use techniques from Chapter 13. Moreover, we apply Stanley's formula to compute the Euler characteristic of the full complex NCB_n. Intriguingly, the well-studied ordered Bell number (see Gross [55] and Wilf [148, p. 175]) shows up as the absolute value of the Euler characteristic of the quotient complex CB_n of connected bipartite graphs. Based on this observation, we present a family of labeled trees counted by this number.

Recall that $\mathsf{B}_{n,p}$ is the subcomplex of B_n consisting of all graphs with balance number at most p. For G connected and $2p < n$, we show in Section 14.3 that $\mathsf{B}_{n,p}(G)$ is homotopy equivalent to a wedge of spheres of dimension $2p - 1$ and that the $(2p - 1)$-skeleton of $\mathsf{B}_{n,p}(G)$ is vertex-decomposable. Moreover, the reduced Euler characteristic $\tilde{\chi}(\mathsf{B}_{n,p}) = \tilde{\chi}(\mathsf{B}_{n,p}(K_n))$ is a polynomial $f_p(n)$ in n of degree at most $2p$ for each fixed p and $n > 2p$. Somewhat surprisingly, $f_p(k) = \tilde{\chi}(\mathsf{B}_{k+1})$ for $0 \leq k \leq p$. Section 14.3 also contains a fairly shallow discussion on hypergraph analogues.

14.1 Bipartite Graphs Without Restrictions

The following two theorems summarize known properties of the complex $B_{\dot{n}}(G)$ of bipartite subgraphs of a given graph G on n vertices.

Theorem 14.1 (Chari [31]). *Let G be a graph on n vertices. Then $B_n(G)$ is homotopy equivalent to a wedge of spheres of dimension $n - c(G) - 1$.* $\square$

Linusson and Shareshian [94] gave an alternate proof of Theorem 14.1 based on discrete Morse theory and provided an explicit description of the unmatched graphs with respect to their acyclic matching.

Theorem 14.2 (Stanley [134, Exercise 5.5]). *The reduced Euler characteristic of $B_n = B_n(K_n)$ satisfies*

$$B(x) := \sum_{n \geq 1} \tilde{\chi}(B_n) \frac{x^n}{n!} = -\sqrt{2e^x - 1} + 1. \ \square$$

Proof. Consider the sum

$$H_2(x) = \sum_{n \geq 1} \frac{x^n}{n!} \sum_{f:[n] \to \{0,1\}} \sum_{G \sim f} (-1)^{|G|+1}; \tag{14.1}$$

$G \sim f$ means that $f(u) \neq f(v)$ whenever $uv \in e$ and $|G|$ is the number of edges in G. If there are vertices x and y such that $f(x) \neq f(y)$, then $(G + xy) \sim f$ if and only if $(G - xy) \sim f$. As a consequence, $\sum_{G \sim f} (-1)^{|G|+1}$ equals zero in this case. If $f(x) = f(y)$ for all x and y, then $G \sim f$ if and only if G is empty, which implies that $\sum_{G \sim f} (-1)^{|G|+1} = -1$. We conclude that $H_2(x) = -2e^x + 2$.

Now, a bipartite graph G with c connected components appears exactly 2^c times in the sum (14.1). Namely, there are two possibilities for the restriction of f to each component. It follows that

$$H_2(x) = \sum_{n \geq 1} \frac{x^n}{n!} \sum_{G \in B_n} (-1)^{|G|+1} 2^{c(G)} = 1 - (1 - B(x))^2,$$

where the last equality is a consequence of Corollary 6.15. $\square$

Table 14.1. The reduced Euler characteristic of B_n for small values on n.

n	1	2	3	4	5	6	7	8	9	10
$\tilde{\chi}(B_n)$	−1	0	−1	3	−16	105	−841	7938	−86311	1062435

See Table 14.1 for the first few terms in this series. For a generalization of

Theorem 14.2 and its proof, see Theorem 15.10, in which we consider the complex $\mathsf{DGr}_{n,p}$ of graded digraphs modulo p. B_n and the trivial extension $\mathsf{DGr}_{n,2}$ of B_n are homotopy equivalent by Proposition 4.5.

Proposition 14.3. *Let* $\{c_n : n \geq 0\}$ *be the unique integers satisfying* $c_0 = 0$, $c_1 = 1$, $c_2 = 0$, *and*

$$c_n = \sum_{b=2}^{n-1} \binom{n-2}{b-2}(c_b + c_{b-1}) \tag{14.2}$$

for $n \geq 3$. *Then*

$$\sum_{n \geq 1} c_n \frac{x^n}{n!} = -\sqrt{2e^{-x} - 1} + 1.$$

Proof. Writing $F(x) = \sum_{n \geq 1} c_n x^n / n!$, one readily verifies from (14.2) that

$$F''(x) = (e^x - 1)(F''(x) + F'(x)).$$

Via a straightforward calculation, one checks that $F(x) = -\sqrt{2e^{-x} - 1} + 1$ is a solution to this equation. The first two terms in the expansion of this function are $0 = c_0$ and $x = c_1 x$, which immediately yields the proposition. $\square$

The recursive identity (14.2) shows up in the analysis of the digraph property DNOCy_n of avoiding odd directed cycles; see Section 15.6.

We want to give a new proof of Chari's result and prove that the $(n - c(G) - 1)$-skeleton of $\mathsf{B}_n(G)$ is VD. However, by Theorem 13.21 (or Theorem 13.24), we know that B_n is a strong pseudo-independence (SPI) complex; see Section 13.3. As an immediate consequence of Theorem 13.10, we thus have the following result.

Corollary 14.4. *Let* G *be a graph on* n *vertices. For any* $H \in \mathsf{B}_n(G)$, *we have that* $\mathrm{lk}_{\mathsf{B}_n(G)}(H)$ *is* $VD^+(c(H) - c(G) - 1)$. *In particular,* $\mathsf{B}_n(G)$ *is* $VD^+(n - c(G) - 1)$ *and hence homotopy equivalent to a wedge of spheres of dimension* $n - c(G) - 1$. $\square$

Finally, we apply Corollary 13.26 to $\mathsf{B}_n(G)$.

Corollary 14.5. *For a graph* G *on* n *vertices, let* $\mathsf{B}_n^*(G)$ *be the complex of subgraphs* H *of* G *such that* $G \setminus H$ *is not in* B_n. *Then* $\mathsf{B}_n^*(G)$ *is* $VD^+(|G| - n + c(G) - 2)$. *In particular,* $\mathsf{B}_n^*(G)$ *is homotopy equivalent to a wedge of spheres of dimension* $|G| - n + c(G) - 2$. $\square$

Note that $\mathsf{B}_n^* = \mathsf{B}_n^*(K_n)$ is the complex of graphs that do not contain two disjoint cliques (complete graphs) of total size n.

14.2 Disconnected Bipartite Graphs

We consider the complex $\mathsf{NCB}_n = \mathsf{B}_n \cap \mathsf{NC}_n$ of disconnected bipartite graphs.

Theorem 14.6. *For $n \geq 2$ and any graph G on n vertices, the complex* $\mathsf{NCB}_n(G)$ *is* $VD^+(n - \max\{c(G), 2\} - 1)$. *In particular,* NCB_n *is* $VD^+(n - 3)$.

Proof. One easily checks that NCB_n is SPI over the one-step truncation of the graphic matroid M_n; thus we are done by Theorem 13.10. $\square$

Theorem 14.7. *The reduced Euler characteristic of* NCB_n *satisfies*

$$G(x) := \sum_{n \geq 2} \tilde{\chi}(\mathsf{NCB}_n) \frac{x^n}{n!} = -\sqrt{2e^x - 1} + 1 + \frac{\ln(2e^x - 1)}{2}.$$

Note that this equals $H(x) + \ln(1 - H(x))$, where $H(x) = \sum_{n \geq 1} \tilde{\chi}(\mathsf{B}_n) \frac{x^n}{n!}$.

Proof. Let $f(n)$ be the reduced Euler characteristic of $\mathsf{CB}_n = \mathsf{B}_n/\mathsf{NCB}_n$; this is the quotient complex of connected bipartite graphs on n vertices. Let $h(n)$ be the Euler characteristic of B_n. It is clear that f and h satisfy Corollary 6.15, which immediately implies that $G(x) - H(x) = \ln(1 - H(x))$. Applying Theorem 14.2, we are done. $\square$

Theorems 14.2 and 14.7 imply that $-\frac{1}{2}\ln(2e^x - 1)$ is the exponential generating function for the Euler characteristic of the quotient complex $\mathsf{CB}_n = \mathsf{B}_n/\mathsf{NCB}_n$ of connected bipartite graphs. As a consequence, this Euler characteristic is up to sign an ordered Bell number (sequence A000670 in Sloane's Encyclopedia [127]); see Gross [55] and Wilf [148, p. 175] for some information.

Theorem 14.8. *For $n \geq 1$, let r_n be the number of spanning trees T on the vertex set $[n]$ with the property that each simple path $(1 = a_1, a_2, \ldots, a_k)$ in T starting at the vertex 1 has the property that $a_i < a_{i+2}$ for $1 \leq i \leq k - 2$. Then*

$$\sum_{n \geq 1} r_n \frac{x^n}{n!} = -\frac{1}{2} \ln(2e^{-x} - 1).$$

Proof. We will define a matching on CB_n such that the unmatched graphs are exactly all spanning trees with properties as in the theorem. Since each pair in the matching turns out to be of the form $(G - e, G + e)$, the theorem will be a consequence of Theorem 14.7. $n = 1$ is obvious; thus assume that $n \geq 2$.

Consider a partition (U, W) of the vertex set $[n]$ such that $1 \in U$ and $W \neq \emptyset$. Let w be minimal in W. Let $\mathcal{F}_{U,W}$ be the family of all graphs G in CB_n with bipartition (U, W). Form a matching on $\mathcal{F}_{U,W}$ by pairing $G - 1w$ with $G + 1w$ whenever possible. A graph G in $\mathcal{F}_{U,W}$ is unmatched with respect to this matching if and only if $1w \in G$ and $G - 1w$ is disconnected.

For any subsets $U_1 \subseteq U$ and $W_1 \subseteq W$ such that $1 \in U_1$ and $w \in W_1$, let $\mathcal{F}_{U,W}(U_1, W_1)$ be the subfamily of $\mathcal{F}_{U,W}$ consisting of all unmatched graphs G such that the two connected components in $G - 1w$ have bipartitions $(U_1, W \setminus W_1)$ and $(W_1, U \setminus U_1)$, respectively. $\mathcal{F}_{U,W}(U_1, W_1)$ can be identified with the join of $\mathcal{F}_{U_1, W \setminus W_1}$ and $\mathcal{F}_{W_1, U \setminus U_1}$, except that we have to add the edge $1w$ to each graph in the join.

By induction on n, $\mathcal{F}_{U_1, W \setminus W_1}$ admits a matching such that the unmatched graphs are exactly those spanning trees on the vertex set $[U_1, W \setminus W_1]$ with properties as in the theorem. The similar property holds for $\mathcal{F}_{W_1, U \setminus U_1}$, except that we have to relabel the smallest vertex w in the first block as 1. Combining these two matchings, the resulting unmatched graphs in $\mathcal{F}_{U,W}(U_1, W_1)$ are exactly the graphs satisfying the properties in the theorem. $\square$

An alternative approach to proving Theorem 14.8 would be to observe the position of the vertex n in a spanning tree as in the theorem. For $1 \le k \le n-1$, let $r_{n,k}$ be the number of such trees such that n has exactly $k - 1$ children with respect to the root 1. Note that all these children must be leaves, as we would otherwise have a vertex two steps away from n and smaller than n. It is a fairly straightforward exercise to prove that $r_{n,k} = r_{n-k}\binom{n-1}{k}$, which yields the identity

$$r_n = \sum_{k=1}^{n-1} r_{n-k} \binom{n-1}{k}$$

for $n \ge 2$. This is the well-known recurrence formula for ordered Bell numbers; see Gross [55, Eq. 9].

Yet another possible approach goes via *ordered partitions* (or *preferential arrangements*) of the set $[n-1]$. Such a partition is defined to be an ordered sequence $(U_1, \ldots, U_k)$ of nonempty sets such that $[n-1]$ is the disjoint union of $U_1, \ldots, U_k$. With s_n equal to the number of ordered partitions of $[n-1]$ (note the shift), we have that $\sum_{n \ge 1} s_n x^n = -\frac{1}{2} \ln(2e^{-x} - 1)$; see Wilf [148, p. 175]. In particular, $r_n = s_n$, where r_n is defined as above. Note that there is a natural bijection between ordered partitions of $[n-1]$ and faces of the barycentric subdivision of $\partial 2^{[n-1]}$; identify $(U_1, \ldots, U_k)$ with $\{\bigcup_{i=1}^{r} U_i : i \in [k-1]\}$.

Problem 14.9. Define a bijection between the family of spanning trees on n vertices with properties as in Theorem 14.8 and the family of ordered partitions of the set $[n-1]$.

14.3 Unbalanced Bipartite Graphs

For a bipartite graph H on n vertices, recall that the balance number $\beta(H)$ is the smallest number r such that H is contained in a copy of $K_{r,n-r}$. For $p \ge 0$ and a graph G on n vertices, we consider the simplicial complex $\mathsf{B}_{n,p}(G)$ of bipartite subgraphs H of G with balance number at most p. Note that

$B_{n,0}(G)$ is the (-1)-simplex. Also, $B_{n,p}(G) = B_n(G)$ for $n \leq 2p + 1$; this case was handled in Section 14.1. From now on, we will assume that $n \geq 2p + 1$ unless otherwise stated.

14.3.1 Depth

We consider the depth of $B_{n,p}$. The two *blocks* of a connected bipartite graph H are the two (unique) maximal vertex sets U and W with the property that each edge in H has one endpoint in U and the other endpoint in W. For $a \leq b$, let us say that a connected bipartite graph H has *type* (a, b) if the two blocks of H have size a and b, respectively. We include the case $(a, b) = (0, 1)$ corresponding to the graph with a single vertex.

Theorem 14.10. *Let $0 \leq p < n/2$. Let G be a graph on n vertices and let Σ be a PI complex over the graphic matroid on G (see Section 13.2). Let H be a graph in $\Sigma \cap B_{n,p}(G)$. Then $\mathrm{lk}_{\Sigma \cap B_{n,p}(G)}(H)$ is $VD(c(H) - c(G) + 2p - n)$. In particular, $\Sigma \cap B_{n,p}(G)$ is $VD(2p - c(G))$.*

Proof. We use induction on n and p. We may assume that $p \geq 1$; the case $p = 0$ is trivial. Write $\Delta = \mathrm{lk}_{\Sigma \cap B_{n,p}(G)}(H)$. If $c(G) = c(H)$, then we are done; $c(H) - c(G) + 2p - n \leq -1$. Thus assume that $c(G) < c(H)$. If $n = 2p + 1$, then Proposition 13.7 and Corollary 13.8 yield the desired result; thus assume that $n \geq 2p + 2$.

First, suppose that there is an edge $e \in G \setminus H$ such that $c(G - e) = c(G)$. Induction on $c(H) - c(G)$ yields that $\mathrm{del}_\Delta(e)$ is $VD(c(H) - c(G) + 2p - n)$ and that $\mathrm{lk}_\Delta(e)$ is $VD(c(H + e) - c(G) + 2p - n)$. Since $c(H + e) \geq c(H) - 1$, we obtain that $\mathrm{lk}_\Delta(e)$ is $VD(c(H) - 1 - c(G) + 2p - n)$ and hence that Δ is $VD(c(H) - c(G) + 2p - n)$. Note that $\mathrm{lk}_\Delta(e)$ is void if $H + e \notin \Sigma \cap B_{n,p}(G)$.

It remains to consider the case that $c(G - e) = c(G) + 1$ for every edge e in $G \setminus H$. Let $H_1, \ldots, H_t$ be the connected components of H. It is clear that we obtain a forest structure on G; view $H_1, \ldots, H_t$ as the vertices and the edges in $G \setminus H$ as the edges. Let H_k be a "leaf" in this forest structure and let e be the edge in G separating H_k from the other connected components in H. By induction, we obtain that $\mathrm{lk}_\Delta(e)$ is $VD(c(H) - 1 - c(G) + 2p - n)$.

For the deletion, we need to work harder. If H_k is of type (c, c), then e is a cone point. Namely, any graph in $\{H\} * \Delta$ remains bipartite when e is added, and the balance number stays the same, as the two blocks of H_k have the same size. Moreover, Σ is PI and hence closed under addition of edges joining connected components. By Lemma 6.11 and induction, we obtain that $\mathrm{del}_\Delta(e)$ is $VD(c(H) - c(G) + 2p - n)$.

Suppose H_k is of type (a, b) for some $a < b$. Define H_0 and G_0 as the graphs obtained from H and G by removing the vertex set of H_k, which we may assume is equal to $[n - a - b + 1, n]$. Write $n' = n - a - b$ and $p' = p - a$. We have that

$$\mathrm{del}_\Delta(e) = \mathrm{lk}_{\Sigma \cap B_{n,p}(G-e)}(H) = \mathrm{lk}_{\Sigma \cap B_{n',p'}(G_0)}(H_0);$$

$\beta(H) = \beta(H_0) + a$. Note that $c(H_0) = c(H) - 1$ and $c(G_0) = c(G - e) - 1 = c(G)$. If $n' \geq 2p' + 1$, then induction yields that $\mathrm{lk}_{\Sigma \cap \mathsf{B}_{n',p'}(G_0)}(H_0)$ is $VD(c(H_0) - c(G_0) + 2p' - n')$. We are done in this case, as

$$c(H_0) - c(G_0) + 2p' - n' = c(H) - 1 - c(G) + 2p - 2a - n + a + b$$
$$= b - a - 1 + (c(H) - c(G) + 2p - n) \geq c(H) - c(G) + 2p - n.$$

If $n' \leq 2p'$, then $\Sigma \cap \mathsf{B}_{n',p'}(G_0) = \Sigma \cap \mathsf{B}_{n'}(G_0)$, which is a PI complex over the graphic matroid on G_0 by Proposition 13.7. Hence Corollary 13.8 implies that $\mathrm{lk}_{\Sigma \cap \mathsf{B}_{n',p'}(G_0)}(H_0)$ is $VD(c(H_0) - c(G_0) - 1)$. This time,

$$c(H_0) - c(G_0) - 1 = c(H) - 1 - c(G) - 1$$
$$= n - 2p - 2 + (c(H) - c(G) + 2p - n) \geq c(H) - c(G) + 2p - n,$$

and we are again done. $\square$

Corollary 14.11. *Let $0 \leq p < n/2$. Let G be a graph on n vertices. Then $\mathsf{B}_{n,p}(G)$ is $VD(2p - c(G))$. The same is true for the subcomplex $\mathsf{F}_n \cap \mathsf{B}_{n,p}(G)$ of "unbalanced" forests.* $\square$

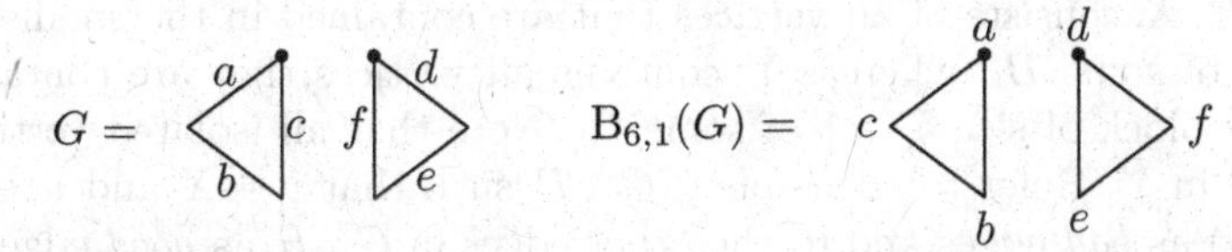

$$G = \quad \overset{a}{\underset{b}{\triangleleft}} c \quad f \overset{d}{\underset{e}{\triangleright}} \qquad \mathsf{B}_{6,1}(G) = \quad c \overset{a}{\underset{b}{\triangleleft}} \quad \overset{d}{\underset{e}{\triangleright}} f$$

Fig. 14.1. The graph G has the property that the corresponding complex $\mathsf{B}_{6,1}(G)$ has nonvanishing homology in two dimensions.

14.3.2 Homotopy Type

Next, we consider the homotopy type of the link $\mathrm{lk}_{\mathsf{B}_{n,p}(G)}(H)$. Note that this complex is not always homotopy equivalent to a wedge of spheres of dimension $c(H) - c(G) + 2p - n$, the quantity in Theorem 14.10. For example, if we add an isolated vertex to each of H and G, then $c(H), c(G)$, and n all increase by one; hence $c(H) - c(G) + 2p - n$ decreases by one. Yet, the corresponding complex is isomorphic to $\mathrm{lk}_{\mathsf{B}_{n,p}(G)}(H)$. For a less trivial example, see Figure 14.1.

These examples suggest that we may not be able to find a general formula for the homotopy type of $\mathrm{lk}_{\mathsf{B}_{n,p}(G)}(H)$. Indeed, in our main result in this section, we restrict our attention to *connected* graphs G and to subgraphs H with a very special structure:

Theorem 14.12. *Let $0 \leq p < n/2$. Let G be a connected graph on n vertices. Let H be a graph in $\mathsf{B}_{n,p}(G)$ such that each connected component is of type (a,b) for some a,b satisfying $a \leq b \leq a+1$. Then $\mathrm{lk}_{\mathsf{B}_{n,p}(G)}(H) \sim dt^{c(H)+2p-1-n}$ for some $d \geq 0$. Hence $\mathrm{lk}_{\mathsf{B}_{n,p}(G)}(H)$ is homotopy equivalent to a wedge of spheres of dimension $c(H) + 2p - 1 - n$. In particular, $\mathsf{B}_{n,p}(G)$ is homotopy equivalent to a wedge of spheres of dimension $2p - 1$.*

Remark. Inspired by Theorem 14.10, one may ask whether Theorem 14.12 would remain true for the intersection of $\mathsf{B}_{n,p}$ with an arbitrary SPI complex (see Section 13.3). This is not the case; $\mathsf{F}_6 \cap \mathsf{B}_{6,2}$ has homology in two different dimensions.

Proof. Quite a few arguments from the proof of Theorem 14.10 will appear once again, but for clarity we keep the proofs separated. We use induction on n and p. We may assume that $p \geq 1$ and $n \geq 2p + 2$; apply Corollary 14.4 for the case $n = 2p + 1$.

Let $B_1, \ldots, B_r, C_1, \ldots, C_s$ be the connected components of H; we assume that each B_i has type $(b_i, b_i + 1)$ for some $b_i \geq 0$ and that each C_j has type (c_j, c_j) for some $c_j \geq 1$.

Divide the vertex set of H (and G) into three sets X, Y, and Z in the following manner: Z consists of all vertices in the connected components $C_1, \ldots, C_s$. X consists of all vertices that are contained in the smaller block of size b_i of some B_i, whereas Y contains all vertices that are contained in the larger block of size $b_i + 1$ of some B_i. Note that all isolated vertices are contained in Y. Refer to edges $vw \in G \setminus H$ such that $v \in X$ and $w \in Y$ (or vice versa) as *bad* edges and to the other edges in $G \setminus H$ as *good* edges.

Clearly, any edge in $G \setminus H$ contained in some connected component B_i or C_j is either a cone point in $\Delta = \mathrm{lk}_{\mathsf{B}_{n,p}(G)}(H)$ or not at all in Δ. We may hence assume that there are no such edges. We divide into a number of cases.

Case 1. There is a good edge $e = vw \in G \setminus H$ such that $G - e$ remains connected. When added to H, e connects two connected components. Note that this implies that $c(H + e) = c(H) - 1$. If $v, w \in X$, $v, w \in Y$, or $v, w \in Z$, then the resulting connected component D is of type (c,c) for some c. If $v \in X \cup Y$ and $w \in Z$, then D is of type $(b, b + 1)$ for some b. By induction, we obtain that

$$\mathrm{del}_\Delta(e) = \mathrm{lk}_{\mathsf{B}_{n,p}(G-e)}(H) \sim d_\emptyset t^{c(H)+2p-1-n};$$

$$\mathrm{lk}_\Delta(e) = \mathrm{lk}_{\mathsf{B}_{n,p}(G)}(H + e) \sim d_e t^{c(H+e)+2p-1-n} = d_e t^{c(H)+2p-2-n};$$

$d_\emptyset, d_e \geq 0$. Note that $\mathrm{lk}_\Delta(e)$ is void if $H + e \notin \mathsf{B}_{n,p}$. As a consequence, the theorem follows in this case; use Lemma 5.22.

Case 2. There are no good edges at all in $G \setminus H$. Then there is no connected component C in H of type (c,c) for any $c > 0$. Namely, C cannot be the only connected component in H; $c \leq \beta(H) \leq p < n/2$. Also, since G is connected, there would be some edge in $G \setminus H$ connecting C with some other connected component, and this edge would be good by definition. As a consequence, the connected components in H are $B_1, \ldots, B_r$ with properties as in the theorem. It is clear that $\beta(H) = \sum_i b_i = |X|$; (X, Y) is the most "unbalanced" bipartition of H. Since each edge in $G \setminus H$ goes from X to Y (all edges are bad), each such edge is a cone point in Δ. As a consequence, Δ is collapsible. Namely, no cone points would mean that H were connected of type $(b, b+1)$ for some $b \leq p$; since $2b + 1 \leq 2p + 1 < n$, we have a contradiction. This, by the way, is the step where a proof for $\mathsf{F}_n \cap \mathsf{B}_{n,p}$ would fail.

Case 3. There are good edges but each good edge e is critical in the sense that $G - e$ is disconnected. Let F be the set of good edges and let $G_1, \ldots, G_t$ be the connected components in $G \setminus F$. It is clear that G has a tree structure; view the connected components in $G \setminus F$ as the vertices and F as the edge set. Let G_k be a "leaf" in this tree structure and let $e \in F$ be the edge separating G_k from the other connected components in $G \setminus F$.

Let H_k be the induced subgraph of H on the vertex set of G_k. Since all edges in $G_k \setminus H_k$ are bad, H_k consists either of one single connected component C_j or of one or several connected components B_i.

Case 3.1. H_k consists of one single connected component C_j; this implies that $H_k = G_k$. We claim that e is a cone point in Δ. Namely, the addition of e to any graph H' such that $H \subseteq H' \subset G$ cannot create an odd cycle, as e separates G. Also, the addition of e does not increase the balance number.

Case 3.2. H_k consists of one or several connected components B_i. As before, we obtain by induction that

$$\mathrm{lk}_\Delta(e) = \mathrm{lk}_{\mathsf{B}_{n,p}(G)}(H + e) \sim d_e t^{c(H) + 2p - 2 - n} \tag{14.3}$$

for some $d_e \geq 0$. The situation for the deletion is slightly more complicated. We divide into two cases depending on the number of components in H_k.

Case 3.2.1. If H_k contains more than one connected component B_i, then each edge in $G_k \setminus H_k$ is a cone point in $\mathrm{del}_\Delta(e)$; the discussion in Case 2 above applies.

Case 3.2.2. If H_k contains one single connected component B_i, then $G_k = H_k = B_i$ and $\beta(G_k) = b_i$. Let G_0 and H_0 be the induced subgraphs of G and H obtained by removing the vertex set of $G_k = B_i$; let W be the vertex set of G_0 and H_0. It is clear that G_0 is connected, $c(H_0) = c(H) - 1$, and $|W| = n - 2b_i - 1$. For simplicity, assume that $W = [n - 2b_i - 1]$. For any graph G' such that $H \subseteq G' \subseteq G - e$, we have that $\beta(G') = \beta(B_i) + \beta(G'(W)) = b_i + \beta(G'(W))$. As a consequence, $G' \in \mathsf{B}_{n,p}$ if and only if $G'(W) \in \mathsf{B}_{|W|, p - \beta(B_i)} = \mathsf{B}_{n - 2b_i - 1, p - b_i}$. Let $n' = n - 2b_i - 1$ and $p' = p - b_i$. Since $\beta(H) \leq p$ and hence $b_i \leq p$, it is clear that $p' \geq 0$. Also, since $2p \leq n - 2$,

$$n' = n - 2b_i - 1 \geq 2p - 2b_i + 1 = 2p' + 1.$$

As a consequence, n' and p' satisfy the conditions in the theorem, which implies by induction that

$$\mathrm{del}_\Delta(e) = \mathrm{lk}_{\mathsf{B}_{n,p}(G-e)}(H) = \mathrm{lk}_{\mathsf{B}_{n',p'}(G_0)}(H_0) \sim d_\emptyset t^{c(H_0)+2p'-1-n'}$$
$$= d_\emptyset t^{c(H)-1+2(p-b_i)-1-(n-2b_i-1)} = d_\emptyset t^{c(H)+2p-1-n}$$

for some $d_\emptyset \geq 0$. Combining this with (14.3), we are done by Lemma 5.22. □

14.3.3 Euler Characteristic

In this section, we restrict our attention to $G = K_n$; write $\mathsf{B}_{n,p} = \mathsf{B}_{n,p}(K_n)$. By Theorem 14.12, $\mathsf{B}_{n,p}$ is homotopy equivalent to a wedge of spheres of dimension $2p - 1$ for $n \geq 2p + 1$. Yet, the proof of Theorem 14.12 does not give much information about the number of spheres in the wedge. The purpose of this section is to demonstrate that this number – which clearly coincides with minus the reduced Euler characteristic – is a polynomial in n for each p.

Let $\mathsf{B}_{n,p}^{P,Q}$ be the simplicial complex defined as follows. The set of 0-cells is the union of the usual set $\binom{[n]}{2}$ of edges and two copies $[n]^P$ and $[n]^Q$ of $[n]$. We denote elements i in the first copy $[n]^P$ as i^P and elements j in the second copy $[n]^Q$ as j^Q; we write $I^P = \{i^P : i \in I\}$ and $I^Q = \{i^Q : i \in I\}$. For a graph G and subsets $I \subseteq [n]$ and $J \subseteq [n]$, the set $G + I^P + J^Q = G \cup I^P \cup J^Q$ is a face of $\mathsf{B}_{n,p}^{P,Q}$ if and only if the following hold:

- $G \in \mathsf{B}_{n,p}$.
- G admits a bipartition (U, W) such that $|U| \leq p$ and such that $I \subseteq U$ and $J \subseteq W$.

In particular, $|I| \leq p$ and $I \cap J = \emptyset$. Note that the definition of $\mathsf{B}_{n,p}^{P,Q}$ makes sense for $p = 0$.

Lemma 14.13. *Let $n \geq 1$ and $p \geq 0$. Let G be a graph on n vertices and let $H \in \mathsf{B}_{n,p}^{P,Q}$ be a subgraph of G. Then $\mathrm{lk}_{\mathsf{B}_{n,p}^{P,Q}(G+[n]^P+[n]^Q)}(H) \sim st^{c(H)-1}$ for some $s \geq 0$, where $c(H)$ is the number of connected components in H; $s = 0$ if $n > 2p$. In particular, $\mathsf{B}_{n,p}^{P,Q} \sim st^{n-1}$ for some $s \geq 0$; $s = 0$ if $n > 2p$.*

Proof. If $p = 0$, then all elements in $[n]^Q$ are cone points. Moreover, $\mathsf{B}_{1,p}^{P,Q} = \{\emptyset, \{1^P\}, \{1^Q\}\}$ whenever $p \geq 1$. Hence we may assume that $n \geq 2$ and $p \geq 1$. Write $\Sigma = \mathrm{lk}_{\mathsf{B}_{n,p}^{P,Q}(G+[n]^P+[n]^Q)}(H)$.

First, suppose that $G \setminus H$ contains some edge e. If e joins two vertices from one and the same connected component of H, then e is either a cone point in Σ or not at all present in Σ. Suppose that e joins vertices from two different components. By induction on $G \setminus H$, $\mathrm{lk}_\Sigma(e) \sim s_e t^{c(H+e)-1} = s_e t^{c(H)-2}$ and $\mathrm{del}_\Sigma(e) \sim s_\emptyset t^{c(H)-1}$ for some integers $s_\emptyset, s_e$, which implies that $\Sigma \sim (s_e + s_\emptyset) t^{c(H)-1}$. For $n > 2p$, induction yields that $s_\emptyset = s_e = 0$ and hence that Σ is nonevasive.

It remains to consider the base case $G = H$. Suppose that G has an isolated vertex i, say $i = n$. We have that $\mathrm{lk}_\Sigma(n^P)$ coincides with

$$\Sigma' = \mathrm{lk}_{\mathsf{B}_{n-1,p-1}^{P,Q}(G([n-1])+[n-1]^P+[n-1]^Q)}(G([n-1])).$$

By induction, Σ' admits a decision tree with all evasive faces of dimension $c(G)-2$. If $n > 2p$, then $n-1 > 2(p-1)$, which yields, again by induction, that Σ' is nonevasive. Moreover, $\mathrm{del}_\Sigma(n^P)$ is a cone with cone point n^Q. Namely, if (U, W) is a bipartition of a face of $\mathrm{del}_\Sigma(n^P)$ such that $n \in U$ and $|U| \le p$, then $(U \setminus \{n\}, W \cup \{n\})$ is another bipartition with the same properties. The desired result follows.

Suppose that G has no isolated vertices. Let $V_1, \ldots, V_r$ be the vertex sets of the connected components in G and let (U_i, W_i) be the unique bipartition of $G(V_i)$ for each i; $|U_i| \le |W_i|$. For each $i \in [r]$, choose $a_i \in U_i$ arbitrarily. Write $A = \{a_i : i \in [r]\}$ and let

$$Y = ([n] \setminus A)^P \cup ([n] \setminus A)^Q.$$

It is clear that $\Sigma(C, Y \setminus C)$ is a cone whenever $C \ne \emptyset$. Namely, if $x^P \in C$ for some $x \in U_i \setminus \{a_i\}$ or $x^Q \in C$ for some $x \in W_i$, then a_i^P is a cone point, whereas a_i^Q is a cone point if $x^Q \in C$ for some $x \in U_i \setminus \{a_i\}$ or $x^P \in C$ for some $x \in W_i$.

The remaining complex is $\Delta = \Sigma(\emptyset, Y)$. Consider $\Delta(M^Q, A^Q \setminus M^Q)$ for each $M \subseteq A$. If $a_i \notin M$, then a_i^P is a cone point in $\Delta(M^Q, A^Q \setminus M^Q)$. Namely, suppose that $I \subseteq A \setminus M$ is a set such that $I^P + M^Q$ belongs to $\Delta(M^Q, A^Q \setminus M^Q)$. Let (U, W) be a bipartition of G such that $I \subseteq U$ and $M \subseteq W$ and such that $|U| \le p$. Clearly, if $W_i \subseteq U$ and $U_i \subseteq W$, then $(U', V') = ((U \setminus W_i) \cup U_i, (W \setminus U_i) \cup W_i)$ is another bipartition of G such that $I \subseteq U'$ and $M \subseteq W'$. Since $|U_i| \le |W_i|$, it follows that $|U'| \le |U| \le p$, which implies that a_i^P is a cone point as desired.

Thus $\Delta(M^Q, A^Q \setminus M^Q)$ is nonevasive unless $M = A$. Now, $\Delta(A^Q, \emptyset)$ is either $\{A^Q\}$ or void, because each element a^P collides with the already present element a^Q. Since $|A| = c(G)$, it follows that $\Delta(A^Q, \emptyset) \sim rt^{c(G)-1}$, where $r \in \{0, 1\}$; hence $\Sigma \sim rt^{c(G)-1}$.

If $n > 2p$, then consider the unique bipartition $(W, U) = (\bigcup_i W_i, \bigcup_i U_i)$ of G such that $A \subseteq U$. Since $|W_i| \ge |U_i|$, it follows that $|W| \ge |U| = n - |W|$. Since $n > 2p$, we have that $|W| > p$, which implies that $\Delta(A^Q, \emptyset) = \emptyset$; thus $\Sigma \sim 0$. $\square$

Let $\mathsf{B}_{n,p}^P$ be the subcomplex of $\mathsf{B}_{n,p}^{P,Q}$ consisting of all elements of the form $G + I^P$; this is the induced subcomplex on the set $\binom{[n]}{2} \cup [n]^P$. Analogously, let $\mathsf{B}_{n,p}^Q$ be the subcomplex of $\mathsf{B}_{n,p}^{P,Q}$ consisting of elements of the form $G + I^Q$; this is the induced subcomplex on the set $\binom{[n]}{2} \cup [n]^Q$.

Proposition 14.14. *The following hold:*

(i) *For $0 \le k \le p$, $\mathsf{B}_{k,p}^P$ and B_{k+1} are isomorphic.*

(ii) *For $n, p \ge 0$, $\tilde{\chi}(\mathsf{B}_{n,p}^P) = \tilde{\chi}(\mathsf{B}_{n,p}) + \tilde{\chi}(\mathsf{B}_{n,p}^{P,Q}) - \tilde{\chi}(\mathsf{B}_{n,p}^Q)$.*

(iii) *For $0 \le 2p \le n$, $\tilde{\chi}(\mathsf{B}_{n,p}^P) = \tilde{\chi}(\mathsf{B}_{n,p})$.*

(iv) *For $n, p \ge 0$,*

$$\tilde{\chi}(\mathsf{B}_{n,p}^{P,Q}) = \sum_{i=0}^{n} \binom{n}{i}(-1)^{n-i}\tilde{\chi}(\mathsf{B}_{i,p}^{P}).$$

Equivalently, $\tilde{\chi}(\mathsf{B}_{n,p}^{P}) = \displaystyle\sum_{i=0}^{2p} \binom{n}{i}\tilde{\chi}(\mathsf{B}_{i,p}^{P,Q}).$

Proof. (i) We obtain an isomorphism $\varphi : \mathsf{B}_{k,p}^{P} \to \mathsf{B}_{k+1}$ by defining $\varphi(G+I) = G + \{i(k+1) : i \in I\}$. Namely, $G + I$ belongs to $\mathsf{B}_{k,p}^{P}$ if and only if G admits a bipartition (U, W) such that $I \subseteq U$ and $|U| \leq p$. Since the latter is always true whenever $p \geq k$, this means exactly that G is bipartite and I is a stable set in G, which is equivalent to $\varphi(G+I)$ being bipartite.

(ii) For a face $G + I^P + J^Q \in \mathsf{B}_{n,p}^{P,Q} \setminus (\mathsf{B}_{n,p}^{P} \cup \mathsf{B}_{n,p}^{Q})$, let $i = \min I$ and $j = \min J$. We obtain a perfect element matching by pairing $(G+ij)+I^P+J^Q$ and $(G-ij) + I^P + J^Q$. As a consequence,

$$\tilde{\chi}(\mathsf{B}_{n,p}^{P,Q}) = \tilde{\chi}(\mathsf{B}_{n,p}^{P} \cup \mathsf{B}_{n,p}^{Q}).$$

Since $\mathsf{B}_{n,p} = \mathsf{B}_{n,p}^{P} \cap \mathsf{B}_{n,p}^{Q}$, the claim follows.

(iii) By (ii), it suffices to prove that $\tilde{\chi}(\mathsf{B}_{n,p}^{P,Q}) = \tilde{\chi}(\mathsf{B}_{n,p}^{Q})$ whenever $n \geq 2p$. Use induction on p; the result is trivial for $p = 0$. Assume that $p > 0$ and consider a face $\sigma = G + I^P + J^Q \in \mathsf{B}_{n,p}^{P,Q}$ such that some $i \in I$ is not isolated in G. Let $x = x(\sigma)$ be minimal such that $ix \in G$. Then $\sigma + x^Q \in \mathsf{B}_{n,p}^{P,Q}$ and $x(\sigma + x^Q) = x(\sigma)$. In particular, we may define an element matching on $\mathsf{B}_{n,p}^{P,Q}$ such that a face $\sigma = G + I^P + J^Q$ is unmatched if and only if all elements in I are isolated in G. For each $I \subseteq [n]$ such that $|I| \leq p$, let $\mathsf{B}_{n,p}^{P,Q}(I)$ be the family of such elements $\sigma = G + I^P + J^Q$.

Clearly, $\mathsf{B}_{n,p}^{P,Q}(\emptyset) = \mathsf{B}_{n,p}^{Q}$. Moreover, for each I such that $1 \leq |I| \leq p$, $\mathsf{B}_{n,p}^{P,Q}(I)$ is isomorphic to $\{I^P\} * \mathsf{B}_{n-|I|,p-|I|}^{Q}$. Since $n-|I| > 2(p-|I|)$, induction yields that $\tilde{\chi}(\mathsf{B}_{n-|I|,p-|I|}^{Q}) = \tilde{\chi}(\mathsf{B}_{n-|I|,p-|I|}^{P,Q})$. By Lemma 14.13, this equals zero, which implies that $\tilde{\chi}(\mathsf{B}_{n,p}^{P,Q}(I)) = 0$; thus we are done.

(iv) We proceed in a manner similar to the proof of claim (iii), except that we swap P and Q. Specifically, we match away all faces $\sigma = G+I^P+J^Q \in \mathsf{B}_{n,p}^{P,Q}$ such that some $j \in J$ is not isolated in G. For each $J \subseteq [n]$, let $\mathsf{B}_{n,p}^{P,Q}(J)$ be the family of elements $\sigma = G + I^P + J^Q$ such that all elements in J are isolated. Clearly, $\mathsf{B}_{n,p}^{P,Q}(\emptyset) = \mathsf{B}_{n,p}^{P}$. Moreover, for each J, $\mathsf{B}_{n,p}^{P,Q}(J)$ is isomorphic to $\{J^Q\} * \mathsf{B}_{n-|J|,p}^{P}$. Summing over all J, we obtain the desired result.

The last claim is just a matrix inversion combined with the fact that $\tilde{\chi}(\mathsf{B}_{i,p}^{P,Q}) = 0$ for $i > 2p$; use Lemma 14.13. □

Theorem 14.15. *For $p \geq 1$ and $n \geq 2p$,*

$$\tilde{\chi}(\mathsf{B}_{n,p}) = \sum_{k=0}^{2p}(-1)^k \binom{n}{k}\binom{n-1-k}{2p-k}\tilde{\chi}(\mathsf{B}_{k,p}^{P}). \tag{14.4}$$

In particular, $\tilde{\chi}(\mathsf{B}_{n,p})$ is a polynomial f_p in n of degree at most $2p$ such that $f_p(0) = -1$ and $f_p(1) = 0$.

Remark. The right-hand side of (14.4) defines a polynomial f_p such that $f_p(k) = \tilde{\chi}(\mathsf{B}_{k+1})$ for $0 \le k \le p$; use Propositions 6.13 and 14.14 (i).

Proof. Let $f_p(n)$ be the unique polynomial of degree at most $2p$ with the property that $f_p(k) = \tilde{\chi}(\mathsf{B}_{k,p}^P)$ for $0 \le k \le 2p$. Then $f_p(n) = \tilde{\chi}(\mathsf{B}_{n,p}^P)$ for all $n \ge 0$. Namely, Lemma 14.13 and Proposition 14.14 (iv) imply that

$$\sum_{i=0}^{n} \binom{n}{i}(-1)^{n-i}\tilde{\chi}(\mathsf{B}_{i,p}^P) = 0 \qquad (14.5)$$

whenever $n \ge 2p + 1$. Hence

$$0 = \sum_{i=0}^{n}\binom{n}{i}(-1)^{n-i}\tilde{\chi}(\mathsf{B}_{i,p}^P)$$

$$= \tilde{\chi}(\mathsf{B}_{n,p}^P) - f_p(n) + \sum_{i=0}^{n}\binom{n}{i}(-1)^{n-i}f_p(i) = \tilde{\chi}(\mathsf{B}_{n,p}^P) - f_p(n).$$

The first equality is (14.5). The second equality is by induction on n starting with $n = 2p + 1$. The third equality is true for any polynomial of degree at most $n-1$. Applying Propositions 6.13 and 14.14 (iii), we obtain (14.4), which concludes the proof. $\square$

In a separate manuscript [73], we generalize a weaker version of Theorem 14.15 to a larger class of monotone graph properties.

Let us examine the polynomial f_p defined by $f_p(n) = \tilde{\chi}(\mathsf{B}_{n,p})$ for $n \ge 2p + 1$. First, consider the case $p = 1$. We refer to graphs in $\mathsf{B}_{n,1}$ as *star graphs*. Theorems 14.12 and 14.15 combined with a direct inspection of $\mathsf{B}_{2,1}$ and $\mathsf{B}_{3,1}$ yield the following result.

Proposition 14.16. *For $n \ge 3$, we have that $\mathsf{B}_{n,1}$ is homotopy equivalent to a wedge of $\binom{n-1}{2}$ spheres of dimension one. In particular, $f_1(n) = -\binom{n-1}{2}$.* $\square$

The situation is much more complicated for $p \ge 2$, but we have enough data to determine f_2 and f_3, and a simple computer calculation yields f_4:

Proposition 14.17. *We have that*

$$f_2(x) = \frac{-x(x-1)(x-3)(x-4) + 2x - 2}{2};$$

$$f_3(x) = \frac{-23x^6 + 393x^5 - 2486x^4 + 7203x^3 - 9425x^2 + 4374x - 36}{36};$$

$$f_4(x) = -\frac{1061x^8}{1152} + \frac{7997x^7}{288} - \frac{65351x^6}{192} + \frac{314935x^5}{144}$$
$$-\frac{1011181x^4}{128} + \frac{4576049x^3}{288} - \frac{4681687x^2}{288} + \frac{51153x}{8} - 1.$$

Proof. Using Theorem 14.2 (see Table 14.1), and Theorem 14.15, we obtain that $f_2(0) = -1$, $f_2(1) = 0$, $f_2(2) = -1$, $f_2(4) = 3$, and $f_2(5) = -16$. There is only one polynomial of degree at most $2p = 4$ with this property. Similarly, we obtain that $f_3(k) = f_2(k)$ for $k \leq 2$, $f_3(3) = 3$, $f_3(6) = 105$, and $f_3(7) = -841$. Moreover, since $B^P_{p+1,p} \cong B_{p+2} \setminus \{K_{[p+1],\{p+2\}}\}$, we have that $\tilde{\chi}(B^P_{p+1,p}) = \tilde{\chi}(B_{p+2}) - (-1)^p$, which implies that $f_3(4) = -\tilde{\chi}(B_5) - (-1)^3 = -15$. We now have seven known values of f_3, and these values determine a unique polynomial of degree at most $2p = 6$. For $p = 4$, proceed in the same manner; we know $f_4(k)$ for $k \in \{0,1,2,3,4,5,8,9\}$ and may easily compute $f_4(6) = \tilde{\chi}(B^P_{6,4}) = 675$ using the computer program homology [42]. Since we have nine known values of f_4, we have a unique polynomial of degree at most eight. In all three cases, we are done by Theorem 14.15. □

Table 14.2. $f_p(n)$ for small values on n and p. We obtained $f_4(6)$ via a computer calculation; all other values are consequences of results in this chapter.

$f_p(n)$	$n = 0$	1	2	3	4	5	6	7	8	9	10
$p = 1$	-1	0	0	-1	-3	-6	-10	-15	-21	-28	-36
2	-1	0	-1	2	3	-16	-85	-246	-553	-1072	-1881
3	-1	0	-1	3	-15	44	105	-841	-5957	-22240	-62661
4	-1	0	-1	3	-16	104	-675	2379	7938	-86311	-763116
5	-1	0	-1	3	-16	105	-840	?	?	?	1062435

Table 14.3. The absolute value of $\tilde{\chi}(B^{P,Q}_{r,p})$ for $p \leq 4$ and for $p = 5$ and $r \leq 6$. By Lemma 14.13, $\tilde{\chi}(B^{P,Q}_{r,p})$ is negative only if r is even and positive only if r is odd.

$|\tilde{\chi}(B^{P,Q}_{r,p})|$	$r = 0$	1	2	3	4	5	6	7	8	9	10
$p = 1$	1	1	1	–	–	–	–	–	–	–	–
2	1	1	2	6	12	–	–	–	–	–	–
3	1	1	2	7	34	160	460	–	–	–	–
4	1	1	2	7	35	225	1615	9975	37135	–	–
5	1	1	2	7	35	226	1786	?	?	?	?

See Table 14.2 for a table with $f_p(n)$ for small n and p. In Table 14.3, we present $\tilde{\chi}(\mathsf{B}_{r,p}^{P,Q})$; recall from Proposition 14.14 (iv) that $f_p(n)$ is equal to the sum $= \sum_{r=0}^{2p} \binom{n}{r} \tilde{\chi}(\mathsf{B}_{r,p}^{P,Q})$.

We have the following intriguing consequences of the results in this section.

Corollary 14.18. *For $p \geq 1$, the coefficients of f_p alternate in sign;*

$$f_p(n) = \sum_{r=0}^{2p} (-1)^{r+1} a_r n^r,$$

where $a_r \geq 0$.

Proof. Since $f_p(n) = \sum_{r=0}^{2p} \binom{n}{r} \tilde{\chi}(\mathsf{B}_{r,p}^{P,Q})$ by Proposition 14.14 (iv), it suffices to prove that $(-1)^{r+1} \tilde{\chi}(\mathsf{B}_{r,p}^{P,Q}) \geq 0$. However, by Lemma 14.13, this is indeed true since all homology is concentrated in dimension $r - 1$. $\square$

Corollary 14.19. *For $p \geq 2$, $f_p(x)$ has at least two roots, counted with multiplicity, in the interval $(0,2)$ and at least one root in each of the intervals $(k, k+1)$ for $2 \leq k \leq p$ and $(2p, 2p+1)$.*

Proof. We have that $f_p(0) = -1$, $f_p(1) = 0$, and $f_p(2) = -1$ for $p \geq 2$; hence there are at least two roots in $(0, 2)$. Moreover, for $2 \leq k \leq p$, $f_p(k) = \tilde{\chi}(\mathsf{B}_{k+1})$ is larger than 0 if k is odd and smaller than 0 if k is even; the inequalities being strict follows from Proposition 14.3. In the same manner, we obtain that $f_p(2p+1) < 0 < f_p(2p)$. Finally, we concluded in the proof of Proposition 14.17 that $f_p(p+1) = \tilde{\chi}(\mathsf{B}_{p+2}) - (-1)^p$. Since $|\tilde{\chi}(\mathsf{B}_{p+2})| > 1$ for $p \geq 2$, it follows that $f_p(p+1) > 0$ if p is even and $f_p(p+1) < 0$ if p is odd. $\square$

Conjecture 14.20. *For $p \geq 1$, f_p is a polynomial of degree exactly $2p$ with only real and positive roots.*

The conjecture clearly holds for $1 \leq p \leq 4$. It would hold for general p if we could prove that $f_p(k)$ alternates in sign all the way from $k = 2$ to $k = 2p-1$; by Corollary 14.19, this would imply that we have $2p$ real roots and hence a polynomial of degree exactly $2p$ with only real roots. One approach to establishing this would be to demonstrate that $\mathsf{B}_{k,p}^P$ is homotopy equivalent to a nonempty wedge of spheres of dimension $k - 1$ for $p + 1 \leq k \leq 2p - 1$; we know that this is true for $2 \leq k \leq p$.

14.3.4 Generalization to Hypergraphs

Given Proposition 14.14 (iii), it is tempting to conjecture that $\mathsf{B}_{n,p}$ and $\mathsf{B}_{n,p}^P$ are homotopy equivalent for $n \geq 2p$. In particular, since $\mathsf{B}_{n,p}^P$ aligns perfectly with the polynomial $f_p(n)$, one may argue that $\mathsf{B}_{n,p}^P$ is a more natural object to study than $\mathsf{B}_{n,p}$.

Now, we may interpret $B_{n,p}^P$ as a complex of hypergraphs with edges of size one and two. It is clear that a hypergraph H belongs to $B_{n,p}^P$ if and only if there is a set T of size at most p such that the intersection of T with each edge in H has size exactly one; note that x must belong to T whenever the singleton edge $\{x\}$ belongs to H. This observation suggests the following generalization:

A hypergraph $H = (V, \mathcal{E})$ admits an *exact r-transversal* if there is a set T of size r such that $|T \cap e| = 1$ for each $e \in \mathcal{E}$. The balance number $\beta(H)$ of a hypergraph H is the size of a smallest exact transversal of H. Note that an ordinary graph G is contained in $B_{n,p}$ and hence has balance number at most p if and only if G admits an exact r-transversal for some $r \le p$; T being a transversal of G is equivalent to $(T, [n] \setminus T)$ being a bipartition of G.

For any $n, p, t \ge 1$, define $HB_{n,p,t}$ as the family of $[t]$-hypergraphs H such that $\beta(H) \le p$, meaning that H admits an exact r-transversal for some $r \le p$. $HB_{n,p,2}$ is exactly the complex $B_{n,p}^P$. While we have not been able to prove very much about $HB_{n,p,t}$ for general t, at least we have the following intriguing observation:

Proposition 14.21. *For any $n, p, t \ge 1$, we have that $HB_{n,p,t} \simeq HB_{n,t,p}$.*

Proof. Assume that $p \ge 1$ and $t \ge 2$ and consider the nerve complex $N_{n,p,t} = N(HB_{n,p,t})$ of $HB_{n,p,t}$; see Section 6.1. We may identify the vertices in $N_{n,p,t}$ with subsets of $[n]$ of size at most p. Namely, for a given set U of size at most p, let H_U be the hypergraph containing all edges e of size at most t such that $|e \cap U| = 1$. Note that H_U contains the singleton set $\{u\}$ and the edge uv for each $u \in U$ and $v \in [n] \setminus U$, which implies that H_U is indeed maximal in $HB_{n,p,t}$; U is the *only* exact transversal of H_U. Conversely, it is easy to see that any maximal hypergraph in $HB_{n,p,t}$ is exactly of this form for some U of size at most p.

Now, a family $\mathcal{W}$ of vertices in $N_{n,p,t}$ forms a face of $N_{n,p,t}$ if and only if the intersection $\bigcap_{W \in \mathcal{W}} H_W$ is nonempty. This means that there is a set S of size at most t such that $|W \cap S| = 1$ for each $W \in \mathcal{W}$. However, this is exactly the condition that there is an exact r-transversal of the hypergraph $([n], \mathcal{W})$ for some $r \le t$. As a consequence, we may identify $N_{n,p,t}$ with $HB_{n,t,p}$. Hence we are done by the Nerve Theorem 6.2, the one remaining case $t = p = 1$ being trivially true. $\square$

For $t = 1$, the situation is very simple:

Corollary 14.22. *For $n, p \ge 1$ and $t = 1$, $HB_{n,p,1} \simeq HB_{n,1,p} \simeq \bigvee_{\binom{n-1}{p}} S^{p-1}$.*

Proof. $HB_{n,p,1}$ coincides with the $(p-1)$-skeleton of an $(n-1)$-simplex. $\square$

Based on Theorem 14.12 and Corollary 14.22, it is tempting to conjecture that $HB_{n,p,t}$ is homotopy equivalent to a wedge of spheres of dimension $pt - 1$ whenever $n \ge pt + 1$. However, we do not have any evidence for such a conjecture when $p, t \ge 3$; the real truth might be substantially more complicated.

Directed Variants of Forests and Bipartite Graphs

We consider complexes of directed variants of forests and bipartite graphs and some relatives thereof.

In Section 15.1, we consider the complex DF_n of directed forests. This is perhaps the most natural variant of the complex F_n of undirected forests. We review the main results about DF_n; these results are due to Kozlov [86]. Most notably, DF_n is vertex-decomposable of dimension $n-2$.

Another variant is the complex $DAcy_n$ of acyclic digraphs. In Section 15.2, we list some important results about $DAcy_n$ due to Björner and Welker [17] and Hultman [64]. The main result is that $DAcy_n$ is homotopy equivalent to a sphere of dimension $n-2$. Using the theory developed in Section 13.2, we show that $DAcy_n$ has a vertex-decomposable $(n-2)$-skeleton.

In Section 15.3, we show that the complex DB_n of bipartite digraphs on n vertices is homotopy equivalent to an $(n-2)$-dimensional sphere and has a vertex-decomposable $(n-2)$-skeleton. In addition, we give a direct proof that DB_n is homotopy equivalent to $DAcy_n$.

In Section 15.4, we proceed with the complex $DGr_{n,p}$ of digraphs on n vertices that are graded modulo p, the most important special case being $DGr_n = DGr_{n,n+1}$. One may view the two special cases DGr_n and $DGr_{n,2}$ as directed variants of the complex B_n of bipartite graphs. We show that $DGr_{n,p}$ is SPI over the digraphic matroid (see Section 13.3), which implies that the complex is homotopy equivalent to a wedge of spheres of dimension $n-2$ and has a vertex-decomposable $(n-2)$-skeleton. Moreover, fixing p, we determine the exponential generating function for the reduced Euler characteristic of $DGr_{n,p}$. We also compute the exponential generating function for $\tilde{\chi}(DGr_n)$.

One of the SPI monotone digraph properties listed at the end of Section 13.3.2 was the complex $DOAC_n$ of digraphs on n vertices with no non-alternating circuits. In Section 15.5, we show that this complex is indeed SPI.

Finally, in Section 15.6, we show that the complex $DNOCy_n$ of digraphs on n vertices without directed cycles of odd length is homotopy equivalent to a wedge of spheres of dimension $2n-3$. The Euler characteristic is, up to sign, the same as for B_n.

15.1 Directed Forests

The following result is a slight generalization of a result due to Kozlov [86] about the complex DF_n of directed forests on n vertices.

Theorem 15.1. *Let D be a digraph on the vertex set $[n]$. Assume that U is a vertex set with the property that, for each $v \in [n] \setminus U$, there is a $u \in U$ such that $uv \in D$. Then $\mathsf{DF}_n(D)$ is $VD(n - 1 - |U|)$.*

Proof. Write $W = [n] \setminus U$. Let Y be the set of edges $ab \in D$ with the property that $ab \notin U \times W$. For each $A \subseteq Y$, we want to show that $\Sigma_A = (\mathsf{DF}_n(D))(A, Y \setminus A)$ is $VD(n - 1 - |U|)$. This is clear if $\Sigma_A = \emptyset$; thus assume that Σ_A is nonvoid.

Let H be the digraph with edge set A and let $F_1, \ldots, F_k$ be the connected components of H that do not contain any element from U. If $k = 0$, then H consists of at most $|U|$ components and hence has at least $n - |U|$ edges, which implies that Σ_A is $VD(n - 1 - |U|)$. Otherwise, let r_i be the root of the component F_i for each i. By assumption, for each i, some $u_i \in U$ has the property that $u_i r_i \in D$. Define

$$Z = D \cap (U \times W) \setminus \{u_i r_i : i \in [1, k]\}.$$

For each $B \subseteq Z$, let us examine $\Sigma_A(B, Z \setminus B)$; assume that the complex is nonvoid. Let H' be the digraph with edge set $A \cup B$ and let $\{F_t : t \in T\}$ be the set of connected components in H' that do not contain any element from U; $T \subseteq [1, k]$. Now, every set in $\Sigma_A(B, Z \setminus B)$ is of the form $A \cup B \cup X$, where X is a subset of $C = \{u_t r_t : t \in T\}$; if $i \notin T$, then $ur_i \in B$ for some $u \in U$. One readily verifies that each $u_t r_t$ is a cone point in $\Sigma_A(B, Z \setminus B)$, which implies that $\Sigma_A(B, Z \setminus B)$ is $VD(|A| + |B| + |C| - 1)$. Since each connected component in the digraph with edge set $A \cup B \cup C$ contains some element from U, it follows that $\Sigma_A(B, Z \setminus B)$ is $VD(n - |U| - 1)$. Hence we are done by Lemma 6.10. $\square$

Corollary 15.2 (Kozlov [86]). *Let D be a digraph with vertex set $[n]$. Suppose that $1i \in D$ for $i \in [2, n]$. Then $\mathsf{DF}_n(D)$ is VD of dimension $n - 2$.* $\square$

With assumptions as in the corollary, Kozlov observed that the Euler characteristic of $\mathsf{DF}_n(D)$ equals (up to sign) the number of directed trees T in D with the property that there are no edges of the form $1i$ in T. We may easily deduce this fact from the proof of Theorem 15.1. As an important special case, Kozlov deduced that the Euler characteristic of DF_n is $-(1 - n)^{n-1}$.

15.2 Acyclic Digraphs

Björner and Welker [17] determined the homotopy type of the complex DAcy_n of acyclic digraphs on n vertices. We give an alternative proof in terms of decision trees.

Theorem 15.3. *For $n \geq 1$, DAcy_n admits a decision tree with one evasive face of dimension $n - 2$. Hence DAcy_n is homotopy equivalent to a sphere of dimension $n - 2$.*

Proof. We use induction on n. For $n = 1$, we have that $\mathsf{DAcy}_1 = \{\emptyset\}$; assume that $n > 1$. Consider the first-hit decomposition of DAcy_n with respect to $(1n, 2n, \ldots, (n-1)n)$; see Definition 5.24. For $r \in [n-1]$, let $A_r = \{in : i \in [r]\}$. Let $\Sigma_r = \mathsf{DAcy}_n(\{rn\}, A_{r-1})$ and $\Sigma_n = \mathsf{DAcy}_n(\emptyset, A_{n-1})$. We want to show that Σ_r is nonevasive for $r \neq n - 1$ and that $\Sigma_{n-1} \sim t^{n-2}$. By Lemma 5.25, it then follows that $\mathsf{DAcy}_n \sim t^{n-2}$.

Clearly, ni is a cone point in Σ_n for any i; if no edges are directed to n, then n cannot be contained in a cycle. In the proof of Theorem 15.4 below, we will need the fact that we may choose $n(n-1)$ as the cone point.

For $r \leq n - 1$, let $B = \{ni : i \in [n-1]\}$. For each $Z \subseteq B$, consider the complex $\Sigma_{r,Z} = \Sigma_r(Z, B \setminus Z)$. If $nr \in Z$, then $\Sigma_{r,Z}$ is void; (nr, rn) is a cycle. If $ni \in Z$ for some $i \neq r$, then ri is a cone point in $\Sigma_{r,Z}$; we already have a directed path from r to i via n.

It remains to consider $Z = \emptyset$. If $r \neq n - 1$, then $(r + 1)n$ is a cone point in $\Sigma_{r,\emptyset}$; n cannot be contained in a cycle since there are no edges directed from n. As a consequence, $\Sigma_r \sim 0$ if $r \neq n - 1$ by Lemma 5.22; $\Sigma_{r,Z} \sim 0$ for all Z. For $r = n - 1$, we have that $\Sigma_{n-1,\emptyset} = \{(n-1)n\} * \mathsf{DAcy}_{n-1}$. Namely, a digraph D containing $(n-1)n$ but no other edges incident to n is clearly acyclic if and only if the digraph obtained from D by removing the vertex n along with the edge $(n-1)n$ is acyclic. By induction, $\mathsf{DAcy}_{n-1} \sim t^{n-3}$, which implies that $\Sigma_{n-1} \sim t^{n-2}$; $\Sigma_{n-1,Z} \sim 0$ if $Z \neq \emptyset$. Hence we are done. $\square$

Theorem 15.4. *For $n \geq 1$, $\tilde{H}_{n-2}(\mathsf{DAcy}_n; \mathbb{Z})$ is generated by the fundamental cycle of the sphere*

$$\Delta_n = \{\emptyset, 12, 21\} * \{\emptyset, 23, 32\} * \ldots \{\emptyset, (n-1)n, n(n-1)\}.$$

Proof. We apply Corollary 4.17 to the decision tree defined in the proof of Theorem 15.3. Let D be a digraph with $n - 1$ edges $e_1, \ldots, e_{n-1}$ such that e_i is either $i(i+1)$ or $(i+1)i$. This implies that D is a maximal face of Δ_n. By Corollary 4.17, it suffices to prove that D is matched with a smaller digraph in the acyclic matching induced by the decision tree unless D is the unmatched digraph D_n with edge set $\{i(i+1) : i \in [n-1]\}$.

First, assume that $e_{n-1} = n(n-1)$. With notation as in the proof of Theorem 15.3, D belongs to Σ_n. In this lifted complex, $n(n-1)$ is a cone point, which means that we may define a perfect matching on Σ_n by adding and deleting $n(n-1)$. In particular, with this matching chosen, D is matched with a smaller digraph.

Second, assume that $e_{n-1} = (n-1)n$. Then D belongs to $\Sigma_{n-1,\emptyset} = \{(n-1)n\} * \mathsf{DAcy}_{n-1}$. By an induction argument, $D - (n-1)n$ is matched with a smaller digraph in DAcy_{n-1} unless $D - (n-1)n = D_{n-1}$ (i.e., $D = D_n$). As a consequence, the same is true for D in $\Sigma_{n-1,\emptyset}$, which concludes the proof. $\square$

Hultman generalized Theorem 15.3 to induced subcomplexes of DAcy_n:

Theorem 15.5 (Hultman [64]). *Let D be a digraph. If every connected component in D is strongly connected (or an isolated vertex), then $\mathsf{DAcy}_n(D)$ is homotopy equivalent to a sphere of dimension $n - c(D) - 1$, where $c(D)$ is the number of connected components in D. Otherwise, $\mathsf{DAcy}_n(D)$ is a cone.* $\square$

Since each minimal nonface of DAcy_n is a directed cycle, which is clearly isthmus-free, DAcy_n is PI. In particular, Corollary 13.8 applies:

Corollary 15.6. *For each digraph D on n vertices, $\mathsf{DAcy}_n(D)$ is $VD(n - c(D) - 1)$.* $\square$

We obtain a closure operator f on $P(\mathsf{DAcy}_n)$ by adding the edge ij whenever there is a directed path from i to j. Björner and Welker [17] examined intervals in the resulting poset $f(P(\mathsf{DAcy}_n))$; this poset is isomorphic to the poset of all posets on n elements, the antichain excluded.

15.3 Bipartite Digraphs

Recall that DB_n is the complex of digraphs on n vertices with the property that each vertex has either zero outdegree or zero indegree; if ij belongs to a given digraph in DB_n, then the digraph contains no edge jk starting in j and no edge ki ending in i.

Theorem 15.7. *For $n \geq 1$, DB_n is $VD^+(n-2)$ and homotopy equivalent to a sphere of dimension $n - 2$.*

Proof. Let $Y = \{ni : i \in [1, n-1]\}$. For each $A \subseteq Y$, we consider the family $\Sigma_A = \mathsf{DB}_n(A, Y \setminus A)$. First, note that $\Sigma_Y = \{Y\}$. Namely, by construction, a digraph in Σ_Y must not contain any edge starting in a vertex in $[1, n-1]$, and all remaining edges have this property.

It remains to prove that Σ_A is $VD(n-2)$ and nonevasive whenever $A \subsetneq Y$. First, consider the case $A = \emptyset$. Let Z be the set of all edges not containing the vertex n. For each $B \subseteq Z$, we want to examine $\Sigma_\emptyset(B, Z \setminus B)$. Let U be the set of vertices u such that B contains no edge of the form vu. Note that the edge un is a cone point whenever $u \in U$; the indegree of u and the outdegree of n remain zero if un is added. Each connected component of the digraph D on $[1, n-1]$ with edge set B has at least one element in U; thus $\Sigma_\emptyset(B, Z \setminus B)$ has at least $c(D)$ cone points. Since $|B| \geq n-1-c(D)$, it follows that $\Sigma_\emptyset(B, Z \setminus B)$ is $VD(n - 2)$ and nonevasive.

Next, assume that $A \neq \emptyset$. Let T be the set of vertices t such that $nt \in A$ and let Z be the set of edges not containing the vertex n or any vertex from

T. Let $B \subseteq Z$ and consider $\Sigma_A(B, Z \setminus B)$. Let W be the set of vertices w such that B contains no edge of the form vw. Note that the edge wt is a cone point whenever $t \in T$ and $w \in W$; the indegree of w and the outdegree of t remain zero if wt is added. Each connected component of the digraph D on $[1, n-1] \setminus T$ with edge set B has at least one element in W; thus $\Sigma_A(B, Z \setminus B)$ has at least $|T| \cdot c(D)$ cone points. Since $|A| = |T|$ and $|B| \geq n - 1 - |T| - c(D)$, we have that

$$|A| + |B| + |T| \cdot c(D) \geq n - 1 - c(D) + |T| \cdot c(D) \geq n - 1.$$

As a consequence, $\Sigma_A(B, Z \setminus B)$ is $VD(n-2)$, which concludes the proof. $\square$

Theorems 15.3 and 15.7 imply that DB_n is homotopy equivalent to the complex DAcy_n of acyclic digraphs on n vertices. This is no coincidence. Specifically, we obtain a collapse from DAcy_n to DB_n in the following manner. For each poset P on the set $[n]$, define $\mathcal{F}(P)$ as the family of digraphs $D \in \mathsf{DAcy}_n$ such that the associated poset $P(D)$ coincides with P; thus $x \leq y$ in P if and only if there is a directed path from x to y in D. It is easy to see that the Cluster Lemma 4.2 applies. Let P be a poset such that there exists a chain $x < y < z$. Then xz is a cone point in $\mathcal{F}(P)$. Hence DAcy_n is collapsible to the union of all $\mathcal{F}(P)$ such that P does not have any chain $x < y < z$. However, this union is exactly DB_n, and we are done. Restricting to posets with no chain of edge length above k, one proves the following result in exactly the same manner.

Theorem 15.8. *For $n \geq 1$ and $1 \leq k \leq n-1$, the complex $\mathsf{DAcy}_{n,k}$ of acyclic digraphs with no directed path of edge length $k+1$ (i.e., vertex length $k+2$) is homotopy equivalent to a sphere of dimension $n - 2$.* $\square$

Consider the face poset $P(\mathsf{DB}_n)$. We obtain a closure operator f on $P(\mathsf{DB}_n)$ by defining

$$f(D) = \{xy : xz \in D \text{ for some } z \text{ and } wy \in D \text{ for some } w\}.$$

By Closure Lemma 6.1, the order complex of the resulting poset $Q_n = f(P(\mathsf{DB}_n))$ has the same homotopy type as DB_n. We may identify a digraph D in Q_n with the pair (X, Y), where X is the set of vertices in D with at least one outgoing edge and Y is the set of vertices with at least one ingoing edge. As it turns out, Q_n is the face poset of a certain cell complex. This complex appears in the work of Babson and Kozlov, who demonstrated that the complex coincides with the boundary complex of a convex polytope [5, §4.2]. Note that this yields yet another proof that DB_n is homotopy equivalent to a sphere.

15.4 Graded Digraphs

We consider the complex $\mathsf{DGr}_{n,p}$ of digraphs on n vertices that are graded modulo p.

Theorem 15.9. *For each* $n \geq 1$ *and* $p \geq 2$, *the complex* $\mathsf{DGr}_{n,p}$ *is SPI over the digraphic matroid* $M_n^{\rightarrow}$. *In particular,* $\mathsf{DGr}_{n,p}$ *is* $VD^+(n-2)$ *and hence homotopy equivalent to a wedge of spheres of dimension* $n - 2$.

Remark. Note that $\mathsf{DGr}_{n,n+1} = \mathsf{DGr}_n$.

Proof. To prove that $\mathsf{DGr}_{n,p}$ is PI, let D be a disconnected digraph in $\mathsf{DGr}_{n,p}$ and let $e = uw \notin D$ be an edge joining two connected components in D; let U and W be the vertex sets of these components; $u \in U$ and $w \in W$. Let $f : [n] \rightarrow [0, p-1]$ be a p-grading of D, meaning that $(f(b) - f(a)) \bmod p = 1$ whenever $ab \in D$. Since no edges in D have one endpoint in W and another endpoint in $[n] \setminus W$, we have, for each integer i, that f_i is a p-grading of D, where

$$f_i(v) = \begin{cases} f(v) & \text{if } v \notin W; \\ (f(v) + i) \bmod p & \text{if } v \in W. \end{cases}$$

In particular, f_x is a p-grading of D, where $x = 1 + f(u) - f(w)$. Since

$$f_x(w) - f_x(u) \equiv (f(w) + 1 + f(u) - f(w)) - f(u) \equiv 1 \pmod{p},$$

it follows that $D + e \in \mathsf{DGr}_{n,p}$ and hence that $\mathsf{DGr}_{n,p}$ is PI.

To prove that $\mathsf{DGr}_{n,p}$ is SPI, let $D \in \mathsf{DGr}_{n,p}$ and let $e = uw \notin D$ be an edge such that $c(D) = c(D + e)$. Let W be the vertex set of the connected component containing e. Let f be a p-grading of D. One readily verifies that if g is another p-grading of D, then $(g(x) - f(x)) \bmod p$ is constant on W. As a consequence, e is a cone point in $\mathrm{lk}_{\mathsf{DGr}_{n,p}}(D)$ if $(f(w) - f(u)) \bmod p = 1$ and not present in $\mathrm{lk}_{\mathsf{DGr}_{n,p}}(D)$ if $(f(w) - f(u)) \bmod p \neq 1$. Thus $\mathsf{DGr}_{n,p}$ is SPI, which concludes the proof. $\square$

Say that a subset X of $[p]$ is *sparse modulo* p if, whenever x belongs to X, the two elements $(x-1) \bmod p$ and $(x+1) \bmod p$ do not belong to X.

Theorem 15.10. *For* $n \geq 1$ *and* $p \geq 2$, *let* $a_{n,p}$ *be the number of functions* $f : [n] \rightarrow [0, p-1]$ *such that* $f([n])$ *is sparse modulo* p. *Then*

$$B_p(x) := \sum_{n \geq 1} \tilde{\chi}(\mathsf{DGr}_{n,p}) \frac{x^n}{n!} = 1 - (1 + A_p(x))^{1/p},$$

where $A_p(x) = \sum_{n \geq 1} a_{n,p} x^n / n!$.

Proof. Consider the sum

$$H_p(x) = \sum_{n \geq 1} \frac{x^n}{n!} \sum_{f:[n] \rightarrow [0,p-1]} \sum_{D \sim f} (-1)^{|D|+1}; \tag{15.1}$$

$D \sim f$ means that f is a p-grading of D and $|D|$ is the number of edges in D. If f is not sparse, then there are vertices x and y satisfying $f(y) \equiv f(x) + 1 \pmod{p}$. In particular, $(D + xy) \sim f$ if and only if $(D - xy) \sim$

f. As a consequence, $\sum_{D \sim f} (-1)^{|D|+1}$ equals zero in this case. If $f([n])$ is sparse modulo p, then $D \sim f$ if and only if D is empty, which implies that $\sum_{D \sim f} (-1)^{|D|+1} = -1$. We conclude that $H_p(x) = -A_p(x)$.

Now, a digraph $D \in \mathsf{DGr}_{n,p}$ with c connected components appears exactly p^c times in the sum (15.1). Namely, let U be a vertex set of size c with one element from each component in D. Then the restriction of f to U uniquely determines the entire function f. Conversely, we may extend any function $U \to [0, p-1]$ to a p-grading of D. It follows that

$$-A_p(x) = H_p(x) = \sum_{n \geq 1} \frac{x^n}{n!} \sum_{D \in \mathsf{DGr}_{n,p}} (-1)^{|D|+1} p^{c(D)} = 1 - (1 - B_p(x))^p,$$

where the last equality is a consequence of Corollary 6.15. $\square$

Theorem 15.11. *With notation as in Theorem 15.10, we have that*

$$\sum_{p \geq 2} A_p(x) y^p = \frac{y + 2(e^x - 1)y^2}{1 - y - (e^x - 1)y^2} - \frac{y}{1-y}.$$

As a consequence,

$$A_p(x) = \frac{(2(e^x - 1) + \beta)\beta^{p-1} - (2(e^x - 1) + \alpha)\alpha^{p-1}}{\sqrt{4e^x - 3}} - 1, \qquad (15.2)$$

where $\alpha = \frac{1 - \sqrt{4e^x - 3}}{2}$ and $\beta = \frac{1 + \sqrt{4e^x - 3}}{2}$ are the two roots of the quadratic polynomial $u^2 - u = e^x - 1$.

Remark. Equivalently, we have that

$$\sum_{p \geq 2} \frac{A_p(x)}{e^x - 1} \cdot y^p = \frac{y^2(2 - y)}{(1 - y - (e^x - 1)y^2)(1 - y)}. \qquad (15.3)$$

Proof. By a simple inclusion-exclusion argument, we obtain that the number $a_{n,p}$ satisfies the identity

$$a_{n,p} = \sum_X \sum_{i=0}^{|X|} (-1)^i \binom{|X|}{i} (|X| - i)^n = \sum_{k=0}^{p} c_{p,k} \sum_{i=0}^{k} (-1)^i \binom{k}{i} (k - i)^n,$$

where the first sum is over all sparse subsets $X \subseteq [0, p-1]$ and $c_{p,k}$ is the number of such subsets of size k. We obtain that

$$A_p(x) = \sum_{k=0}^{p} c_{p,k} \sum_{i=0}^{k} (-1)^i \binom{k}{i} (e^{(k-i)x} - 1)$$

$$= -1 + \sum_{k=0}^{p} c_{p,k}(e^x - 1)^k = -1 + C_p(e^x - 1), \qquad (15.4)$$

where $C_p(t) = \sum_{k=0}^{p} c_{p,k} t^k$. It is easy to prove that $c_{p,k} = c_{p-1,k} + c_{p-2,k-1}$ for $p \geq 3$. As a consequence, defining $C = C(t,y) = \sum_{p \geq 1} C_p(y) t^p$, we conclude that

$$C - y - y^2(1 + 2t) = (C - y)y + Cty^2 \iff C = \frac{y + 2ty^2}{1 - y - ty^2}.$$

Combining this with (15.4), we obtain the first statement of the theorem.

The second statement of the theorem is a straightforward consequence of the first. □

Table 15.1. The function $A_p(x)/(e^x - 1)$ in (15.3) for small values on p.

p	$A_p(x)/(e^x - 1)$
2	2
3	3
4	$2e^x + 2$
5	$5e^x$
6	$2e^{2x} + 5e^x - 1$
7	$7e^{2x}$
8	$2e^{3x} + 10e^{2x} - 6e^x + 2$
9	$9e^{3x} + 3e^{2x} - 6e^x + 3$
10	$2e^{4x} + 17e^{3x} - 13e^{2x} + 2e^x + 2$

In Table 15.1, we provide a closed formula for $A_p(x)/(e^x - 1)$ for $p \leq 10$.

We now proceed with the problem of determining the Euler characteristic of DGr_n.

Theorem 15.12. *We have that* $H(x) := \sum_{n \geq 1} \tilde{\chi}(\mathsf{DGr}_n) \dfrac{x^n}{n!} = \dfrac{1 - \sqrt{4e^x - 3}}{2}$.

Proof. Since $\mathsf{DGr}_n = \mathsf{DGr}_{n,p}$ whenever $p > n$, it is clear that

$$H(x) = \lim_{p \to \infty} B_p(x)$$

coefficient-wise, where $B_p(x)$ is defined as in Theorem 15.10. Let notation be as in Theorem 15.11. A close examination of (15.2) yields that we may find a neighborhood U of the origin such that $(1 + A_p(x))^{1/p}$ converges uniformly to $\beta = \frac{1 + \sqrt{4e^x - 3}}{2}$ for all $x \in U$. Namely, choosing the neighborhood sufficiently small, we have that $|\alpha/\beta| < \delta$ for some fixed $\delta < 1$ for all x in this region. Since the convergence is uniform, the MacLaurin expansion of the limit coincides with the coefficient-wise limit; hence Theorem 15.10 implies that $H(x)$ is equal to $1 - \beta = \alpha = \frac{1 - \sqrt{4e^x - 3}}{2}$. □

For a different proof of Theorem 15.12, see the author's thesis [71].

Remark. $|\tilde{\chi}(\mathsf{DGr}_n)|$ is the number of semiorders on n elements with a connected incomparability graph; see sequence A048287 in Sloane's Encyclopedia [127].

15.5 Digraphs with No Non-alternating Circuits

We proceed with the complex DOAC_n of digraphs on n vertices with no non-alternating circuits.

Theorem 15.13. *For each $n \geq 1$, the complex DOAC_n is SPI over the digraphic matroid $M_n^{\rightarrow}$. In particular, DOAC_n is $VD^+(n-2)$ and hence homotopy equivalent to a wedge of spheres of dimension $n-2$.*

Proof. Let Ω_n be the set of edges in the complete digraph $K_n^{\rightarrow}$. Define $\varphi : \Omega_n \to \mathbb{Z}_2^n$ by $\varphi(pq) = \mathbf{e}_p + \mathbf{e}_q$; $\mathbf{e}_p$ is the pth unit vector in $\mathbb{Z}_2^n$. This is clearly a representation of the matroid $M_n^{\rightarrow}$. Define $\psi : \Omega_n \to \mathbb{Z}_2^{2n}$ by $\psi(pq) = \mathbf{e}_p + \mathbf{e}_{q+n}$. With notation as in Section 13.3.1, we want to prove that $\mathsf{DOAC}_n = \mathsf{B}_{M_n^{\rightarrow},\varphi,\psi}$; by Theorem 13.21, this will yield the desired result. By Corollary 13.22, we need only prove that the alternating circuits are exactly those circuits that belong to $\mathsf{B}_{M_n^{\rightarrow},\varphi,\psi}$. Now, every vertex incident to some edge in a circuit is incident to exactly two edges. This means that a circuit $\{a_1, \ldots, a_r\}$ satisfies $\sum \psi(a_i) = 0$ if and only if each relevant vertex is either the tail of two edges or the head of two edges in the circuit. This means exactly that the circuit is alternating. $\square$

Problem 15.14. Compute the Euler characteristic of DOAC_n.

15.6 Digraphs Without Odd Directed Cycles

We examine the complex DNOCy_n of digraphs on n vertices without directed cycles of odd length. One readily verifies that if a digraph contains a directed cycle of odd length, then there is a *simple* directed cycle of odd length. We may hence define DNOCy_n as the complex of digraphs avoiding simple directed cycles of odd length.

Theorem 15.15. *For $n \geq 1$, $\mathsf{DNOCy}_n \sim c_n t^{2n-3}$, where $c_n = |\tilde{\chi}(\mathsf{B}_n)|$. In particular, DNOCy_n is homotopy equivalent to a wedge of c_n spheres of dimension $2n-3$, and*

$$\sum_{n \geq 1} c_n \frac{x^n}{n!} = -\sqrt{2e^{-x} - 1} + 1.$$

Proof. One easily checks the theorem for $n \in \{1, 2\}$; thus assume that $n \geq 3$. Let $Y = \{in, ni : i \in [2, n-1]\}$. For each $A \subseteq Y$, we consider the family $\Sigma_A = \mathsf{DNOCy}_n(A, Y \setminus A)$. Let A^+ be the set of vertices i such that $ni \in A$ and $in \notin A$. Let A^- be the set of vertices i such that $in \in A$ and $ni \notin A$. Let $A^\pm$ be the set of vertices i such that $in, ni \in A$.

We have a number of cases:

- $A = \emptyset$. We have that $1n$ and $n1$ are cone points in $\Sigma_\emptyset$; n has no other neighbors than 1, which means that $1n$ and $n1$ cannot be contained in any simple directed cycle of odd length.
- $A^- = A^\pm = \emptyset$ and $A^+ \neq \emptyset$. Decompose Σ_A with respect to $1n$.
  - We have that $n1$ is a cone point in $\Sigma_A(\emptyset, 1n)$; no directed cycles contain the vertex n, as there are no edges ending in n.
  - Consider $\Sigma_A(1n, \emptyset)$ and let i be a vertex in A^+; thus $ni \in A$. Let $Z = \{ij : j \in [n-1] \setminus \{i\}\}$. For $B \subseteq Z$ such that $B \neq \emptyset$, let j be such that $ij \in B$. If $i1 \in B$, then $\Sigma_A(B + 1n, Z \setminus B)$ is void; $(1, n, i)$ forms an odd directed cycle. In particular, we may assume that $j \neq 1$. We claim that $1j$ is a cone point in $\Sigma_A(B + 1n, Z \setminus B)$. Namely, if the path $(1, j)$ is part of an odd directed cycle, then so is the path $(1, n, i, j)$. We refer to this property as the *4-vertex property*: Given a directed graph in DNOCy_n containing a simple directed path (a, b, c, d), we may add the edge ad without introducing odd directed cycles.

    The remaining case is $B = \emptyset$. Digraphs in $\Sigma_A(1n, Z)$ have the property that there are no edges starting in i. In particular, no cycles contain i, which immediately implies that $1i$ is a cone point.
- $A^+ = A^\pm = \emptyset$ and $A^- \neq \emptyset$. By symmetry, this case is analogous to the previous case.
- $A^+ \neq \emptyset$ and $A^- \cup A^\pm \neq \emptyset$. Let $i \in A^+$ and let $Z = \{ij : j \in [n-1] \setminus \{i\}\}$. Let j be a vertex such that $jn \in A$. For $B \subseteq Z$ such that $B \neq \emptyset$, we have that $\Sigma_A(B, Z \setminus B)$ is a cone by the 4-vertex property. Namely, we obtain the path (j, n, i, r), where $ir \in B$. For $B = \emptyset$, there are no edges starting in i, which implies that $1i$ is a cone point in $\Sigma_A(\emptyset, Z)$.
- $A^- \neq \emptyset$ and $A^+ \cup A^\pm \neq \emptyset$. Again by symmetry, this case is analogous to the previous case.
- $A^\pm \neq \emptyset$ and $A^- = A^+ = \emptyset$. Pick some $j \in A^\pm$ and let Z be the set of edges containing some vertex in $A^\pm \setminus \{j\}$, already checked edges excluded. If $B \neq \emptyset$, then $\Sigma_A(B, Z \setminus B)$ is void or a cone by the 4-vertex property. Namely, suppose that $ir \in B$ for some $i \in A^\pm$ and $r \neq i, j, n$. Then (j, n, i, r) is a directed path, and jr remains to be checked. Analogously, if $ri \in B$, then (r, i, n, j) is a directed path.

  The remaining case is $B = \emptyset$. The edges remaining to be checked are exactly all edges between vertices in $[n-1] \setminus (A^\pm \setminus \{j\})$ and the two edges $1n$ and $n1$. Decompose with respect to $1n$ and $n1$. There are four subfamilies of $\Sigma_A(\emptyset, Z)$ to consider:

- $\Sigma_A(1n, Z+n1)$. Let W be the set of edges ending in 1, already checked edges excluded. For $C \subseteq W$ such that $C \neq \emptyset$, we have that $\Sigma_A(C + 1n, (Z+n1) \cup (W \setminus C))$ is a cone by the 4-vertex property; rj is a cone point whenever r satisfies $r1 \in C$. For $C = \emptyset$, there are no edges ending in 1, which implies that $1j$ is a cone point in $\Sigma_A(1n, (Z+n1) \cup W)$.

- $\Sigma_A(n1, Z+1n)$. By symmetry, this case is analogous to the previous case.

- $\Sigma_A(\emptyset, Z \cup \{1n, n1\})$. We have that j is a cut point separating $(A^{\pm} \setminus \{j\}) \cup \{n\}$ from $[n-1] \setminus A^{\pm}$. Moreover, the edges remaining to be checked are exactly all edges between vertices in $[n-1] \setminus (A^{\pm} \setminus \{j\})$. As a consequence, $\Sigma_A(\emptyset, Z \cup \{1n, n1\})$ is isomorphic to $\{A\} * \mathsf{DNOCy}_{n-a}$, where $a = |A|/2 = |A^{\pm}|$. By induction on n, we obtain that

$$\Sigma_A(\emptyset, Z \cup \{1n, n1\}) \sim c_{n-a} t^{|A|+2(n-a)-3} = c_{n-a} t^{2n-3},$$

where $c_k = |\tilde{\chi}(\mathsf{B}_k)|$.

- $\Sigma_A(\{1n, n1\}, Z)$. Let W be the set of edges starting or ending in 1, already checked edges excluded. If $C \subseteq W$ is nonempty, then the family $\Sigma_A(\{1n, n1\} \cup C, Z \cup (W \setminus C))$ is a cone. For example, if C contains an edge $1r$ such that $j \neq r$, then jr is a cone point by the 4-vertex property; $(j, n, 1, r)$ is a simple path. The remaining case is $C = \emptyset$. We have that j is a cut point in every digraph in $\Sigma_A(\{1n, n1\}, Z \cup W)$ separating $(A^{\pm} \setminus \{j\}) \cup \{1, n\}$ from $[2, n-1] \setminus A^{\pm}$. The conclusion is that $\Sigma_A(\{1n, n1\}, Z \cup W)$ is isomorphic to $\{A\} * \{1n, n1\} * \mathsf{DNOCy}_{n-a-1}$, where again $a = |A|/2 = |A^{\pm}|$. By induction on n, we obtain that

$$\Sigma_A(\{1n, n1\}, Z \cup W) \sim c_{n-a-1} t^{|A|+2+2(n-a-1)-3} = c_{n-a-1} t^{2n-3}.$$

Note that there are $\binom{n-2}{a}$ sets $A^{\pm} \subseteq [2, n-1]$ such that $|A^{\pm}| = a$. Applying Lemma 5.22, we thus obtain that $\mathsf{DNOCy}_n \sim c'_n t^{2n-3}$, where

$$c'_n = \sum_{a=1}^{n-2} \binom{n-2}{a}(c_{n-a} + c_{n-a-1}) = \sum_{b=2}^{n-1} \binom{n-2}{b-2}(c_b + c_{b-1}).$$

This is exactly (14.2), which implies that $c'_n = c_n$ by Proposition 14.3. $\square$

Conjecture 15.16. DNOCy_n is $VD(2n-3)$ or at least has a Cohen-Macaulay $(2n-3)$-skeleton.

16

Noncrossing Graphs

Recall that the associahedron A_n is the complex of graphs on the vertex set $[n]$ without crossings and boundary edges. The associahedron was introduced by Stasheff [136]. We discuss the associahedron and some related dihedral properties, all defined in terms of crossing avoidance.

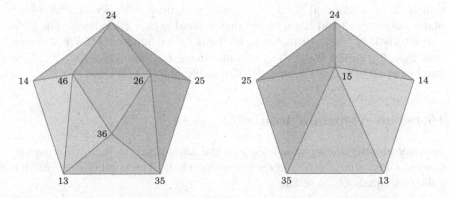

Fig. 16.1. Geometric realization of the two-dimensional sphere A_6 with the front of the sphere to the left and the back of the sphere to the right.

In Section 16.1, we provide an overview of the basic topological properties of A_n, the main property being that A_n is the boundary complex of a convex polytope of dimension $n - 3$; see Haiman [57] or Lee [90]. In particular, A_n is a shellable sphere. See Figure 16.1 for the case $n = 6$. We also provide a simple argument that A_n is semi-nonevasive.

In Section 16.2, we consider a certain well-known shelling of the n-fold cone NX_n over A_n with respect to $Bd_n = \{12, 23, \ldots, (n-1)n, 1n\}$. This shelling has the attractive property that every minimal face of the shelling is a forest. We use this shelling to compute the homotopy type of certain dihedral subcomplexes of NX_n. The most important example is the complex

NXF_n of noncrossing forests discussed below. We also show how to collapse certain dihedral complexes to their subcomplexes of noncrossing graphs. The crucial condition to be satisfied is that if a face contains a 2-crossing $\{ac, bd\}$, then we may add the edges ab, bc, cd, and da and remain inside the complex. We use this result in Chapter 21 to compute the homotopy type of the complexes $NCR_n^{0,0}$ and $NCR_n^{1,0}$ of graphs with a disconnected or separable polygon representation.

In Section 16.3, we analyze the complex $NXM_n = M_n \cap NX_n$ of noncrossing matchings. We show that NXM_n is $VD(\lceil \frac{n-4}{3} \rceil)$ and that this bound on the depth is sharp. In fact, NXM_n is semi-nonevasive and has homology in dimension d if and only if $\lceil \frac{n-4}{3} \rceil \leq d \leq \lfloor \frac{2n-5}{5} \rfloor$. As a consequence, the depth of NXM_n coincides with that of the full matching complex M_n; see Section 11.2. We also give a formula for the Euler characteristic.

In Section 16.4, we examine the complex $NXF_n = F_n \cap NX_n$ of noncrossing forests on the vertex set $[n]$. NXF_n inherits all nice topological properties from the full complex F_n of forests (see Section 13.1), except that NXF_n is not a matroid complex.

In Section 16.5, we consider the complex $NXB_n = B_n \cap NX_n$ of noncrossing bipartite graphs. Using properties of the associahedron established in Section 16.2, we prove that NXB_n is homotopy equivalent to a wedge of spheres of dimension $n-2$ and has a vertex-decomposable $(n-2)$-skeleton. The Euler characteristic of NXB_n turns out to be the nth Fine number [44]. We also show that the subcomplex $B_{n,p} \cap NX_n$ of noncrossing bipartite graphs with balance number at most p has a vertex-decomposable $(2p-1)$-skeleton.

16.1 The Associahedron

We may identify the maximal faces of the associahedron A_n with triangulations of the n-gon (the boundary edges excluded). The number of such triangulations equals C_{n-2}, where

$$C_m = \frac{1}{m+1} \binom{2m}{m};$$ (16.1)

C_m is the mth *Catalan number*. For a few different proofs of this fact, see Lovász [96, Ex. 1.38-40]. For an extensive list of other objects counted by Catalan numbers, see Stanley [134, 135].

The most important topological result about A_n is as follows.

Theorem 16.1 (Haiman [57], Lee [90]). *For $n \geq 3$, the associahedron A_n coincides with the boundary complex of an $(n-3)$-dimensional polytope and is hence shellable and homeomorphic to a sphere of dimension $n-4$.* $\square$

See Haiman [57], Lee [90], and Ziegler [152, Ex. 0.10 and 9.11] for further discussion and references.

Proving that A_n is homotopy equivalent to a sphere is not difficult:

Proposition 16.2. *For $n \geq 3$, $A_n \sim t^{n-4}$. In particular, A_n has the homotopy type of a sphere of dimension $n - 4$.*

Proof. Let $E_1 = \{1i : i \in [3, n-1]\}$ and consider the family $A_n(Y, E_1 \setminus Y)$ for each $Y \subseteq E_1$. First, assume that $Y \neq E_1$. Let $x_1 \geq 2$ be minimal such that $1(x_1 + 1) \notin Y$. Such an $x_1 \leq n - 2$ exists, as $Y \neq E_1$. Let $x_2 \geq x_1 + 2$ be minimal such that $1x_2 \in Y$ or $x_2 = n$. Then $x_1 x_2$ is a cone point in $A_n(Y, E_1 \setminus Y)$. Namely, $x_1 x_2 \in \mathrm{Int}_n$, because $2 \leq x_2 - x_1 \leq n - 2$. Moreover, any edge crossing $x_1 x_2$ contains the vertex 1 or crosses either $1x_1$ or $1x_2$. The remaining family is $A_n(E_1, \emptyset)$, which contains the single graph with edge set E_1; every edge in $\mathrm{Int}_n - E_1$ crosses some edge in E_1. Since $|E_1| = n - 3$, we are done by Lemma 5.22. $\square$

There are many proposed generalizations and variants of the associahedron in the literature:

- Kapranov's permutoassociahedron [79]; see also Reiner and Ziegler [112].
- The cyclohedron and its relatives; see Bott and Taubes [20], Simion [125], Markl [99], and Fomin et al. [47, 29].
- Complexes of "p-divisble" polygon dissections; see Przytycki and Sikora [109] and Tzanaki [140].
- Complexes of graphs avoiding $(k + 1)$-sets of mutually crossing edges; see Capoyleas and Pach [27], Nakamigawa [102], and Dress et al. [38, 39, 40]

16.2 A Shelling of the Associahedron

To facilitate analysis of many of the dihedral complexes to be studied in this paper, we will make use of a specific shelling of A_n due to Lee [90]. This shelling turns out to be a VD-shelling (see Section 6.3) with certain quite nice properties. Recall that NX_n is the complex of all noncrossing graphs on $[n]$; NX_n is the n-fold cone over A_n with respect to Bd_n.

Theorem 16.3. *For each $n \geq 3$, A_n admits a VD-shelling such that each minimal face G in the shelling has the property that G is a forest and hence cycle-free. Equivalently, the n-fold cone NX_n over A_n has the same property.*

Proof. Given a triangulation T of the n-gon, every interior edge e is a diagonal in the quadrilateral obtained by removing e from T; let e' be the other diagonal in the same quadrilateral. Aligning with Lee [90], refer to an interior edge e in T as *red* if e is smaller than e', the edges ordered lexicographically; refer to e as *green* if e is larger than e'.[1] All boundary edges are referred to as green. For a given triangulation T, let $R(T)$ be the set of red edges in T. We want to present a vertex decomposition such that the corresponding shelling $(\emptyset = \Delta_0, \ldots, \Delta_r = A_n)$ has the following property:

---

[1] Actually, Lee labeled the edges the other way around.

(i) For each $i \in [1, r]$, we obtain the minimal face σ_i of $\Delta_i \setminus \Delta_{i-1}$ from the maximal face τ_i of $\Delta_i \setminus \Delta_{i-1}$ by removing all green edges (including the boundary edges); thus $\sigma_i = R(\tau_i)$.

First, we prove that such a shelling has the desired property. Consider a cycle C in a maximal face τ_i, and let e be the edge joining the two largest vertices v and w in C. The edge e is either a boundary edge or a diagonal in a certain quadrilateral Q in τ_i. In the former case, e is clearly not part of σ_i; hence assume the latter case. Since τ_i is crossing-free, one of the two vertices joined by the other diagonal e' in Q belongs to C. However, this endpoint must then be strictly smaller than v and w, which implies that e' is smaller than e and hence that e is green in τ_i. It follows that C is not contained in the corresponding minimal nonface σ_i.

For $i \in [3, n-1]$, define

$$\Sigma_i = \mathsf{A}_n(1i, \{1j : j \in [3, i-1]\}).$$

In addition, let $\Sigma_n = \Delta(\emptyset, \{1j : j \in [3, n-1]\})$. In any triangulation, edges with one endpoint in 1 will always be red; hence the given decomposition does not violate the desired condition (i).

First, consider Σ_n. It is clear that $2n$ is a cone point, as no edges except the removed ones cross this edge. Also, in any triangulation in Σ_n, $2n$ is a green edge, as the other diagonal in the corresponding quadrilateral must contain the vertex 1. For any set $S \subseteq [n]$, define NX_S as the induced subcomplex of NX_S on the set $\binom{S}{2}$; define A_S as the induced subcomplex obtained from NX_S by removing the boundary edges in the convex polygon spanned by the vertices in S. Study the $(n-1)$-gon with vertex set $[2, n]$ and the corresponding complex $\mathsf{A}_{[2,n]}$. By induction on n, there is a vertex decomposition of $\mathsf{A}_{[2,n]}$ satisfying (i). Since Σ_n is a cone over $\mathsf{A}_{[2,n]}$ with cone point $2n$, we easily translate the given decomposition of $\mathsf{A}_{[2,n]}$ into a decomposition of Σ_n satisfying (i).

Next, consider Σ_i for $i \leq n-1$. Note that $2i$ is a cone point in Σ_i and clearly green in any triangulation in Σ_i. This time, we are interested in the two polygons on the vertex sets $[2, i]$ and $[i, 1] = \{1\} \cup [i, n]$, respectively. By induction, each of $\mathsf{A}_{[2,i]}$ and $\mathsf{A}_{[i,1]}$ admits a vertex decomposition satisfying (i). Now, we have that Σ_i is a cone over the join $\{1i\} * \mathsf{A}_{[2,i]} * \mathsf{A}_{[i,1]}$ with cone point $2i$ for $i \neq 3$ and that $\Sigma_3 = \{13\} * \mathsf{A}_{[3,1]}$. In particular, we obtain a decomposition of Σ_i satisfying (i) by first applying the given decomposition of $\mathsf{A}_{[2,i]}$ and then applying the given decomposition of $\mathsf{A}_{[i,1]}$. $\square$

Corollary 16.4. *Let Σ be a subcomplex of NX_n. Let d be a fixed integer.*

(i) *If $\Sigma(\sigma, \mathrm{Int}_n \setminus \tau)$ is void or shellable (VD-shellable) of dimension d for each cycle-free and noncrossing set σ of interior edges and each triangulation τ of the n-gon containing σ, then so is Σ. Analogously, if $\Sigma(\sigma, \mathrm{Int}_n \setminus \tau)$ is $VD(d)$ for each σ and τ as above, then so is Σ.*

(ii) *If $\Sigma(\sigma, \mathrm{Int}_n \setminus \tau)$ admits an acyclic matching (decision tree) with all critical (evasive) sets of the same dimension d for each cycle-free and noncrossing set σ of interior edges and each triangulation τ of the n-gon containing σ, then so does Σ.*

Proof. The shelling pairs (σ, τ) with respect to the VD-shelling in Theorem 16.3 satisfy the conditions in the corollary. Thus the corollary follows from Lemma 6.12. $\square$

Corollary 16.5. *Let M be a matroid on the edge set of K_n such that each triangulation of the n-gon has full rank r. If Δ is a PI complex over M (see Section 13.2), then $\Delta \cap \mathsf{NX}_n$ is $VD(r-1)$. If Δ is an SPI complex over M (see Section 13.3), then $\Delta \cap \mathsf{NX}_n$ is $VD^+(r-1)$.*

Proof. For each cycle-free and noncrossing set σ of interior edges and each triangulation τ of the n-gon containing σ, we have that $(\Delta \cap \mathsf{NX}_n)(\sigma, \mathrm{Int}_n \setminus \tau) = \Delta(\sigma, \mathrm{Int}_n \setminus \tau)$. Using Theorem 13.5 and the fact that pseudo-independence is closed under taking links and deletions, we obtain that $\Delta(\sigma, \mathrm{Int}_n \setminus \tau)$ is $VD(r-1)$. By Corollary 16.4 (i), it follows that $\Delta \cap \mathsf{NX}_n$ is $VD(r-1)$. If Δ is an SPI complex, then Theorem 13.10 implies that $\Delta(\sigma, \mathrm{Int}_n \setminus \tau)$ is $VD^+(r-1)$. As a consequence, $\Delta \cap \mathsf{NX}_n$ is $VD^+(r-1)$ by Corollary 16.4 (i)-(ii). $\square$

Given that certain conditions are satisfied, we may collapse a complex containing graphs with crossing edges to its subcomplex of noncrossing graphs:

Theorem 16.6. *Let Σ be a simplicial complex on $[n]$. Suppose that whenever a face $\sigma \in \Sigma$ contains two crossing edges ac and bd, the face $\sigma \cup \{ab, bc, cd, ad\}$ belongs to Σ. Then Σ admits a collapse to $\Sigma \cap \mathsf{NX}_n$.*

Proof. We define a perfect acyclic matching on $\Sigma \setminus \mathsf{NX}_n$ as follows. Define a linear order $\leq_L$ on the edges in $\binom{[n]}{2}$ in the following manner: An edge ij such that $i < j$ is smaller than an edge kl such that $k < l$ if $j < l$ or if $j = l$ and $i > k$. For a given graph $G \in \Sigma \setminus \mathsf{NX}_n$, let $e(G) = uv$ be maximal with respect to this order such that there are crossing edges xu and vy in G satisfying $x < y < u < v$. For $e \in \binom{[n]}{2}$, let $\mathcal{F}(e)$ be the family of graphs $G \in \Sigma \setminus \mathsf{NX}_n$ such that $e(G) = e$. We obtain a poset map $\Sigma \setminus \mathsf{NX}_n \to (\binom{[n]}{2}, \leq_L)$ by sending a graph in $\mathcal{F}(e)$ to the edge e. Namely, if we add edges to a graph G, then $e(G)$ cannot decrease. As a consequence, we may apply the Cluster Lemma 4.2 to the families $\mathcal{F}(e)$.

Let $e = uv$ with $u < v$ and let $G \in \mathcal{F}(e)$. Let x and y be such that $xu, yv \in G$ and $x < y < u < v$. We claim that $e(G + uv) = uv$. Namely, otherwise we must have an edge $rs \in G$ with $r < s$ such that rs and uv cross and such that $sv >_L uv$. If $r < u < s < v$, then $sv <_L uv$. If $u < r < v < s$, then yv and rs cross, which implies that $e(G) \geq_L vs >_L uv$. In both cases, we obtain a contradiction. As a consequence, it follows that $e(G + uv) = uv$.

Now, by assumption, we have that $G + e(G) \in \Sigma$ whenever $G \in \Sigma$. In addition, we have just demonstrated that $e(G) = e(G+e(G))$. In particular, e is a cone point in the family $\mathcal{F}(e)$. By the Cluster Lemma 4.2, it follows that $\Sigma \setminus \mathsf{NX}_n$ admits a perfect acyclic matching and hence that we have a collapse from Σ to $\Sigma \cap \mathsf{NX}_n$. $\square$

The generic example of a complex as in Theorem 16.6 is the complex of graphs with a separable polygon representation; see Section 21.3. Another example is the complex of graphs with a disconnected polygon representation; see Section 21.2.

16.3 Noncrossing Matchings

We discuss the complex NXM_n of noncrossing matchings on n vertices. It is well-known [134] that the number of perfect noncrossing matchings on $2m$ vertices is equal to the Catalan number $C_m = \frac{1}{m+1}\binom{2m}{m}$.

First, we consider the homology and depth of NXM_n. By convention, we define $\mathsf{NXM}_0 = \{\emptyset\}$. For $0 \le k \le n$, define $\mathsf{NXM}_{n,k} = \mathrm{del}_{\mathsf{NXM}_n}(\mathrm{Int}_k^*)$, where Int_k^* is the set of edges ij such that $i \ge 1$ and $i+2 \le j \le k$; $\mathrm{Int}_k^* = \emptyset$ for $k \le 2$ and $\mathrm{Int}_k^* = \mathrm{Int}_k + 1k$ for $k \ge 3$. Note that $\mathsf{NXM}_{n,k} = \mathsf{NXM}_n$ whenever $0 \le k \le 2$.

Theorem 16.7. *For* $0 \le k \le n$*, we have that* $\mathsf{NXM}_{n,k}$ *is* $VD(\nu_n)$*, where* $\nu_n = \lceil \frac{n-4}{3} \rceil$.

Proof. We use double induction on n and $n - k$. The case $n \le 3$ is easy to check by hand. Assume that $n \ge 4$.

The base case is that $k = n$; we have that $\mathsf{NXM}_{n,n}$ is the induced subcomplex on the set $\{i(i+1) : i \in [n-1]\}$. This is the edge set of the graph Pa_n discussed in Section 11.4; hence $\mathsf{NXM}_{n,n}$ coincides with $\mathsf{M}(\mathsf{Pa}_n)$. By Proposition 11.42, $\mathsf{NXM}_{n,n}$ is $VD(\nu_n)$.

Now, assume that $k < n$. Decompose $\mathsf{NXM}_{n,k}$ with respect to the 0-cells that are contained in $\mathsf{NXM}_{n,k}$ but not in $\mathsf{NXM}_{n,k+1}$; these 0-cells are the edges $i(k+1)$ for $i \in [k-1]$. Since at most one of these edges can be present in a matching, the order in which we decompose $\mathsf{NXM}_{n,k}$ is irrelevant. Note that the deletion of $\mathsf{NXM}_{n,k}$ with respect to $\{i(k+1) : i \in [k-1]\}$ is exactly $\mathsf{NXM}_{n,k+1}$, which is $VD(\nu_n)$ by induction on $n - k$.

It remains to consider the link with respect to $i(k+1)$ for each $i \in [k-1]$. The edge $i(k+1)$ divides the vertex set into the two intervals $[i+1,k]$ and $[k+2,i-1]$. Since there are no edges between those two intervals in a noncrossing matching containing $i(k+1)$, we have that $\mathrm{lk}_{\mathsf{NXM}_{n,k}}(i(k+1))$ is a join of two complexes. The first complex is isomorphic to $\mathsf{M}(\mathsf{Pa}_{k-i}) = \mathsf{NXM}_{k-i,k-i}$. Namely, the only remaining edges in $[i+1,k]$ are the boundary edges. The second complex is isomorphic to $\mathsf{NXM}_{n-k+i-2,i-1}$. Namely, all edges in $[k+2,i-1]$ remain to be checked except for the ones between vertices in $[i-1]$ on distance at least two. By induction, $\mathrm{lk}_{\mathsf{NXM}_{n,k}}(i(k+1))$ is hence $VD(\gamma)$, where

$$\gamma = \nu_{k-i} + \nu_{n-k+i-2} \geq \frac{k-i-4}{3} + \frac{n-k+i-6}{3} + 1$$

$$= \frac{n-4}{3} - 1. \tag{16.2}$$

Since the last expression rounded up is equal to $\nu_n - 1$, the conclusion is that $\mathrm{lk}_{\mathsf{NXM}_{n,k}}(i(k+1))$ is $VD(\nu_n - 1)$.

By induction, we obtain that $\mathsf{NXM}_{n,k}$ is $VD(\nu_n)$. $\square$

One may view Theorem 16.7 as a dihedral analogue of Athanasiadis' Theorem 11.7. Indeed, the vertex decomposition of NXM_n in the proof is inspired by Athanasiadis' decomposition [2] of HM_n^k.

Theorem 16.8. *For $0 \leq k < n$, we have that $\mathsf{NXM}_{n,k}$ is semi-nonevasive. In fact, $\tilde{H}_d(\mathsf{NXM}_{n,k}; \mathbb{Z})$ is isomorphic to the group*

$$\tilde{H}_d(\mathsf{NXM}_{n,k+1}; \mathbb{Z}) \oplus \bigoplus_i \tilde{H}_{d-\nu_i-2}(\mathsf{NXM}_{n-i-2,k-i-1}; \mathbb{Z}),$$

where the direct sum is over all $i \in [1, k-1]$ such that $i \bmod 3 \in \{0, 1\}$.

Proof. By Corollary 5.10, it suffices to prove that $\mathsf{NXM}_{n,k}$ is semi-nonevasive over $\mathbb{Q}$. First note that $\mathsf{NXM}_{n,n} = \mathsf{M}(\mathsf{Pa}_n)$ is semi-nonevasive by Proposition 11.42. For $k < n$, observe that the vertex decomposition in the proof of Theorem 16.7 partitions $\mathsf{NXM}_{n,k}$ into the subfamilies $\mathsf{NXM}_{n,k+1}$ and $\{i(k+1)\} * \mathrm{lk}_{\mathsf{NXM}_{n,k}}(i(k+1))$ for $1 \leq i \leq k+1$. By the same proof,

$$\mathrm{lk}_{\mathsf{NXM}_{n,k}}(i(k+1)) = \mathsf{M}(\mathsf{Pa}_{[i+1,k]}) * \Lambda_{n,k,i},$$

where $\mathsf{Pa}_{[i+1,k]}$ is the graph with edge set $\{j(j+1) : j \in [i+1, k-1]\}$ and $\Lambda_{n,k,i} \cong \mathsf{NXM}_{n-k+i-2,i-1}$ is a graph complex defined on the vertex set $[1, i-1] \cup [k+2, n]$. Note that

$$\tilde{H}_d(\mathsf{NXM}_{n,k}/\mathsf{NXM}_{n,k+1}; \mathbb{Q}) \tag{16.3}$$

$$\cong \bigoplus_{i=1}^{k-1} [i(k+1)] \otimes \tilde{H}_{\nu_{k-i}}(\mathsf{M}(\mathsf{Pa}_{[i+1,k]}); \mathbb{Q}) \otimes \tilde{H}_{d-\nu_{k-i}-2}(\Lambda_{n,k,i}; \mathbb{Q});$$

apply Corollary 4.23 and use the fact that $\mathsf{M}(\mathsf{Pa}_r)$ has homology only in dimension ν_r.

To settle semi-nonevasiveness over $\mathbb{Q}$, it suffices to show that the natural map $f_i : \tilde{H}_{\nu_{k-i}}(\mathsf{M}(\mathsf{Pa}_{[i+1,k]}); \mathbb{Q}) \otimes \tilde{H}_{d-\nu_{k-i}-2}(\Lambda_{n,k,i}; \mathbb{Q}) \to \tilde{H}_{d-1}(\mathsf{NXM}_{n,k+1}; \mathbb{Q})$ is zero for each i. Namely, this will imply that the exact sequence over $\mathbb{Q}$ for the pair $(\mathsf{NXM}_{n,k}, \mathsf{NXM}_{n,k+1})$ has the property that the natural map

$$f : \tilde{H}_d(\mathsf{NXM}_{n,k}/\mathsf{NXM}_{n,k+1}; \mathbb{Q}) \to \tilde{H}_{d-1}(\mathsf{NXM}_{n,k+1}; \mathbb{Q})$$

is zero. By (16.3), an induction argument yields the desired result.

Noting that the complex $\mathsf{M}(\mathsf{Pa}_{[i,k+1]}) * \Lambda_{n,k,i}$ is contained in $\mathsf{NXM}_{n,k+1}$ and contains $\mathsf{M}(\mathsf{Pa}_{[i+1,k]}) * \Lambda_{n,k,i}$, we may decompose f_i as

$$\tilde{H}_{\nu_{k-i}}(\mathsf{M}(\mathsf{Pa}_{[i+1,k]}); \mathbb{Q}) \otimes \tilde{H}_{d-\nu_{k-i}-2}(\Lambda_{n,k,i}; \mathbb{Q})$$

$$\downarrow$$

$$\tilde{H}_{\nu_{k-i}}(\mathsf{M}(\mathsf{Pa}_{[i,k+1]}); \mathbb{Q}) \otimes \tilde{H}_{d-\nu_{k-i}-2}(\Lambda_{n,k,i}; \mathbb{Q})$$

$$\downarrow$$

$$\tilde{H}_{d-1}(\mathsf{NXM}_{n,k+1}; \mathbb{Q}).$$

Since $\tilde{H}_{\nu_{k-i}}(\mathsf{M}(\mathsf{Pa}_{[i,k+1]}); \mathbb{Q}) \cong \tilde{H}_{\nu_{k-i}}(\mathsf{M}(\mathsf{Pa}_{k-i+2}); \mathbb{Q}) = 0$ for all $k - i$ by Proposition 11.42, it follows that f is indeed zero.

The final statement in the theorem is an immediate consequence of (16.3), Proposition 11.42, and Corollary 5.10. $\square$

Theorem 16.9. *For $n \geq 1$, NXM_n has homology in dimension d if and only if*

$$\left\lceil \frac{n-4}{3} \right\rceil \leq d \leq \left\lfloor \frac{2n-5}{5} \right\rfloor \iff \left\lceil \frac{5d+5}{2} \right\rceil \leq n \leq 3d+4.$$

Remark. Note that the upper bound $\lfloor \frac{2n-5}{5} \rfloor$ is significantly less than the nearly trivial upper bound $\lfloor \frac{n-3}{2} \rfloor$ for M_n (which is best possible by Theorem 11.12).

Proof. First, we show that there is indeed homology in the given dimensions. This is clear if $n \leq 4$; thus assume that $n \geq 5$. Iterating the homology formula in Theorem 16.8 for increasing values of k, starting with $k = 2$, we deduce that $\tilde{H}_d(\mathsf{NXM}_n; \mathbb{Z}) = \tilde{H}_d(\mathsf{NXM}_{n,2}; \mathbb{Z})$ is nonzero whenever $\tilde{H}_{d-\nu_i-2}(\mathsf{NXM}_{n-i-2,k-i-1}; \mathbb{Z})$ is nonzero for some $1 \leq i < k < n$ such that $i \bmod 3 \in \{0, 1\}$. We may hence conclude the following:

- For $i = 1$ and $k = 2$, we obtain $d - \nu_i - 2 = d - 1$ and $(n - i - 2, k - i - 1) = (n - 3, 0)$. By induction, $\tilde{H}_{d-1}(\mathsf{NXM}_{n-3}; \mathbb{Z})$ is nonzero whenever $\frac{5(d-1)+5}{2} \leq n-3 \leq 3(d-1)+4$, which implies that $\tilde{H}_d(\mathsf{NXM}_n; \mathbb{Z})$ is nonzero whenever $\frac{5d+6}{2} \leq n \leq 3d+4$.
- For $i = 3$ and $k = 4$, we obtain $d - \nu_i - 2 = d - 2$ and $(n - i - 2, k - i - 1) = (n - 5, 0)$. By induction, $\tilde{H}_{d-2}(\mathsf{NXM}_{n-5}; \mathbb{Z})$ is nonzero whenever $\frac{5(d-2)+5}{2} \leq n-5 \leq 3(d-2)+4$, which implies that $\tilde{H}_d(\mathsf{NXM}_n; \mathbb{Z})$ is nonzero whenever $\frac{5d+5}{2} \leq n \leq 3d+3$.

It remains to prove that there is no homology in NXM_n above dimension $\beta_n = \lfloor \frac{2n-5}{5} \rfloor$. We show that this is true for $\mathsf{NXM}_{n,k}$ for all k. To achieve this, we again use the homology formula in Theorem 16.8 combined with induction on n and $n - k$.

For $k = n$, we obtain $\mathsf{M}(\mathsf{Pa}_n)$, which has homology only in dimension ν_n by Proposition 11.42. One easily checks that $\nu_n \leq \beta_n$ for all n except $n = 2$, in which case $\mathsf{M}(\mathsf{Pa}_n)$ is collapsible.

For $k < n$, we need to show that $2n \geq 5d + 5$ whenever the homology group $\tilde{H}_{d-\nu_i-2}(\mathsf{NXM}_{n-i-2,k-i-1}; \mathbb{Z})$ is nonzero and $i \bmod 3 \in \{0, 1\}$. By induction, $2(n - i - 2) \geq 5(d - \nu_i - 2) + 5$. This yields that

$$0 \leq 2(n - i - 2) - 5((d - \nu_i - 2) + 1) = 2n - 5(d + 1) + 5\nu_i - 2i + 6.$$

Defining $\epsilon = i \bmod 3$, we obtain that

$$5\nu_i - 2i + 6 = \frac{5(i - 3 - \epsilon)}{3} - 2i + 6 = \frac{-i + 3 - 5\epsilon}{3} \leq 0$$

for all relevant i, which concludes the proof. $\square$

Corollary 16.10. *For $n \geq 1$, the homotopical depth of NXM_n is $\lceil \frac{n-4}{3} \rceil$.* $\square$

Computational evidence suggests the following conjecture:

Conjecture 16.11. *For each $k \geq 0$, we have that the rank of the homology group $\tilde{H}_{k-1}(\mathsf{NXM}_{3k+1}; \mathbb{Z})$ is $\frac{1}{k+1}\binom{4k+2}{k} = \frac{1}{3k+2}\binom{4k+2}{k+1}$.*

Table 16.1. Euler characteristic of NXM_n for small n.

$n = 1$	2	3	4	5	6	7	8	9	10	11	12
$\tilde{\chi}(\mathsf{NXM}_n)$ -1	0	2	3	-1	-11	-15	13	77	86	-144	-595

Finally, we compute the exponential generating function for the Euler characteristic of NXM_n; see Table 16.1 for the first few values.

Theorem 16.12. *Adopting the convention that $\mathsf{NXM}_0 = \{\emptyset\}$, the Euler characteristic of NXM_n satisfies*

$$F(x) := \sum_{n \geq 0} \tilde{\chi}(\mathsf{NXM}_n)x^n = \frac{1 - x - \sqrt{1 - 2x + 5x^2}}{2x^2}.$$

Proof. It is easy to see that

$$\mathsf{NXM}_n = \mathsf{NXM}_{n-1} \cup \bigcup_{i=1}^{n-1} \mathsf{NXM}_n(\{in\}, \emptyset); \tag{16.4}$$

for any $G \in \mathsf{NXM}_n$, the vertex n is adjacent to at most one other vertex. Moreover, the edge in divides the vertex set into the two intervals $[1, i-1]$ and $[i+1, n-1]$. In particular,

$$\mathrm{lk}_{\mathsf{NXM}_n}(in) \cong \mathsf{NXM}_{i-1} * \mathsf{NXM}_{n-i-1}.$$

With $a_k = \tilde{\chi}(\mathsf{NXM}_k)$, this implies that

$$\tilde{\chi}(\mathrm{lk}_{\mathsf{NXM}_n}(in)) = -a_{i-1}a_{n-i-1}. \tag{16.5}$$

Combining (16.4) and (16.5), we obtain that

$$a_n = a_{n-1} + \sum_{i=1}^{n-1} a_{i-1}a_{n-i-1};$$

here, we use the fact that $\tilde{\chi}(\mathsf{NXM}_n(\{in\}, \emptyset)) = -\tilde{\chi}(\mathrm{lk}_{\mathsf{NXM}_n}(in))$. Summing over n, we get

$$F(x) + 1 = \sum_{n \geq 1} a_n x^n = \sum_{n \geq 1} a_{n-1}x^n + \sum_{n \geq 1}\sum_{i=1}^{n-1} a_{i-1}a_{n-i-1}x^n$$
$$= xF(x) + x^2 F^2(x),$$

and we are done. $\square$

16.4 Noncrossing Forests

We discuss the complex NXF_n of noncrossing forests on the vertex set $[n]$. The number of noncrossing spanning trees is known to be $\frac{1}{2n-1}\binom{3(n-1)}{n-1}$ [41, 45, 104].

Table 16.2. Absolute value of the Euler characteristic of NXF_n for small n.

	$n = 1$	2	3	4	5	6	7	8	9	10	11
$\|\tilde{\chi}(\mathsf{NXF}_n)\|$	1	0	1	3	11	43	176	745	3235	14331	64516

Before stating our main result about NXF_n, we introduce some notation. A noncrossing cycle is *interior* if all edges in the cycle belong to Int_n and *non-interior* otherwise. In a subdivision of the n-gon into regions, a region is *interior* if the cycle forming its boundary is interior and *non-interior* otherwise.

Theorem 16.13. *For $n \geq 1$, NXF_n is VD and has the homotopy type of a wedge of τ_n spheres of dimension $n-2$, where τ_n is the number of dissections of the n-gon without any interior regions (equivalently, the number of noncrossing forests using only interior edges). The generating function $F(x) = \sum_{n \geq 1} \tau_n x^n$ satisfies the equation*

$$F^3(x) + (x^2 + x)F^2(x) - (2x^2 + x)F(x) + x^2 = 0;$$

note that the inverse of $F(x)$ is equal to $G(y) = \dfrac{y(1 + \sqrt{1 - 4y})}{2(1 - y)}.$

Remark. In Table 16.2, we present the Euler characteristic of NXF_n for small values on n.

Proof. Since $\mathsf{NXF}_n = \mathsf{NX}_n \cap \mathsf{F}_n$ and F_n is the independence complex of M_n and hence SPI of dimension $n - 2$, Corollary 16.5 immediately implies that NXF_n is VD of dimension $n - 2$.

To compute $\tilde{\chi}(\mathsf{NXF}_n)$, we prove that the Euler characteristic of $\Lambda_Y = \{Y\} * \mathsf{NXF}_n(Y, \mathrm{Int}_n \setminus Y)$ equals $(-1)^{n-2}$ for each edge set $Y \subset \mathrm{Int}_n$ forming a noncrossing forest. This will imply that $\tilde{\chi}(\mathsf{NXF}_n) = (-1)^{n-2}\tau_n$ as desired.

Now, all subgraphs of $Y \cup \mathrm{Bd}_n$ are noncrossing. In particular, Λ_Y coincides with $\{Y\} * \mathrm{lk}_{\mathsf{F}_n(Y \cup \mathrm{Bd}_n)}(Y)$, where $\mathsf{F}_n(Y \cup \mathrm{Bd}_n)$ is the independence complex of the graphic matroid on the edge set $Y \cup \mathrm{Bd}_n$. Since the rank of this matroid is $n - 1$, it follows that Λ_Y has homology only in top dimension $n - 2$.

It remains to prove that $|\tilde{\chi}(\Lambda_Y)| = 1$. Let $R_1, \ldots, R_k$ be the regions in the graph with edge set $Y \cup \mathrm{Bd}_n$. For $i \in [k]$, let E_i be the set of edges in Bd_n that are on the boundary of R_i. Since all regions R_i are non-interior, each E_i has size at least 1. We obtain that

$$\Lambda_Y = \{Y\} * \partial 2^{E_1} * \cdots * \partial 2^{E_k}.$$

Namely, a graph G such that $Y \subseteq G \subseteq Y \cup \mathrm{Bd}_n$ contains a cycle if and only if G contains the entire set E_i for some $i \in [k]$. Hence $|\tilde{\chi}(\Lambda_Y)| = 1$ as desired.

Now, we examine the Euler characteristic $a_n = \tilde{\chi}(\mathsf{NXF}_n)$ in greater detail. For a given graph G on $[n]$, let $v(G)$ be the smallest integer v such that $1v \in G$; if no such v exists, we define $v(G) = n + 1$. Let $\mathcal{F}(v)$ be the family of graphs $G \in \mathsf{NXF}_n$ such that $v(G) = v$. It is clear that $\tilde{\chi}(\mathcal{F}(n+1)) = \tilde{\chi}(\mathsf{NXF}_{n-1}) = a_{n-1}$.

For $v \leq n$, we have that the edge $1v$ divides the n-gon into the two parts $[2, v]$ and $[v, 1]$; we exclude the vertex 1 from the first part, as there are no edges from 1 to $(1, v)$. A graph G with $v(G) = v$ belongs to $\mathcal{F}(v)$ if and only if the induced subgraphs on $[2, v]$ and $[v, 1]$ both are noncrossing forests and $1v \in G$. Now, the family of noncrossing forests on $[2, v]$ is $\mathsf{NXF}_{[2,v]}$, whereas the family of noncrossing forests on $[v, 1]$ is $\mathsf{NXF}_{[v,1]}$. In the latter case, we should consider the subfamily $\mathsf{NXF}_{[v,1]}(1v, \emptyset)$, because the edge $1v$ is always present. Hence

$$\mathcal{F}(v) = \mathsf{NXF}_{[2,v]} * \mathsf{NXF}_{[v,1]}(1v, \emptyset).$$

Observing that one may identify $\mathsf{NXF}_{[v,1]}(1v, \emptyset)$ with $\mathsf{NXF}_{n-v+2}(12, \emptyset)$, we obtain that

$$\tilde{\chi}(\mathcal{F}(v)) = -a_{v-1}b_{n-v+2},$$

where $b_i = \tilde{\chi}(\mathsf{NXF}_i(12, \emptyset))$.

Summing, we obtain that

$$A(x) = \sum_{n \geq 1} a_n x^n = a_1 x + \sum_{n \geq 2} a_{n-1} x^n - \sum_{n \geq 2} \sum_{v=2}^{n} a_{v-1} b_{n-v+2} x^n$$
$$= -x + xA(x) - A(x)B(x)/x, \tag{16.6}$$

where $B(x) = \sum_{n \geq 2} b_n x^n$.

For a graph G containing 12, let $w = w(G) \geq 3$ be minimal such that $1w \in G$; we define $w(G) = n + 1$ if no such w exists. Let $\mathcal{G}(w)$ be the family of graphs G in $\mathsf{NXF}_n(12, \emptyset)$ such that $w(G) = w$. We observe that $\tilde{\chi}(\mathcal{G}(n + 1)) = -\tilde{\chi}(\mathsf{NXF}_{n-1}) = -a_{n-1}$; a graph such that 1 is adjacent only to 2 is a noncrossing forest if and only if this is true for the induced subgraph on $[2, n]$.

For $w \leq n$, $1w$ divides the n-gon into the two parts $[1, w]$ and $[w, 1]$; as we will see, this time we cannot exclude 1 from the first interval. A graph G with $w(G) = w$ belongs to $\mathcal{G}(w)$ if and only if the induced subgraphs G_1 on $[w, 1]$ and G_2 on $[1, w]$ both are noncrossing forests. This is true if and only if $G_1 \in \mathsf{NXF}_{[w,1]}(1w, \emptyset)$ and $G_2 \in \mathsf{NXF}_w(\{12, 1w\}, 1 \times (2, w))$; hence

$$\mathcal{G}(w) = \mathrm{lk}_{\mathsf{NXF}_{[w,1]}}(1w) * \mathsf{NXF}_w(\{12, 1w\}, 1 \times (2, w)).$$

However, $G_2 \in \mathsf{NXF}_w(\{12, 1w\}, 1 \times (2, w))$ if and only if the graph obtained from G_2 by removing the vertex 1 and adding the edge $2w$ belongs to $\mathsf{NXF}_{[2,w]}(2w, \emptyset)$. Thus

$$\mathcal{G}(w) \cong \mathrm{lk}_{\mathsf{NXF}_{n-w+2}}(12) * \mathrm{lk}_{\mathsf{NXF}_{w-1}}(12) * \{12, 1w\}.$$

The conclusion is that

$$\tilde{\chi}(\mathcal{G}(w)) = -b_{n-w+2} b_{w-1}.$$

Summing, we obtain that

$$B(x) = \sum_{n \geq 2} b_n x^n = -\sum_{n \geq 2} a_{n-1} x^n - \sum_{n \geq 2} \sum_{w=3}^{n} b_{w-1} b_{n-w+2} x^n$$
$$= -xA(x) - B^2(x)/x. \tag{16.7}$$

From (16.6), we derive that $B(x)/x = x - 1 - x/A(x)$. Inserting this in (16.7), we obtain with $A = A(x)$ that

$$x - 1 - \frac{x}{A} = -A - \left(x - 1 - \frac{x}{A}\right)^2$$
$$\iff A^3 + (x^2 - x)A^2 - (2x^2 - x)A + x^2 = 0.$$

Since $F(x) = A(-x)$, we are done. $\square$

Table 16.3. Absolute value of the Euler characteristic of NXB_n for small n. This value equals the Fine number F_n.

$n = 1$	2	3	4	5	6	7	8	9	10	11	12
1	0	1	2	6	18	57	186	622	2120	7338	25724

(row label: $|\tilde{\chi}(NXB_n)|$)

16.5 Noncrossing Bipartite Graphs

Recall that NXB_n is the complex of noncrossing bipartite graphs on $[n]$.

Theorem 16.14. *For $n \geq 1$, NXB_n is $VD^+(n-2)$ and homotopy equivalent to a wedge of F_n spheres of dimension $n-2$, where F_n is the nth Fine number defined by*

$$\sum_{n \geq 1} F_n x^n = \frac{1 + 2x - \sqrt{1 - 4x}}{4 + 2x} = \frac{xC(x) + x}{2 + x}.$$

Here, $C(x) = \sum_{n \geq 0} C_n x^n$, where C_n is the Catalan number $\frac{1}{n+1}\binom{2n}{n}$.

Remark. The Fine number F_n satisfies $2F_n + F_{n-1} = C_{n-1}$ for $n \geq 2$. See Fine [44] and Deutsch [35, App. C] for more information about Fine numbers and Table 16.3 for the first few values.

Proof. Since $NXB_n = NX_n \cap B_n$ and B_n is an SPI complex over M_n, Corollary 16.5 immediately implies that NXB_n is $VD^+(n-2)$.

It remains to determine the Euler characteristic $a_n = \tilde{\chi}(NXB_n)$. The procedure is very similar to that in the proof of Theorem 16.13: For a given graph G on $[n]$, let $v(G)$ be the smallest integer v such that $1v \in G$; if no such v exists, we define $v(G) = n + 1$. Let $\mathcal{F}(v)$ be the family of graphs $G \in NXB_n$ such that $v(G) = v$. As in the proof of Theorem 16.13, we obtain that $\tilde{\chi}(\mathcal{F}(n+1)) = \tilde{\chi}(NXB_{n-1}) = a_{n-1}$ and that

$$\tilde{\chi}(\mathcal{F}(v)) = -a_{v-1}b_{n-v+2},$$

where $b_i = \tilde{\chi}(NXB_i(12, \emptyset))$.

Summing as in (16.6), we obtain that

$$A(x) = -x + xA(x) - A(x)B(x)/x, \qquad (16.8)$$

where $B(x) = \sum_{n \geq 2} b_n x^n$.

Again as in the proof of Theorem 16.13, for a graph G containing 12, let $w = w(G) \geq 3$ be minimal such that $1w \in G$; we define $w(G) = n + 1$ if no such w exists. Let $\mathcal{G}(w)$ be the family of graphs $G \in NXB_n(12, \emptyset)$ such that $w(G) = w$. We observe that $\tilde{\chi}(\mathcal{G}(n+1)) = -\tilde{\chi}(NXB_{n-1}) = -a_{n-1}$; a graph such that 1 is adjacent only to 2 is bipartite and noncrossing if and only if this is true for the induced subgraph on $[2, n]$.

For $w \leq n$, $1w$ divides the n-gon into the two parts $[1, w]$ and $[w, 1]$. A graph G with $w(G) = w$ belongs to $\mathcal{G}(w)$ if and only if the induced subgraphs G_1 on $[w, 1]$ and G_2 on $[1, w]$ are both noncrossing and bipartite. This is true if and only if $G_1 \in \mathsf{NXB}_{[w,1]}(1w, \emptyset)$ and $G_2 \in \mathsf{NXB}_w(\{12, 1w\}, 1 \times (2, w))$; hence

$$\mathcal{G}(w) = \mathrm{lk}_{\mathsf{NXB}_{[w,1]}}(1w) * \mathsf{NXB}_w(\{12, 1w\}, 1 \times (2, w)).$$

We now arrive at a point where the proof no longer aligns with the proof of Theorem 16.13; the family $\mathcal{B}_w = \mathsf{NXB}_w(\{12, 1w\}, 1 \times (2, w))$ does not easily reduce to a family in $\mathsf{NXB}_{[2,w]}$. Instead, we claim that $\tilde{\chi}(\mathcal{B}_w) = \tilde{\chi}(\mathsf{NXB}_{w-2})$.

To prove this claim, proceed as follows. For a graph G in $\mathcal{B}_w$, let $x = x(G) \geq 3$ be minimal such that $xw \in G$. If no such x exists, define $x(G) = w$. Define $\mathcal{H}(x)$ as the family of graphs G in $\mathcal{B}_w$ satisfying $x(G) = x$. Now, if $x \leq w - 1$ and $G \in \mathcal{H}(x)$, then the graph H obtained by adding $2x$ to G remains in $\mathcal{H}(x)$. Namely, H remains bipartite, as 2 and x belong to different blocks in any bipartition of G; xw, $1w$, and 12 are present edges. Also, no edges cross $2x$, as there are no edges from 1 to $(2, w) \supset (2, x)$ and no edges from w to $(2, x)$. As a consequence, we have a perfect matching on $\mathcal{H}(x)$ given by pairing $G - 2x$ with $G + 2x$; hence $\tilde{\chi}(\mathcal{H}(x)) = 0$. The remaining set $\mathcal{H}(w)$ has the property that a graph G belongs to it if and only if the induced subgraph on $[2, w - 1]$ is bipartite; hence $\tilde{\chi}(\mathcal{H}(w)) = \tilde{\chi}(\mathsf{NXB}_{w-2})$. We conclude that

$$\tilde{\chi}(\mathcal{G}(w)) = b_{n-w+2} \cdot \tilde{\chi}(\mathsf{NXB}_w(\{12, 1w\}, 1 \times (2, w))) = a_{w-2}b_{n-w+2}.$$

Summing, we obtain that

$$B(x) = \sum_{n \geq 2} b_n x^n = -\sum_{n \geq 2} a_{n-1}x^n + \sum_{n \geq 2} \sum_{w=3}^{n} a_{w-2}b_{n-w+2}$$
$$= -xA(x) + A(x)B(x). \tag{16.9}$$

From this equation, we conclude that $B(x) = -xA(x)/(1 - A(x))$. Inserting this in (16.8), we obtain with $A = A(x)$ that

$$A + x = xA + \frac{A^2}{1 - A} \iff A = \frac{1 - 2x - \sqrt{1 + 4x}}{4 - 2x}.$$

Since $F(x) = A(-x)$, we are done. $\square$

For $p \geq 1$, recall that $\mathsf{B}_{n,p}$ is the subcomplex of B_n consisting of those bipartite graphs that admit a bipartition (U, W) such that one of U and W has size at most p.

Theorem 16.15. *For $p \geq 1$ and $n \geq 2p+1$, $\mathsf{NXB}_{n,p} = \mathsf{B}_{n,p} \cap \mathsf{NX}_n$ is $VD(2p-1)$.*

Proof. Let σ and τ be as in Corollary 16.4; τ is a maximal face of NX_n and σ is a cycle-free subset of $\tau \cap \mathrm{Int}_n$. By Corollary 16.4, it suffices to prove that

$\Sigma_{\sigma,\tau} = \mathsf{NXB}_n(\sigma, \mathrm{Int}_n \setminus \tau)$ is $VD(2p-1)$. Since $T = \tau \cup \mathrm{Bd}_n$ is noncrossing, $\Sigma_{\sigma,\tau} = \mathsf{NXB}_{n,p}(\sigma, \mathrm{Int}_n \setminus \tau)$ coincides with $\mathsf{B}_{n,p}(\sigma, \mathrm{Int}_n \setminus \tau)$. By Theorem 14.10, $\mathsf{B}_{n,p}(\sigma, \mathrm{Int}_n \setminus \tau)$ is $VD(r)$, where

$$r = c(\sigma) - c(\tau) + 2p - n + |\sigma|.$$

However, σ is a forest, which implies that $c(\sigma) - n + |\sigma| = 0$. Moreover, τ is connected; hence $c(\tau) = 1$. It follows that $r = 2p - 1$ as desired. $\square$

17

Non-Hamiltonian Graphs

[1] Using discrete Morse theory, we obtain some information about NHam_n, the simplicial complex of non-Hamiltonian graphs on n vertices. More precisely, in Section 17.1, we show that NHam_n is homotopy equivalent to

$$\bigvee_{(n-2)!} S^{2n-5} \vee \Sigma_n, \tag{17.1}$$

where Σ_n is a certain subcomplex of NHam_n.

For small values of n (at least for $n \leq 7$), the homology of Σ_n vanishes. However, this nice property does not seem to hold in general. Specifically, we show that $\tilde{H}_{14}(\Sigma_{10}, \mathbb{Z})$ contains a free subgroup of rank $8!/2$. It seems reasonable to conjecture that the homology of Σ_n is always nontrivial when n is at least ten.

In Section 17.2, we examine the homology of the quotient complex $\mathsf{Ham}_n = 2^{K_n}/\mathsf{NHam}_n$. The above result implies that the group $\tilde{H}_{2n-4}(\mathsf{Ham}_n, \mathbb{Z})$ contains a free subgroup of rank at least $(n-2)!$. We show that this subgroup has a basis with the property that each element coincides with a simple transformation of the fundamental cycle of the associahedron A_n (see Sections 16.1-16.2). In Section 19.1, we will see that this basis coincides with Shareshian's basis [118] for the homology of C_n^2, where C_n^2 is the quotient complex of 2-connected graphs.

In Section 17.3, we consider a directed variant of NHam_n. Specifically, define DNHam_n as the complex of non-Hamiltonian digraphs on n vertices. We prove that the shifted connectivity degree of DNHam_n is at least $2n - 3$; this bound is likely to be far from sharp.

---

[1] This chapter is a revised and extended version of Sections 5 and 8 in a paper [67] published in *Journal of Combinatorial Theory, Series A*.

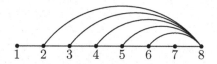

Fig. 17.1. A graph of the first kind when $n = 8$.

17.1 Homotopy Type

We provide an acyclic matching on the complex NHam_n of non-Hamiltonian graphs on the vertex set $[n]$. The matching has the property that the unmatched graphs are of two kinds:

1. Graphs with edge set

$$\{1\rho_2, \rho_2\rho_3, \ldots, \rho_{n-2}\rho_{n-1}, \rho_{n-1}n, \rho_2 n, \rho_3 n, \ldots, \rho_{n-2}n\},$$

   where $\{\rho_2, \ldots, \rho_{n-1}\} = \{2, \ldots, n-1\}$; see Figure 17.1 for an example.
2. Graphs G with a Hamiltonian path from 1 to n such that $G + 1n$ is 3-connected.

We denote the family of graphs of the first kind as $\mathcal{V}$ and the family of graphs of the second kind as $\mathcal{W}$. As it turns out, all graphs in $\mathcal{V}$ satisfy the conditions in Corollary 4.13. In particular, (17.1) holds, where Σ_n is the complex $(\mathsf{NHam}_n)_{\mathcal{W}}$ defined as in (4.2) with respect to the matching yet to be defined.

We divide the description of the acyclic matching into several steps.

Step 1: *Matching with the edge $1n$ to obtain the family* NHam'_n.

Match with $1n$ whenever possible, meaning that we pair $G - 1n$ and $G + 1n$ whenever $G + 1n$ is non-Hamiltonian. Let NHam'_n be the family of critical graphs with respect to this matching. Note that NHam'_n consists of all non-Hamiltonian graphs with a Hamiltonian path from 1 to n. By Lemma 4.1, any acyclic matching on NHam'_n together with the matching just defined yields an acyclic matching on NHam_n.

Step 2: *Defining the set* HP_G *and partitioning* NHam'_n *into subfamilies* $\mathsf{NHam}'_n(\mathcal{H})$.

For any graph G, let HP_G be the set of Hamiltonian paths from 1 to n in G. For $\mathcal{H} \subseteq \mathrm{HP}_{K_n}$, let

$$\mathsf{NHam}'_n(\mathcal{H}) = \{G :\in \mathsf{NHam}'_n : \mathrm{HP}_G = \mathcal{H}\}.$$

It is clear that the family $\{\mathsf{NHam}'_n(\mathcal{H}) : \mathcal{H} \subseteq \mathrm{HP}_{K_n}\}$ satisfies the conditions in the Cluster Lemma 4.2; thus it suffices to find an acyclic matching on $\mathsf{NHam}'_n(\mathcal{H})$ for each $\mathcal{H}$ such that $\mathsf{NHam}'_n(\mathcal{H})$ is nonvoid.

Step 3: *Defining the family $X(G)$ and the critical family $\mathcal{W}$.*

Let $\mathcal{H} \subseteq \mathrm{HP}_{K_n}$ be fixed. Consider the smallest Hamiltonian path

$$(1, \rho_2, \rho_3, \ldots, \rho_{n-1}, n)$$

in $\mathcal{H}$ with respect to lexicographic order (from left to right). Define $\rho_1 = 1$ and $\rho_n = n$.

For $G \in \mathsf{NHam}'_n(\mathcal{H})$, let $X(G)$ be the family of 2-sets $\{a, b\}$ of vertices such that $G([n] \setminus \{a, b\}) + 1n$ is disconnected. If an edge $ab \in X(G)$ is added to or deleted from G, then the family of Hamiltonian paths from 1 to n remains the same. Note that the pairs $1n, 1\rho_2, \rho_2\rho_3, \ldots, \rho_{n-2}\rho_{n-1}, \rho_{n-1}n$ do not belong to $X(G)$. Let $\mathcal{W}(\mathcal{H}) \subseteq \mathsf{NHam}'_n(\mathcal{H})$ be the family of graphs G such that $\mathrm{HP}_G = \mathcal{H}$ and $X(G) = \emptyset$; let

$$\mathcal{W} = \bigcup_{\mathcal{H}} \mathcal{W}(\mathcal{H}).$$

$X(G) = \emptyset$ means that $G + 1n$ is 3-connected; hence the graphs in $\mathcal{W}$ are exactly the graphs of the second kind as defined at the beginning of this section. We leave them unmatched and concentrate on the graphs G with nonempty $X(G)$; let

$$\mathcal{F}(\mathcal{H}) = \mathsf{NHam}'_n(\mathcal{H}) \setminus \mathcal{W}(\mathcal{H}).$$

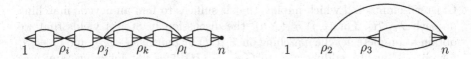

Fig. 17.2. The case $\rho_j\rho_l = S_G \neq \rho_i\rho_k = S_{G-\rho_j\rho_l}$ in Step 4; the situation turns out to be as in the picture to the right with $i = \rho_i = 1$, $j = 2$, $k = 3$, and $l = \rho_l = n$.

Step 4: *Proceeding with the family $\mathcal{F}(\mathcal{H})$ and defining the pair S_G.*

For a graph $G \in \mathcal{F}(\mathcal{H})$, we obtain a total order $\prec$ on pairs $\rho_r\rho_s$ in $X(G)$ such that $r < s$ by defining

$$\rho_i\rho_j \prec \rho_k\rho_l \iff (j < l) \text{ or } (j = l \text{ and } i < k);$$

this is lexicographic order from right to left. Let the *closed interval* $[\rho_i, \rho_j]$ be the set of all elements ρ_k such that $i \leq k \leq j$. The *half-open interval* $(\rho_i, \rho_j]$ is obtained from $[\rho_i, \rho_j]$ by removing ρ_i, while the *open interval* (ρ_i, ρ_j) is obtained by removing both endpoints ρ_i and ρ_j.

Let S_G be the smallest member of $X(G)$ with respect to $\prec$; assume that $S_G = \rho_j \rho_l$, $j < l$. If $\rho_j \rho_l \notin G$, then it is clear that $S_{G+\rho_j \rho_l} = S_G$. Suppose that $\rho_j \rho_l \in G$ and that

$$S_{G-\rho_j \rho_l} = \rho_i \rho_k \prec \rho_j \rho_l;$$

$i < k$. We claim that this implies that $i = 1$, $j = 2$, $k = 3$, and $l = n$. Namely, one readily verifies that

$$i < j < k < l;$$

the other possibilities would imply that $\rho_i \rho_k \in S_G$. Also, since $\rho_j \rho_l \in S_G$, we must have that $\rho_i \rho_k \notin G$. Since $(G + 1n)([n] \setminus \{\rho_j, \rho_l\})$ is disconnected, there are no edges in G between the open interval (ρ_j, ρ_l) and the union $[1, \rho_j) \cup (\rho_l, n]$ of half-open intervals. Similarly, there are no edges besides $x_j x_l$ between (ρ_i, ρ_k) and $[1, \rho_i) \cup (\rho_k, n]$. As a conclusion, there are no edges between (ρ_i, ρ_l) and $[1, \rho_i) \cup (\rho_l, n]$. Since $\rho_i \rho_l$ is smaller than $\rho_j \rho_l$ with respect to $\prec$, we must have $i = 1$ and $l = n$, because otherwise S_G would not be equal to $\rho_j \rho_l$. In the same manner, one shows that $j = 2$ and $k = 3$; otherwise either $\rho_i \rho_j$ or $\rho_j \rho_k$ (both smaller than $\rho_j \rho_l$) would be in $X(G)$. The situation is illustrated in Figure 17.2.

Step 5: *Partitioning $\mathcal{F}(\mathcal{H})$ into subfamilies $\mathcal{F}_{ij}(\mathcal{H})$.*

For $i < j$, let

$$\mathcal{F}_{ij}(\mathcal{H}) = \{G \in \mathcal{F}(\mathcal{H}) : S_G = \rho_i \rho_j\}.$$

It is clear that the family $\{\mathcal{F}_{ij}(\mathcal{H}) : i < j\}$ satisfies the conditions in the Cluster Lemma 4.2, which implies that it suffices to find an acyclic matching on each $\mathcal{F}_{ij}(\mathcal{H})$. For $(i, j) \neq (2, n)$, the discussion in Step 4 yields that we obtain a complete acyclic matching on $\mathcal{F}_{ij}(\mathcal{H})$ by pairing $G - \rho_i \rho_j$ and $G + \rho_i \rho_j$ for all G. Namely, adding $\rho_i \rho_j$ to $G \in \mathcal{F}_{ij}(\mathcal{H})$ does not introduce any new Hamiltonian paths from 1 to n and removing the same edge does not eliminate any such paths, which means that $G - \rho_i \rho_j$ and $G + \rho_i \rho_j$ belong to the same set $\mathcal{F}_{ij}(\mathcal{H})$.

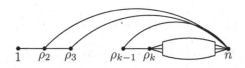

Fig. 17.3. The situation in Step 6; $k = k_G$.

Step 6: *Defining an optimal acyclic matching on* $\mathcal{F}_{2n}(\mathcal{H})$.

It remains to study $\mathcal{F}_{2n}(\mathcal{H})$. For $G \in \mathcal{F}_{2n}(\mathcal{H})$, let

$$k_G = \max\{k : \rho_2 n, \ldots, \rho_k n \in X(G)\};$$

clearly, $2 \leq k_G \leq n-2$. Note that $\rho_j n \in G$ for $2 \leq j < k_G$; otherwise, we would have $\rho_{j-1}\rho_{j+1} \in X(G)$, contradicting the minimality of $2n$ in $X(G)$. Moreover, if $k_G < n - 2$, then there is an m such that $k_G + 2 \leq m < n$ and $\rho_{k_G}\rho_m \in G$. Namely, otherwise $\rho_{k_G+1} n \in X(G)$, which would contradict the maximality of k_G. In particular, with $e = \rho_{k_G} n$ we have that $k_G = k_{G-e} = k_{G+e}$. See Figure 17.3 for an illustration.

Let

$$\mathcal{F}_{2n}^k(\mathcal{H}) = \{G \in \mathcal{F}_{2n}(\mathcal{H}) : k_G = k\}$$

for $2 \leq k \leq n - 2$. Another application of the Cluster Lemma 4.2 yields that it suffices to find an acyclic matching on each of the families $\mathcal{F}_{2n}^k(\mathcal{H})$. By the above discussion, the matching defined by pairing $G - \rho_k n$ with $G + \rho_k n$ for all G is a complete acyclic matching on $\mathcal{F}_{2n}^k(\mathcal{H})$ when $k < n-2$. If $k_G = n-2$, then G is a graph of the first kind as defined at the beginning of this section, which implies that $\mathcal{V}$ is exactly the union of all $\mathcal{F}_{2n}^{n-2}(\mathcal{H})$. As a consequence, taking the union of all matchings mentioned in the construction, we obtain an acyclic matching $\mathcal{M}$ on NHam_n whose critical graphs are the graphs in $\mathcal{V} \cup \mathcal{W}$.

Step 7: *Examining paths in the digraph associated to the given acyclic matching.*

By Corollary 4.13, (17.1) is a consequence of the following lemma.

Lemma 17.1. *If G and H are critical graphs in NHam_n with respect to the provided matching, then $H \longrightarrow G$ if and only if $G \subseteq H$ and $G, H \in \mathcal{W}$.*

Proof. By the maximality of $\mathcal{W}$ in NHam_n and the fact that all graphs in $\mathcal{V}$ have the same size, we need only show that

$$H \in \mathcal{W}, G \in \mathcal{V} \Longrightarrow H \not\longrightarrow G.$$

Assume that the Hamiltonian path from 1 to n in G is

$$P = (1, \rho_2, \rho_3, \ldots, \rho_{n-1}, n).$$

Suppose that

$$(G_1, G_2, \ldots, G_{r-1}, G_r = G)$$

is a directed path of graphs in the directed graph D corresponding to our acyclic matching. By a simple induction argument it follows immediately that

$P \in \mathrm{HP}_{G_i}$ for $1 \leq i \leq r$. Namely, our matching has the property that two graphs $G, H \in \mathsf{NHam}'_n$ cannot be matched unless the corresponding sets HP_G and HP_H are the same. When we go from G_r and backwards, we only add edges distinct from $1n$; hence all graphs are in NHam'_n. Moreover, we remove only edges that are used in the matching on NHam'_n; thus HP_{G_i} grows (weakly) as i decreases.

We claim the following:

(i) *For each $k \in [2, n-2]$ and each $i \in [1, r]$, there is an $m \geq k+2$ such that $\rho_k \rho_m \in G_i$.*

(ii) *For each $k \geq 3$, if (i) holds for a graph G containing the Hamiltonian path P, then there is a Hamiltonian path in G from 1 to ρ_k containing $1\rho_2$.*

Before proving the claims, we note that (ii) implies that $1\rho_k \notin G_i$ for $k > 2$; thus $\rho_2 n \in X(G_i)$. This implies that $G_i \notin \mathcal{W}$ as desired.

Proof of claim (i). Use induction: Assume that G_{i+1} satisfies (i). If $G_{i+1} \subset G_i$, then G_i trivially satisfies (i). Otherwise, G_i and G_{i+1} are matched. In particular, $X(G_i) \neq \emptyset$. If $l+2 \leq j < n$, then $\rho_l \rho_j \notin X(G_{i+1})$, because there is an edge from ρ_{j-1} to some vertex ρ_m, $m > j$. Hence we must have $S_{G_{i+1}} = \rho_2 n$ and $G_{i+1} = G_i + \rho_k n$ for some $k \geq 2$. Whatever the minimal Hamiltonian path P' from 1 to n in G_i looks like, the first k elements in P' must be $\{1, \rho_2, \ldots, \rho_k\}$ with ρ_k on position k. Namely, by construction, $\rho_k n \in X(G_i)$. Since $P \in \mathrm{HP}_{G_i}$, this implies that there are no edges from $\{1, \rho_2, \ldots, \rho_{k-1}\}$ to $\{\rho_{k+1}, \ldots, \rho_{n-1}\}$. If ρ_k is not followed by ρ_{k+1} in P' in G_i, then there is trivially an edge $\rho_k \rho_m \in G_i$ such that $k+2 \leq m < n$. By the discussion in Step 6 above, this is also true if ρ_k is followed by ρ_{k+1} in P'.

Proof of claim (ii). To simplify notation, assume that $\rho_j = j$ for all j. We use induction on n, $n \geq 3$. For $n = 3$, the statement is trivial. Assume that $n \geq 4$. By assumption, there is an edge $(k-1)m$ such that $k+1 \leq m \leq n$. If $k = n-1$, then we are finished; thus assume that $k \leq n-2$.

First, assume that $m > k+1$. Then, by induction hypothesis, there is a Hamiltonian path

$$(m_1 = k, m_2 = k+1, m_3, \ldots, m_{n-k+1} = m)$$

from k to m in $G([k, n])$. Hence

$$(1, \ldots, k-1, m_{n-k+1} = m, m_{n-k}, \ldots, m_1 = k)$$

is a Hamiltonian path in G.

Next, assume that $m = k+1$. Let $m' \geq k+2$ be such that $km' \in G$. Again by induction hypothesis, there is a Hamiltonian path

$$(m_1 = k, m_2 = k+1, m_3, \ldots, m_{n-k+1} = m')$$

from k to m' in $G([k,n])$. This time,

$$(1,\ldots,k-1,m_2=k+1,m_3,\ldots,m_{n-k+1}=m',k)$$

is a Hamiltonian path in G, which concludes the proof of claim (ii).

Conclusion.

We have proved the following:

Theorem 17.2. *The complex* NHam_n *is homotopy equivalent to*

$$\bigvee_{(n-2)!} S^{2n-5} \vee (\mathsf{NHam}_n)_\mathcal{W},$$

where $(\mathsf{NHam}_n)_\mathcal{W}$ *is defined as in* (4.2) *with respect to the matching given in Steps 1-6.* $\square$

Corollary 17.3. *For* $n \geq 6$, NHam_n *has shifted connectivity degree at least* $\lceil \frac{3n}{2} \rceil - 2$.

Proof. Each graph G in $\mathcal{W}$ has the property that $G+1n$ is 3-connected, which implies that G contains at least $3n/2 - 1$ edges. Since $3n/2 - 1$ is at most $2n - 4$ for $n \geq 6$, we are done by Theorem 4.7. $\square$

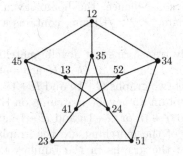

Fig. 17.4. The Petersen graph.

Corollary 17.4. NHam_{10} *is homotopy equivalent to a wedge of some simplicial complex and*

$$\bigvee_{8!/2} S^{14} \vee \bigvee_{8!} S^{15}. \qquad (17.2)$$

The spheres of dimension 14 correspond to Petersen graphs (see Figure 17.4) in which $\{1,10\}$ *is not an edge, while the spheres of dimension 15 correspond to the graphs in* $\mathcal{V}$.

Proof. It is a straightforward exercise to check that the Petersen graph is non-Hamiltonian and 3-connected. By Corollary 4.13 and Lemma 17.1, we need only show that any critical Petersen graph P is isolated among the critical graphs in $\mathcal{W}$. Since P is 3-regular, $X(G)$ is nonempty whenever $G = P - e$ for some $e \in P$. Moreover, between any pair of nonadjacent vertices ab, bc in the Petersen graph, there are several Hamiltonian paths, one example being

$$(ab, cd, ea, bd, ce, da, eb, ac, de, bc)$$

(notation as in Figure 17.4; $\{a, b, c, d, e\} = \{1, 2, 3, 4, 5\}$). Thus P is a maximal non-Hamiltonian graph, which concludes the proof. $\square$

While we have obtained partial information about the homotopy type of NHam_n, we have failed in our attempts to provide a more complete description of the complex. In particular, the following problem remains open:

Question 1. What is the homotopy type of $(\mathsf{NHam}_n)_\mathcal{W}$ in Theorem 17.2?

17.2 Homology

Unfortunately, we have not been able to give a complete description of the homology of NHam_n; this open problem is probably very difficult to solve. However, at least we know that $\tilde{H}_{2n-5}(\mathsf{NHam}_n)$ contains a free subgroup of rank $(n-2)!$. As a consequence, the homology in dimension $2n-4$ of the quotient complex $\mathsf{Ham}_n = 2^{K_n}/\mathsf{NHam}_n$ contains a free subgroup of rank $(n-2)!$.

For $n \geq 4$, one easily transforms our acyclic matching on NHam_n into an acyclic matching on Ham_n. Namely, if $G \subset H$ are matched in NHam_n, then either $H = G+1n$ or the two graphs $G+1n$ and $H+1n$ are both Hamiltonian. As a consequence, we obtain an acyclic matching on Ham_n by pairing G and H if either H is equal to $G+1n$ or $G-1n$ and $H-1n$ are matched in NHam_n.

The critical graphs of the matching are the graphs obtained by adding the edge $1n$ to each of the graphs in the families $\mathcal{V}$ and $\mathcal{W}$ described at the beginning of Section 17.1. Therefore, modify these families by adding $1n$ to every graph in the families. This means that $\mathcal{V}$ contains graphs $G(\rho) = G(1, \rho_2, \ldots, \rho_{n-1}, n)$ with edge sets

$$\{1\rho_2, \rho_2\rho_3, \ldots, \rho_{n-2}\rho_{n-1}, \rho_{n-1}n, \rho_2 n, \rho_3 n, \ldots, \rho_{n-2}n\} \cup \{1n\},$$

where $\{\rho_2, \ldots, \rho_{n-1}\} = \{2, \ldots, n-1\}$.

It is of great importance for us that Lemma 17.1 remains true for Ham_n. One readily verifies that no path starting and ending in $\mathcal{V} \cup \mathcal{W}$ contains any graph matched using $1n$. This implies that the only thing we have to show is again that $H \not\longrightarrow G$ if $H \in \mathcal{W}$ and $G \in \mathcal{V}$. However, following the proof of Lemma 17.1, we easily obtain the desired result.

In particular, we may apply Theorem 4.19 to conclude that there is a $b_G \in B$ for every $G \in \mathcal{V}$ such that $\{G - b_G : G \in \mathcal{V}\}$ is a basis for $H_{2n-4}(\mathsf{V})$, where B is the group generated by all graphs matched with smaller graphs and V is defined as in Theorem 4.19. We will show that the elements $G - b_G$ are "lifted" fundamental cycles in the associahedron A_n. As we will describe in Section 19.1, Shareshian [118] earlier proved that these elements also generate the homology of the quotient complex $\mathsf{C}_n^2 = 2^{K_n}/\mathsf{NC}_n^2$ of 2-connected graphs.

We will need the following lemma.

Lemma 17.5. *With notation as in Section 16.1, let* $G \in \mathsf{NX}_n$. *Then* G *is Hamiltonian if and only if all edges in* Bd_n *belong to* G.

Proof. It suffices to prove that the only Hamiltonian cycle in a triangulated n-gon is the n-gon itself. However, this is obvious, because a graph in NX_n that does not contain the entirety of Bd_n cannot be 2-connected; see Shareshian [118, Lemma 4.1] for details. $\square$

For a given permutation $\rho = [\rho_1, \rho_2, \ldots, \rho_{n-1}, \rho_n]$ of $[n]$, write

$$\rho(\mathrm{Bd}_n) = \{\rho_1\rho_2, \rho_2\rho_3, \ldots, \rho_{n-1}\rho_n, \rho_n\rho_1\} = \{\rho(i)\rho(j) : ij \in \mathrm{Bd}_n\}.$$

Let π be the fundamental cycle of A_n and let $\rho(\pi)$ be the cycle obtained from π by relabeling the vertex i as ρ_i. Define

$$\rho(\pi^*) = \rho(\pi) \wedge [\rho(\mathrm{Bd}_n)]; \tag{17.3}$$

we obtain $\rho(\pi^*)$ from $\rho(\pi)$ by replacing each summand $[\sigma]$ with $[\sigma] \wedge [\rho(\mathrm{Bd}_n)]$. It is clear that $\rho(\pi^*)$ is an element in the chain complex of Ham_n.

Theorem 17.6. *With notation as above, the set* $\{\rho(\pi^*) : \rho \in \mathfrak{S}_{[n]}, \rho_1 = 1, \rho_n = n\}$ *is linearly independent in* $\tilde{H}_{2n-4}(\mathsf{Ham}_n)$.

Proof. As an immediate consequence of Lemma 17.5, the boundary of $\rho(\pi^*)$ in the chain complex of Ham_n equals $\partial(\rho(\pi)) \wedge [\rho(\mathrm{Bd}_n)]$, which is zero.

Let $[T]$ be a summand in $\rho(\pi^*)$; $T = \rho(T')$ for some triangulation T' of the n-gon. It remains to show that T is matched with a smaller graph unless $T = G(\rho)$. Consider the family $X(T)$ of all pairs $\{a, b\}$ separating T as defined in Step 3 in Section 17.1, and let S_T be as in Step 4 in the same section. Note that there is a unique Hamiltonian path from 1 to n in $T - 1n$. Suppose that S_T is not an edge in T. Since T is a triangulation, this means that S_T crosses some edge if S_T is drawn in the interior of the Hamiltonian cycle. That is, if $S_T = \rho_i\rho_j$ with $i < j$, then there is a k between i and j such that that there is an edge $\rho_k\rho_l$ with $l > j$ or $l < i$. In particular, $\rho_i\rho_j \notin X(T)$, which is a contradiction. Thus $S_T \in T$.

If $S_T \neq \rho_2 n$, then Step 5 shows that T is matched with a smaller graph. If $S_T = \rho_2 n$, then turn to Step 6 and consider the element k_T. The same argument as above shows that $k_T n$ is contained in T. Thus again T is matched with a smaller graph unless $T = G(\rho^*)$. $\square$

17.3 Directed Variant

Finally, we consider the complex DNHam_n of directed non-Hamiltonian graphs. Our only result is the following simple bound on the connectivity degree:

Proposition 17.7. *For $n \geq 3$, the shifted connectivity degree of DNHam_n is at least $2n - 3$.*

Proof. For a given digraph D in DNHam_n, let $X(D)$ be the set of vertices x such that there is a directed Hamiltonian path from x to n. Define $\mathsf{DNHam}_n(X)$ to be the family of digraphs D in DNHam_n such that $X(D) = X$. One easily checks that the families $\mathsf{DNHam}_n(X)$ satisfy the Cluster Lemma 4.2.

If $X \subsetneq [n-1]$, then let x be minimal in $[n-1] \setminus X$. It is clear that we can add or delete nx to or from a digraph in $\mathsf{DNHam}_n(X)$ without ending up outside $\mathsf{DNHam}_n(X)$. It follows that $\mathsf{DNHam}_n(X)$ is a cone.

The remaining case is that $X = [n-1]$. Let $D \in \mathsf{DNHam}_n([n-1])$. We claim that every vertex x in $[n-1]$ has two outgoing edges. This will imply that every digraph in $\mathsf{DNHam}_n([n-1])$ contains at least $2n-2$ edges, which in turn will imply the proposition by Theorem 4.7.

Now, x has at least one outgoing edge xy for some $y \in [n-1] \setminus \{x\}$, because there is a directed Hamiltonian path starting in x and ending in n. However, there is also a directed Hamiltonian path starting in y and ending in n. In this path, the edge with tail x cannot be xy, which concludes the proof. $\square$

Table 17.1. The homology of DNHam_n for $n \leq 5$.

$\tilde{H}_i(\mathsf{DNHam}_n, \mathbb{Z})$	$i = 0$	1	2	3	4	5	6	7	8	9	10
$n = 2$	$\mathbb{Z}$	-	-	-	-	-	-	-	-	-	-
3	-	-	-	$\mathbb{Z}$	-	-	-	-	-	-	-
4	-	-	-	-	-	-	$\mathbb{Z}^2$	-	-	-	-
5	-	-	-	-	-	-	-	-	-	$\mathbb{Z}^6$	-

Computer calculations give some indications that the actual bound on the shifted connectivity degree of DNHam_n might be as large as $3(n-2)$; see Table 17.1. By the table, DNHam_n has the homology of a wedge of $(n-2)!$ spheres of dimension $3(n-2)$ for $n \leq 5$. Given the situation in the undirected case, such a nice formula is not likely to hold in general. However, one may suspect that $\tilde{H}_{3(n-2)}(\mathsf{DNHam}_n, \mathbb{Z})$ does contain a free subgroup isomorphic to $\mathbb{Z}^{(n-2)!}$ for all $n \geq 2$.

Part VI

Connectivity

Disconnected Graphs

We examine the complex NC_n of disconnected graphs on n vertices. We also consider subcomplexes consisting of graphs with certain restrictions on the vertex size of the connected components.

Due to its interpretation in terms of matroid theory, NC_n has a very simple topological structure; recall Corollary 13.4. Specifically, NC_n is homotopy equivalent to a wedge of spheres of dimension $n - 3$. Intriguingly, the number of spheres equals $(n - 1)!$. We discuss NC_n along with induced subcomplexes in Section 18.1.

In Section 18.2, we consider the complex $NLC_{n,k}$ of graphs with the property that each connected component contains at most k vertices. For $k = 2$, we obtain the matching complex M_n, whereas $k = n-1$ yields NC_n. Sundaram [137] proved the somewhat surprising result that $NLC_{n,k}$ is homotopy equivalent to NC_{n-1} for $3 \leq k + 2 \leq n \leq 2k + 1$. For general n and k, the situation is much more complicated, but we have been able to prove that the shifted connectivity degree and the depth of $NLC_{n,k}$ are at least $\frac{(k-1)(n-1+r/k)}{k+1} - 1$, where $r = (n-1) \bmod (k+1)$. For $k = 3$ and $n \geq 4$, we prove that this bound is sharp; there *is* homology in the given dimension. The homology is finite for $n = 4t+1$ whenever $t \geq 2$ and infinite otherwise; in the finite case, we have an elementary 2-group. For general n and k, we do not know whether our bound on the connectivity degree is sharp.

In Section 18.3, we proceed with the complex $SSC_n^{k,s}$ of graphs with at least s connected components of vertex size at most k. Generalizing to a larger family of complexes, we show that $SSC_n^{k,s}$ is homotopy equivalent to a wedge of spheres of dimension $n - s - 2$ and that the $(n - s - 2)$-skeleton is vertex-decomposable whenever $n > ks$. We also show that the nice topological properties of $SSC_n^{k,s}$ are preserved under intersection with strong pseudo-independence complexes (see Section 13.3).

In Section 18.4, we summarize our results for the complex $NC_{n,p}$ of graphs such that the size of some connected component is not divisible by p.

Specifically, this complex is homotopy equivalent to a wedge of spheres of dimension $n-3$ and has a vertex-decomposable $(n-3)$-skeleton.

Finally, in Section 18.5, we provide an overview of known properties of the complex $\mathsf{HNC}_{n,k}$ of disconnected k-hypergraphs.

The complexes considered in this chapter are closely related to certain sublattices of the partition lattice Π_n. More precisely, NC_n is homotopy equivalent to the order complex of the proper part of the full lattice, whereas $\mathsf{NLC}_{n,k}$ corresponds to the sublattice consisting of all partitions in which all sets have size at most k. In the same manner, $\mathsf{SSC}_n^{k,s}$ corresponds to partitions with at least s parts of size at most k, while $\mathsf{NC}_{n,p}$ corresponds to partitions with at least one set of size not divisible by p. Restricting to partitions in which all sets have size one or at least k, we obtain the lattice $\Pi_n^{1,\geq k}$ corresponding to $\mathsf{HNC}_{n,k}$. Björner and Welker [16] studied $\Pi_n^{1,\geq k}$ in the context of subspace arrangements; see Section 1.1.4 for some details.

18.1 Disconnected Graphs Without Restrictions

We devote this section to the complex NC_n. This complex is well-known to have very attractive topological properties:

Proposition 18.1. *For $n \geq 2$ and any graph G on n vertices with at most two components, $\mathsf{NC}_n(G)$ is $VD^+(n-3)$ and homotopy equivalent to a wedge of spheres of dimension $n-3$. For the full complex NC_n, the number of spheres in the wedge equals $(n-1)!$.*

Proof. If G consists of two components, then $\mathsf{NC}_n(G)$ is a cone of dimension at least $n-3$. If G is connected, then the first part of the proposition is an immediate consequence of Theorem 13.25; $\mathsf{NC}_n(G)$ is SPI* over the graphic matroid on G. Define C_n as the quotient complex of connected graphs on n vertices. The nonzero Betti number being $(n-1)!$ is a consequence of Corollary 6.15. Namely, with notation as in the corollary and with $f(n) = \tilde{\chi}(\mathsf{C}_n)$ and $h(n) = \tilde{\chi}(2^{K_n})$, we obtain that $H(x) = -x = 1 - e^{-F(x)}$; thus $-F(x) = \ln(1+x)$. $\square$

See Babson et al. [3] for more information and references. There are at least two natural explanations for the simplicity of the topology of NC_n:

The first explanation is the one used in the above proof; a graph is disconnected if and only if the graph does not have full rank with respect to the graphic matroid on the complete graph. In particular, NC_n and its induced subcomplexes are all Alexander duals of independence complexes.

The second explanation is that there is a simple homotopy equivalence between NC_n and the the proper part $\overline{\Pi}_n$ of the partition lattice Π_n on the set $[n]$. Namely, we may define a closure operator on the face poset of NC_n with image $\overline{\Pi}_n$ by mapping a graph with connected components $V_1,\ldots,V_k$ to the partition $\{V_1,\ldots,V_k\}$. Applying Lemma 6.1, we obtain the desired homotopy equivalence. Now, Π_n is a geometric lattice of rank $(n-1)$ with Möbius function $(-1)^{n-1} \cdot (n-1)!$; see Rota [116] or Stanley [131]. By a result of Björner [7] about geometric lattices (see also Folkman [46]), this implies the following:

Theorem 18.2. *The order complex of* $\overline{\Pi}_n$ *is shellable and homotopy equivalent to a wedge of* $(n-1)!$ *spheres of dimension* $n-3$. $\square$

Let us give an explicit decision tree on NC_n; note that we already know that such a tree exists by Proposition 18.1.

Proposition 18.3. *For* $n \geq 2$, $\mathsf{NC}_n \sim (n-1)! \cdot t^{n-3}$.

Proof. If $n = 2$, then NC_n is the (-1)-simplex. Assume that $n \geq 3$. Let $A = \{in : i \in [n-1]\}$ and consider the complex $\Sigma_Y = \mathsf{NC}_n(Y, A \setminus Y)$ for each $Y \subseteq A$. If Y contains two edges in and jn, then ij is a cone point in Σ_Y; i and j are contained in the same component. Hence $\Sigma_Y \sim 0$. If $Y = \emptyset$, then Σ_Y is the full simplex on $\binom{[n-1]}{2}$ and hence nonevasive. The case remaining is $\Sigma_{\{kn\}}$ for each $k \in [n-1]$. It is clear that a graph G belongs to $\Sigma_{\{kn\}}$ if and only if $G([n-1])$ belongs to NC_{n-1}. This yields by induction that $\Sigma_{\{kn\}} \sim (n-2)! \cdot t^{n-3}$. Using Lemma 5.22, we obtain that

$$\mathsf{NC}_n \sim \sum_{k=1}^{n-1} (n-2)! \cdot t^{n-3} = (n-1)! \cdot t^{n-3}. \quad \square$$

Corollary 18.4. *The quotient complex* C_n *admits a decision tree such that a graph is unmatched if and only if its edge set is of the form*

$$\sigma_\rho := \{\rho_i \rho_{i+1} : i \in [n-1]\}$$

for some permutation $\rho = \rho_1 \ldots \rho_n$ *in* $\mathfrak{S}_{[n]}$ *such that* $\rho_1 = 1$.

Proof. This is an immediate consequence of the proof of Proposition 18.3. $\square$

Corollary 18.5. *For* $n \geq 2$, *the set* $\{[\sigma_\rho] : \rho \in \mathfrak{S}_{[n]}, \rho_1 = 1\}$ *forms a basis for* $\tilde{H}_{n-2}(\mathsf{C}_n; \mathbb{Z})$. *As a consequence, the set* $\{\partial([\sigma_\rho]) : \rho \in \mathfrak{S}_{[n]}, \rho_1 = 1\}$ *forms a basis for* $\tilde{H}_{n-3}(\mathsf{NC}_n; \mathbb{Z})$.

Proof. For the first statement, use Theorem 5.2 and Corollary 4.17. The second statement is an immediate consequence of Theorem 3.3. $\square$

18.2 Graphs with No Large Components

We discuss the complex $\mathsf{NLC}_{n,k}$ of graphs on n vertices with all connected components of vertex size at most k. The reader may want to keep in mind that $\mathsf{NLC}_{n,2}$ is the matching complex M_n discussed in Section 11.2. One easily checks that $\mathsf{NLC}_{n,k}$ has the same homotopy type as the lattice $\varPi_n^{\leq k}$ of partitions in which all sets have size at most k; compare to the discussion in the previous section. As a consequence, for any result about the homotopy type or the homology of $\mathsf{NLC}_{n,k}$, we have an equivalent result about the topology of the order complex of the proper part of $\varPi_n^{\leq k}$.

18.2.1 Homotopy Type and Depth

In the language of partition lattices, Sundaram [137] proved the following; we give a new proof in terms of discrete Morse theory.

Theorem 18.6 (Sundaram [137]). *For $3 \leq k + 2 \leq n \leq 2k + 1$, $\mathsf{NLC}_{n,k}$ is homotopy equivalent to NC_{n-1} and hence has the homotopy type of a wedge of $(n-2)!$ spheres of dimension $n-4$.*

Proof. First, we use discrete Morse theory to prove that $\mathsf{NLC}_{n,k}$ is homotopy equivalent to $\mathsf{NLC}_{n,n-2}$ for $\frac{n-1}{2} \leq k < n-2$. For any set W of size at least $k+1$ and at most $n-2$, define $\mathcal{F}(W)$ as the family of graphs in $\mathsf{NLC}_{n,n-2}$ containing a connected component with vertex set W. It is clear that each graph G in $\mathsf{NLC}_{n,n-2} \setminus \mathsf{NLC}_{n,k}$ belongs to exactly one family $\mathcal{F}(W)$; $|W| \geq k + 1 \geq \frac{n+1}{2}$, so there is no other component in G with a vertex set of size at least $k + 1$. In particular, the families $\mathcal{F}(W)$ satisfy the Cluster Lemma 4.2. Now, any edge between two vertices in the set $[n] \setminus W$ is a cone point in $\mathcal{F}(W)$, which yields a perfect acyclic matching on $\mathcal{F}(W)$. Using the Cluster Lemma 4.2, we obtain a perfect acyclic matching on $\mathsf{NLC}_{n,n-2} \setminus \mathsf{NLC}_{n,k}$; hence we may collapse $\mathsf{NLC}_{n,n-2}$ to $\mathsf{NLC}_{n,k}$.

It remains to compute the homotopy type of $\mathsf{NLC}_{n,n-2}$. Again, we use discrete Morse theory. For each set $X \subseteq [n-1]$, let $\mathcal{H}(X)$ be the family of graphs such that $X \cup \{n\}$ is the vertex set of the connected component containing n. By construction, $\mathcal{H}(X)$ is void if X is of size at least $n-2$, which implies that $[n-1] \setminus X$ has size at least two – and hence contains an edge – whenever $\mathcal{H}(X)$ is nonvoid. Moreover, $\mathcal{H}(\emptyset)$ coincides with NC_{n-1}. It is clear that the families $\mathcal{H}(X)$ satisfy the Cluster Lemma 4.2. Now, if X is nonempty, then any edge between two vertices in the set $[n-1] \setminus X$ is a cone point in $\mathcal{H}(X)$; the largest component we may achieve by adding such an edge has size $n-2$, which is allowed. Hence $\mathsf{NLC}_{n,n-2}$ admits an acyclic matching that is perfect outside $\mathcal{H}(\emptyset) = \mathsf{NC}_{n-1}$, which implies that we may collapse $\mathsf{NLC}_{n,n-2}$ to NC_{n-1}. $\square$

Theorem 18.7. *Let $k, n \geq 1$. Write $r = (n-1) \bmod (k+1)$. Then $\mathsf{NLC}_{n,k}$ has homology in dimension d only if*

$$\alpha_{n,k} := \frac{(k-1)(n-1+r/k)}{k+1} - 1 \leq d \leq \frac{(k-1)(n-1)}{k} - 1 =: \beta_{n,k}.$$

Moreover, $\mathsf{NLC}_{n,k}$ is $(\lceil \alpha_{n,k} \rceil - 1)$-connected.

Table 18.1. The bound $\lceil \alpha_{n,k} \rceil$ in Theorem 18.7 on the shifted connectivity degree of $\mathsf{NLC}_{n,k}$ for $n \leq 19$ and $k \leq 8$. For entries in bold, $\mathsf{NLC}_{n,k}$ has the homotopy type of a wedge of spheres in the given dimension.

$\lceil \alpha_{n,k} \rceil$	$n=3$	4	5	6	7	8	9	10	11	12	13	14	15	16	17	18	19	20
$k=2$	**0**	**0**	**1**	**1**	1	**2**	2	2	3	3	3	4	4	4	5	5	5	6
3	–	**1**	**1**	2	**3**	**3**	3	4	**5**	5	5	6	7	7	7	8	9	9
4	–	–	**2**	**2**	3	**4**	**5**	5	5	6	7	**8**	8	8	9	10	11	11
5	–	–	–	**3**	3	4	5	**6**	**7**	**7**	7	8	9	10	**11**	11	11	12
6	–	–	–	–	**4**	**4**	5	**6**	**7**	**8**	**9**	**9**	9	10	11	12	13	**14**
7	–	–	–	–	–	**5**	**5**	6	**7**	**8**	**9**	**10**	**11**	**11**	11	12	13	14
8	–	–	–	–	–	–	**6**	**6**	**7**	**8**	**9**	**10**	**11**	**12**	**13**	**13**	13	14

Remark. The upper bound appears in the work of Sundaram [137], as does the lower bound for $n \leq 3k + 4$. See Table 18.1 for the value of $\lceil \alpha_{n,k} \rceil$ for small n and k.

Proof. The cases $k = 1$ and $n = 1$ are trivial; thus assume that $k, n \geq 2$. We use discrete Morse theory and induction on n; our goal is to find an acyclic matching such that the dimension of each unmatched graph is in the desired interval. We have three base cases:

- $2 \leq n \leq k$. In this case, $\mathsf{NLC}_{n,k}$ is collapsible; all graphs belong to $\mathsf{NLC}_{n,k}$, as the size of the largest component is at most $n \leq k$. Indeed, $\alpha_{n,k} = \beta_{n,k} = \frac{(n-1)(k-1)}{k} - 1$, and this is not an integer unless k divides $n - 1$.
- $n = k + 1$. Then $\mathsf{NLC}_{n,k} = \mathsf{NC}_{k+1}$, which has all homology in dimension $k - 2$ by Proposition 18.1; note that $\alpha_{k+1,k} = \beta_{k+1,k} = k - 2$.
- $k + 2 \leq n \leq 2k + 1$. Note that $n = k + r + 2$. By Theorem 18.6, $\mathsf{NLC}_{n,k}$ is homotopy equivalent to NC_{n-1}, which has all its homology concentrated in dimension $n - 4 = k + r - 2$. This yields that

$$\alpha_{n,k} = \frac{(k-1)(k+1+r+r/k)}{k+1} - 1 = k - 2 + \frac{(k-1)r}{k};$$

$$\beta_{n,k} = \frac{(k-1)(k+1+r)}{k} - 1 = k - 2 + \frac{(k-1)(r+1)}{k}.$$

Since

$$r - 1 < \frac{(k-1)r}{k} \leq r \leq \frac{(k-1)(r+1)}{k} < r + 1$$

whenever $0 \leq r \leq k - 1$, the theorem follows for this case.

We now proceed by induction on n. Assume that $n \geq 2k + 2$. Match with the edge 12 whenever possible. A graph G is unmatched if and only if 1 and 2 belong to different components U_1 and U_2 in G and the number of vertices in $U_1 \cup U_2$ is at least $k+1$. By Lemma 4.1, it suffices to find an acyclic matching on the family $\mathcal{F}$ of unmatched graphs.

For any two disjoint sets U_1 and U_2 of size at most k such that $1 \in U_1$ and $2 \in U_2$ and such that $|U_1 \cup U_2| > k$, let $\mathcal{F}(U_1, U_2)$ be the family of graphs $G \in \mathsf{NLC}_{n,k}$ such that U_i is the vertex set of the connected component containing i for $i = 1, 2$. It is clear that the families $\mathcal{F}(U_1, U_2)$ satisfy the conditions in the Cluster Lemma 4.2.

Write $u_i = |U_i|$. Note that $\mathcal{F}(U_1, U_2)$ is isomorphic to the join of the complexes $\mathsf{NLC}_{n-u_1-u_2,k}$, C_{u_1}, and C_{u_2}, where C_u is the quotient complex of connected graphs on u vertices. By Proposition 18.3 (and Proposition 5.36), C_{u_i} admits a decision tree with all evasive sets of dimension $u_i - 2$; this decision tree can be translated into an acyclic matching via Theorem 5.2. Using Theorem 5.29, we conclude that the join of C_{u_1} and C_{u_2} admits an acyclic matching such that all unmatched sets have dimension $u_1 + u_2 - 3$. This implies that $\mathcal{F}(U_1, U_2)$ admits an acyclic matching such that the family of unmatched sets is the disjoint union of families of the form $\{G\} * \mathsf{NLC}_{n-u_1-u_2,k}$, where G is a graph with $u_1 + u_2 - 2$ edges. These families clearly satisfy the conditions in the Cluster Lemma 4.2, as they form an antichain.

Write $u_1 + u_2 = t$; note that $k+1 \leq t \leq 2k$ and that $n - t \geq 2k + 2 - t \geq 2$. By induction, $\mathsf{NLC}_{n-t,k}$ admits an acyclic matching such that a graph is unmatched only if its dimension d satisfies

$$\alpha_{n-t,k} \leq d \leq \beta_{n-t,k}.$$

As a consequence, $\{G\} * \mathsf{NLC}_{n-t,k}$ admits an acyclic matching such that a graph is unmatched only if its dimension d satisfies

$$\alpha_{n-t,k} + t - 2 \leq d \leq \beta_{n-t,k} + t - 2.$$

Now,

$$\beta_{n,k} - \beta_{n-t,k} = \frac{(k-1)t}{k} = t - \frac{t}{k} \geq t - 2;$$

the last inequality is because $t \leq 2k$. This proves the upper bound in the theorem. For the lower bound, write $r_0 = (n - t - 1) \bmod (k+1)$. Note that $(r - r_0) \bmod (k+1) = t$. Since $k+1 \leq t \leq 2k$ and since $r - r_0 \leq k$, we have that

$$r - r_0 \leq t - k - 1.$$

Thus

$$
\begin{aligned}
\alpha_{n,k} - \alpha_{n-t,k} &= \frac{(k-1)(t + (r - r_0)/k)}{k+1} \\
&\leq \frac{(k-1)(t + (t - k - 1)/k)}{k+1} = \frac{(k-1)(t-1)}{k} \\
&= t - 1 - \frac{t-1}{k} \leq t - 2;
\end{aligned}
$$

the last inequality is because $t \geq k + 1$. This proves the lower bound in the theorem, and we are done. □

Corollary 18.8 (Sundaram [137]). *For $k \geq 1$, we have that $\mathsf{NLC}_{2k+2,k}$ is homotopy equivalent to a wedge of $\frac{(2k)!k}{k+1}$ spheres of dimension $2k - 3$, whereas $\mathsf{NLC}_{3k+2,k}$ is homotopy equivalent to a wedge of spheres of dimension $3k - 4$.*

Proof. With $n = 2k + 2$, we obtain that $r = k$ and hence that

$$\alpha_{2k+2} = \frac{(k-1)(2k+2)}{k+1} - 1 = 2k - 3;$$

$$\beta_{2k+2} = \frac{(k-1)(2k+1)}{k+1} - 1 = 2k - 3 + \frac{k-1}{k}.$$

With $n = 3k + 2$, we obtain that $r = k - 1$ and hence that

$$\alpha_{3k+2} = \frac{(k-1)(3k+1+(k-1)/k)}{k+1} - 1 = 3k - 4 - \frac{k-1}{k};$$

$$\beta_{3k+2} = \frac{(k-1)(3k+1)}{k+1} - 1 = 3k - 4 + \frac{k-1}{k}.$$

It follows that all homology is contained in one dimension. By Theorem 4.8, the complexes are hence wedges of spheres. For the nonvanishing Betti number of $\mathsf{NLC}_{2k+2,k}$, see Sundaram [137] or apply Corollary 18.10 below. □

Proposition 18.9. *Let $k \geq 1$. Then the reduced Euler characteristic $h_k(n) = \tilde{\chi}(\mathsf{NLC}_{n,k})$ satisfies*

$$H_k(x) := \sum_{n \geq 1} \frac{h_k(n)}{n!} x^n = 1 - \exp\left(-\sum_{r=1}^{k} \frac{(-x)^r}{r}\right).$$

Proof. We apply Corollary 6.15. Define $f_k(n) = 0$ for $n > k$ and $f_k(n) = (-1)^n \cdot (n-1)!$ for $1 \leq n \leq k$; the latter is the Euler characteristic of C_n. Then f_k and h_k satisfy the conditions in Corollary 6.15, which immediately implies the proposition. □

Corollary 18.10. *Let $2k + 2 \leq n \leq 3k + 2$. Then the reduced Euler characteristic $h_k(n) = \tilde{\chi}(\mathsf{NLC}_{n,k})$ satisfies*

$$\frac{(-1)^{n+1} h_k(n)}{n!} = \frac{1}{n} - \frac{1}{n-1} + \sum_{i=k+1}^{n-k-1} \frac{1}{2i(n-i)} - \sum_{i=k+1}^{n-k-2} \frac{1}{2i(n-i-1)}. \quad (18.1)$$

In particular, for each $t \geq 2$, $(-1)^{t+1} \frac{h_k(2k+t)}{(2k+t)!}$ is a rational function in k for $k \geq t - 2$.

Proof. Write $\theta(x) = \sum_{r=k+1}^{n} \frac{(-x)^r}{r}$. By Proposition 18.9, we have that

$$H_k(x) = 1 - \exp\left(\ln(1+x) + \theta(x) + x^{n+1}R(x)\right)$$
$$= 1 - (1+x)\exp\left(\theta(x) + x^{n+1}R(x)\right)$$

for some polynomial $R(x)$. If $3(k+1) > n$, then this equals

$$1 - (1+x)(1 + \theta(x) + \frac{\theta^2(x)}{2}) + x^{n+1}Q(x)$$

for some polynomial $Q(x)$. This is easily seen to imply (18.1). For the last claim, simply note that the two sums in the right-hand side of (18.1) contain exactly $t-1$ and $t-2$ terms, respectively, each of which is a rational function in k. □

By the next result, $\alpha_{n,k}$ is a lower bound on the depth.

Theorem 18.11. *Let $k, n \geq 1$ and let σ be an edge set. Then the lifted complex $\mathsf{NLC}_{n,k}(\sigma, \emptyset)$ is $(\lceil \alpha_{n,k} \rceil - 1)$-connected. In particular, the $\lceil \alpha_{n,k} \rceil$-skeleton of $\mathsf{NLC}_{n,k}$ is Cohen-Macaulay; hence the depth of $\mathsf{NLC}_{n,k}$ is at least $\lceil \alpha_{n,k} \rceil$.*

Proof. Defining $\mathsf{NLC}_{0,k} = \{\emptyset\}$, we may extend the theorem to $n = 0$; note that $\alpha_{0,k} = -1$. The cases $k = 1$ and $n \leq 1$ are trivial; thus assume that $k, n \geq 2$. As in the proof of Theorem 18.7, we will define an acyclic matching such that the dimension of each unmatched graph is in the desired interval. This time, our base case is $2 \leq n \leq k$. Since $\mathsf{NLC}_{n,k}$ is the full simplex, it follows that $\mathsf{NLC}_{n,k}(\sigma, \emptyset)$ is collapsible unless σ is the full edge set $\binom{[n]}{2}$. Now, $|\binom{[n]}{2}| = \binom{n}{2}$ is at least $\alpha_{n,k} + 1 = \frac{(k-1)(n-1)}{k}$; thus we are done with the base case.

Now, assume that $n \geq k + 1$. $\mathsf{NLC}_{n,k}(\sigma, \emptyset)$ is void if $\sigma = \binom{[n]}{2}$; thus we may assume that some edge ab, say 12, is not in σ. Match with the edge 12 whenever possible. As in the proof of Theorem 18.7, a graph G is unmatched if and only if 1 and 2 belong to different components U_1 and U_2 in G and the number of vertices in $U_1 \cup U_2$ is at least $k + 1$. Define $\mathcal{F}(U_1, U_2)$ as in the proof of Theorem 18.7, except that we restrict to graphs in $\mathsf{NLC}_{n,k}(\sigma, \emptyset)$. As before, the Cluster Lemma 4.2 applies.

For $i \in \{1, 2\}$, define σ_i to be the restriction of σ to the set U_i. Define σ_0 to be the restriction of σ to $\sigma \setminus (U_1 \cup U_2)$. The family $\mathcal{F}(U_1, U_2)$ is nonvoid only if $\sigma = \sigma_0 \cup \sigma_1 \cup \sigma_2$.

Write $u_i = |U_i|$ and $t = u_1 + u_2$. We have that $\mathcal{F}(U_1, U_2)$ is isomorphic to the join of $\mathsf{NLC}_{n-t,k}(\sigma_0, \emptyset)$, $\mathsf{C}_{u_1}(\sigma_1, \emptyset)$, and $\mathsf{C}_{u_2}(\sigma_2, \emptyset)$. Now, every graph in $\mathsf{C}_{u_i}(\sigma_i, \emptyset)$ has size at least $u_i - 1$. Moreover, by induction on n, $\mathsf{NLC}_{n-t,k}(\sigma_0, \emptyset)$ admits an acyclic matching such that all critical faces have dimension at least $\alpha_{n-t,k}$. This implies that $\mathcal{F}(U_1, U_2)$ admits an acyclic matching such that a graph is unmatched only if its dimension d satisfies

$$d \geq \alpha_{n-t,k} + t - 2.$$

As in the proof of Theorem 18.7, we obtain that $\alpha_{n,k} \leq \alpha_{n-t,k} + t - 2$, which concludes the proof. □

18.2.2 Bottom Nonvanishing Homology Group

The main object of this section is to examine the homology of $\mathsf{NLC}_{n,3}$; we generalize to larger odd k whenever reasonably straightforward. Unfortunately, we have no results in the case when k is even and at least four. We remind the reader that all results in this section apply to the order complex of the proper part of the lattice $\Pi_{\bar{n}}^{\leq k}$.

Table 18.2. The homology of $\mathsf{NLC}_{n,3}$ for $n \leq 9$ and $n = 11$.

$\tilde{H}_i(\mathsf{NLC}_{n,3}; \mathbb{Z})$	$i = 0$	1	2	3	4	5	6
$n = 4$	-	$\mathbb{Z}^6$	-	-	-	-	-
5	-	$\mathbb{Z}^6$	-	-	-	-	-
6	-	-	$\mathbb{Z}^{24}$	-	-	-	-
7	-	-	-	$\mathbb{Z}^{120}$	-	-	-
8	-	-	-	$\mathbb{Z}^{540}$	-	-	-
9	-	-	-	$\mathbb{Z}_2^2$	$\mathbb{Z}^{1764}$	-	-
10	-	-	-	-	?	?	-
11	-	-	-	-	-	$\mathbb{Z}^{68256}$	-

Recall from Section 11.2 that the bottom nonvanishing homology group of $\mathsf{M}_n = \mathsf{NLC}_{n,2}$ is finite for almost all n. Using computer, we have been able to verify that the bottom nonvanishing homology group of $\mathsf{NLC}_{9,3}$ is also finite; see Table 18.2. This might suggest that the general situation for $k = 2$ generalizes to $k = 3$. Indeed, in Theorem 18.14 below, we show that the relevant homology group is an elementary 2-group whenever $n = 4t + 1$ for some $t \geq 2$. Nevertheless, for *all* other values of $n \geq 4$, it turns out that the group is infinite. In our first theorem, we consider two thirds of these cases, postponing the case $n \bmod 4 = 3$ until later.

Theorem 18.12. *For $m \geq 1$ and $t \geq 2$, let $n = mt$ and $k = 2m - 1$. Then $\tilde{H}_{\lceil \alpha_{n,k} \rceil}(\mathsf{NLC}_{n,k}; \mathbb{Z})$ is infinite; note that $\lceil \alpha_{n,k} \rceil = t(m - 1) - 1$.*

Remark. For $k = 3$, this specializes to $n = 2t$ and $\lceil \alpha_{n,k} \rceil = t - 1$.

Proof. For $j \in [t]$, let $S_j = [(j - 1)m + 1, jm]$; we have that $\{S_1, \ldots, S_t\}$ is a partition of the vertex set $[n] = [mt]$ and $|S_j| = m = \frac{k+1}{2}$. Let $\Gamma_{n,k}$ be the family of graphs G in $\mathsf{NLC}_{n,k}$ such that the vertex sets of the connected components in G are exactly $S_1, \ldots, S_t$. It is clear that

$$\Gamma_{n,k} = \mathsf{C}_{S_1} * \cdots * \mathsf{C}_{S_t},$$

where C_{S_j} is the quotient complex of connected graphs on the vertex set S_j. By Corollary 18.4, C_{S_j} admits a decision tree such that a graph is evasive if and only if its edge set equals $\{\rho_i\rho_{i+1} : i \in [m-1]\}$, where $\{\rho_1, \ldots, \rho_m\} = S_j$ and $\rho_1 = (j-1)m+1$. Applying Theorem 5.29, we obtain that $\Gamma_{n,k}$ admits a decision tree such that a graph is evasive if and only if its edge set is a union of sets $\{\rho_i\rho_{i+1} : i \in [m-1]\}$ with properties as above.

Let $C_{n,k}$ be the family of of evasive graphs and write

$$\Sigma_{n,k} = \mathsf{NLC}_{n,k} \setminus (\Gamma_{n,k} \setminus C_{n,k}).$$

This is a simplicial complex, because each graph in $C_{n,k}$ is minimal in $\Gamma_{n,k}$ and no graph in $\mathsf{NLC}_{n,k} \setminus \Gamma_{n,k}$ contains any graph in $\Gamma_{n,k}$. By Theorem 4.4, $\mathsf{NLC}_{n,k}$ and $\Sigma_{n,k}$ are homotopy equivalent.

For integers a, b such that $a < b$, define

$$\pi_{a,b} = [a(a+1)] \wedge [(a+1)(a+2)] \wedge \cdots \wedge [(b-2)(b-1)] \wedge [(b-1)b].$$

Define $\omega_{m,k} = \pi_{1,m}$ and, recursively,

$$\omega_{mt,k} = \omega_{m(t-1),k} \wedge \pi_{m(t-1)+1,mt}. \tag{18.2}$$

For all $t \geq 0$ except $t = 1$, we claim that there is a cycle $z_{mt,k}$ in $\tilde{C}_{\lceil \alpha_{mt,k} \rceil}(\Sigma_{mt,k}; \mathbb{Z})$ such that the coefficient of $\omega_{mt,k}$ in $z_{mt,k}$ is nonzero. Since $\omega_{mt,k}$ is a maximal face of $\Sigma_{mt,k}$, this will imply that $\tilde{H}_{\lceil \alpha_{mt,k} \rceil}(\Sigma_{mt,k}; \mathbb{Z})$ is infinite.

To prove the claim, we use induction on t. As it turns out, the base step consists of the cases $t = 0$ and $t = 3$. The case $t = 0$ is trivially true. We postpone the case $t = 3$ until later and consider the induction step; assume that $t \geq 2$ and $t \neq 3$. By induction, we have a cycle $z_{m(t-2),k}$ in $\tilde{C}_{\lceil \alpha_{m(t-2),k} \rceil}(\Sigma_{m(t-2),k}; \mathbb{Z})$ with desired properties. Now, define

$$z_{mt,k} = z_{m(t-2),k} \wedge \partial(\pi_{m(t-2)+1,mt}). \tag{18.3}$$

Since $\pi_{m(t-2)+1,mt}$ is a tree on $2m = k+1$ vertices, its boundary is a sum of graphs in which all components have size at most k. Moreover, the only way to split $\pi_{m(t-2)+1,mt}$ into two components of size m is to remove the edge in the middle, which yields the face $\pi_{m(t-2)+1,m(t-1)} \wedge \pi_{m(t-1)+1,mt}$. Joining this face to any face of $C_{m(t-2),k}$, we clearly obtain a face of $C_{mt,k}$. In particular, by properties of $z_{m(t-2),k}$, every face of $z_{mt,k}$ belongs to $\Sigma_{mt,k}$. In addition, it is clear that the coefficient of $\omega_{mt,k}$ is nonzero.

It remains to consider the case $t = 3$. Define

$$x = \partial(\pi_{1,2m}) \wedge \partial(\pi_{2m,3m}).$$

While x is not an element in $\tilde{C}_{\lceil \alpha_{3m,k} \rceil}(\Sigma_{3m,k}; \mathbb{Z})$, x admits a unique decomposition $x = x_0 + x_1$ such that $x_0 \in \tilde{C}_{\lceil \alpha_{3m,k} \rceil}(\Sigma_{3m,k}; \mathbb{Z})$ and $x_1 \in \tilde{C}_{\lceil \alpha_{3m,k} \rceil}(2^{K_{3m}}/\Sigma_{3m,k}; \mathbb{Z})$. Define

$$z = x_0 + [1(3m)] \wedge \partial(x_1).$$

This is a cycle, because $\partial(x_0) + \partial(x_1) = 0$. Moreover, the coefficient of $\omega_{3m,k}$ in z is nonzero.

It remains to prove that $z \in \tilde{C}_{\lceil \alpha_{3m,k} \rceil}(\Sigma_{3m,k}; \mathbb{Z})$. Since $z = x_0 + [1(3m)] \wedge \partial(x_1)$, any element $[\sigma]$ with nonzero coefficient in $\partial(x_1)$ has the property that σ belongs to $\Sigma_{3m,k}$. Let τ be such that $\sigma \subset \tau$ and τ appears in x_1 with nonzero coefficient. This means that we obtain $[\tau]$ from $\pi_{1,3m}$ by removing one element $[a(a+1)]$ such that $a \in [1, 2m-1]$ and one element $[(b-1)b]$ such that $b \in [2m+1, 3m]$; hence

$$[\tau] = \pi_{1,a} \wedge \pi_{a+1,b-1} \wedge \pi_{b,3m}.$$

Moreover, one of the three resulting components must contain at least $2m = k+1$ vertices. This is possible only for the component $\pi_{a+1,b-1}$, which implies that the two other components contain a total of at most $3m - (b - a - 1) \leq m \leq k$ elements. In particular, the component in $\tau + 1(3m)$ containing 1 and $3m$ has size at most k. Since $\sigma \in \Sigma_{3m,k}$ and $\sigma \subset \tau$, it follows that $\sigma + 1(3m) \in \Sigma_{3m,k}$ as desired. $\square$

We now restrict our attention to the cases $k \in \{3, 7\}$; we do not know whether the following result generalizes to other values of k.

Theorem 18.13. *For $t \geq 0$, the following hold:*

(i) $\tilde{H}_{2t-1}(\mathsf{NLC}_{4t+1,3}; \mathbb{Z}_2)$ *is nonzero.*

(ii) $\tilde{H}_{6t-1}(\mathsf{NLC}_{8t+1,7}; \mathbb{Z}_2)$ *is nonzero.*

Proof. (i) Define a homomorphism $\varphi : \tilde{C}_{2t-1}(\mathsf{NLC}_{4t+1,3}; \mathbb{Z}_2) \to \mathbb{Z}_2$ by $\varphi([\sigma]) = 1$ if and only if σ is a matching. It is clear that $\varphi(z_{4t,3}) = 1$, where $z_{4t,3}$ is defined as in (18.3); the unique element appearing in $z_{4t,3}$ that corresponds to a matching is $\omega_{4t,3}$ defined in (18.2). We want to show that $\varphi(\partial([\tau])) = 0$ for every face τ of $\mathsf{NLC}_{4t+1,3}$ of dimension $2t$; this will imply that $z_{4t,3}$ is nonzero in $\tilde{H}_{2t-1}(\mathsf{NLC}_{4t+1,3}; \mathbb{Z}_2)$. Now, since there is not room for a matching of size $2t + 1$ on $4t + 1$ vertices, the boundary of a face τ contains a matching of size $2t$ if and only if the face consists of $2t - 1$ components of size two and one component of size three (with two edges). Since the boundary of such a face contains exactly two matchings, the claim follows.

(ii) Define a homomorphism $\varphi : \tilde{C}_{6t-1}(\mathsf{NLC}_{8t+1,7}; \mathbb{Z}_2) \to \mathbb{Z}_2$ by $\varphi([\sigma]) = 1$ if and only if every connected component but one in σ is a graph isomorphic to the four-path $P_4 = ([4], \{12, 23, 34\})$; the remaining component is then necessarily an isolated vertex. We observe that $\varphi(z_{8t,7}) = 1$. It suffices to prove that $\varphi(\partial([\tau])) = 0$ for every τ of dimension $6t$ in $\mathsf{NLC}_{8t+1,7}$. This is trivially true unless τ has the property that there are $2t - 1$ components isomorphic to P_4 in τ. Let H be the graph on the remaining five vertices. We need only prove that H contains an even number of subgraphs isomorphic to P_4 plus an isolated vertex. There are four possibilities for H:

- H consists of a square graph and an isolated vertex.
- H is isomorphic to the graph $([5], \{12, 13, 23, 34\})$.
- H is a path of vertex length five.
- H is isomorphic to the graph $([5], \{12, 23, 34, 35\})$.

In each case, we have an even number of subgraphs of the desired shape, which concludes the proof. $\square$

Theorem 18.14. *For $t \geq 2$, $\tilde{H}_{2t-1}(\mathsf{NLC}_{4t+1,3}; \mathbb{Z}) \cong \mathbb{Z}_2^{e_t}$ for some $e_t \geq 1$. Moreover, for all $t \geq 1$, $\tilde{H}_{2t+1}(\mathsf{NLC}_{4t+3,3}; \mathbb{Z})$ is infinite.*

Proof. For $t = 2$, we are done by the computation in Table 18.2. To prove the general case, we consider two exact sequences.

For $n \geq 5$, let $\mathsf{NLC}_{n,3}^2$ be the subcomplex of $\mathsf{NLC}_{n,3}$ consisting of all graphs such that the component containing the vertex 1 has size at most two and also all graphs such that 1 and 2 belong to the same component. One easily checks that

$$\mathsf{NLC}_{n,3}/\mathsf{NLC}_{n,3}^2 \cong \bigvee_U \mathsf{C}_U * \mathsf{NLC}_{[n] \setminus U, 3},$$

where the wedge is over all U such that $1 \in U$, $2 \notin U$, and $|U| = 3$; $\mathsf{NLC}_{[n] \setminus U, 3}$ is defined in the obvious manner. In particular, since C_U is homotopy equivalent to a nonempty wedge of spheres of dimension one, we have that

$$\tilde{H}_i(\mathsf{NLC}_{n,3}/\mathsf{NLC}_{n,3}^2; \mathbb{Z}) \cong \bigoplus \tilde{H}_{i-2}(\mathsf{NLC}_{n-3,3}; \mathbb{Z});$$

apply Corollary 4.23. This yields our first exact sequence:

$$\tilde{H}_i(\mathsf{NLC}_{n,3}^2) \longrightarrow \tilde{H}_i(\mathsf{NLC}_{n,3}) \longrightarrow \bigoplus \tilde{H}_{i-2}(\mathsf{NLC}_{n-3,3}) \longrightarrow \tilde{H}_{i-1}(\mathsf{NLC}_{n,3}^2).$$

Next, let $\mathsf{NLC}_{n,3}^{2,4}$ be the subcomplex of $\mathsf{NLC}_{n,3}^2$ consisting of all graphs such that the union of the components containing the vertices 1 and 2 has vertex size at most four. One easily checks that

$$\mathsf{NLC}_{n,3}^2/\mathsf{NLC}_{n,3}^{2,4} \cong \bigvee_{U_1, U_2} \mathsf{C}_{U_1} * \mathsf{C}_{U_2} * \mathsf{NLC}_{[n] \setminus (U_1 \cup U_2), 3},$$

where the wedge is over all disjoint U_1 and U_2 such that $i \in U_i$, $|U_1| = 2$, and $|U_2| = 3$. In particular, we have that

$$\tilde{H}_i(\mathsf{NLC}_{n,3}^2/\mathsf{NLC}_{n,3}^{2,4}; \mathbb{Z}) \cong \bigoplus \tilde{H}_{i-3}(\mathsf{NLC}_{n-5,3}; \mathbb{Z}).$$

Finally, let $\mathsf{NLC}_{n,3}^{2,3}$ be the subcomplex of $\mathsf{NLC}_{n,3}^{2,4}$ consisting of all graphs such that the union of the components containing the vertices 1 and 2 has vertex size at most three. This means that $\mathsf{NLC}_{n,3}^{2,3}$ is a cone with cone point 12. In particular, $\mathsf{NLC}_{n,3}^{2,4} \simeq \mathsf{NLC}_{n,3}^{2,4}/\mathsf{NLC}_{n,3}^{2,3}$. One easily checks that

$$\mathrm{NLC}_{n,3}^{2,4}/\mathrm{NLC}_{n,3}^{2,3} \cong \bigvee_{U_1,U_2} C_{U_1} * C_{U_2} * \mathrm{NLC}_{[n]\setminus(U_1\cup U_2),3},$$

where the wedge is over all disjoint U_1 and U_2 such that $i \in U_i$, $|U_1|+|U_2| = 4$, and $|U_1| \le 2$. As a consequence, we have that

$$\tilde{H}_i(\mathrm{NLC}_{n,3}^{2,4};\mathbb{Z}) \cong \tilde{H}_i(\mathrm{NLC}_{n,3}^{2,4}/\mathrm{NLC}_{n,3}^{2,3};\mathbb{Z}) \cong \bigoplus \tilde{H}_{i-2}(\mathrm{NLC}_{n-4,3};\mathbb{Z}).$$

We obtain our second exact sequence:

$$\bigoplus \tilde{H}_{i-2}(\mathrm{NLC}_{n-4,3}) \longrightarrow \tilde{H}_i(\mathrm{NLC}_{n,3}^2) \longrightarrow \bigoplus \tilde{H}_{i-3}(\mathrm{NLC}_{n-5,3})$$
$$\longrightarrow \bigoplus \tilde{H}_{i-3}(\mathrm{NLC}_{n-4,3}).$$

First, consider $n = 4t+1$ for $t \ge 3$. We have that $\alpha_{4t-4,3} = \alpha_{4t-3,3} = 2t-3$. In particular, the tail end of our second exact sequence becomes

$$\bigoplus \tilde{H}_{2t-3}(\mathrm{NLC}_{4t-3,3}) \longrightarrow \tilde{H}_{2t-1}(\mathrm{NLC}_{4t+1,3}^2) \longrightarrow 0.$$

Moreover, $\alpha_{4t-2,3} = 2t-2$. In particular, the tail end of our first exact sequence becomes

$$\tilde{H}_{2t-1}(\mathrm{NLC}_{4t+1,3}^2) \longrightarrow \tilde{H}_{2t-1}(\mathrm{NLC}_{4t+1,3}) \longrightarrow 0.$$

By induction on t, $\tilde{H}_{2t-3}(\mathrm{NLC}_{4t+1,3})$ is an elementary 2-group. Combining the above two tail ends, we obtain that the same is true for $\tilde{H}_{2t-1}(\mathrm{NLC}_{4t+1,3}^2)$ and $\tilde{H}_{2t-1}(\mathrm{NLC}_{4t+1,3})$. The group $\tilde{H}_{2t-1}(\mathrm{NLC}_{4t+1,3})$ being nonzero is a consequence of Theorem 18.13.

Next, consider $n = 4t + 3$ for $t \ge 1$. We have that $\alpha_{4t-2,3} = 2t - 2$ and $\alpha_{4t-1,3} = 2t - 1$. In particular, the tail end of our second exact sequence becomes

$$\tilde{H}_{2t+1}(\mathrm{NLC}_{4t+3,3}^2) \longrightarrow \bigoplus \tilde{H}_{2t-2}(\mathrm{NLC}_{4t-2,3}) \longrightarrow 0.$$

Most importantly, $\tilde{H}_{2t}(\mathrm{NLC}_{4t+3,3}^2) = 0$. Since $\alpha_{4t,3} = 2t - 1$, the tail end of our first exact sequence becomes

$$\tilde{H}_{2t+1}(\mathrm{NLC}_{4t+3,3}) \longrightarrow \bigoplus \tilde{H}_{2t-1}(\mathrm{NLC}_{4t,3}) \longrightarrow 0.$$

Since $\tilde{H}_{2t-1}(\mathrm{NLC}_{4t,3})$ is infinite by Theorem 18.12, $\tilde{H}_{2t+1}(\mathrm{NLC}_{4t+3,3})$ is also infinite, which concludes the proof. $\square$

Remark. The pair of sequences in the above proof is similar, but not exactly the same, as a certain pair of sequences appearing in the work of Sundaram [137]. Specifically, while we relate $\mathrm{NLC}_{n,3}^2$ to $\mathrm{NLC}_{n-4,3}$ and $\mathrm{NLC}_{n-5,3}$ in our second sequence, Sundaram relates $\mathrm{NLC}_{n,3}^2$ to $\mathrm{NLC}_{n-1,3}$ and $\mathrm{NLC}_{n-2,3}$ (or rather their partition lattice counterparts). One easily generalizes our construction of exact sequences to any k; compare to Sundaram's general construction [137].

Corollary 18.15. *For $n \geq 4$, the shifted connectivity degree and the depth of $\mathsf{NLC}_{n,3}$ are equal to $\lceil \alpha_{n,3} \rceil$. Moreover, the bottom nonvanishing homology group is finite if and only if $n \bmod 4 = 1$ and $n \geq 9$. In this case, the group is an elementary 2-group.* □

We have collected some evidence for the following conjecture:

Conjecture 18.16. *For $n \geq k + 1 \geq 2$, the shifted connectivity degree of $\mathsf{NLC}_{n,k}$ is equal to $\lceil \alpha_{n,k} \rceil$.*

By Corollaries 11.13 and 18.15, the conjecture is true for $k \in \{2, 3\}$. Moreover, by the results of this section, the conjecture holds for $n \in [k+2, 2k+2] \cup \{3k+2\}$ for all k and also for $n = \frac{t(k+1)}{2}$ whenever $t \geq 2$ and k is odd.

18.3 Graphs with Some Small Components

Let $\mu = (\mu_1, \ldots, \mu_m)$ be a weakly increasing sequence of positive integers. Define SSC_n^μ as the simplicial complex of graphs G on n vertices such that G has at least m connected components and such that the ith smallest connected component (with respect to vertex size) has at most μ_i vertices for $i \in [1, m]$. Note that $\mathsf{SSC}_n^{k^s} = \mathsf{SSC}_n^{k,\ldots,k}$ coincides with the complex $\mathsf{SSC}_n^{k,s}$ introduced in Section 7.1. In this section, we show that SSC_n^μ inherits many of the nice properties of NC_n.

To facilitate analysis, we generalize the definition of SSC_n^μ further, introducing weights on the vertices. For weakly increasing sequences $\lambda = (\lambda_1, \ldots, \lambda_l)$ and $\mu = (\mu_1, \ldots, \mu_m)$, say that $\lambda \leq \mu$ if $l \geq m$ and $\lambda_i \leq \mu_i$ for $1 \leq i \leq m$. Let $\omega = (\omega_1, \ldots, \omega_n)$ be a sequence of positive integers; for $S \subseteq [n]$, let

$$\omega_S = \sum_{s \in S} \omega_s.$$

For a graph G on the vertex set $[n]$ consisting of k connected components, let the corresponding vertex sets $V_1, \ldots, V_k$ be ordered such that $\omega_{V_1} \leq \ldots \leq \omega_{V_k}$. Define

$$\omega(G) = (\omega_{V_1}, \ldots, \omega_{V_k}).$$

Let $\mu = (\mu_1, \ldots, \mu_m)$ be a (not necessarily nonempty) weakly increasing sequence of positive integers such that $\sum_i \mu_i < \sum_j \omega_j$; we say that (μ, ω) is a *permitted pair* on n vertices if this condition is satisfied. Let SSC_ω^μ be the simplicial complex of graphs G on the vertex set $[n]$ satisfying $\omega(G) \leq \mu$. Let $L(\mu)$ be the length of the sequence μ; $L(\mu_1, \ldots, \mu_m) = m$.

Theorem 18.17. *Let Σ be an SPI complex over the graphic matroid on K_n (see Section 13.3) and let (μ, ω) be a permitted pair on n vertices. Write $\Lambda = \Sigma \cap \mathsf{SSC}_\omega^\mu$. Let $H \subseteq G$ be graphs such that every two connected components in H are joined by at least one edge from $G \setminus H$. Then $\mathrm{lk}_{\Lambda(G)}(H)$ is $VD^+(c(H) - 2 - L(\mu))$. In particular, Λ is $VD^+(n - 2 - L(\mu))$.*

Proof. We use induction on $G \setminus H$. Write $d(H) = c(H) - 2 - L(\mu)$. If $H = G$, then $\mathrm{lk}_{\Lambda(G)}(H)$ is void if $\mu \neq \emptyset$ and equal to the (-1)-simplex if $\mu = \emptyset$.

Suppose that $G \setminus H$ contains an edge e that joins two vertices from the same connected component in H. Then e is either not present at all or a cone point in $\mathrm{lk}_{\Lambda(G)}(H)$. Namely, this holds for $\mathrm{lk}_{\Sigma(G)}(H)$ since Σ is SPI, and e is obviously a cone point in $\mathrm{lk}_{\mathrm{SSC}^\mu_\omega(G)}(H)$. It follows that $\mathrm{lk}_{\Lambda(G)}(H)$ is a cone over or equal to $\mathrm{lk}_{\Lambda(G-e)}(H)$; by induction, $\mathrm{lk}_{\Lambda(G-e)}(H)$ is $VD^+(d(H))$, which implies that the same is true for $\mathrm{lk}_{\Lambda(G)}(H)$.

Next, suppose that all edges in $G \setminus H$ join vertices from different connected components in H. Suppose that there are edges e, e' joining the same pair of components. By induction, $\mathrm{lk}_{\Lambda(G)}(H + e)$ is $VD^+(c(H + e) - 2 - L(\mu)) = VD^+(d(H) - 1)$ and $\mathrm{lk}_{\Lambda(G-e)}(H)$ is $VD^+(d(H))$. By definition, we obtain that $\mathrm{lk}_{\Lambda(G)}(H)$ is $VD^+(d(H))$.

The remaining case is that every two connected components in H are joined by exactly one edge in G. Let $V_1, \ldots, V_k$ be the connected components ordered such that $\omega_{V_1} \leq \omega_{V_j}$ for $j \in [2, k]$. If $\omega_{V_1} > \mu_1$, then $\mathrm{lk}_{\Lambda(G)}(H)$ is void. Thus assume that $\omega_{V_1} \leq \mu_1$. For $j \in [2, k]$, let e_j be the edge joining V_1 and V_j. Write $E_1 = \{e_j : j \in [2, k]\}$ and $\Gamma = \mathrm{lk}_{\Lambda(G)}(H)$. Consider the lifted complex $\Gamma_Y = \Gamma(Y, E_1 \setminus Y)$ for each $Y \subseteq E_1$. We need to prove that each Γ_Y is $VD^+(d(H))$.

First, consider the case $Y \neq \emptyset$. Then $\Gamma_Y = \{Y\} * \mathrm{lk}_{\Lambda(G-(E_1 \setminus Y))}(H + Y)$, which by induction is $VD^+(r)$, where

$$r = |Y| + c(H + Y) - 2 - L(\mu) = c(H) - 2 - L(\mu) = d(H);$$

here we use the fact that $c(H + Y) = c(H) - |Y|$.

Next, consider the case $Y = \emptyset$. For simplicity, assume that $V_1 = [n' + 1, n]$ for some $n' \geq 1$. We have that $\Gamma_\emptyset = \mathrm{lk}_{\Lambda(G-E_1)}(H)$. In $G - E_1$, we have two connected components, one with vertex set $[n']$ and the other with vertex set $V_1 = [n' + 1, n]$. A set Z such that $H \subseteq H + Z \subseteq G - E_1$ belongs to $\mathrm{lk}_{\Lambda(G-E_1)}(H)$ if and only if $H([n']) + Z$ belongs to $\Lambda' := \Sigma(K_{n'}) \cap \mathrm{SSC}^{\mu'}_{\omega'}$, where $\mu' = (\mu_2, \ldots, \mu_m)$ and $\omega' = (\omega_1, \ldots, \omega_{n'})$; $m = L(\mu)$. Note that (μ', ω') is a permitted pair on n' vertices. Namely,

$$\omega_{[n']} = \omega_{[n]} - \omega_{V_1} > \sum_{i=1}^{m} \mu_i - \mu_1 = \sum_{i=2}^{m} \mu_i.$$

Now, induced subcomplexes of SPI complexes remain SPI over the corresponding induced matroid; in particular, $\Sigma(K_{n'})$ is SPI. Moreover, $\mathrm{lk}_{\Lambda(G-E_1)}(H)$ coincides with $\mathrm{lk}_{\Lambda'(G')}(H')$, where $G' = G([n'])$ and $H' = H([n'])$. G' and H' satisfy the conditions in the theorem; we have exactly one edge in $G' \setminus H'$ between any two connected components V_i and V_j in H'. By induction, $\mathrm{lk}_{\Lambda'(G')}(H')$ is hence $VD^+(r)$, where

$$r = c(H') - 2 - L(\mu') = c(H) - 1 - 2 - L(\mu) + 1 = d(H).$$

This concludes the proof. $\square$

By Theorem 13.24, the full simplex, the complex F_n of forests, and the complex B_n of bipartite graphs on n vertices are all SPI. As a consequence, we have the following corollary.

Corollary 18.18. *Let (μ, ω) be a permitted pair on n vertices. Then the complexes SSC_ω^μ, $\mathsf{F}_n \cap \mathsf{SSC}_\omega^\mu$, and $\mathsf{B}_n \cap \mathsf{SSC}_\omega^\mu$ are $VD^+(n - 2 - L(\mu))$. $\square$*

Remark. This yields a new proof that $\mathsf{F}_n \cap \mathsf{NC}_n$ and $\mathsf{B}_n \cap \mathsf{NC}_n$ are $VD^+(n-3)$; see Corollary 13.2 and Theorem 14.6.

Let us consider the special case $\mathsf{SSC}_n^{k,s}$; recall that this is the complex of graphs on n vertices with at least s connected components of vertex size at most k.

Theorem 18.19. *Let $k, s \geq 1$. For each $n > ks$, we have that $\mathsf{SSC}_n^{k,s}$ is $VD^+(n - 2 - s)$. Moreover, write $F_k(x) = \sum_{r=1}^{k} (-1)^r \frac{x^r}{r}$. Then $H_{k,s}(x) = \sum_{n \geq 1} \tilde{\chi}(\mathsf{SSC}_n^{k,s}) \frac{x^n}{n!}$ satisfies*

$$H_{k,s}(x) = (1 + x)\left(e^{F_k(x)} \cdot \sum_{r=0}^{s-1} \frac{(-F_k(x))^r}{r!} - 1 \right).$$

Proof. The first claim is an immediate consequence of Theorem 18.17. For the second claim, recall that C_n is the family of all connected graphs on n vertices; we know by Proposition 18.1 that $\tilde{\chi}(\mathsf{C}_n) = (-1)^n(n-1)!$. Let Σ_n^k be the family of graphs in which each component has size at least $k + 1$; define $\Sigma_0^k = \{\emptyset\}$. By Corollary 6.15, we have that

$$\sum_{n \geq 0} \tilde{\chi}(\Sigma_n^k) = -\exp\left(-\sum_{r > k} (-1)^r \frac{x^r}{r} \right) = -\exp\left(\ln(1 + x) + F_k(x) \right)$$
$$= -(1 + x)e^{F_k(x)}.$$

Next, let $\Gamma_n^{k,s}$ be the family of graphs with at least s components such that each component has size at most k. By Corollary 6.16,

$$\sum_{n \geq 0} \tilde{\chi}(\Gamma_n^{k,s}) = \sum_{r=0}^{s-1} \frac{(-F_k(x))^r}{r!} - e^{-F_k(x)}.$$

For any graph G in $\mathsf{SSC}_n^{k,s}$, let $G_{>k}$ be the induced subgraph consisting of all components of size at least $k + 1$ and let $G_{\leq k}$ be the induced subgraph consisting of all components of size at most k. It is clear that $G_{>k}$ is isomorphic to some graph in Σ_r^k for some $r \in [0, n]$ (in fact, $r \leq n - s$) and that $G_{\leq k}$ is isomorphic to some graph in $\Gamma_{n-r}^{k,s}$. Indeed, this property defines $\mathsf{SSC}_n^{k,s}$. As a consequence,

$$-\tilde{\chi}(\mathsf{SSC}_n^{k,s}) = \sum_{r=0}^{n} \binom{n}{r} \tilde{\chi}(\Sigma_r^k)\tilde{\chi}(\Gamma_{n-r}^{k,s}).$$

It follows that

$$H_{k,s}(x) = (1+x)e^{F_k(x)} \cdot \left(\sum_{r=0}^{s-1} \frac{(-F_k(x))^r}{r!} - e^{-F_k(x)} \right),$$

which concludes the proof. $\square$

Corollary 18.20. *For $1 \le s \le n-1$, we have that*

$$|\tilde{\chi}(\mathsf{SSC}_n^{1,s})| = n\binom{n-2}{s-1} - \binom{n-1}{s-1}.$$

Proof. By Theorem 18.19, we have that

$$H_{1,s}(x) = (1+x)\left(e^{-x} \cdot \sum_{r=0}^{s-1} \frac{x^r}{r!} - 1 \right).$$

For $n \ge s+1$, the coefficient of $\frac{x^n}{n!}$ is

$$\sum_{r=0}^{s-1}(-1)^{n-r}\left(\binom{n}{r} - n\binom{n-1}{r} \right) = (-1)^{n+s-1}\left(\binom{n-1}{s-1} - n\binom{n-2}{s-1} \right),$$

which concludes the proof. $\square$

A weaker variant of Theorem 18.17 is the following result:

Theorem 18.21. *Let Σ be a PI complex over the graphic matroid on K_n (see Section 13.2) and let (μ, ω) be a permitted pair on n vertices. Write $\Lambda = \Sigma \cap \mathsf{SSC}_\omega^\mu$. Let $H \subseteq G$ be graphs such that every two connected components in H are joined by at least one edge from $G \setminus H$. Then $\mathrm{lk}_{\Lambda(G)}(H)$ is $VD(c(H) - 2 - L(\mu))$. In particular, Λ is $VD(n-2-L(\mu))$.*

Proof. The proof is identical to the proof of Theorem 18.17, except for the case that $G \setminus H$ contains an edge e that joins two vertices from the same connected component in H. In this case, e may be present in $\mathrm{lk}_{\Lambda(G)}(H)$ without being a cone point if Σ is not SPI. However, it is still true then that we may use induction to conclude that $\mathrm{lk}_{\Lambda(G)}(H+e)$ is $VD(c(H) - 2 - L(\mu))$ and hence $VD(c(H) - 3 - L(\mu))$. Since $\mathrm{lk}_{\Lambda(G-e)}(H)$ is $VD(c(H) - 2 - L(\mu))$, it follows that $\mathrm{lk}_{\Lambda(G)}(H)$ is $VD(c(H) - 2 - L(\mu))$ as desired.

For the final case in the proof of Theorem 18.17, we need the fact that induced subcomplexes of PI complexes remain PI over the corresponding induced matroid, but this is obvious by Theorem 13.6. $\square$

18.4 Graphs with Some Component of Size Not Divisible by p

Let n and p be positive integers such that p divides n. Recall that $\mathsf{NC}_{n,p}$ is the complex of graphs such that some component has a vertex set of size not divisible by p.

Corollary 18.22. *For any graph G on n vertices, $\mathsf{NC}_{n,p}(G)$ is $VD^+(n - c(G) - 2)$, where $c(G)$ is the number of connected components in G. Moreover, for $p > 1$, the Euler characteristic of $\mathsf{NC}_{n,p}$ satisfies*

$$H(x) = \sum_{k \geq 1} \tilde{\chi}(\mathsf{NC}_{kp,p}) \frac{x^{kp}}{(kp)!} = (1 - (-x)^p)^{1/p} - 1.$$

Proof. The first claim is a consequence of Theorem 13.25 and the fact that $\mathsf{NC}_{n,p}$ is a SPI* complex; use Theorem 13.32. For the second claim, we know by Proposition 18.1 that $\tilde{\chi}(\mathsf{C}_n) = (-1)^n(n-1)!$. Hence by Corollary 6.15, we have that

$$-H(x) = 1 - \exp\left(\sum_{k \geq 1} \frac{(-x)^{kp}}{kp}\right) = 1 - \exp\left(\frac{1}{p}\ln(1 - (-x)^p)\right)$$

$$= 1 - (1 - (-x)^p)^{1/p},$$

which concludes the proof. $\square$

18.5 Disconnected Hypergraphs

The *k-equal partition lattice* $\varPi_n^{1,\geq k}$ is the sublattice of $\varPi_n$ consisting of all partitions in which each set has either size one or size at least k. Analogously to the correspondence between $\varPi_n$ and NC_n described earlier in this section, one easily proves that the order complex of the proper part of $\varPi_n^{1,\geq k}$ has the same homotopy type as the complex $\mathsf{HNC}_{n,k}$ of disconnected k-uniform hypergraphs. By a result of Björner and Welker about $\varPi_n^{1,\geq k}$, this yields the following theorem:

Theorem 1 (Björner and Welker [16]). $\mathsf{HNC}_{n,k}$ *is homotopy equivalent to a wedge of spheres in various dimensions; there is homology in dimension d if and only if $d = n - 3 - t(k - 2)$ for some integer $t \in [1, \lfloor \frac{n}{k} \rfloor]$. For $k \geq 3$, the rank of the top nonvanishing homology group $\tilde{H}_{n-k-1}(\mathsf{HNC}_{n,k}; \mathbb{Z})$ is $\binom{n-1}{k-1}$. Indeed, the rank of $\tilde{H}_d(\mathsf{HNC}_{n,k}; \mathbb{Z})$ is a multiple of $\binom{n-1}{k-1}$ for all d.*

Björner and Wachs [13] proved that $\varPi_n^{1,\geq k}$ is nonpure shellable. See Sundaram and Wachs [138] for detailed information about the homology of $\varPi_n^{1,\geq k}$.

Not 2-connected Graphs

We examine the complex NC_n^2 of not 2-connected graphs on n vertices. One of the most well-known and celebrated results in the field of graph complex topology is the theorem that NC_n^2 is homotopy equivalent to a wedge of $(n-2)!$ spheres of dimension $2n-5$; Babson, Björner, Linusson, Shareshian, and Welker [3] and, independently, Turchin [139] proved this theorem. In Section 19.1, we outline a third proof due to Shareshian [118] and present some consequences and related results. We also discuss the important lattice of block-closed graphs introduced by Babson et al. [3]. This lattice is Cohen-Macaulay [117] and its proper part is homotopy equivalent to NC_n^2. In Section 19.2, we proceed with a result by Shareshian about a concrete basis for the homology of the quotient complex $\mathsf{C}_n^2 = 2^{K_n}/\mathsf{NC}_n^2$ of 2-connected graphs. In Section 19.3, we show that NC_n^2 is semi-nonevasive and derive from that a new proof of Shareshian's homology basis result. The chapter is concluded in Section 19.4 with a generalization of NC_n^2 defined in terms of a sequence $\mathsf{A} = (A_1, \ldots, A_r)$ of subsets of $[n]$. A graph G belongs to this generalized complex if and only if each A_i contains a vertex a with the property that $G([n] \setminus \{a\})$ is disconnected.

The complex C_n^2 appears in Vassiliev's analysis of the homology and cohomology of certain spaces of knots [143]; see Section 1.1.5 for some discussion.

19.1 Homotopy Type

Babson, Björner, Linusson, Shareshian, and Welker [3] and, independently, Turchin [139] proved the following about NC_n^2:

Theorem 19.1. NC_n^2 *is homotopy equivalent to a wedge of* $(n-2)!$ *spheres of dimension* $2n-5$. $\square$

Babson et al. [3] also introduced the lattice of block-closed graphs. A graph G is *block-closed* if the edge vw belongs to G for any vertices v and w that are contained in one and the same simple cycle in G. The maximal cliques in a block-closed graph G are exactly the "2-connected components" in G; see Babson et al. [3] and Shareshian [118] for more information. A *block* in a block-closed graph G is a maximal clique containing at least two vertices; define $b(G)$ as the number of blocks in G. Define $\Pi_{n,2}$ as the poset of block-closed graphs.

Theorem 19.2 (Babson et al. [3]). *$\Pi_{n,2}$ is a lattice with rank function $\rho(G) = 2n - 2c(G) - b(G)$. In particular, a block-closed graph G contains at least $2n - 2c(G) - b(G)$ edges.* $\square$

Define the operator $f : P(\mathsf{NC}_n^2) \to P(\mathsf{NC}_n^2)$ by mapping a graph G to the graph obtained from G by adding all edges vw such that v and w are contained in one and the same cycle. We refer to $f(G)$ as the *block closure* of G. It is easy to see that f is a closure operator with image the proper part of $\Pi_{n,2}$. This implies the following:

Theorem 19.3 (Babson et al. [3]). *For $n \geq 2$, the proper part of $\Pi_{n,2}$ is homotopy equivalent to NC_n^2 and hence has the homotopy type of a wedge of $(n-2)!$ spheres of dimension $2n - 5$.* $\square$

We now outline a proof of a generalization of Theorem 19.1 based on discrete Morse theory. This alternate proof is, in all essence, due to Shareshian [118], who exploited a matching on NC_n^2 used by Rodica Simion to determine the Euler characteristic of NC_n^2. Shareshian's proof is to our knowledge the first example of a specific problem in topological combinatorics that was solved using discrete Morse theory in its full strength.

Theorem 19.4 (Shareshian [118, 117]). *If H is a block-closed graph on n vertices, then the lifted complex $\mathsf{NC}_n^2(H, \emptyset)$ is homotopy equivalent to a wedge of spheres of dimension $\delta(H) := |H| + 2c(H) + b(H) - 5$. If H is not block-closed, then $\mathsf{NC}_n^2(H, \emptyset)$ is a cone. As a consequence, the $(2n - 5)$-skeleton of NC_n^2 is Cohen-Macaulay. Moreover, NC_n^2 is homotopy equivalent to a wedge of $(n-2)!$ spheres of dimension $2n - 5$.*

Proof. We want to prove that $\mathsf{NC}_n^2(H, \emptyset)$ admits an acyclic matching such that all unmatched graphs contain $|H| + 2c(H) + b(H) - 4$ edges. First, assume that H is not block-closed. Let v and w be any nonadjacent vertices contained in a simple cycle in H. It is clear that vw is a cone point in $\mathsf{NC}_n^2(H, \emptyset)$, because there is no way to separate v from w by removing a vertex in a graph containing H.

From now on, assume that H is block-closed. If H is the complete graph, then the theorem is trivial. Moreover, the case $n = 2$ is easy to check. Thus assume that H is not complete and that $n \geq 3$. We define an acyclic matching on the quotient complex $\mathsf{C}_n^2(H, \emptyset)$ of 2-connected graphs containing H such that all unmatched graphs contain $\delta(H) + 2$ edges. The matching turns out to be straightforward to translate into an acyclic matching on $\mathsf{NC}_n^2(H, \emptyset)$; compare to the discussion at the beginning of Section 17.1.

Let a and b be any nonadjacent vertices in H; we may assume that $a = 1$ and $b = n$. Define a matching on $\mathsf{C}_n^2(H, \emptyset)$ by pairing $G + 1n$ with $G - 1n$ whenever possible. Let Δ be the family of unmatched graphs. By Lemma 4.1, we obtain an acyclic matching on $\mathsf{C}_n^2(H, \emptyset)$ by combining any acyclic matching on Δ with the matching just defined.

Δ consists of all 2-connected graphs G containing H such that $G - 1n$ is not 2-connected. Let G be a graph in Δ and let $\hat{G}$ be the block closure of $G - 1n$. We have that the vertex 1 is contained in a unique block in $\hat{G}$; let M_G be the vertex set of this block. To prove uniqueness, note that if 1 were contained in more than one block, then 1 would be a cut point in $\hat{G}$ and hence in G; this is a contradiction. Note that M_G is the set of neighbors of 1 in $\hat{G}$ and that M_G does not contain the vertex n.

We claim that M_G contains a unique vertex $x_G \neq 1$ such that x_G is a cut point in $\hat{G}$.

For existence, let x be a vertex in M_G such that $xy \in \hat{G}$ for some y not in M_G. There must be some x with this property, because otherwise there would be no path from 1 to n in $\hat{G}$. If x were not a cut point in $\hat{G}$, then there would be a path from 1 to y in $\hat{G}([n] \setminus \{x\})$. Since $1x, xy \in \hat{G}$, this would imply that there is a simple cycle in $\hat{G}$ containing 1 and y. However, $\hat{G}$ is block-closed, which yields that $1y \in \hat{G}$ and hence that $y \in M_G$, a contradiction.

For uniqueness, suppose that x' is another vertex such that $x'z \in \hat{G}$ for some z not in M_G; by symmetry, we obtain that x' is a cut point in $\hat{G}$ separating 1 and z. Now, since x' is not a cut point in $\hat{G} + 1n$, there must be a path from z to n in $\hat{G}([n] \setminus \{x'\})$. This path cannot use any vertex in M_G, as this would yield a path from z to 1 in $\hat{G}([n] \setminus \{x'\})$. By symmetry, there is a path from y to n in $G([n] \setminus M_G)$. As a consequence, there is a path from z to y in $G([n] \setminus M_G)$. However, this is a contradiction, as we can extend this to a path from z to 1 in $\hat{G}([n] \setminus \{x'\})$; yx and $x1$ are both present in this graph.

Let $\Delta(M, x)$ be the family of graphs G in Δ such that $M = M_G$ and $x = x_G$. Since M_G can only increase if we add edges to G, it is clear that the poset map sending $\Delta(M, x)$ to (M, x) satisfies the Cluster Lemma 4.2; we consider (M, x) as smaller than (M', x') if $M \subsetneq M'$. It remains to prove that each $\Delta(M, x)$ admits an acyclic matching such that each unmatched graph contains $\delta(H) + 2$ edges. If $\Delta(M, x)$ is void, then we are done. From now on, assume that $\Delta(M, x)$ is nonvoid. In particular, there are no edges in H from $M \setminus \{x\}$ to $[n] \setminus (M \setminus \{x\})$.

We divide into two cases depending on whether or not xn belongs to H. First, assume that $xn \notin H$. Define a matching on $\Delta(M, x)$ by pairing $G - xn$ and $G + xn$ whenever possible. Let $\Sigma(M, x)$ be the family of unmatched graphs. We claim that a graph G in $\Delta(M, x)$ belongs to $\Sigma(M, x)$ if and only if $G([n-1])$ is 2-connected and the neighborhood of n in G is $\{1, x\}$. In particular, $\Sigma(M, x)$ is void unless $M = [n - 1]$.

To prove the above statement, we first note that one direction is obvious. For the other direction, assume that $G \in \Sigma(M, x)$; this means that G and $G(M)$ are 2-connected, whereas $G - 1n$ and $G - xn$ are not. The only possible cut point in $G - xn$ is 1, because any cut point in $G - xn$ must separate x and n, and 1 cannot be separated from either of these vertices. Moreover, by construction, x separates 1 from n in $G - 1n$. As a consequence, n cannot be in the same connected component as 1 and x in $G \setminus \{1n, xn\}$. Since n is not a cut point in G, it follows that n is isolated in $G \setminus \{1n, xn\}$ as desired. Moreover, again by construction, 1 and x belong to the same 2-connected component in $G - 1n$, which implies that we must have that $M = [n - 1]$.

The conclusion is that $\Sigma(M, x)$ is void unless $M = [n - 1]$ and n is isolated in H, in which case we have that

$$\Sigma([n - 1], x) = \{\{1n, xn\}\} * \mathsf{C}^2_{n-1}(H([n - 1]), \emptyset).$$

By induction, $\mathsf{C}^2_{n-1}(H([n - 1]), \emptyset)$ admits an acyclic matching such that all unmatched graphs contain exactly

$$\delta(H[n - 1]) + 2 = |H| + b(H([n - 1])) + 2c(H([n - 1])) - 3$$
$$= |H| + b(H) + 2c(H) - 5 = \delta(H)$$

edges. It follows that $\Sigma([n - 1], x)$ admits an acyclic matching such that all unmatched graphs contain $\delta(H) + 2$ edges.

At this point, note that if H is the empty graph, then there are $n - 2$ choices for x; thus an induction argument yields that the reduced Euler characteristic of C^2_n is $(n - 2) \cdot \tilde{\chi}(\mathsf{C}^2_{n-1}) = (n - 2)!$ as desired.

Next, suppose that $xn \in H$. We claim that a graph G containing H belongs to $\Delta(M, x)$ if and only if the two induced subgraphs on M and $P = ([n] \setminus M) \cup \{x\}$ are 2-connected. One direction is immediate. For the other direction, suppose that G belongs to $\Delta(M, x)$ but that the induced subgraph $G(P)$ is not 2-connected. Then there is a cut point y in $G(P)$ separating x from some vertex z. We cannot have that $y = n$, because then n would separate M from z in G. Moreover, since $xn \in G(P)$, we obtain that n belongs to the same connected component as x in $G(P \setminus \{y\})$. Since $1n$ is the only edge between $M \setminus \{x\}$ and $P \setminus \{x\}$ in G, it follows that y separates the whole of M from z in G. Hence y is a cut point in G, which is a contradiction.

As a consequence,

$$\Delta(M, x) = \{\{1n\}\} * \mathsf{C}^2_M(H(M), \emptyset) * \mathsf{C}^2_P(H(P), \emptyset);$$

C^2_X is the quotient complex of 2-connected graphs on the vertex set X. Induction yields that $\mathsf{C}^2_M(H(M), \emptyset)$ and $\mathsf{C}^2_P(H(P), \emptyset)$ admit acyclic matchings such that all unmatched graphs contain $\delta(H(M)) + 2$ and $\delta(H(P)) + 2$ edges, respectively. This yields an acyclic matching on $\Delta(M, x)$ such that all unmatched graphs have

$$1 + \delta(H(M)) + 2 + \delta(H(P)) + 2$$
$$= |H| + b(H(M)) + b(H(P)) + 2c(H(M)) + 2c(H(P)) - 5$$

edges. One readily verifies that $b(H(M)) + b(H(P)) = b(H)$ and $c(H(M)) + c(H(P)) = c(H) + 1$; hence

$$|H| + b(H(M)) + b(H(P)) + 2c(H(M)) + 2c(H(P)) - 5$$
$$= |H| + b(H) + 2c(H) - 3 = \delta(H) + 2$$

as desired.

To prove that the $(2n - 5)$-skeleton of NC_n^2 is Cohen-Macaulay, use Theorem 19.2 to conclude that $\mathrm{lk}_{\mathsf{NC}_n^2}(H)$ is $(2n - 6 - |H|)$-connected;

$$\delta(H) = |H| + 2c(H) + b(H) - 5 \geq 2n - 5. \ \square$$

As an immediate consequence of the proof of Theorem 19.4, we have the following result:

Corollary 19.5. *For $n \geq 2$, NC_n^2 admits an acyclic matching with $(n-2)!$ critical faces of dimension $2n-5$. In particular, NC_n^2 is semi-collapsible. Moreover, C_n^2 admits an acyclic matching with $(n-2)!$ critical faces of dimension $2n - 4$. $\square$*

Using Theorem 19.4, Shareshian [117] established the following result about the lattice $\Pi_{n,2}$:

Theorem 19.6 (Shareshian [117]). *For $n \geq 2$, $\Pi_{n,2}$ is Cohen-Macaulay.*

Proof. Recall from Theorem 19.2 that the rank function on $\Pi_{n,2}$ is given by $\rho(G) = 2n - 2c(G) - b(G)$. For each interval (H, G), we prove that the order complex $\Delta(H, G)$ is homotopy equivalent to a wedge of spheres of dimension $\rho(G) - \rho(H) - 2$. By Theorem 5.29, this is sufficient to prove the theorem; every link in the order complex of a poset is a join of intervals.

First, assume that $G = K_n$. By the Closure Lemma 6.1, we have that $\Delta(H, K_n)$ is homotopy equivalent to $\mathrm{lk}_{\mathsf{NC}_n^2}(H)$. By Theorem 19.4, this link is homotopy equivalent to a wedge of spheres of dimension $2c(H) + b(H) - 5 = \rho(K_n) - \rho(H) - 2$ as desired.

Suppose that $G \neq K_n$ and let $V_1, \ldots, V_r$ be the blocks in G. Write $G_i = G(V_i)$ and $H_i = H(V_i)$. Let $\Sigma_{G,H}$ be the lifted complex consisting of all graphs G' such that $H \subseteq G' \subseteq G$ and such that $G'(V_i)$ is not 2-connected for at least one i. By the Closure Lemma 6.1, it is clear that $\mathrm{lk}_{\Sigma_{G,H}}(H)$ is homotopy equivalent to (H, G). Define an acyclic matching on $\Sigma_{G,H}$ in the following manner:

We may assume that $G_i \neq H_i$ for some i, say $i = r$. Let ab be an edge in $G_r \setminus H_r$. Let $\Sigma_{G,H}^0$ be the subfamily of $\Sigma_{G,H}$ consisting of all graphs G' such that $G'(V_j) \in \mathsf{NC}_{V_j}^2$ for some $j < r$; $\mathsf{NC}_{V_j}^2$ is the complex of not 2-connected graphs on the vertex set V_j. It is clear that ab is a cone point in $\Sigma_{G,H}^0$.

Let $\Sigma_{G,H}^1$ be the remaining family; $\Sigma_{G,H}^1$ is the join of the families $\mathsf{C}_{V_1}^2(H_1, \emptyset), \ldots, \mathsf{C}_{V_{r-1}}^2(H_{r-1}, \emptyset)$, and $\mathsf{NC}_{V_r}^2(H_r, \emptyset)$. By Theorems 5.29 and 19.4, $\Sigma_{G,H}^1$ admits an acyclic matching such that all unmatched graphs contain

$$\sum_{i=1}^{r}(|H_i| + b(H_i) + 2c(H_i) - 3) - 1$$

edges. One readily verifies that $b(H) = \sum_i b(H_i)$ and $b(G) = r$. Moreover, it is easy to check that

$$c(H) = \sum_{i=1}^{r} c(H_i) + n - \sum_{i=1}^{r} |V_i|$$

and $c(G) = r + n - \sum_i |V_i|$. For any unmatched graph G_0, it follows that the number of edges in $G_0 \setminus H$ equals

$$b(H) + 2c(H) - 2n + 2\sum_i |V_i| - 3r - 1 = 2\sum_i |V_i| - 3r - 1 - \rho(H).$$

Now,

$$\rho(G) = 2n - r - 2r - 2n + 2\sum_i |V_i| = 2\sum_i |V_i| - 3r.$$

As a consequence, the number of edges in $G_0 \setminus H$ is $\rho(G) - \rho(H) - 1$, which implies that $\mathrm{lk}_{\Sigma_{G,H}}(H)$ is homotopy equivalent to a wedge of spheres of dimension $\rho(G) - \rho(H) - 2$. $\square$

Finally, let us say a few words about the complex $\mathsf{HNC}^2_{n,t}$ of not 2-connected t-uniform hypergraphs. A t-uniform hypergraph H on a vertex set V is k-connected if the induced subhypergraph $H(V \setminus W)$ is connected for each set $W \subset V$ of size less than k. Analyzing the lattice $\Pi_{n,2}$ in greater detail, Shareshian [117] was able to prove that $\mathsf{HNC}^2_{n,3}$ is homotopy equivalent to a wedge of spheres of dimension $n-4$. In addition, he computed the exponential generating function for $\tilde{\chi}(\mathsf{HNC}^2_{n,3})$. For $t \geq 4$, the homotopy type of $\mathsf{HNC}^2_{n,t}$ remains an open problem except in trivial cases.

19.2 Homology

Shareshian [118] computed an explicit basis for the homology of the quotient complex $\mathsf{C}^2_n = 2^{K_n}/\mathsf{NC}^2_n$ in terms of the fundamental cycle of the associahedron:

Theorem 19.7 (Shareshian [118]). *Let $n \geq 3$. Define $\rho(\pi^*)$ as in (17.3) in Section 17.2; we have that $\rho(\pi^*) = \rho(\pi) \wedge [\rho(\mathrm{Bd}_n)]$. Then the set $\{\rho(\pi^*) : \rho \in \mathfrak{S}_{[n]}, \rho_1 = 1, \rho_n = n\}$ is a basis for the homology of C^2_n.* $\square$

Note that the set in Theorem 19.7 coincides with that in Theorem 17.6. As a consequence, there is a natural embedding of $\tilde{H}_{2n-4}(\mathsf{C}^2_n)$ into $\tilde{H}_{2n-4}(\mathsf{Ham}_n)$. In Section 19.3, we give a new proof of Theorem 19.7.

The observation that one may express Shareshian's basis in terms of the fundamental cycle of the associahedron is due to Shareshian and Wachs [118].

Define
$$\rho(\hat{\pi}^*) = \rho(\pi) \wedge \partial([\rho(\mathrm{Bd}_n)]). \tag{19.1}$$

Note that $\hat{\pi}^*$ is the fundamental cycle of the sphere obtained by taking the join of the associahedron A_n and the boundary complex of the simplex on the set Bd_n. This sphere is of importance in the analysis of the complex $\mathsf{NCR}_n^{1,0}$ of graphs with a separable polygon representation; see Section 21.3.

Corollary 19.8. *The set $\{\rho(\hat{\pi}^*) : \rho \in \mathfrak{S}_{[n]}, \rho_1 = 1, \rho_n = n\}$ is a basis for the homology of NC_n^2.*

Proof. This is an immediate consequence of Theorem 19.7 and the long exact sequence for the pair $(2^{K_n}, \mathsf{NC}_n^2)$; see Theorem 3.3. $\square$

19.3 A Decision Tree

We show that NC_n^2 is semi-nonevasive, thereby strengthening Shareshian's result in Corollary 19.5 that NC_n^2 is semi-collapsible. Moreover, we show how to use this result to reestablish Theorem 19.7.

Theorem 19.9. *For $n \geq 3$, $\mathsf{NC}_n^2 \sim (n-2)! \cdot t^{2n-5}$. In particular, NC_n^2 is semi-nonevasive.*

Proof. Let $E_n = \{in : i \in [n-1]\}$ and consider the family $\Sigma_Y = \mathsf{NC}_n^2(Y, E_n \setminus Y)$ for each $Y \subseteq E_n$. $|Y| \leq 1$ means that the degree of the vertex n is at most one. In particular, any edge ij such that $i, j \neq n$ is a cone point in Σ_Y, which implies that Σ_Y is nonevasive.

From now on, assume that $|Y| \geq 2$. First, we claim that Σ_{E_n} coincides with $\{E_n\} * \mathsf{NC}_{n-1}$, where NC_{n-1} is the complex of disconnected graphs on $n-1$ vertices. Namely, since n is adjacent to all other vertices in a graph G containing E_n, n is the only possible cut point. Clearly, n is a cut point if and only if $G([n-1])$ is disconnected. By Proposition 18.1, $\mathsf{NC}_{n-1} \sim (n-2)! \cdot t^{n-4}$. As a consequence, if we can prove that Σ_Y is nonevasive whenever $Y \subsetneq E_n$, then it follows that $\mathsf{NC}_n^2 \sim (n-2)! \cdot t^{|E_n|} \cdot t^{n-4} = (n-2)! \cdot t^{2n-5}$ by Lemma 5.22.

Now, for a given set $Y \subsetneq E_n$ such that $|Y| \geq 2$, define $\epsilon(Y) = \{i : in \in Y\}$ and $B_Y = \binom{[n-1]}{2} \setminus \binom{\epsilon(Y)}{2}$. Consider the family $\Sigma_{Y,Z} = \Sigma_Y(Z, B_Y \setminus Z)$ for each possible edge set $Z \subseteq B_Y$. $\Sigma_{Y,Z}$ consists of graphs containing the edge set $E' = Y \cup Z$ but not any edges from $E_n \setminus Y$ or $B_Y \setminus Z$. There are three possibilities for the graph $G = ([n], E')$.

- G *is disconnected.* Since any two vertices $w_1, w_2 \in \epsilon(Y)$ already belong to the same component in G, $w_1 w_2$ is a cone point in $\Sigma_{Y,Z}$.
- G *is connected, and some cycle contains the vertex n.* Let $w_1, w_2 \in \epsilon(Y)$ be the neighbors of n in this cycle. It is clear that adding or deleting $w_1 w_2$ to or from a face of $\Sigma_{Y,Z}$ does not affect the 2-connectivity of the corresponding graph; thus $w_1 w_2$ is a cone point.

- G *is connected, and no cycle contains the vertex* n. Let $w_1 \in \epsilon(Y)$ be such that n is not the only neighbor of w_1 in G; such a w_1 exists since G is connected and fewer than $n - 1$ vertices are adjacent to n. Let $v \neq n$ be a neighbor of w_1 in G. We claim that w_1 is a cut point separating v from $\{n\} \cup (\epsilon(Y) \setminus \{w_1\})$ in G, and hence also in $G + \binom{\epsilon(Y)}{2}$. Namely, if there were a path from v to n not using w_1, then this path would form a cycle together with w_1. In particular, $w_1 w_2$ is a cone point in $\Sigma_{Y,Z}$ for any $w_2 \in \epsilon(Y) \setminus \{w_1\}$.

As a consequence, $\Sigma_{Y,Z}$ is always nonevasive, which by Lemma 5.22 implies that Σ_Y is nonevasive; thus we are done. $\square$

Corollary 19.10. *Let* $G \neq K_n$ *be a graph on* n *vertices. Then the induced subcomplex* $\mathsf{NC}_n^2(G)$ *has no homology above dimension* $2n - 6$.

Proof. By Theorem 19.4, for any nonempty graph H, $\mathrm{lk}_{\mathsf{NC}_n^2}(H)$ has no homology above dimension $2c(H) + b(H) - 5$. This is at most $2n - 6$; the block-closure of H has rank at least one in the lattice $\Pi_{n,2}$, meaning that $2n - 2c(H) - b(H) \geq 1$ by Theorem 19.2. As we just concluded, $\mathrm{del}_{\mathsf{NC}_n^2}(e)$ is nonevasive for every e; hence we are done by Proposition 6.7. $\square$

Remark. Corollary 19.10 implies the well-known and easily proved result that every minimal 2-connected graph on $n \geq 4$ vertices has at most $2n - 4$ edges. Namely, if $G = ([n], E)$ is a minimal nonface of NC_n^2, then $\mathsf{NC}_n^2(G)$ coincides with the boundary of the full simplex on E.

New proof of Theorem 19.7. We may view the decision tree just given as a decision tree on $\mathsf{C}_n^2 = 2^{K_n}/\mathsf{NC}_n^2$ with $(n-2)!$ evasive graphs of dimension $2n - 4$. First, we claim that we can define the decision tree such that each evasive graph is of the form

$$T_\rho = E_n \cup \{\rho_i \rho_{i+1} : i \in [1, n-2]\},$$

where $E_n = \{in : i \in [n-1]\}$ and $\{\rho_1, \ldots, \rho_{n-1}, \rho_n\} = [n]$; $\rho_1 = 1$ and $\rho_n = n$. Namely, by the proof of Theorem 19.9, given any optimal decision tree on $\mathsf{C}_{n-1} = 2^{K_{n-1}}/\mathsf{NC}_{n-1}$, we obtain an optimal decision tree on $\mathsf{C}_n^2 = 2^{K_n}/\mathsf{NC}_n^2$ with one evasive set $E_n \cup \sigma$ for each evasive set σ in C_{n-1}. By the proof of Proposition 18.3, we may define an optimal decision tree on C_{n-1} such that the evasive faces are exactly of the form $\{\rho_i \rho_{i+1} : i \in [1, n-2]\}$ with $\{\rho_1, \ldots, \rho_{n-1}\} = [n-1]$ and $\rho_1 = 1$.

It remains to show that we can define the decision tree such that the underlying acyclic matching has the property that all graphs appearing in the cycle π_ρ^* except T_ρ are matched with smaller graphs. By Corollary 4.17 and the fact that π_ρ^* is a cycle in the chain complex of C_n^2, it then follows that the $(n-2)!$ cycles π_ρ^* are exactly the homology cycles generated by the acyclic matching.

Let T be a triangulated n-gon with boundary edges $\rho_1\rho_2, \rho_2\rho_3, \ldots, \rho_n\rho_1$; $\rho_1 = 1$ and $\rho_n = n$; assume that $T \neq T_\rho$. With notation as in the proof of Theorem 19.9, T belongs to a family $\Sigma_{Y,Z}$ for some $Y \subsetneq E_n$ containing $1n$ and $\rho_{n-1}n$ and some $Z \subset B_Y = \binom{[n-1]}{2} \setminus \binom{\epsilon(Y)}{2}$.

As in the proof of Theorem 19.9, let G be the graph with edge set $Y \cup Z$. Now, we have to be careful, because different T and ρ may give rise to the same G. Specifically, we need to find a cone point in $\Sigma_{Y,Z}$ such that this cone point is present in any triangulated n-gon T in $\Sigma_{Y,Z}$; it is not sufficient to find a cone point that is present in a given fixed T.

Let x be minimal such that $xn \in Y$ and x is contained in a cycle in G containing n. Such an x exists: we obtain G from T by removing all $\rho_i\rho_{i+1}$ such that $\rho_i n, \rho_{i+1}n \in Y$. Since $Y \neq E_n$ and $1n, \rho_{n-1}n \in Y$, there must exist some a, b with $b - a \geq 2$ such that $\rho_a n, \rho_b n \in Y$ and $\rho_c n \notin Y$ if $a < c < b$; compare to the situation in the proof of Proposition 16.2. Let y be arbitrarily chosen such that x and y are contained in a minimal cycle in G containing n and such that $yn \in G$.

By the proof of Theorem 19.9 (the second of the three possibilities for G), xy is a cone point in $\Sigma_{Y,Z}$. Since x and y are defined without reference to T or ρ, it suffices to prove that $xy \in T$; by symmetry, this will then hold for all other relevant triangulated n-gons in $\Sigma_{Y,Z}$. We have that $x = \rho_i$ and $y = \rho_j$ for some i, j; $j - i \geq 2$ (or $i - j \geq 2$). Since $xn, yn \in T$, it is clear that xy does not cross any edge in T with both endpoints in $[n-1]$. Namely, such an edge would cross either xn or yn. Suppose that xy crosses some edge $\rho_k n$; $i < k < j$. Since x and y are part of the same minimal cycle in G, we have vertices

$$i = i_0 < i_1 < \ldots < i_{r-1} < i_r = j$$

such that $\rho_{i_s}\rho_{i_{s+1}} \in G$ for all s and such that the cycle

$$(n, \rho_i, \rho_{i_1}, \ldots, \rho_{i_{r-1}}, \rho_j, n)$$

is minimal. If some i_s equals k, then we have a contradiction to the minimality of the cycle. Hence $i_s < k < i_{s+1}$ for some s, but then $\rho_k n$ and $\rho_{i_s}\rho_{i_{s+1}}$ cross in T, another contradiction. Thus we are done. $\square$

19.4 Generalization and Yet Another Proof

Finally, we present a generalization of NC_n^2 that yields another approach to proving Theorem 19.1. Let Δ_{n-1} be a monotone graph property on $n - 1$ vertices. For any vertex set V of size $n-1$, we define Δ_V as the monotone graph property on V naturally isomorphic to Δ_{n-1}. For a sequence $\mathsf{A} = (A_1, \ldots, A_r)$ of subsets of $[n]$, let $\Delta_n^2(\mathsf{A})$ be the complex of graphs G on n vertices such that the following holds:

- For each $i \in [r]$, there is an $a \in A_i$ such that $G([n] \setminus \{a\})$ belongs to $\Delta_{[n]\setminus\{a\}}$.

For example, if $\Delta_{n-1} = \mathsf{NC}_{n-1}$, then $\Delta_n^2([n]) = \mathsf{NC}_n^2$.

Define $\Delta_n^2(a_1, \ldots, a_r) = \Delta_n^2(\{a_1\}, \ldots, \{a_r\})$.

Lemma 19.11. *Suppose that* $\Delta_n^2(1, \ldots, r)$ *is contractible whenever* $r \leq n-1$. *Let* $\mathsf{A} = (A_1, \ldots, A_r)$ *be a sequence of pairwise disjoint and nonempty subsets of* $[n]$. *Then*

$$\Delta_n^2(\mathsf{A}) \simeq \begin{cases} \mathrm{Susp}^{n-r}(\Delta_n^2(1, \ldots, n)) & \text{if } \sum_i |A_i| = n; \\ \text{point} & \text{otherwise.} \end{cases}$$

Moreover, if $\Delta_n^2(1, \ldots, r)$ *is buildable for* $r \leq n-1$ *and* $\Delta_n^2(1, \ldots, n)$ *is semi-buildable, then* $\Delta_n^2(\mathsf{A})$ *is semi-buildable.*

Proof. We use induction on the number r of components in A. If each A_i has size one, then we are done by assumption and symmetry. Otherwise, assume that $|A_r| \geq 2$ and write A_r as a disjoint union of two nonempty sets $A_{r,1}$ and $A_{r,2}$. Write $\mathsf{A}' = (A_1, \ldots, A_{r-1})$; hence $\mathsf{A} = (\mathsf{A}', A_r)$. One easily checks that

$$\Delta_n^2(\mathsf{A}', A_{r,1}) \cup \Delta_n^2(\mathsf{A}', A_{r,2}) = \Delta_n^2(\mathsf{A}', A_r);$$
$$\Delta_n^2(\mathsf{A}', A_{r,1}) \cap \Delta_n^2(\mathsf{A}', A_{r,2}) = \Delta_n^2(\mathsf{A}', A_{r,1}, A_{r,2}).$$

By induction, $\Delta_n^2(\mathsf{A}', A_{r,1})$ and $\Delta_n^2(\mathsf{A}', A_{r,2})$ are both contractible. We hence obtain that

$$\Delta_n^2(\mathsf{A}', A_r) \simeq \frac{\Delta_n^2(\mathsf{A}', A_r)}{\Delta_n^2(\mathsf{A}', A_{r,1})} = \frac{\Delta_n^2(\mathsf{A}', A_{r,2})}{\Delta_n^2(\mathsf{A}', A_{r,1}, A_{r,2})} \simeq \mathrm{Susp}(\Delta_n^2(\mathsf{A}', A_{r,1}, A_{r,2}));$$

use the Contractible Subcomplex Lemma 3.16 for the first homotopy equivalence and Lemma 3.18 for the second. Since $\sum_{i=1}^r |A_i| = \sum_{i=1}^{r-1} |A_i| + |A_{r,1}| + |A_{r,2}|$, we are done. $\square$

For $\mathsf{A} = (A_1, \ldots, A_r)$, we have that $\mathsf{NC}_n^2(\mathsf{A})$ is the complex of graphs G such that each A_i contains a vertex a with the property that $G([n] \setminus \{a\})$ is disconnected. Note that this definition does not take into account whether G itself is connected. For $n \geq 3$, $\mathsf{NC}_n^2([n])$ coincides with NC_n^2.

Theorem 19.12. *Let* $n \geq 3$ *and let* $\mathsf{A} = (A_1, \ldots, A_r)$ *be a sequence of pairwise disjoint and nonempty subsets of* $[n]$. *Then*

$$\mathsf{NC}_n^2(\mathsf{A}) \simeq \begin{cases} \bigvee_{(n-2)!} S^{2n-r-4} & \text{if } \sum_i |A_i| = n; \\ \text{point} & \text{otherwise.} \end{cases}$$

Proof. We prove that $\mathsf{NC}_n^2(1, \ldots, r) \sim 0$ whenever $0 \leq r \leq n-1$ and $n \geq 3$ and that $\mathsf{NC}_n^2(1, \ldots, n) \sim (n-2)! \cdot t^{n-4}$. By Lemma 19.11, this implies the theorem.

First, consider $\Sigma_{n,r} = \mathsf{NC}_n^2(1, \ldots, r)$ for $0 \leq r \leq n-1$. This complex is trivially nonevasive if $n = 3$; thus assume that $n \geq 4$. If $r \leq 1$, then the edge $1n$ is a cone point; thus assume that $r \geq 2$.

Let E_n be the set of edges incident to the vertex n. If B contains two edges an and bn, then ab is a cone point in $\Sigma_{n,r}(B, E_n \setminus B)$. If B is empty, then any edge in K_{n-1} is a cone point. The remaining case is that $B = \{kn\}$ for some $k \in [n-1]$. One readily verifies that $\Sigma_{n,r}(\{kn\}, E_n \setminus \{kn\})$ coincides with $\{\{kn\}\} * \Sigma_{n-1,r}$ if $k > r$ and with $\{\{kn\}\} * \mathsf{NC}^2_{n-1}(1, \ldots, k-1, k+1, \ldots, r) \cong \{\{kn\}\} * \Sigma_{n-1,r-1}$ if $k \le r$. By induction, we obtain the desired statement.

It remains to consider $\Sigma_{n,n}$. In a connected graph, not all vertices are cut points; take a leaf in a spanning tree. In particular, $\Sigma_{n,n}$ is exactly the complex of disconnected graphs, except that $\Sigma_{n,n}$ does not contain graphs with an isolated vertex and a connected component of size $n - 1$. Again, consider the set E_n. As before, ab is a cone point in $\Sigma_{n,n}(B, E_n \setminus B)$ whenever $an, bn \in B$. Moreover, $\Sigma_{n,n}(\{kn\}, E_n \setminus \{kn\})$ coincides with

$$\{\{kn\}\} * \mathsf{NC}^2_{n-1}(1, \ldots, k-1, k+1, \ldots, n-1) \cong \{\{kn\}\} * \Sigma_{n-1,n-2} \sim 0.$$

The remaining case is $B = \emptyset$. Since $\Sigma_{n,n}(\emptyset, E_n) = \mathsf{NC}_{n-1}$, Proposition 18.3 implies that $\Sigma_{n,n} \sim \Sigma_{n,n}(\emptyset, E_n) \sim t^{n-4}$ as desired. $\square$

Remark. Lemma 19.11 and the above proof imply that NC^2_n is buildable, a strictly weaker property than the semi-nonevasiveness established in Theorem 19.9.

Not 3-connected Graphs and Beyond

[1] We verify a conjecture due to Babson, Björner, Linusson, Shareshian, and Welker [3] about the complex NC_n^3 of not 3-connected graphs on n vertices:

$$\mathsf{NC}_n^3 \simeq \bigvee_{(n-3)\frac{(n-2)!}{2}} S^{2n-4}. \tag{20.1}$$

We obtain this result via a certain acyclic matching on NC_n^3 such that exactly $(n-3) \cdot (n-2)!/2$ graphs, each containing $2n-3$ edges, remain unmatched. See Section 20.1 for details.

While NC_n^3 is hence semi-collapsible, we have not been able to establish semi-nonevasiveness. By Proposition 18.3 and Theorem 19.9, both NC_n and NC_n^2 are semi-nonevasive.

In Section 20.2, we use the given acyclic matching to determine a basis for the homology of the quotient complex $\mathsf{C}_n^3 = 2^{K_n}/\mathsf{NC}_n^3$. In Section 20.3, we analyze the elements in this basis in greater detail. As it turns out, the corresponding elements in the homology of NC_n^3 coincide with the fundamental cycles of certain spheres with the property that maximal faces correspond to "disconnected" lattice paths.

In Section 20.4, we give an overview of known results and some open problems related to the complex NC_n^k of not k-connected graphs for $k > 3$. The main result is a formula for the Euler characteristic of NC_n^{n-3} due to Babson et al. [3].

20.1 Homotopy Type

We establish the homotopy equivalence (20.1), thereby verifying the conjecture of Babson et al. [3]. Our proof involves an acyclic matching on NC_n^3 such

---

[1] This chapter is a revised and extended version of Sections 4 and 7 in a paper [67] published in *Journal of Combinatorial Theory, Series A*.

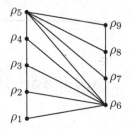

Fig. 20.1. $G(\rho_1, \rho_2, \rho_3, \rho_4, \rho_5 | \rho_6, \rho_7, \rho_8, \rho_9)$.

that there are $(n-3) \cdot (n-2)!/2$ critical graphs, each of which has $2n-3$ edges. The graphs are easy to describe: For $2 \le k \le n-2$ and a permutation $\rho = \rho_1 \rho_2 \ldots \rho_n \in \mathfrak{S}_{[n]}$, let $G(\rho_1, \ldots, \rho_k | \rho_{k+1}, \ldots, \rho_n)$ be the graph with edge set

$$\{\rho_1 \rho_2, \rho_2 \rho_3, \ldots, \rho_{k-1} \rho_k\} \cup \{\rho_{k+1} \rho_{k+2}, \rho_{k+2} \rho_{k+3}, \ldots, \rho_{n-1} \rho_n\} \cup$$
$$\{\rho_1 \rho_{k+1}, \rho_2 \rho_{k+1}, \ldots, \rho_k \rho_{k+1}, \rho_k \rho_{k+2}, \rho_k \rho_{k+3}, \ldots, \rho_k \rho_n\}.$$

The graph $G(\rho_1, \ldots, \rho_k | \rho_{k+1}, \ldots, \rho_n)$ consists of two "walls" with $\rho_1, \ldots, \rho_k$ forming a path on the left-hand side and $\rho_{k+1}, \ldots, \rho_n$ forming a path on the right-hand side. All vertices on the left wall are connected to ρ_{k+1}, whereas all vertices on the right wall are connected to ρ_k. See Figure 20.1 for an example.

Let

$$\mathcal{U}_k = \{G(\rho_1, \rho_2, \ldots, \rho_k | \rho_{k+1}, \rho_{k+2}, \ldots, \rho_n) : \rho_1 = 1, \rho_n = n, \rho_2 < \rho_{k+1}\}.$$

The family of critical graphs in the acyclic matching is

$$\mathcal{U} = \bigcup_{k=2}^{n-2} \mathcal{U}_k.$$

For each of the $n-3$ choices of k, we have $(n-2)!/2$ valid permutations ρ, which implies that $|\mathcal{U}| = (n-3) \cdot (n-2)!/2$. Since all graphs in $\mathcal{U}$ have the same number $2n-3$ of edges, (20.1) is a consequence of Theorem 4.8.

Before proceeding, we give a brief description of the acyclic matching to be defined: First, we match with respect to the edge $1n$ whenever possible; the remaining graphs are those with the property that $1n$ cannot be added without creating a 3-connected graph. Second, for any remaining graph G, we show that there are two unique vertices x, y separating G such that the connected component in $G([n] \setminus \{x, y\})$ containing 1 is minimal. Matching with the edge xy, we get rid of all graphs but those of the kind illustrated in Figure 20.2.

In the final step, we proceed by induction on n to obtain a matching with the desired properties. This step is technical in nature; roughly speaking,

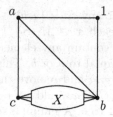

Fig. 20.2. A graph in $\Lambda'_n(\{1\}, x, y)$; $\{a, b\} = \{x, y\}$ and $n \in X$.

we consider the graph in NC^3_{n-1} obtained from the graph in Figure 20.2 by removing the vertex 1 and the edges $1a$ and $1b$.

We divide the description of the acyclic matching into several steps.

Step 1: *Matching with the edge $1n$ to obtain the family Λ_n.*

Match with $1n$ whenever possible, meaning that we pair $G - 1n$ and $G + 1n$ whenever $G + 1n$ is not 3-connected. Let Λ_n be the family of critical graphs with respect to this matching. By Lemma 4.1, any acyclic matching on Λ_n together with the matching just defined yields an acyclic matching on NC^3_n. One readily verifies that Λ_4 consists of the graph $K_4 - 14 = G(1, 2|3, 4)$ and nothing more. Hence from now on we may assume that $n \geq 5$.

Step 2: *Defining the set $X(G)$.*

For any graph G (3-connected or not), let $X(G)$ be the set of pairs $\{x, y\}$ such that $G([n] \setminus \{x, y\})$ is disconnected. Let $G \in \Lambda_n$. Since $G + 1n$ is 3-connected, the set $X(G + 1n)$ is empty. This implies that any $\{x, y\} \in X(G)$ separates 1 and n in G. In particular, if $\{x, y\} \in X(G)$, then $x, y \in [2, n-1]$.

Step 3: *Defining the pair S_G.*

For any $G \in \Lambda_n$ and $S = \{a, b\} \in X(G)$, note that the induced subgraph $G([n] \setminus S)$ consists of exactly two connected components

$$M_1(S, G) \text{ and } M_n(S, G),$$

where $1 \in M_1(S, G)$ and $n \in M_n(S, G)$. Namely, since $G + 1n$ is 3-connected, $G([n] \setminus S) + 1n$ is connected. By the following lemma, there is a unique set $S = S_G$ in $X(G)$ such that $M_1(S, G)$ is minimal.

Lemma 20.1. *Let $G \in \Lambda_n$. Then there is a set $S_G \in X(G)$ such that*

$$M_1(S_G, G) \subsetneq M_1(S, G)$$

for all $S \in X(G) \setminus \{S_G\}$.

Proof. Let $G \in \Lambda_n$. First, for any $S = \{a, b\} \in X(G)$, we claim that there is a simple path from 1 to each $x \in M_1(S) \cup \{a, b\}$ ($M_1(S) = M_1(S, G)$) such that the path does not contain any element in $M_n(S) \cup \{a, b\}$, except that the endpoint might be equal to a or b. This is true by definition for all $x \in M_1(S)$. Thus consider $x = a$ and remove the vertex b from G. The new graph is connected (otherwise $G + 1n$ would not be 3-connected), which implies that it contains a path from 1 to a. A minimal such path cannot contain any element from $M_n(S)$, as this would imply that there is a path not containing either of a and b from 1 to this element. By symmetry, the same property holds for b when a is removed, and the claim follows.

As a consequence, if $S_{ab} = \{a, b\}$ and $S_{cd} = \{c, d\}$ are distinct but not necessarily disjoint sets in $X(G)$ such that $S_{cd} \subset M_n(S_{ab}) \cup S_{ab}$, then

$$M_1(S_{cd}) \supseteq M_1(S_{ab}) \cup (S_{ab} \setminus S_{cd}) \supsetneq M_1(S_{ab}).$$

Note that this implies that

$$M_1(S_{cd}) \cup S_{cd} \supsetneq M_1(S_{cd}) \cup (S_{ab} \cap S_{cd}) \supseteq M_1(S_{ab}) \cup S_{ab}.$$

In particular, by symmetry (swap 1 and n), if $S_{cd} \subset M_1(S_{ab}) \cup S_{ab}$, then

$$M_1(S_{cd}) \subsetneq M_1(S_{ab}).$$

It remains to consider the case $a \in M_1(S_{cd})$, $b \in M_n(S_{cd})$, $c \in M_1(S_{ab})$, and $d \in M_n(S_{ab})$. Since there are no edges between $M_1' = M_1(S_{ab}) \cap M_1(S_{cd})$ and $M_n' = M_n(S_{ab}) \cup M_n(S_{cd}) = [n] \setminus (M_1' \cup \{a, c\})$, it is clear that $S_{ac} = \{a, c\} \in X(G)$ and that $M_1(S_{ac}) = M_1'$ and $M_n(S_{ac}) = M_n'$. $M_1(S_{ac})$ is properly included in $M_1(S_{ab})$ (which contains c) and $M_1(S_{cd})$ (which contains c), which concludes the proof. $\square$

Step 4: *Partitioning Λ_n into subfamilies $\Lambda_n(M, x, y)$.*

For any $M \subset [n-1]$ and $x, y \notin M$, let

$$\Lambda_n(M, x, y) = \{G \in \Lambda_n : S_G = \{x, y\}, M = M_1(S_G, G)\}.$$

This yields a partition of Λ_n into smaller families. Consider the map $f : P(\Lambda_n) \to P(2^{[n-1]})$ defined by $f(G) = M$ whenever $G \in \Lambda_n(M, x, y)$. We want to show that the Cluster Lemma 4.2 applies, meaning that f is a poset map. Now, if $G \subseteq H$, then

$$M_1(S_G, G) \subseteq M_1(S_H, G) = M_1(S_H, H)$$

with equality if and only if $S_G = S_H$ by Lemma 20.1. In particular, the family

$$\{\Lambda_n(M, x, y) : M \subset [n], x < y\}$$

does satisfy the conditions in the Cluster Lemma 4.2. The condition $x < y$ is necessary for avoiding double-counting; $\Lambda_n(M, x, y) = \Lambda_n(M, y, x)$.

Step 5: *Matching with the edge* xy *in* $\Lambda_n(M,x,y)$ *to obtain the family* $\Lambda'_n(M,x,y)$.

In $\Lambda_n(M,x,y)$, match with xy whenever possible. Let $\Lambda'_n(M,x,y)$ be the family of critical graphs in $\Lambda_n(M,x,y)$ with respect to this matching. The following lemma implies that we need only consider $M = \{1\}$.

Lemma 20.2. $\Lambda'_n(M,x,y)$ *is nonvoid if and only if* $M = \{1\}$. *A graph* G *is in* $\Lambda'_n(\{1\},x,y)$ *if and only if the following conditions are satisfied.*

(i) *$G + 1n$ is 3-connected, whereas G is not.*
(ii) *One of the elements x and y, denoted a, has degree 3.*
(iii) *Let b be the element in $\{x,y\} \setminus \{a\}$. The neighborhood $N_G(a)$ of a equals $\{1,b,c\}$ for some $c \in [n] \setminus \{1,a,b,n\}$, whereas $N_G(b)$ contains 1 and a.*

Remark. By the lemma, graphs in $\Lambda'_n(\{1\},x,y)$ are of the form illustrated in Figure 20.2.

Proof. One easily checks that a graph satisfying conditions (i)-(iii) belongs to $\Lambda'_n(\{1\},x,y)$. For the other direction, first note that $\Lambda'_n(M,x,y)$ is the family of all graphs G in $\Lambda_n(M,x,y)$ containing the edge xy and having the property that $(G - xy) + 1n$ is not 3-connected. Namely, suppose that $(G - xy) + 1n$ is 3-connected. We need to show that $S_G = S_{G-xy} = \{x,y\}$. Suppose not; hence there is a set S in $X(G - xy)$ such that

$$M_1(S, G - xy) \subsetneqq M_1(S_G, G - xy).$$

This means that x and y are not contained in $M_1(S, G - xy)$; hence they are both contained in $M_n(S, G - xy) \cup S$. However, this means that $S \in X(G)$ and $M_1(S,G) \subsetneqq M_1(S_G, G)$, which is a contradiction to the minimality of $M_1(S_G, G)$.

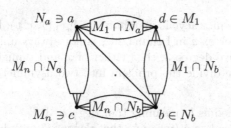

Fig. 20.3. A graph in $\Lambda'_n(M,x,y)$; $ab = xy$ and $1, n \in N_b \cup \{c,d\}$.

For a graph $G \in \Lambda'_n(M,x,y)$, let

$$M_1 = M_1(\{x, y\}, G) \ni 1;$$
$$M_n = M_n(\{x, y\}, G) \ni n.$$

Consider a pair $\{c, d\} \in X((G - xy) + 1n)$, and let N_a and N_b be the two components in $G' = (G + 1n)([n] \setminus \{c, d\}) - xy$. We have that $\{c, d\} \cap \{x, y\} = \emptyset$ for any $\{c, d\} \in X((G - xy) + 1n)$, because $G + 1n$ is 3-connected. Since 1 and n are adjacent in $(G - xy) + 1n$, they must be in the same component in G' unless one of them is contained in $\{c, d\}$; assume that

$$1, n \in N_b \cup \{c, d\}.$$

Furthermore, assume that $d \in M_1$ and $c \in M_n$. Let a, b be such that $\{a, b\} = \{x, y\}$, $a \in N_a$, and $b \in N_b$. The situation for G is as in Figure 20.3; there are no edges between M_1 and M_n, and the only edge between N_a and N_b is ab.

Since there is no edge between $M_1 \cap N_a$ and $M_n \cup N_b$ in $G + 1n$ and since

$$(M_1 \cap N_a) \cup M_n \cup N_b = (M_1 \cup M_n \cup N_b) \cap (M_n \cup N_a \cup N_b) = [n] \setminus \{a, d\},$$

$\{a, d\}$ separates $G + 1n$ unless $M_1 \cap N_a = \emptyset$ (recall that $1, n \notin N_a$). Since $G + 1n$ is 3-connected, we must indeed have that $M_1 \cap N_a = \emptyset$. By the same argument, $M_n \cap N_a = \emptyset$. Note that if $1 \in M_1 \cap N_b$, then

$$M_1(\{b, d\}, G) = M_1 \cap N_a \subsetneqq M_1,$$

which is a contradiction to the fact that $S_G = \{a, b\}$; hence $d = 1$. Moreover, since $\{1, b\} \notin X(G)$, we must have that $M_1 \cap N_b = \emptyset$. In particular, $M_1(S_G, G) = \{1\}$.

To conclude the proof, note that we have just demonstrated that the neighborhood of a is $\{1, b, c\}$. If $c = n$, then we have $M_n \cap N_b = \emptyset$, which implies that $n = 4$. This is a contradiction; hence $c \neq n$, and we are done. $\square$

Step 6: *Partitioning the family $\Lambda'_n(\{1\}, x, y)$ into subfamilies $\mathcal{F}^n_i(x, y, z)$.*

It remains to find a nice acyclic matching on $\Lambda'_n(\{1\}, x, y)$. Before proceeding, we note that the vertex a in Lemma 20.2 is not always uniquely determined; if both x and y have degree 3, then either of x and y might be chosen as a. For this reason, we divide the problem into two asymmetric (as opposed to symmetric) cases:

1. x *may* be defined as b (meaning that $\deg y = 3$).
2. x *must* be defined as a (meaning that $\deg y > 3$ and $\deg x = 3$).

For the first case, introduce the family

$$\mathcal{F}^n_1(x, y, z) = \{G : N_G(y) = \{1, x, z\}\} \cap \Lambda'_n(\{1\}, x, y)$$

for each $z \in [n] \setminus \{1, x, y, n\}$. For the second case, introduce the family

$$\mathcal{F}_2^n(x,y,z) = \{G : \deg y > 3, N_G(x) = \{1, y, z\}\} \cap \Lambda_n'(\{1\}, x, y)$$

for each $z \in [n] \setminus \{1, x, y, n\}$. The partition

$$\{\mathcal{F}_1^n(x,y,z), \mathcal{F}_2^n(x,y,z) : z \in [n] \setminus \{1, x, y, n\}\}$$

of $\Lambda_n'(\{1\}, x, y)$ satisfies the conditions in the Cluster Lemma 4.2; the inclusion relation between two graphs in $\Lambda_n'(\{1\}, x, y)$ not in the same family $\mathcal{F}_i^n$ can hold only if the smaller set is contained in some $\mathcal{F}_1^n$ and the larger set is contained in some $\mathcal{F}_2^n$.

Step 7: *Defining optimal acyclic matchings on $\mathcal{F}_i^n(x, y, z)$.*

The final step is the following Lemma, which implies (20.1).

Lemma 20.3.

(i) *There is an acyclic matching on $\mathcal{F}_1^n(x, y, z)$ with critical graphs*

$$G(1, x | y, z, \rho_5, \dots, \rho_{n-1}, n),$$

where $\{x, y, z, \rho_5, \dots, \rho_{n-1}\} = \{2, \dots, n-1\}$.

(ii) *There is an acyclic matching on $\mathcal{F}_2^n(x, y, z)$ with critical graphs*

$$G(1, x, z, \rho_4, \dots, \rho_k | y, \rho_{k+2}, \dots, \rho_{n-1}, n), \quad 3 \le k \le n-2,$$

where $\{x, z, \rho_4, \dots, \rho_k, y, \rho_{k+2}, \dots, \rho_{n-1}\} = \{2, \dots, n-1\}$.

Proof of (i). Consider a graph $G \in \mathcal{F}_1^n(x, y, z)$. One readily verifies that there is a unique maximal simple path

$$P_G = (v_1, v_2, \dots, v_t)$$

with $v_1 = 1$, $v_2 = y$, and $v_3 = z$ such that

$$N_G(v_k) = \{v_{k-1}, v_{k+1}, x\}$$

for all $k \in \{2, \dots, t-1\}$. Specifically, add one vertex v_k at a time and continue as long as $N_G(v_k)$ is of the form $\{v_{k-1}, w, x\}$ for some $w \notin \{v_1, \dots, v_k\}$. For example, with $1 = \rho_1$, $x = \rho_6$, $y = \rho_2$, and $z = \rho_3$ in Figure 20.1, we have $P_G = (\rho_1, \rho_2, \rho_3, \rho_4, \rho_5)$. Note that if $k < t$, then $v_k \ne n$, because otherwise $\{x, n\}$ would separate $G + 1n$.

If $v_t = n$, then (for the same reason) all vertices in $[n] \setminus \{x\}$ are contained in the path, which implies that $t = n - 1$. By construction, n is adjacent to v_{n-2} in G but not to v_i for any $i < n-2$. This implies that n must be adjacent to x in G; $G + 1n$ is 3-connected. As a consequence, we have that

$$G = G(1, x | y, z, v_4, \dots, v_{n-2}, n).$$

For $t < n - 1$, denote by

$$\mathcal{F}_1^n(x, y, z, v_4, \ldots, v_t)$$

those graphs G in $\mathcal{F}_1^n(x, y, z)$ with $P_G = (1, y, z, v_4, \ldots, v_t)$. Since $v_t \neq n$ and $G + 1n$ is 3-connected, the degree of v_t in G must be at least three. In fact, by the maximality of P_G, if v_t is adjacent to x, then v_t is adjacent to a total of at least four vertices. In particular, the families $\mathcal{F}_1^n(x, y, z, v_4, \ldots, v_t)$ satisfy the conditions in the Cluster Lemma 4.2; t cannot increase if we add an edge.

Let $G \in \mathcal{F}_1^n(x, y, z, v_4, \ldots, v_t)$ be a graph containing the edge xv_t. Suppose that $G - xv_t$ is not contained in $\mathcal{F}_1^n(x, y, z, v_4, \ldots, v_t)$. Then $K = (G + 1n) - xv_t$ is not 3-connected, which implies that there are $p, q \in [n]$ with the property that $K' = K([n] \setminus \{p, q\})$ is disconnected. Since $K' + xv_t$ is connected, v_t and x belong to different components in K'. This means that (say) $p = v_{t-1}$, because x and v_t are both adjacent to v_{t-1}. We concluded above that $\deg v_t > 3$ in G, which implies that $\deg v_t \geq 3$ in K. Hence the component in K' containing v_t must contain something more than v_t, and it does not contain x or v_{t-2}, the other neighbors of $p = v_{t-1}$. Therefore, $K([n] \setminus \{v_t, q\}) = (G + 1n)([n] \setminus \{v_t, q\})$ is disconnected, which is a contradiction to the fact that $G + 1n$ is 3-connected. It follows that $G - xv_t \in \mathcal{F}_1^n(x, y, z, v_4, \ldots, v_t)$.

As a consequence, we may use the edge xv_t to obtain a complete matching on the family $\mathcal{F}_1^n(x, y, z, v_4, \ldots, v_t)$. Namely, if a graph G not containing xv_t belongs to $\mathcal{F}_1^n(x, y, z, v_4, \ldots, v_t)$, then the same is certainly true for $G + xv_t$.

Proof of (ii). We use induction on n. This requires a base step, but we already provided an acyclic matching on NC_4^3. For $n > 4$, let $\hat{\Lambda}_{n-1}(x, z, y)$ be the family of graphs H on the vertex set $[2, n]$ such that

$$H + xn \text{ is 3-connected and } N_G(x) = \{y, z\}.$$

Obviously $\hat{\Lambda}_{n-1}(x, z, y) = \hat{\Lambda}_{n-1}(x, y, z)$, but we prefer having z before y for reasons that will be explained later. We want to prove that

$$\mathcal{F}_2^n(x, y, z) = \{\{1x, 1y\}\} * \hat{\Lambda}_{n-1}(x, z, y). \tag{20.2}$$

Before proving (20.2), we show how it implies the second part of Lemma 20.3. After some relabeling (such as replacing x with 1), we easily see that $\hat{\Lambda}_{n-1}(x, z, y)$ can be identified with the family $\Lambda_{n-1}(\{1\}, z, y)$ introduced in step 4 above. In particular, by induction on n, we may apply steps 4-7 on $\hat{\Lambda}_{n-1}(x, z, y)$ to obtain an acyclic matching such that the unmatched graphs are

$$G(x, z, \rho_4, \ldots, \rho_k | y, \rho_{k+2}, \ldots, \rho_{n-1}, n)$$

with $2 \leq k \leq n - 3$ and $\{x, z, \rho_4, \ldots, \rho_k, y, \rho_{k+2}, \ldots, \rho_{n-1}\} = \{2, \ldots, n-1\}$. Note that if we add 1, $1x$, and $1y$ to $G(x, z, \rho_4, \ldots, \rho_k | y, \rho_{k+2}, \ldots, \rho_{n-1}, n)$, then we obtain the graph

$$G(1, x, z, \rho_4, \ldots, \rho_k | y, \rho_{k+2}, \ldots, \rho_{n-1}, n);$$

this is the reason why we wanted to have z before y. Thus choosing the acyclic matching on $\mathcal{F}_2^n(x,y,z)$ corresponding in the natural way to the chosen acyclic matching on $\hat{\Lambda}_{n-1}(x,z,y)$, we obtain Lemma 20.3.

To obtain (20.2), let $H \in \hat{\Lambda}_{n-1}(x,z,y)$ and let G be the graph obtained from H by adding the vertex 1 and the edges $1x$ and $1y$. Clearly, (ii) and (iii) in Lemma 20.2 hold with $(a,b,c) = (x,y,z)$, and the degree of y in G is at least 4. To prove that $G \in \mathcal{F}_2^n(x,y,z)$, it remains to show that (i) in Lemma 20.2 holds. It is clear that G is not 3-connected; $\{x,y\}$ is a cut point in G. Moreover, we claim that $G + 1n$ is 3-connected. To prove this, note that the 3-connected graph $H + xn$ is obtained from $G + 1n$ by contracting the edge $1x$. Thus any $S \in X(G + 1n)$ contains either 1 or x. Now, the former is impossible, as $H = (G + 1n)([2,n])$ is 2-connected. For the latter, suppose that $S = \{x,q\}$ separates $G + 1n$. Since 1 is adjacent to y and n, 1 is not isolated in $K = (G + 1n)([n] \setminus \{x,q\})$. However, this implies that the graph $(H + xn)([2,n] \setminus \{x,q\})$ obtained from K by removing 1 is disconnected, which is a contradiction to $H + xn$ being 3-connected. It follows that $G + 1n$ is 3-connected and hence that $G \in \mathcal{F}_2^n(x,y,z)$.

Conversely, let $G \in \mathcal{F}_2^n(x,y,z)$ and write $H = G([2,n])$. Clearly, the neighborhood of x in H is $\{y,z\}$. It remains to verify that $H + xn$ is 3-connected. Suppose that $S = \{p,q\}$ separates $H + xn$. If $x \notin S$, then S also separates $G + 1n$; n and y belong to the same connected component as x in the separated graph. As this is a contradiction, we must have that $S = \{x,q\}$ for some q. Now, the two components A and B in $H([2,n] \setminus \{x,q\})$ have the property that y and n belong to different components, say $y \in A$ and $n \in B$. Namely, otherwise, $\{x,q\}$ would separate $G + 1n$. Also, z is in B; otherwise, q would be a cut point in H separating B from $\{x,y,z\}$ and hence a cut point in G. Now, A contains something more than just y; the degree of y is at least four in G and hence at least three in H. However, this means that $A \setminus \{y\}$ is a nonempty connected component of $(G + 1n)([n] \setminus \{y,q\})$. Namely, there are no edges from $A \setminus \{y\}$ to $B \cup \{1,x\}$ in $G + 1n$, because $z, n \in B$. This is a contradiction; hence (20.2) follows.

Conclusion.

Let us summarize the main achievement of this section:

Theorem 20.4. NC_n^3 *is homotopy equivalent to a wedge of* $(n-3) \cdot (n-2)!/2$ *spheres of dimension* $2n - 4$. $\square$

20.2 Homology

Translate the acyclic matching given in Section 20.1 into an acyclic matching on the quotient complex $\mathsf{C}_n^3 = 2^{K_n}/\mathrm{NC}_n^3$; follow the procedure described at the beginning of Section 17.2. Clearly, the homology of C_n^3 vanishes except in dimension $2n - 3$, and $\tilde{H}_{2n-3}(\mathsf{C}_n^3; \mathbb{Z})$ is free of rank $(n - 3) \cdot (n - 2)!/2$.

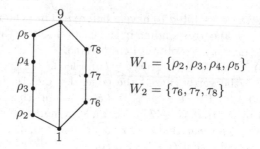

Fig. 20.4. $R(1, \rho_2, \rho_3, \rho_4, \rho_5 | \tau_6, \tau_7, \tau_8, 9)$ and the walls W_1 and W_2.

Our goal is to find a basis for $\tilde{H}_{2n-3}(\mathrm{C}_n^3; \mathbb{Z})$. With notation as in Section 4.4.1, it suffices to find an element $b_G \in B$ with the same boundary as G for each critical graph G. Namely, then $\{[G] - b_G : G \text{ is critical}\}$ will form a basis for $\tilde{H}_{2n-3}(\mathrm{C}_n^3; \mathbb{Z})$ by Corollary 4.17.

The critical graphs are of the form

$$G(1, \rho_2, \ldots, \rho_k | \tau_{k+1}, \tau_{k+2}, \ldots, \tau_{n-1}, n) + 1n,$$

where $\rho_2 < \tau_{k+1}$, $\{\rho_2, \ldots, \rho_k, \tau_{k+1}, \tau_{k+2}, \ldots, \tau_{n-1}\} = \{2, \ldots, n-1\}$, and $2 \leq k \leq n - 2$. Note that we use two different symbols ρ_* and τ_* to denote the vertices, as opposed to the previous single symbol ρ_*; this is to make it easier to distinguish between the two kinds of vertices. To simplify notation, write

$$\rho^* = (1, \rho_2, \ldots, \rho_{k-1}, \rho_k);$$
$$\tau^* = (\tau_{k+1}, \ldots, \tau_{n-1}, n);$$
$$G(\rho^* | \tau^*) = G(1, \rho_2, \ldots, \rho_k | \tau_{k+1}, \ldots, \tau_{n-1}, n).$$

Consider the graph $R(\rho^* | \tau^*)$ obtained from $G(\rho^* | \tau^*) + 1n$ by removing the edge set

$$\{\rho_i \tau_{k+1} : i = 2, \ldots, k\} \cup \{\rho_k \tau_{k+i} : i = 2, \ldots, n - k - 1\}.$$

The graph $R(\rho^* | \tau^*) - 1n$ is the unique Hamiltonian cycle in $G(\rho^* | \tau^*)$ (see Figure 20.4). In $R(\rho^* | \tau^*) - 1n$, there are exactly two simple paths $(1, \rho_2, \ldots, \rho_k, n)$ and $(1, \tau_{k+1}, \ldots, \tau_{n-1}, n)$ from 1 and n. The sets $W_1 = \{\rho_2, \ldots, \rho_k\}$ and $W_2 = \{\tau_{k+1}, \ldots, \tau_{n-1}\}$ are the two *walls* of $R(\rho^* | \tau^*)$. We say that ρ_m (τ_m) is *above* ρ_l (τ_l) if $m > l$ and that z is *between* x and y if y is above z and z is above x.

Let $\mathcal{S}(\rho^* | \tau^*)$ be the family of edge sets $S = \{\rho_{i_m} \tau_{j_m} : 1 \leq m \leq n - 3\}$ satisfying

$$\begin{cases} 2 = i_1 \leq i_2 \leq \ldots \leq i_{n-3} = k; \\ k + 1 = j_1 \leq j_2 \leq \ldots \leq j_{n-3} = n - 1; \\ i_m + j_m = m + k + 2, \; 1 \leq m \leq n - 3. \end{cases} \tag{20.3}$$

Condition (20.3) means that S is a triangulation of the n-gon $R(\rho^*|\tau^*) - 1n$ with the property that every interior edge contains one element from each wall. In particular, every $v \neq 1, n$ is contained in at least one edge in S, whereas no edge in S contains 1 or n.

Let $\mathcal{T}(\rho^*|\tau^*)$ be the family of graphs $R(\rho^*|\tau^*) \cup S$ such that $S \in \mathcal{S}(\rho^*|\tau^*)$. Note that $G(\rho^*|\tau^*) + 1n$ is contained in $\mathcal{T}(\rho^*|\tau^*)$. Each of the group elements $[G] - b_G$ discussed at the beginning of this section turns out to be a linear combination of elements from some $\mathcal{T}(\rho^*|\tau^*)$.

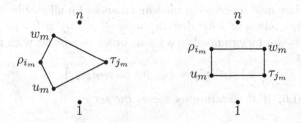

Fig. 20.5. The two possibilities for w_m given that $u_m \in W_1$; $e_m = \rho_{i_m}\tau_{j_m}$.

Lemma 20.5. *Every graph in $\mathcal{T}(\rho^*|\tau^*)$ is 3-connected. Let u_m be the vertex in the set $e_{m-1} \setminus e_m$ and let w_m be the vertex in the set $e_{m+1} \setminus e_m$ for $2 \leq m \leq n - 4$; see Figure 20.5. Then $T - e_m$ is 3-connected if and only if the vertices u_m and w_m belong to different walls. Finally, $T - e$ is not 3-connected whenever $e \in R(\rho^*|\tau^*) \cup \{e_1, e_{n-3}\}$.*

Proof. Consider $T \in \mathcal{T}(\rho^*|\tau^*)$. First, note that if we remove 1 or n from T, then the remaining graph is 2-connected. Namely, $e_1 = \rho_2\tau_{k+1}$ and $e_{n-3} = \rho_k\tau_{n-1}$, which implies that each of $T([2, n])$ and $T([n-1])$ contains a Hamiltonian cycle. In particular, any pair separating T must be contained in $W_1 \cup W_2$.

For the vertex $\tau_{k+i} \in W_2$, let e_m be an edge containing τ_{k+i}; this means that $e_m = \rho_{m+2-i}\tau_{k+i}$. There are three disjoint simple paths from 1 to τ_{k+i}:

$$(1, \tau_{k+1}, \ldots, \tau_{k+i}), (1, n, \tau_{n-1}, \ldots, \tau_{k+i+1}, \tau_{k+i}), (1, \rho_2, \ldots, \rho_{m+2-i}, \tau_{k+i}).$$

By symmetry, there are also three disjoint simple paths from 1 to each $\rho_i \in W_1$. Thus for any pair $\{v, w\}$ of elements from $W_1 \cup W_2$, all remaining vertices in $T([n] \setminus \{v, w\})$ must be contained in the same connected component as 1. As a consequence, T is 3-connected.

For the second statement in the lemma, consider the graph $T - e_m$. We still have three disjoint paths from 1 to any element $x \in W_1 \cup W_2$ with the property that x is contained in some $e_{m'} \neq e_m$. In particular, $T - e_m$ is 3-connected if and only if each of the two vertices contained in e_m is contained in some other edge $e_{m'}$ (if not, then one of the vertices would have degree two in $T - e_m$). This is exactly the property that one of the vertices is contained in

e_{m+1} and the other is contained in e_{m-1}, which is equivalent to the condition that u_m and w_m as defined in the lemma are on different walls.

If e_1 is removed from T, then one of the vertices ρ_2 and τ_{k+1} has degree two, which implies that $T-e_1$ is not 3-connected. For similar reasons, $T-e_{n-3}$ is not 3-connected. For the remaining part of the last statement in the lemma, see Shareshian [118, Lemma 4.1]; compare to the proof of Lemma 17.5. $\square$

We need to define an orientation of each graph $T = S \cup R(\rho^*|\tau^*) \in \mathcal{T}(\rho^*|\tau^*)$, which amounts to defining an order of the edges in T. Let the edges in $R(\rho^*|\tau^*)$ be the first edges in T (ordered in the same manner for all graphs in $\mathcal{T}(\rho^*|\tau^*)$). Order the other edges in T by defining $\rho_{i_l}\tau_{j_l} \leq \rho_{i_m}\tau_{j_m}$ if $l \leq m$ (notation as in (20.3)); this can be extended to a linear order consistent with all graphs in $\mathcal{T}(\rho^*|\tau^*)$. Let

$$V_0(\rho^*|\tau^*) = \{\rho_2, \rho_4, \dots, \rho_{2\lfloor k/2 \rfloor}\}.$$

Theorem 20.6. *With notation as above, the set*

$$\{\sigma(1, \rho_2, \dots, \rho_k|\tau_{k+1}, \dots, \tau_{n-1}, n) : 2 \leq k \leq n-2, \rho_2 < \tau_{k+1}\}$$

is a basis for $\tilde{H}_{2n-3}(\mathsf{C}_n^3; \mathbb{Z})$*, where*

$$\sigma(\rho^*|\tau^*) = \sum_{T \in \mathcal{T}(\rho^*|\tau^*)} \operatorname{sgn}(T) T$$

and

$$\operatorname{sgn}(T) = \prod_{v \in V_0(\rho^*|\tau^*)} (-1)^{\deg_T(v)};$$

$\deg_T(v)$ *is the number of neighbors of* v *in* T.

Proof. The theorem is obvious for $n = 4$; assume $n > 4$. By Corollary 4.17, we have to prove two things:

(i) The element $\sigma(\rho^*|\tau^*)$ is a cycle.
(ii) Every $T \in \mathcal{T}(\rho^*|\tau^*) \setminus \{G(\rho^*|\tau^*) + 1n\}$ is matched with a smaller graph.

Proof of (i). Consider a graph $T \in \mathcal{T}(\rho^*|\tau^*)$; $T = R(\rho^*|\tau^*) \cup \{e_m = \rho_{i_m}\tau_{j_m} : 1 \leq m \leq n-3\}$, where $e_1 < \dots < e_{n-3}$. By Lemma 20.5, $U = T - e_m$ is 3-connected if and only if $m \in [2, n-4]$, $i_{m+1} = i_{m-1}+1$, and $j_{m+1} = j_{m-1}+1$. In this case, there are exactly two graphs in $\mathcal{T}(\rho^*|\tau^*)$ containing U; the graphs are $T_1 = U + \rho_{i_{m-1}}\tau_{j_{m+1}}$ and $T_2 = U + \rho_{i_{m+1}}\tau_{j_{m-1}}$. With the given orientations of T_1 and T_2, it follows that

$$\langle \partial(T_1), U \rangle = \langle \partial(T_2), U \rangle, \tag{20.4}$$

where $\langle \cdot, \cdot \rangle$ is the standard inner product (see Section 4.4). Exactly one of the indices i_{m-1} and i_{m+1} is even, which implies that $\operatorname{sgn}(T_1) = -\operatorname{sgn}(T_2)$; hence (i) is a consequence of (20.4).

Proof of (ii). Our acyclic matching might appear as somewhat implicitly defined, as we used induction in the proof of Theorem 20.4. However, using the very same induction again, we may succeed anyway. Let $T \in \mathcal{T}(\rho^*|\tau^*)$. In T, either ρ_2 is adjacent to τ_{k+2} or τ_{k+1} is adjacent to ρ_3. In either case, one of ρ_2 and τ_{k+1} has degree three in T, while the other has degree at least four.

If $\deg \tau_{k+1} = 3$, then $T - 1n \in \mathcal{F}_1^n(\rho_2, \tau_{k+1}, \tau_{k+2})$ (notation as in Step 6 in Section 20.1). Let $v_1, \ldots, v_t$ be defined as in the proof of (i) in Lemma 20.3 for $G = T - 1n$; $v_1 = 1$, $v_2 = \tau_{k+1}$, and $v_3 = \tau_{k+2}$. We have to show that $\rho_2 v_t \in T$; this is the edge used in the matching unless $t = n - 2$, in which case T is equal to $G(1, \rho_2|\tau^*) + 1n$ and hence critical. Clearly $v_j = \tau_{k+j-1}$ for $2 \le j \le t$, which implies that either $\rho_2 v_t \in T$ or $\rho_3 v_{t-1} \in T$; recall (20.3). However, the only neighbors of v_{t-1} are v_{t-2}, v_t, and ρ_2; hence $\rho_2 v_t \in T$.

Next, consider the case $\deg \tau_{k+1} > 3$ and $\deg \rho_2 = 3$. This means that $T - 1n$ belongs to the set $\mathcal{F}_2^n(\rho_2, \tau_{k+1}, \rho_3)$ defined in Step 6 in Section 20.1. Removing the vertex 1 as in the proof of (ii) in Lemma 20.3 and adding the edge $\rho_2 n$, we obtain a graph contained in

$$\mathcal{T}(\rho_2, \rho_3, \ldots, \rho_k|\tau^*) \subset \{G + \rho_2 n : G \in \hat{\Lambda}_{n-1}(\rho_2, \rho_3, \tau_{k+1})\}.$$

An induction argument concludes the proof. $\square$

We obtain a basis for $\tilde{H}_{2n-4}(\mathsf{NC}_n^3; \mathbb{Z})$ by applying the boundary operator in the chain complex of 2^{K_n} to each basis element $\sigma(\rho^*|\tau^*)$ in $\tilde{H}_{2n-3}(\mathsf{C}_n^3; \mathbb{Z})$; compare to Corollary 19.8. We have that the coefficient of a graph G in $\partial(\sigma(\rho^*|\tau^*))$ is ± 1 if one of the following two conditions holds and zero otherwise.

- $G + e \in \mathcal{T}(\rho^*|\tau^*)$ for some $e \in R(\rho^*|\tau^*)$.
- $G + e \in \mathcal{T}(\rho^*|\tau^*)$ for some $e = \rho_i \tau_j$ such that either $\deg_G(\rho_i) = 2$ or $\deg_G(\tau_j) = 2$.

To see this, use Lemma 20.5.

20.3 A Related Polytopal Sphere

The purpose of this section is to analyze the cycle $\sigma(\rho^*|\tau^*)$ in greater detail for each relevant ρ^* and τ^*. Specifically, we consider the family $\mathcal{S}(\rho^*|\tau^*)$. To facilitate analysis, write $\rho^* = (1, \rho_1, \ldots, \rho_r)$ and $\tau^* = (\tau_1, \ldots, \tau_s, n)$; $r + s = n - 2$. If we identify the edge $\rho_i \tau_j$ with the pair (i, j), we may identify a set $S \in \mathcal{S}(\rho^*|\tau^*)$ with a subset of $I_{r,s} = [1, r] \times [1, s]$. We view $I_{r,s}$ as a chessboard and use matrix notation; (i, j) is the element in the ith row and the jth column. List the edges in S in increasing order as

$$S = \{\rho_{i_1} \tau_{i_1}, \rho_{i_2} \tau_{i_2}, \ldots, \rho_{i_{n-3}} \tau_{i_{n-3}}\}.$$

This means that $(i_{m+1} - i_m, j_{m+1} - j_m) \in \{(1, 0), (0, 1)\}$. As a consequence, S forms a path from $(1, 1)$ to (r, s) with the property that each step is either one step down or one step to the right.

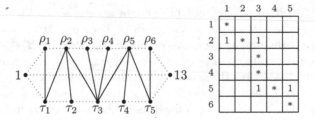

Fig. 20.6. To the left an edge set in $\mathcal{S}(\rho^*|\tau^*)$. The underlying set $R(\rho^*|\tau^*)$ is dashed with dots. To the right the corresponding face σ of $\mathsf{L}_{6,5}$. The removal of any element marked with a star yields a maximal face of the complex $\mathsf{NCL}_{6,5}$.

Let $\mathsf{L}_{r,s}$ be the simplicial complex of all sets σ such that σ is contained in such a path. Equivalently, if $(a,b),(c,d) \in \sigma$, then either we have that $a \leq c$ and $b \leq d$ or we have that $a \geq c$ and $b \geq d$. See Figure 20.6 for an example. We refer to (a,b) and (c,d) as *inconsistent* if $a > c$ and $b < d$ (or vice versa); this means that $\{(a,b),(c,d)\} \notin \mathsf{L}_{r,s}$. Obviously, $\mathsf{L}_{r,s}$ is pure of dimension $r+s-2$, and each maximal face of $\mathsf{L}_{r,s}$ contains exactly one element (a,b) such that $a+b=k$ for each $k \in [2, r+s]$. As a side note, we mention that $\mathsf{L}_{r,s}$ appears in the analysis of ideals of 2×2 determinants [62, 25]; see Section 1.1.6.

Recall the definition of $R(\rho^*|\tau^*)$ from the previous section; $R(\rho^*|\tau^*)$ consists of a Hamiltonian cycle plus the edge $1n$. Let π be a face of $\mathsf{L}_{r,s}$ and let $\hat{\pi}$ be the corresponding edge set $\{\rho_a\tau_b : (a,b) \in \pi\}$. By the proof of Lemma 20.5, a face $\pi \in \mathsf{L}_{r,s}$ has the property that $\hat{\pi} \cup R(\rho^*|\tau^*)$ is 3-connected if and only if each vertex except 1 and n is contained in at least one edge in $\hat{\pi}$. This means exactly that each row and each column in $I_{r,s}$ contains at least one element from π. Let $\mathsf{NCL}_{r,s}$ be the subcomplex of $\mathsf{L}_{r,s}$ consisting of all sets π such that some row or some column does not contain any element from π. See Figure 20.6 for an illustration.

Theorem 20.7. *Let $r, s \geq 1$. Then $\mathsf{NCL}_{r,s}$ is a polytopal sphere of dimension $r+s-3$. Indeed, $\mathsf{NCL}_{r,s}$ coincides with $\partial \mathsf{L}_{r,s}$.*

Proof. We obtain a poset structure L on the 0-cell set of $\mathsf{L}_{r,s}$ by defining (a,b) to be smaller than (c,d) whenever $a \leq c$ and $b \leq d$. One easily checks that L is a distributive lattice [133] and that $\mathsf{L}_{r,s}$ coincides with the order complex of L. As a consequence, $\mathsf{L}_{r,s}$ is a shellable ball and the boundary of $\mathsf{L}_{r,s}$ is polytopal; see Björner [7] or Provan [107] for a proof of the former fact and Björner and Farley [10] for a proof outline of the latter fact.

It remains to prove that $\mathsf{NCL}_{r,s}$ coincides with the boundary of $\mathsf{L}_{r,s}$. First, we prove that $\mathsf{NCL}_{r,s}$ is pure of dimension $r+s-3$. Let π be a face of $\mathsf{NCL}_{r,s}$. By construction, some row or column, say row a, is empty in π. Forgetting about row a and relabeling rows $a+1$ through r in the natural manner, we may view π as a face π' of $\mathsf{L}_{r-1,s}$. Extending π' to a maximal face of $\mathsf{L}_{r-1,s}$

and adding back row a again, we obtain a face of $\mathsf{NCL}_{r,s}$ of dimension $r+s-3$ as desired.

Now, let π be a face of $\mathsf{L}_{r,s}$ of codimension one (i.e., dimension $r+s-3$). We need to prove that π is contained in one single maximal face of $\mathsf{L}_{r,s}$ if and only if π is a maximal face of $\mathsf{NCL}_{r,s}$. We have a number of cases:

- $(1,1)$ does not belong to π. Then the only maximal face containing π is $\pi + (1,1)$. It is clear that either row 1 or column 1 is empty in π.
- (r,s) does not belong to π. This case is analogous to the previous case.
- $(a-1,b)$ and $(a+1,b)$ belong to π, whereas (a,b) does not belong to π. Then π does not contain any element from row a, because such an element would be inconsistent with either $(a-1,b)$ or $(a+1,b)$. It is clear that $\pi + (a,b)$ is the only maximal face of $\mathsf{L}_{r,s}$ containing π.
- $(a,b-1)$ and $(a,b+1)$ belong to π, whereas (a,b) does not belong to π. This case is analogous to the previous case.
- (a,b) and $(a+1,b+1)$ belong to π, whereas $(a,b+1)$ and $(a+1,b)$ do not belong to π. In this case, $\pi + (a,b+1)$ and $\pi + (a+1,b)$ are both maximal faces of $\mathsf{L}_{r,s}$ containing π. Clearly, all rows and columns in π are nonempty.

This concludes the proof. □

Define $\mathsf{CL}_{r,s} = \mathsf{L}_{r,s}/\mathsf{NCL}_{r,s}$. Since $\mathsf{L}_{r,s}$ is contractible, $\mathsf{CL}_{r,s}$ is homotopy equivalent to the suspension of $\mathsf{NCL}_{r,s}$; use Lemma 3.18. By Lemma 3.19 and Theorem 20.7, it follows that $\mathsf{CL}_{r,s}$ is homotopy equivalent to a sphere of dimension $r+s-2$. In fact, since $\partial\mathsf{L}_{r,s}$ coincides with $\mathsf{NCL}_{r,s}$, $\|\mathsf{CL}_{r,s}\|$ is homeomorphic to a sphere. It is clear that the fundamental cycle z of this sphere, viewed as an element in $\tilde{H}_{r+s-2}(\mathsf{CL}_{r,s};\mathbb{Z})$, has the property that $z \wedge [R(\rho^*|\tau^*)]$ is isomorphic to the cycle $\sigma(1,\rho_1,\ldots,\rho_r|\tau_1,\ldots,\tau_s,n)$ in $\tilde{H}_n(\mathsf{C}_n^3;\mathbb{Z})$.

20.4 Not k-connected Graphs for $k > 3$

We have seen that the complex NC_n^k of not k-connected graphs on n vertices has a nice structure for $1 \le k \le 3$. A natural question to ask is whether any of this structure is preserved for larger k. Unfortunately, the answer is likely to be negative. For example, it is not always the case that NC_n^k is homotopy equivalent to a wedge of spheres. One counterexample is $(n,k) = (7,5)$, and there are infinitely many other counterexamples of the form $(n, n-2)$.

To see this, note that NC_n^{n-2} is the Alexander dual of the matching complex with respect to the ground set $\binom{[n]}{2}$; this observation is due to Babson et al. [3]. Namely, a graph is $(n-2)$-connected if and only if every induced subgraph on three vertices is connected, which is equivalent to every induced subgraph on three vertices of the complement graph being empty or consisting of a single edge. Equivalently, the complement graph is a matching. As described in Section 11.2, the work of Bouc [21] and Shareshian and Wachs [122] implies that there is 3-torsion in the homology of M_n for infinitely many n. As a consequence, the same is true for the Alexander dual NC_n^{n-2}; use Theorem 3.4.

Applying Theorem 11.6, Alexander duality, and Theorem 3.8, one easily proves that the shifted connectivity degree of NC_n^{n-2} is $\left\lceil \frac{(n+1)(n-3)}{2} \right\rceil$ for $n \geq 3$.

Table 20.1. The homology of $(NC_n^{n-3})^*$ for $n \leq 7$.

$\tilde{H}_i((NC_n^{n-3})^*, \mathbb{Z})$	$i = 1$	2	3	4	5	6	7
$n = 4$	-	$\mathbb{Z}^6$	-	-	-	-	-
5	-	$\mathbb{Z}^6$	-	-	-	-	-
6	-	-	-	$\mathbb{Z}^{36}$	-	-	-
7	-	-	-	$\mathbb{Z}$	$\mathbb{Z}^{181}$	-	-

Let us proceed with the case $k = n - 3$. The Alexander dual of NC_n^{n-3} nearly coincides with the complex BD_n^2 of graphs on n vertices such that the degree of each vertex is at most 2. The only difference is that $(NC_n^{n-3})^*$ does not contain squares $\{ab, bc, cd, ad\}$; the complement of such a square is not $(n-3)$-connected. See Table 20.1 for information about the homology of $(NC_n^{n-3})^*$ for small n; the homology for general n remains unsettled.

Theorem 20.8 (Babson et al. [3]). *We have that*

$$\sum_{n \geq 4} \tilde{\chi}((NC_n^{n-3})^*) \frac{x^n}{n!} = 1 + x - \frac{\exp(\frac{x}{2+2x} + x - \frac{x^2}{4} - \frac{x^4}{8})}{\sqrt{1+x}}.$$

Proof. There are exactly three squares on four vertices; hence we are done by Corollary 6.15 and Theorem 12.23. □

While it seems difficult to adapt this proof to NC_n^{n-k} when $k > 3$, an interesting question is whether it is at least possible to establish the existence of a "nice" closed formula for the series

$$\sum_{n \geq k+1} \tilde{\chi}((NC_n^{n-k})^*) \frac{x^n}{n!}.$$

One may also examine the series

$$\sum_{n \geq k+1} \tilde{\chi}(NC_n^k) \frac{x^n}{n!}.$$

For $k \leq 3$, this series is of the form $p(x) \ln(1 \pm x) + q(x)$ for some polynomials $p(x)$ and $q(x)$; use Proposition 18.1 and Theorems 19.1 and 20.4. Such a nice result is very unlikely to hold in general, but again even a proof of existence of a closed formula for $k \geq 4$ would be of interest.

Dihedral Variants of k-connected Graphs

We examine dihedral variants of the concept of k-connectivity. More precisely, we consider graphs with a disconnected, separable, or two-separable representation.

$NCR_n^{0,0}$ is the complex of all graphs on the vertex set $[n]$ with a disconnected polygon representation. As we saw in Chapter 18, the monotone graph property NC_n of being disconnected as an abstract graph is homotopy equivalent to a wedge of $(n-1)!$ spheres of dimension $n-3$. In Section 21.2, we establish a similar result, demonstrating that $NCR_n^{0,0}$ is homotopy equivalent to a wedge of a Catalan number of spheres of the same dimension $n-3$. We also make the observation that $NCR_n^{0,0}$ is not an entirely new object. Specifically, the complex relates to the well-studied lattice of noncrossing partitions on $[n]$ analogously to the way NC_n relates to the full partition lattice as outlined in Chapter 18. As a consequence, we are able to deduce that the $(n-3)$-skeleton of $NCR_n^{0,0}$ is Cohen-Macaulay.

$NCR_n^{1,0}$ is the complex of graphs with a separable polygon representation. In Section 21.3, we show that $NCR_n^{1,0}$ is homotopy equivalent to a sphere of dimension $2n-5$. One may compare this to the result that the monotone graph property NC_n^2 of being not 2-connected is homotopy equivalent to a wedge of $(n-2)!$ spheres of the same dimension $2n-5$; see Theorem 19.1. In fact, we obtain a generator for the homology of the quotient complex $CR_n^{1,0} = 2^{K_n}/NCR_n^{1,0}$ by picking a certain member of the basis for the homology of $C_n^2 = 2^{K_n}/NC_n^2$ in Theorem 19.7. In addition, we introduce the lattice $NX\Pi_{n,2}$ of graphs with a "block-closed" representation; this is a dihedral analogue of the lattice $\Pi_{n,2}$ of block-closed graphs discussed in Section 19.1. $NX\Pi_{n,2}$ turns out to be Gorenstein* and homotopy equivalent to $NCR_n^{1,0}$.

$NCR_n^{1,1}$ is the complex of graphs with a two-separable polygon representation. In Section 21.4, we show that $NCR_n^{1,1}$ is an n-fold cone over a complex $\overline{NCR}_n^{1,1}$, which is homotopy equivalent to a sphere of dimension $n-4$. In fact, $\overline{NCR}_n^{1,1}$ is collapsible to the associahedron A_n. This time, the related monotone graph property is the complex NC_n^3 of not 3-connected graphs, which is homotopy equivalent to a wedge of $(n - 3) \cdot (n - 2)!/2$ spheres of dimension $n + n - 4 = 2n - 4$; see Chapter 20.

As the above discussion indicates, it makes sense to think about these complexes as "light-weight" versions of NC_n^k for $1 \leq k \leq 3$.

21.1 A General Observation

Recall the definition of $\mathsf{NCR}_n^{k,l}$ from Section 8.1. Before proceeding, we make the following crucial observation for the cases $(k, l) = (0, 0), (1, 0), (1, 1)$:

Lemma 21.1.   (i) *Let* $k, l \in \{0, 1\}$. *Let* G *be a graph in* $\mathsf{NCR}_n^{k,l}$ *containing two crossing edges* rt *and* su; $r < s < t < u$. *Then the graph obtained by adding the edges* rs, st, tu, ru *to* G *is contained in* $\mathsf{NCR}_n^{k,l}$.

(ii) *Let* G *be a graph in* $\mathsf{NCR}_n^{0,0}$ *containing two edges* rt *and* ru *with a common endpoint* r. *Then the graph obtained by adding the edge* tu *to* G *is contained in* $\mathsf{NCR}_n^{0,0}$.

Remark. The lemma does not remain true if either k or l is at least two.

Proof. (i) We show that rs can be added to G; the other three cases are analogous. Let a and b be such that $(a, b - l]$ and $(b, a - k]$ are nonempty intervals with no edges from G between them. If $G + rs$ is not in $\mathsf{NCR}_n^{k,l}$, then we must have that $r \in (a, b - l]$ and $s \in (b, a - k]$ (or vice versa). Since $rt \in G$, we have that $t \notin (b, a - k]$ and hence that $t \in (a - k, b]$. Since $t \in (s, r)$, this implies that

$$t \in (s, r) \cap (a - k, b] = ((s, a - k] \cup (a - k, r)) \cap (a - k, b] = (a - k, r).$$

Since $u \in (t, r)$, this implies that

$$u \subseteq (a - k + 1, r) \subseteq (a, b - l),$$

which is a contradiction, as su would then be an edge between $(b, a - k]$ and $(a, b - l]$.

(ii) Let a and b be such that $(a, b]$ and $(b, a]$ are nonempty intervals with no edges from G between them. Suppose that $G + tu$ is not in $\mathsf{NCR}_n^{0,0}$. Without loss of generality, we may assume that $t \in (a, b]$ and $u \in (b, a]$. Now, r is contained in either $(b, a]$ or $(a, b]$, which implies that either rt or ru violates the assumption on a and b. This is a contradiction; hence we are done. $\square$

As a consequence of (i), Theorem 16.6 applies to $\mathsf{NCR}_n^{k,l}$ for $k, l \in \{0, 1\}$.

21.2 Graphs with a Disconnected Polygon Representation

We consider the complex $\mathsf{NCR}_n^{0,0}$ of graphs on the vertex set $[n]$ with a disconnected polygon representation.

Theorem 21.2. *Let $n \geq 3$. Then*

$$\mathsf{NCR}_n^{0,0} \simeq \mathsf{NCR}_n^{0,0} \cap \mathsf{NX}_n \simeq \bigvee_{C_{n-1}} S^{n-3},$$

where C_{n-1} is the Catalan number $\frac{1}{n}\binom{2n-2}{n-1}$.

Proof. Write $\mathsf{CR}_n^{0,0} = 2^{K_n}/\mathsf{NCR}_n^{0,0}$; we adopt the convention that $\mathsf{CR}_1^{0,0} = \{\emptyset\}$. We want to form a decision tree on $\mathsf{CR}_n^{0,0}$ with C_{n-1} evasive faces of dimension $n - 2$. For $n = 1, 2$, we have that $\mathsf{CR}_n^{0,0}$ consists of one single face of dimension $n - 2$. For $n \geq 3$, check all edges in the set $A = \{1i : i \in [2, n]\}$ and consider the family $\Sigma_Y = \mathsf{CR}_n^{0,0}(Y, A \setminus Y)$ for $Y \subseteq A$. If $Y = \emptyset$, then Σ_Y is void. If $|Y| \geq 2$, let r and s be any vertices such that $1r, 1s \in Y$. By Lemma 21.1 (ii), rs is a cone point in Σ_Y, which implies that Σ_Y is nonevasive.

The remaining case is that $|Y| = 1$; let $Y = \{1r\}$ and write $\Sigma_r = \Sigma_Y$. Check all edges in the set $B_r = \{ij : i \in (1, r), j \in (r, 1)\}$; this is the set of edges crossing the edge $1r$. If $Z \subseteq B_r$ is nonempty, then $\Sigma_r(Z, B_r \setminus Z)$ is nonevasive. Namely, if $ij \in Z$, then ir and jr are cone points in $\Sigma_r(Z, B_r \setminus Z)$ by Lemma 21.1 (i); ij and $1k$ cross.

It remains to consider the case $Z = \emptyset$; write $\Gamma_r = \Sigma_r(\emptyset, B_r)$. The set of edges remaining to be checked is the union of the complete set of edges on $[2, r]$ and the complete set of edges on $[r, n]$. It is clear that G belongs to Γ_r if and only if the induced subgraph $G([2, r])$ belongs to $\mathsf{CR}_{[2,r]}^{0,0}$ and the induced subgraph $G([r, n])$ belongs to $\mathsf{CR}_{[r,n]}^{0,0}$, where $\mathsf{CR}_{[a,b]}^{0,0}$ is defined in the natural manner. As a consequence,

$$\Gamma_r \cong \{1r\} * \mathsf{CR}_{r-1}^{0,0} * \mathsf{CR}_{n-r+1}^{0,0}.$$

We want to prove that $\mathsf{CR}_n^{0,0} \sim C_{n-1}t^{n-2}$. By induction on n, $\mathsf{CR}_{r-1}^{0,0} \sim C_{r-2}t^{r-3}$ and $\mathsf{CR}_{n-r+1}^{0,0} \sim C_{n-r}t^{n-r-1}$. By Theorem 5.29, this implies that

$$\Gamma_r \sim t \cdot (C_{r-2}t^{r-3}) \cdot (C_{n-r}t^{n-r-1}) \cdot t = C_{r-2}C_{n-r}t^{n-2}.$$

The conclusion is that

$$\mathsf{CR}_n^{0,0} \sim \sum_{r=2}^{n} C_{r-2}C_{n-r}t^{n-2} = \sum_{i=0}^{n-2} C_i C_{n-i-2}t^{n-2} = C_{n-1}t^{n-2},$$

and we are done.

To prove that $\mathsf{NCR}_n^{0,0} \simeq \mathsf{NCR}_n^{0,0} \cap \mathsf{NX}_n$, we simply note that $\mathsf{NCR}_n^{0,0}$ satisfies the conditions in Theorem 16.6 by Lemma 21.1. $\square$

It is easy to see that each evasive graph with respect to the given decision tree is a noncrossing spanning tree. In particular, each proper subgraph is contained in $\mathsf{NCR}_n^{0,0}$. As a consequence, we may describe the homology of

$\mathsf{NCR}_n^{0,0}$ and $\mathsf{CR}_n^{0,0}$ in exactly the same manner as we described the homology of NC_n and C_n in Corollary 18.5.

A partition of the set $[n]$ is *noncrossing* if, for every two edges $e_i \subseteq B_i$ and $e_j \subseteq B_j$ such that $i \neq j$, we have that e_i and e_j are noncrossing. The set of noncrossing partitions of $[n]$ forms a lattice $\mathsf{NX}\varPi_n$ with elements ordered by refinement. Kreweras [87] introduced this lattice, which has appeared in a variety of settings; see Simion [124] for a nice survey.

Björner [7] showed that the proper part $\overline{\mathsf{NX}\varPi_n}$ of $\mathsf{NX}\varPi_n$ is shellable and that $\Delta(\overline{\mathsf{NX}\varPi_n})$ is homotopy equivalent to a wedge of C_{n-1} spheres of dimension $n-3$. We obtain a poset map φ from $P(\mathsf{NCR}_n^{0,0})$ to $\overline{\mathsf{NX}\varPi_n}$ by defining $\varphi(G)$ as the family of connected components in the polygon representation of G; each component is identified with its underlying vertex set. It is easy to see that this map preserves homotopy type. Indeed, we may identify a noncrossing partition with the graph obtained by replacing each set in the partition with the complete graph on this set. Hence, by Closure Lemma 6.1, Björner's result implies Theorem 21.2.

In fact, we may deduce more information about $\mathsf{NCR}_n^{0,0}$ from Björner's result. Namely, consider a graph G in $\mathsf{NCR}_n^{0,0}$. The link of $\mathsf{NCR}_n^{0,0}$ with respect to G is either collapsible or homotopy equivalent to the order complex of the poset of all partitions strictly above $\varphi(G)$ in $\overline{\mathsf{NX}\varPi_n}$. The latter case occurs precisely when all connected components in the polygon representation of G are complete graphs. Since $\overline{\mathsf{NX}\varPi_n}$ is Cohen-Macaulay of dimension $n-3$, it follows that $\mathrm{lk}_{\mathsf{NCR}_n^{0,0}}(G)$ is $(n-4-\rho(\varphi(G)))$-connected, where $\rho(G)$ is the rank of G in $\overline{\mathsf{NX}\varPi_n}$. Obviously, $\rho(G)$ cannot exceed the size $|G|$ of the edge set of G, which implies that $\mathrm{lk}_{\mathsf{NCR}_n^{0,0}}(G)$ is $(n-4-|G|)$-connected. As a consequence, we have the following result.

Theorem 21.3. *The $(n-3)$-skeleton of $\mathsf{NCR}_n^{0,0}$ is Cohen-Macaulay.* $\square$

We conjecture that this skeleton is also vertex-decomposable.

21.3 Graphs with a Separable Polygon Representation

We consider the complex $\mathsf{NCR}_n^{1,0}$ of graphs with a separable polygon representation.

Theorem 21.4. *Let $n \geq 3$. Then*

$$\mathsf{NCR}_n^{1,0} \simeq \mathsf{NCR}_n^{1,0} \cap \mathsf{NX}_n \simeq S^{2n-5}.$$

Moreover, $\mathsf{NCR}_n^{1,0} \cap \mathsf{NX}_n$ is the join of the associahedron A_n and the boundary of an $(n-1)$-simplex.

Proof. By Lemma 21.1 (i), $\mathsf{NCR}_n^{1,0}$ satisfies the conditions in Theorem 16.6, which implies that $\mathsf{NCR}_n^{1,0}$ is collapsible to $\mathsf{NCR}_n^{1,0} \cap \mathsf{NX}_n$.

We claim that $G \in \mathrm{NCR}_n^{1,0} \cap \mathrm{NX}_n$ if and only if G is noncrossing and does not contain the full polygon $\mathrm{Bd}_n = \{12, 23, \ldots, (n-1)n, 1n\}$. The "only if" statement is clear; Bd_n is not in $\mathrm{NCR}_n^{1,0} \cap \mathrm{NX}_n$. For the "if" statement, suppose that G is noncrossing and does not contain all boundary edges, say $12 \notin G$. Let $j > 2$ be minimal such that $1j$ does not cross any edge in G. Such a j exists, as $1n$ does not cross any edge. Now, there is no edge from $(1, j)$ to $(j, 1)$ in G, as such an edge would cross $1j$. Moreover, by assumption, there is no edge from 1 to $[2, j-1]$. The conclusion is that j is a cut point in G, and it follows that $G \in \mathrm{NCR}_n^{1,0} \cap \mathrm{NX}_n$ (see also Shareshian [118, Lemma 4.1]).

As a consequence, $\mathrm{NCR}_n^{1,0} \cap \mathrm{NX}_n = \mathsf{A}_n * \partial 2^{\mathrm{Bd}_n}$, and we are done. $\square$

Define π_n^* as in (17.3) in Section 17.2; $\pi_n^* = \pi_n \wedge [\mathrm{Bd}_n]$, where π_n is the fundamental cycle of A_n. By Theorem 19.7, π_n^* is a cycle in the chain group $\tilde{C}_{2n-4}(\mathsf{C}_n^2)$, where C_n^2 is the quotient complex of 2-connected graphs on n vertices. Define $\hat{\pi}_n^*$ as in (19.1); $\hat{\pi}_n^* = \pi_n \wedge \partial([\mathrm{Bd}_n])$.

Theorem 21.5. *Let $n \geq 3$. Then the cycle π_n^* generates the homology of $\mathrm{CR}_n^{1,0} = 2^{K_n}/\mathrm{NCR}_n^{1,0}$ and $\mathrm{CR}_n^{1,0} \cap \mathrm{NX}_n = \mathrm{NX}_n/(\mathrm{NCR}_n^{1,0} \cap \mathrm{NX}_n)$. Moreover, $\hat{\pi}_n^*$ generates the homology of $\mathrm{NCR}_n^{1,0}$ and $\mathrm{NCR}_n^{1,0} \cap \mathrm{NX}_n$.*

Proof. By Theorem 21.4, $\hat{\pi}_n^*$ is the fundamental cycle of $\mathrm{NCR}_n^{1,0} \cap \mathrm{NX}_n$. Theorem 16.6 yields that there exists a perfect acyclic matching on $\mathrm{NCR}_n^{1,0} \setminus \mathrm{NX}_n$, which implies that $\hat{\pi}_n^*$ also generates the homology of $\mathrm{NCR}_n^{1,0}$.

To prove that π_n^* generates the homology of the corresponding quotient complexes, observe that $\partial(\pi_n^*) = \pm\hat{\pi}_n^*$ and apply the long exact sequences for the pairs $(\mathrm{NX}_n, \mathrm{NCR}_n^{1,0} \cap \mathrm{NX}_n)$ and $(2^{K_n}, \mathrm{NCR}_n^{(1,0)})$; see Theorem 3.3. $\square$

Our next goal is to analyze links in $\mathrm{NCR}_n^{1,0}$, the overall goal being to prove that the $(2n-5)$-skeleton of $\mathrm{NCR}_n^{1,0}$ is Cohen-Macaulay. Before proceeding, we need to introduce some concepts. First, let us extend the definition of $\mathrm{NCR}_n^{1,0}$ to $n = 2$; we set $\mathrm{NCR}_2^{1,0}$ equal to $\{\emptyset\}$ and hence consider the graph with two vertices and one edge as having a non-separable polygon representation. We say that a graph G has a *block-closed representation* if V is a clique whenever the induced subgraph $G(V)$ has a non-separable representation. If G has a block-closed representation, then G is block-closed in the sense of Section 19.1. Let $b(G)$ be defined as in Section 19.1; $b(G)$ is the number of blocks in G. We now prove an analogue of Shareshian's Theorem 19.4.

Theorem 21.6. *Let $n \geq 2$. If $H \neq K_n$ is a graph on n vertices with a block-closed representation, then the lifted complex $\mathrm{NCR}_n^{1,0}(H, \emptyset)$ is homotopy equivalent to a sphere of dimension $\delta(H) := |H| + 2c(H) + b(H) - 5$. If H does not have a block-closed representation, then $\mathrm{NCR}_n^{1,0}(H, \emptyset)$ is a cone. As a consequence, the $(2n-5)$-skeleton of $\mathrm{NCR}_n^{1,0}$ is Cohen-Macaulay.*

Proof. We want to prove that $\mathrm{NCR}_n^{1,0}(H, \emptyset)$ admits an acyclic matching such that there is either one unmatched graph with $|H| + 2c(H) + b(H) - 4$ edges or no unmatched graph at all.

First, assume that H does not have a block-closed representation. Let V be a vertex set such that $H(V)$ is non-separable and such that V is not a clique in H. Let v and w be nonadjacent vertices in V. Then vw is a cone point in $\mathsf{NCR}_2^{1,0}(H, \emptyset)$. Namely, let a be a cut point in the representation of a graph G in $\mathsf{NCR}_2^{1,0}(H, \emptyset)$ and let b be such that a separates G into the two vertex sets $[a+1, b]$ and $[b+1, a-1]$. We must have that V is a subset of either $[a, b]$ or $[b+1, a]$. Namely, otherwise either $G(V)$ would have a disconnected representation or a would be a cut point in $G(V)$. As a consequence, a remains a cut point in $G + vw$, which proves that vw is indeed a cone point.

Next, assume that H does have a block-closed representation. The case $n = 2$ is easy to check; thus assume that $n \geq 3$. We will define an acyclic matching on the quotient complex $\mathsf{CR}_n^{1,0}(H, \emptyset)$ of graphs with a non-separable representation. As we will see, there will be exactly one unmatched graph, and the number of edges in this graph is $\delta(H) + 2$ edges. Note that this remains true if H is the complete graph K_n. The matching turns out to be straightforward to translate into an acyclic matching on $\mathsf{NCR}_n^{1,0}(H, \emptyset)$.

As before, assume that $H \neq K_n$. Let e be an edge in $\mathrm{Bd}_n \setminus H$. Such an edge exists, because otherwise the full set Bd_n of exterior edges would be contained in H, which would imply that H is the complete graph. By symmetry, we may assume that $e = 1n$. Define a matching on $\mathsf{CR}_n^{1,0}(H, \emptyset)$ by pairing $G + 1n$ with $G - 1n$ whenever possible. Let Δ be the family of unmatched graphs. Lemma 4.1 yields that it suffices to define an appropriate acyclic matching on Δ.

A graph G containing H belongs to Δ if and only if G, but not $G - 1n$, has a non-separable representation. Let G be a graph in Δ and let $x = x_G$ be minimal such that the vertex x separates the representation of $G - 1n$. It is clear that $x_G \in [2, n-1]$. Moreover, if we remove x_G from the representation of $G - 1n$, then the resulting topological space X_G consists of two connected components. One of the components contains all vertices in the set $[1, x_G - 1]$ and the other contains the remaining vertices in the set $[x_G + 1, n]$. Namely, if we add the edge $1n$, the resulting representation is connected.

Let $\Delta(x)$ be the family of graphs G in Δ such that $x = x_G$; obviously, the families $\Delta(x)$ satisfy the Cluster Lemma 4.2. Let y be minimal such that $yn \in H$. If no such y exists, we define $y = n - 1$. It is clear that $y \geq x_G$, because n and y belong to the same component in X_G unless $y = x_G$.

Let us examine $\Delta(x)$. First, assume that $x < y$. We are done if $\Delta(x)$ is void. Otherwise, match with the edge xn; pair $G - xn$ with $G + xn$ whenever possible. This is possible, because $xn \notin H$ by assumption. We claim that this is a perfect matching. Namely, x_G remains the same if we add $x_G n$ to G, because this edge does not cross any edges in G by the properties of the space X_G. Conversely, suppose that $G - xn$ is not in $\Delta(x)$; this means that $G - xn$ is separable. In particular, there is a cut point z somewhere in the interval $[x+1, n-1]$ separating x and n. However, this means that z separates 1 and x. Since there are no edges from $[x_G + 1, n-1]$ to $[1, x_G - 1]$, it follows that n also separates 1 and x in $G - xn$ and hence also in G, which is a contradiction.

Next, we consider $\Delta(x)$ in the case that $x = y$. We claim that a graph G containing H belongs to $\Delta(x)$ if and only if the induced subgraphs $G([1,x])$ and $G([x,n])$ have non-separable representations and $1n$ is the only edge in G from $[1, x - 1] \times [x + 1, n]$. The last claim is obvious, as x is a cut point in $G - 1n$ separating 1 and n.

First, we prove that $G([1,x])$ has a non-separable representation. Assume the opposite and let z be a cut point in $G([1,x])$. Consider the space X obtained from the representation of $G([1,x])$ by removing z. For $i \in \{1, x\}$, let W_i be a connected component in X that does not contain the vertex i. Note that we may have that $W_1 = W_x$. Suppose that $z < x$. Since there are no edges from W_x to $[x + 1, n]$, we obtain that z is a cut point in $G - 1n$, which is a contradiction to the minimality of x. Suppose instead that $z = x$. Then x separates W_1 from $([n] \setminus \{x\}) \setminus W_1$ in G, another contradiction.

Next, we prove that $G([x,n])$ has a non-separable representation. Assume the opposite and let z be a cut point in $G([x,n])$. Let X be the space obtained from the representation of $G([x,n])$ by removing z. Let W be a connected component in X that does not contain the vertices x and n; these vertices are adjacent, which implies that such a component does exist. However, then z separates W from $([n] \setminus \{z\}) \setminus W$ in G, which is a contradiction.

For the converse, assume that $G([1,x])$ and $G([x,n])$ have non-separable representations and that $1n$ is the only other edge in G. One readily verifies that G does not have any cut points. We obtain that

$$\Delta(x) = \{\{1n\}\} * \mathsf{CR}_{[1,x]}^{1,0}(H([1,x]), \emptyset) * \mathsf{CR}_{[x,n]}^{1,0}(H([x,n]), \emptyset),$$

where $\mathsf{CR}_U^{1,0}$ is the quotient complex of graphs on the vertex set U with a non-separable representation. By induction, we obtain that $\mathsf{CR}_{[1,x]}^{1,0}(H([1,x]), \emptyset)$ and $\mathsf{CR}_{[x,n]}^{1,0}(H([x,n]), \emptyset)$ admit acyclic matchings with exactly one unmatched graph containing $\delta(H([1,x]))+2$ and $\delta(H([x,n]))+2$ edges, respectively. Using exactly the same approach as in the proof of Theorem 19.4, we obtain the desired result about $\Delta(x)$.

To prove that the $(2n - 5)$-skeleton of $\mathsf{NCR}_n^{1,0}$ is Cohen-Macaulay, again apply the proof of Theorem 19.4. $\square$

Just as for $\mathsf{NCR}_n^{0,0}$, there is a potentially interesting lattice $\mathsf{NX}\Pi_{n,2}$ closely related to $\mathsf{NCR}_n^{1,0}$. Specifically, recall the definition of the lattice $\Pi_{n,2}$ of block-closed graphs discussed in Section 19.1; by Theorem 19.6, $\mathsf{NX}\Pi_{n,2}$ is Cohen-Macaulay. Restricting to graphs with a block-closed representation, we obtain a sublattice, which we denote by $\mathsf{NX}\Pi_{n,2}$. There is an obvious closure map from the face poset of $\mathsf{NCR}_n^{1,0}$ to $\overline{\mathsf{NX}\Pi_{n,2}}$. Namely, map a graph G to the graph obtained by adding the edge ab whenever there is a set V containing a and b such that $G(V)$ is non-separable. This implies the following result.

Corollary 21.7. *For $n \geq 3$, $\Delta(\overline{\mathsf{NX}\Pi_{n,2}})$ is homotopy equivalent to a sphere of dimension $2n - 5$.* $\square$

We want to prove a stronger result. A ranked poset P with rank function ρ is *homotopically Gorenstein** if $\Delta(I)$ is homotopy equivalent to a sphere of dimension $\rho(G) - \rho(H) - 2$ for each nonempty interval $I = (x, y) = \{z : x < z < y\}$ such that $x, y \in P \cup \{\hat{0}, \hat{1}\}$. Since every link in the order complex of a poset is a join of order complexes of intervals, every link in a homotopically Gorenstein* poset is homotopy equivalent to a sphere in top dimension; apply Lemma 3.6. As a consequence, Gorenstein* posets are Cohen-Macaulay.

Theorem 21.8. *For $n \geq 3$, $\overline{NX\Pi}_{n,2}$ is homotopically Gorenstein* with rank function $\rho(G) = 2n - 2c(G) - b(G)$.*

Proof. The proof is very similar to the proof of Theorem 19.6. Consider an interval (H, G) in the lattice. For $G = K_n$, we have that $\Delta(H, K_n)$ is homotopy equivalent to $\mathrm{lk}_{\mathsf{NCR}_n^{1,0}}(H)$. By Theorem 21.6, this link is homotopy equivalent to a sphere of dimension $2c(H) + b(H) - 5 = \rho(K_n) - \rho(H) - 2$ as desired.

Suppose that $G \neq K_n$ and let $V_1, \ldots, V_r$ be the blocks in G. Write $G_i = G(V_i)$ and $H_i = H(V_i)$. Let $\Sigma_{G,H}$ be the lifted complex consisting of all graphs G' such that $H \subseteq G' \subset G$ and such that $G'(V_i)$ has a separable representation for at least one i. By the Closure Lemma 6.1, we have that $\mathrm{lk}_{\Sigma_{G,H}}(H)$ is homotopy equivalent to the order complex of (H, G). Using exactly the same approach as in the proof of Theorem 19.6, we obtain an acyclic matching on $\Sigma_{G,H}$ such that the remaining family is the join of $\mathsf{CR}_{V_1}^{1,0}(H_1)$, ..., $\mathsf{CR}_{V_{r-1}}^{1,0}(H_{r-1})$, and $\mathsf{NCR}_{V_r}^{1,0}(H_r)$. Since each of these families admits an acyclic matching with one single unmatched graph, the same is true for their join and hence for $\Sigma_{G,H}$. The remainder of the proof is identical to the proof of Theorem 19.6. $\square$

21.4 Graphs with a Two-separable Polygon Representation

We consider the complex $\mathsf{NCR}_n^{1,1}$ of graphs with a two-separable polygon representation.

Theorem 21.9. *Let $n \geq 4$. Then the complex $\overline{\mathsf{NCR}}_n^{1,1}$ obtained from $\mathsf{NCR}_n^{1,1}$ by removing the n cone points $i(i + 1)$ for $i \in [n]$ satisfies*

$$\overline{\mathsf{NCR}}_n^{1,1} \simeq S^{n-4}.$$

Proof. While $\mathsf{NCR}_n^{1,1}$ satisfies the conditions in Theorem 16.6, this is not true for the smaller complex $\overline{\mathsf{NCR}}_n^{1,1}$ that we are interested in. In fact, the proof of Theorem 16.6 does not apply, as the edge e defining the perfect matching on the complex $\Sigma(e)$ is not necessarily contained in $\mathrm{Int}_n = \binom{[n]}{2} \setminus \mathrm{Bd}_n$. To prove that $\overline{\mathsf{NCR}}_n^{1,1}$ is indeed homotopy equivalent to $\overline{\mathsf{NCR}}_n^{1,1} \cap \mathsf{NX}_n = \overline{\mathsf{NCR}}_n^{1,1} \cap \mathsf{A}_n$, we have to work a bit more. Note that we are done as soon as such a homotopy equivalence is established. Namely, $A_n \subset \overline{NCR}_n^{1,1}$, because all noncrossing graphs are two-separable.

For a given graph G, define $X(G)$ as the family of cut sets $\{i, j\}$ in $G \cup Bd_n$; i and j separate the representation of $G - ij$. Define $Cl(G)$ as the maximal graph K containing Bd_n with the property that $X(K) = X(G)$. K is easily seen to be unique, because the union $G \cup H$ of two graphs G and H with $X(G) = X(H)$ is readily seen to satisfy $X(G \cup H) = X(G)$.

Define a linear order $\leq_L$ on the family of subsets of $[n]$ in the following manner. A set S is smaller than a set T if $|S| < |T|$ or if $|S| = |T|$ and S is smaller than T with respect to a given fixed linear order. For a graph $G \in \overline{NCR}_n^{1,1} \setminus A_n$, let $Q = Q(G)$ be maximal with respect to the order $\leq_L$ such that Q is a clique in $Cl(G)$. By the properties of $\leq_L$, it follows that Q is a maximal clique in $Cl(G)$. For $Q \subset [1, n]$, define $\mathcal{F}(Q)$ as the family of graphs $G \in \overline{NCR}_n^{1,1} \setminus A_n$ such that $Q(G) = Q$. The families $\mathcal{F}(Q)$ satisfy the Cluster Lemma 4.2; if we add an edge to G, then Q cannot decrease.

Let $G \in \mathcal{F}(Q)$. Since G contains crossings, Q has size at least four; the four vertices in a crossing form a clique in $Cl(G)$. Since $X(G)$ is nonempty, we have that $Q \subsetneq [n]$. In particular, there are vertices $i, j \in Q$ such that $ij \in Int_n$ and such that $Q \cap (i, j) = \emptyset$; this means that i and j are adjacent on the polygon with vertex set Q but not on the big n-gon. Choose (i, j) with this property such that i is minimal and $i < j$.

We claim that

$$Q(G - ij) = Q(G + ij). \qquad (21.1)$$

If we can prove this, we obtain that ij is a cone point in $\mathcal{F}(Q)$, which concludes the proof.

(21.1) obviously holds if $ij \notin G$, because $ij \in Cl(G)$; Q is a clique in $Cl(G)$. Suppose that $ij \in G$. First, we show that ij does not cross any edge in $Cl(G)$. Assume the opposite; there are $a \in (i, j)$ and $b \in (j, i)$ such that $ab \in Cl(G)$. Note that $a \notin Q$. By a straightforward adaptation of the proof of Lemma 21.1, we have, for any graph H, that $X(H + rs) = X(H)$ if rt and su are crossing edges in H; by the proof, every cut set in H remains a cut set in $H + rs$. In particular, $ai, aj \in Cl(G)$, as ab and ij cross in $Cl(G)$. Moreover, if $r \in Q \setminus \{i, j, b\}$, then $ar \in Cl(G)$ as ab crosses either ir or rj. However, this means that a is adjacent to all vertices in Q, which contradicts the fact that Q is a maximal clique in $Cl(G)$.

If (21.1) does not hold, then $X(G - ij) \neq X(G)$. Let $\{a, b\} \in X(G - ij) \setminus X(G)$; we must have that $a \in (i, j)$ and $b \in (j, i)$ (or vice versa). Observe that each of $\{b, i\}$ and $\{b, j\}$ is either a cut set in $Cl(G)$ or a boundary edge. Namely, we learned above that $\{i, j\}$ is a cut set in $Cl(G)$ separating (j, b) and (b, i) from (i, j), whereas $\{a, b\}$ is a cut set in $Cl(G - ij)$ separating (j, b) from (b, a) and (b, i) from (a, b). Since $(i, j) \cup (b, a) = (b, j)$ and $(i, j) \cup (a, b) = (i, b)$, the claim follows.

Now, since Q has size at least four, the set $Q \cap ((j,b) \cup (b,i)) = Q \setminus \{b,i,j\}$ is nonempty. The conclusion is that at least one of $\{b,j\}$ and $\{b,i\}$, say $\{b,i\}$, is indeed a cut set in $G \cup \mathrm{Bd}_n$ and hence in $\mathrm{Cl}(G)$, separating some element $q \in Q \setminus \{i,j\}$ from j. However, $\mathrm{Cl}(G)$ contains the edge qj, which implies that $\{b,i\}$ is not a cut set in $\mathrm{Cl}(G)$. This is a contradiction; hence $X(G - ij) = X(G)$, and we are done. $\square$

Corollary 21.10. *The homology of* $\overline{\mathrm{NCR}}_n^{1,1}$ *is generated by the fundamental cycle of the associahedron* A_n.

Proof. By the above proof, $\overline{\mathrm{NCR}}_n^{1,1}$ is collapsible to A_n; hence the corollary follows immediately. $\square$

Directed Variants of Connected Graphs

We devote this chapter to some directed variants of connectivity.

In Section 22.1, we review some known results about the complex DNSC_n of not strongly connected digraphs, the main result being that DNSC_n is homotopy equivalent to a wedge of $(n-1)!$ spheres of dimension $2n-4$. This result is due to Björner and Welker [17]; Hultman [64] generalized it to certain induced subcomplexes of DNSC_n. In addition, we prove that DNSC_n has a Cohen-Macaulay $(2n-4)$-skeleton; the proof is inspired by Shareshian's proof of the corresponding result for NC_n^2 (see Theorem 19.1).

In Section 22.2, we proceed with complexes of not strongly 2-connected digraphs, for which very little is known.

In Section 22.3, we consider the complex DNSp_n of non-spanning digraphs on n vertices and show that the homotopy type coincides with that of NC_n^2, thus a wedge of $(n-2)!$ spheres of dimension $2n-5$.

22.1 Not Strongly Connected Digraphs

Björner and Welker [17] determined the homotopy type of the complex DNSC_n of not strongly connected digraphs on n vertices. We generalize their result to the complex $\mathsf{DNSC}_{n,k}$ of digraphs D such that the associated poset $P(D)$ has at least $k+1$ elements.

Theorem 22.1. *For* $1 \le k \le n-1$, $\mathsf{DNSC}_{n,k} \sim c_{n,k} \cdot t^{2n-k-3}$, *where*

$$P_n(x) := \sum_{k=1}^{n-1} c_{n,k} x^k = x(2+x)(3+x) \cdots (n-1+x) = x \cdot \prod_{i=2}^{n-1} (i+x).$$

In particular, DNSC_n *is homotopy equivalent to a wedge of* $(n-1)!$ *spheres of dimension* $2n-4$.

Proof. We use induction on n. First, note that $\mathsf{DNSC}_{n,n-1} = \mathsf{DAcy}_n \sim t^{n-2}$ by Theorem 15.3. Thus we may assume that $1 \le k \le n-2$; in particular, $n \ge 3$. Consider the first-hit decomposition of $\mathsf{DNSC}_{n,k}$ with respect to

$(1n, 2n, \ldots, (n-1)n)$; see Definition 5.24. For $r \in [n-1]$, let $A_r = \{in : i \in [r]\}$. Let $\Sigma_r = \mathsf{DNSC}_{n,k}(\{rn\}, A_{r-1})$ and $\Sigma_n = \mathsf{DNSC}_{n,k}(\emptyset, A_{n-1})$. We want to show that

$$\Sigma_r \sim \begin{cases} c_{n-1,k} \cdot t^{2n-k-3} & \text{if } r \in [n-2]; \\ (c_{n-1,k-1} + c_{n-1,k}) \cdot t^{2n-k-3} & \text{if } r = n-1; \\ 0 & \text{if } r = n. \end{cases} \qquad (22.1)$$

Lemma 5.25 will then yield that

$$\mathsf{DNSC}_{n,k} \sim (c_{n-1,k-1} + (n-1)c_{n-1,k}) \cdot t^{2n-k-3}.$$

Since this implies that

$$\sum_{k=1}^{n-1} |\tilde{\chi}(\mathsf{DNSC}_{n,k})| \cdot x^k = P_{n-1}(x) \cdot ((n-1)+x) = P_n(x),$$

the theorem will follow.

Clearly, ni is a cone point in Σ_n for any i; if no edges are directed to n, then n cannot be contained in a cycle, which means that n forms an element on its own in $P(D)$ whenever $D \in \Sigma_n$.

Consider Σ_r for $r \leq n-1$. Let $B = \{ni : i \in [n-1]\}$. For each $Z \subseteq B$, consider the complex $\Sigma_{r,Z} = \Sigma_r(Z, B \setminus Z)$. Note that $A_{n-1} \setminus A_r$ is the set of edges incident to n that remain to be checked. There are three cases:

- $ni \in Z$ for some $i \neq r$. Then ri is a cone point in $\Sigma_{r,Z}$; we already have a directed path from r to i via n.
- $Z = \{nr\}$. For each $Y \subseteq A_{n-1} \setminus A_r$, consider the complex $\Sigma_{r,\{nr\},Y} = \Sigma_{r,\{nr\}}(Y, (A_{n-1} \setminus A_r) \setminus Y)$. If $Y \neq \emptyset$, then yr is a cone point in $\Sigma_{r,\{nr\},Y}$ for each $y \in Y$; we already have a directed path from y to r via n. The remaining case is $Y = \emptyset$. Now, a digraph D on $n-1$ vertices belongs to $\mathsf{DNSC}_{n-1,k}$ if and only if D' belongs to $\Sigma_{r,\{nr\},\emptyset}$, where D' is the digraph obtained by adding the vertex n and the edges rn and nr to D. By induction, $\mathsf{DNSC}_{n-1,k} \sim c_{n-1,k} \cdot t^{2n-k-5}$, which implies that $\Sigma_{r,\{nr\},\emptyset} \sim c_{n-1,k} \cdot t^{2n-k-3}$. It follows that

$$\Sigma_{r,\{nr\}} \sim c_{n-1,k} \cdot t^{2n-k-3}.$$

- $Z = \emptyset$. This means that no edge begins in n, which implies that n is not contained in any cycle. If $r < n-1$, then $(n-1)n$ is hence a cone point in $\Sigma_{r,\emptyset}$; thus $\Sigma_{r,\emptyset} \sim 0$. The remaining case is $r = n-1$. Now, a digraph D on $n-1$ vertices belongs to $\mathsf{DNSC}_{n-1,k-1}$ if and only if D' belongs to $\Sigma_{n-1,\emptyset}$, where D' is the digraph obtained by adding the vertex n and the edge $(n-1)n$ to D. By induction, $\mathsf{DNSC}_{n-1,k-1} \sim c_{n-1,k-1} \cdot t^{2n-k-4}$, which implies that

$$\Sigma_{n-1,\emptyset} \sim c_{n-1,k-1} \cdot t^{2n-k-3}.$$

As a consequence, (22.1) follows. $\square$

For the remainder of the section, we confine ourselves to $\mathsf{DNSC}_n = \mathsf{DNSC}_{n,1}$. Let DSC_n be the quotient complex of strongly connected digraphs on n vertices. As an immediate consequence of the above proof, we have a decision tree on DSC_n such that D is evasive if and only if D is the digraph D_μ with edge set

$$\sigma_\mu := \{\mu_j j, j\mu_j : j \in [2, n]\}$$

for some $(\mu_2, \ldots, \mu_n)$ such that $1 \leq \mu_j < j$ for all j. One readily verifies that every proper subdigraph of D_μ belongs to DNSC_n. As a consequence, we may conclude the following; apply Corollary 4.17.

Corollary 22.2. *For* $n \geq 2$, $\{[\sigma_\mu] : 1 \leq \mu_j < j \text{ for } j \in [2, n]\}$ *is a basis for* $\tilde{H}_{2n-3}(\mathsf{DSC}_n; \mathbb{Z})$. *As a consequence,* $\{\partial([\sigma_\mu]) : 1 \leq \mu_j < j \text{ for all } j\}$ *is a basis for* $\tilde{H}_{2n-4}(\mathsf{DNSC}_n; \mathbb{Z})$. $\square$

Note that $\partial([\sigma_\mu])$ is the fundamental cycle of the sphere $\partial 2^{\sigma_\mu}$.

A digraph D is 2-*dense* if every directed cycle in D contains two vertices u and v such that (u, v) forms a 2-cycle in D; $(u, v), (v, u) \in D$. Define $\hat{D}$ as the ordinary undirected graph on the same vertex set as D and with one edge for each 2-cycle in D. Hultman generalized the result of Björner and Welker in the following manner:

Theorem 22.3 (Hultman [64]). *Let D be a digraph on n vertices. If D is 2-dense and strongly connected, then $\mathsf{DNSC}_n(D)$ is homotopy equivalent to a wedge of $(2n - 4)$-dimensional spheres. The number of spheres in the wedge equals $|\chi'_{\hat{D}}(0)|$, where $\chi_{\hat{D}}(t)$ is the chromatic polynomial of $\hat{D}$.* $\square$

Finally, we show that the $(2n - 4)$-skeleton of DNSC_n is Cohen-Macaulay. For a digraph D, we define $b(D)$ as the number of elements in the associated poset $P(D)$. Thus if $A_1, A_2, \ldots, A_b$ are the elements in $P(D)$, then $b(D) = b$. We refer to the sets A_i as the *blocks* of D. For vertices x and y in a digraph D, let $x \xrightarrow{D} y$ mean that there is a directed path from x to y in D; if the underlying digraph is clear from context, we simply write $x \longrightarrow y$. D is *block-closed* if $xy \in D$ whenever $x \xrightarrow{D} y$.

Theorem 22.4. *If H is a block-closed digraph on n vertices, then the lifted complex $\mathsf{DNSC}_n(H, \emptyset)$ is homotopy equivalent to a wedge of spheres of dimension $\delta(H) := |H| + b(H) + c(H) - 4$. If H is not block-closed, then $\mathsf{DNSC}_n(H, \emptyset)$ is a cone. As a consequence, the $(2n - 4)$-skeleton of DNSC_n is Cohen-Macaulay.*

Proof. We want to prove that $\mathsf{DNSC}_n(H, \emptyset)$ admits an acyclic matching such that all unmatched digraphs contain $\delta(H) + 1$ edges. First, assume that H is not block-closed. Let v and w be such that $vw \notin H$ and $v \xrightarrow{H} w$. It is clear that vw is a cone point in $\mathsf{DNSC}_n(H, \emptyset)$.

From now on, assume that H is block-closed. If H is the complete digraph, then the theorem is trivial. Moreover, the cases $n = 1$ and $n = 2$ are easy to check. Thus assume that H is not complete and that $n \geq 3$. We will define an acyclic matching on the quotient complex $\mathsf{DSC}_n(H, \emptyset)$ of strongly connected digraphs containing H such that all unmatched digraphs contain $\delta(H) + 2$ edges. The matching turns out to be straightforward to translate into an acyclic matching on $\mathsf{DNSC}_n(H, \emptyset)$.

First, assume that there are vertices v and w in H such that neither vw nor wv belongs to H; we may assume that $v = 1$ and $w = n$. Define a matching on $\mathsf{DSC}_n(H, \emptyset)$ by pairing $D + n1$ with $D - n1$ whenever possible. Let Δ be the family of unmatched digraphs. By Lemma 4.1, we obtain an acyclic matching on $\mathsf{DSC}_n(H, \emptyset)$ by combining any acyclic matching on Δ with the matching just defined.

Δ consists of all strongly connected digraphs D containing H such that $D - n1$ is not strongly connected. Equivalently, for all $x \in [n]$, we have that $1 \longrightarrow x$ and $x \longrightarrow n$ in both D and $D - n1$, whereas $n \longrightarrow 1$ in D but not in $D - n1$. Now, define a matching on Δ by pairing $D + 1n$ with $D - 1n$ whenever possible. Let Σ be the family of unmatched digraphs. It is clear that a digraph D in Δ belongs to Σ if and only if $1 \not\longrightarrow n$ in $D - 1n$.

For a digraph D in Σ, write $\hat{D} = D \setminus \{1n, n1\}$. Let X_D be the set of vertices x such that $1 \xrightarrow{\hat{D}} x$; note that $1 \in X_D$ and $n \notin X_D$. For each set $X \subseteq [n-1]$, let $\Sigma(X)$ be the family of digraphs D such that $X_D = X$. Write $Y_D = [n] \setminus X_D$. We claim that $\hat{D}$ is the disjoint union of the two induced subdigraphs $D(X_D)$ and $D(Y_D)$ and that these subdigraphs are strongly connected.

To see this, first note that $x \not\longrightarrow y$ in $\hat{D}$ if $x \in X_D$ and $y \in Y_D$. Namely, otherwise we would have that $1 \xrightarrow{\hat{D}} y$. Moreover, $x \xrightarrow{\hat{D}} 1$, because $x \xrightarrow{D} 1$, whereas $x \not\longrightarrow n$ in $\hat{D}$. It follows that there are no edges from X_D to Y_D in $\hat{D}$ and that $D(X_D)$ is strongly connected.

Next, note that $n \xrightarrow{\hat{D}} y$ if $y \in Y_D$. Namely, $n \longrightarrow y$ in D and hence in $D - 1n$. Since $1 \not\longrightarrow y$ in $\hat{D}$, no directed path from n to y in $D - 1n$ contains the edge $n1$, and the claim follows. As a consequence, we have that $y \not\longrightarrow 1$ in $\hat{D}$, because $n \not\longrightarrow 1$ in $\hat{D}$. Finally, we have that $y \xrightarrow{\hat{D}} n$, because $y \longrightarrow n$ in $\hat{D} + 1n$ and $y \not\longrightarrow 1$ in $\hat{D}$. It follows that there are no edges from Y_D to X_D and that $D(Y_D)$ is strongly connected.

As an immediate consequence, we have that the families $\Sigma(X)$ satisfy the Cluster Lemma 4.2; they form an antichain with respect to inclusion. Moreover, with $Y = [n] \setminus X$, we have that

$$\Sigma(X) = \{\{1n, n1\}\} * \mathsf{DSC}_X(H(X), \emptyset) * \mathsf{DSC}_Y(H(Y), \emptyset);$$

DSC_X is the quotient complex of strongly connected digraphs on the vertex set X. Induction yields that $\mathsf{DSC}_X(H(X), \emptyset)$ and $\mathsf{DSC}_Y(H(Y), \emptyset)$ admit acyclic matchings such that all unmatched digraphs contain $\delta(H(X)) + 2$ and $\delta(H(Y)) + 2$ edges, respectively. This yields an acyclic matching on $\Sigma(X)$

with

$$2 + \delta(H(X)) + 2 + \delta(H(Y)) + 2$$
$$= |H| + b(H(X)) + b(H(Y)) + c(H(X)) + c(H(Y)) - 2$$

edges. One readily verifies that $b(H(X)) + b(H(Y)) = b(H)$ and $c(H(X)) + c(H(Y)) = c(H)$; hence

$$|H| + b(H(X)) + b(H(Y)) + c(H(X)) + c(H(Y)) - 2$$
$$= |H| + b(H) + c(H) - 2 = \delta(H) + 2$$

as desired.

It remains to consider the case that $vw \in H$ or $wv \in H$ for all vertices v and w in H. Then the associated poset $P(H)$ is a linear order. If $P(H)$ consists of one single element, then H is the complete digraph; thus assume that $P(H)$ consists of at least two elements. Let $A_1, \ldots, A_b$ be the blocks of H and assume that A_i is smaller than A_j in $P(H)$ if and only if $i < j$. Since H is block-closed, this means that $xy \in H$ if and only if $x \in A_i$ and $y \in A_j$ for some $i \le j$. Pick an element a_1 from A_1 and some element $a_2 \in A_2$; we may assume that $a_1 = 1$ and $a_2 = n$.

We obtain a matching on $\mathsf{DSC}_n(H, \emptyset)$ by pairing $D + n1$ with $D - n1$ whenever possible. D remains unmatched if and only if D is strongly connected and $D - n1$ is not. Write $A = A_1$ and $B = [n] \setminus A_1$.

Let D be unmatched. First, we claim that there is no edge xy in $D - n1$ such that $x \in B$ and $y \in A$. Namely, either $n = x$ or $nx \in H$; similarly, either $y = 1$ or $y1 \in H$. In particular, $n \longrightarrow 1$ in $D - n1$, a contradiction. Next, we claim that the induced subdigraph of D on B is strongly connected. Namely, $x \overset{D}{\longrightarrow} n$ for all $x \in B$, and since the only edge from B to A in D is $n1$, it follows that $x \longrightarrow n$ in $D(B)$.

As a consequence,

$$\mathsf{DSC}_n(H, \emptyset) = \{K_A^{\rightarrow} \cup (A \times B)\} * \{n1\} * \mathsf{DSC}_B(H(B), \emptyset);$$

$K_A^{\rightarrow}$ is (the edge set of) the complete digraph on the set A. By induction, we have that $\mathsf{DSC}_B(H(B), \emptyset)$ admits an acyclic matching such that all unmatched digraphs contain $\delta(H(B)) + 2$ edges. This yields an acyclic matching on $\mathsf{DSC}_n(H, \emptyset)$ with

$$|K_A^{\rightarrow} \cup (A \times B)| + 1 + \delta(H(B)) + 2 = |H| + b(H(B)) + c(H(B)) - 1$$

edges. It is clear that $c(H(B)) = 1 = c(H)$ and $b(H(B)) = b(H) - 1$; we lose the poset element $A = A_1$. Hence

$$|H| + b(H(B)) + c(H(B)) - 1 = |H| + b(H) + c(H) - 2 = \delta(H) + 2$$

as desired.

Finally, we have to prove that the $(2n-4)$-skeleton of DNSC_n is Cohen-Macaulay. It suffices to prove that

$$|H| \geq 2n - b(H)' - c(H).$$

Namely, this will yield the desired inequality $\delta(H) \geq 2n - 4$.

Now, let $A_1, \ldots, A_b$ be the blocks in H. Since H is block-closed, $H(A_i)$ is a complete digraph for each A_i. In particular, $H(A_i)$ contains at least $2(|A_i|-1)$ edges; we have equality for $|A_i| \leq 2$. Moreover, since the poset $P(H)$ contains $c(H)$ components and $b(H)$ elements, there are at least $b(H) - c(H)$ covering relations in $P(H)$. For each covering relation $A_i < A_j$ in $P(H)$, H contains the set $A_i \times A_j$, which has size at least 1; thus there are at least $b(H) - c(H)$ edges in H between different blocks. As a consequence, H contains at least

$$\sum_{i=1}^{b(H)} 2(|A_i| - 1) + b(H) - c(H) = 2n - 2b(H) + b(H) - c(H)$$

edges, which concludes the proof. □

22.2 Not Strongly 2-connected Digraphs

Recall that DNSC_n^2 is the simplicial complex of not strongly 2-connected digraphs on n vertices. Not much is known about DNSC_n^2. In particular, the following conjecture due to Björner and Welker [17] remains unsettled:

Conjecture 22.5. *For $n \geq 3$, DNSC_n^2 is homotopy equivalent to a wedge of $(n-2) \cdot (n-2)!$ spheres of dimension $3n - 5$.*

Computer calculations yield the conjecture for $n \leq 5$ [17]. In addition, using computer, we have been able to verify that the reduced Euler characteristic of DNSC_6^2 equals the conjectured value -96.

While we have not made any further progress on Conjecture 22.5, we have discovered a rather unexpected correspondence between DNSC_n^2 and the complex of digraphs D such that $D([n] \setminus \{x\})$ is not strongly connected for *any* $x \in [n]$. Specifically, for any sequence $\mathsf{A} = (A_1, \ldots, A_r)$ of subsets of $[n]$, let $\mathsf{DNSC}_n^2(\mathsf{A})$ be the complex of digraphs D such that each A_i contains a vertex a such that $D([n] \setminus \{a\})$ is not strongly connected. For $n \geq 3$, $\mathsf{DNSC}_n^2([n])$ coincides with DNSC_n^2.

Theorem 22.6. *Let $n \geq 3$ and let $\mathsf{A} = (A_1, \ldots, A_r)$ be a sequence of pairwise disjoint and nonempty subsets of $[n]$. Then*

$$\mathsf{DNSC}_n^2(\mathsf{A}) \simeq \begin{cases} \mathrm{Susp}^{n-r}(\mathsf{DNSC}_n^2(1, \ldots, n)) & \text{if } \sum_i |A_i| = n; \\ \text{point} & \text{otherwise.} \end{cases}$$

Moreover, if $\mathsf{DNSC}_n^2(1, \ldots, n)$ is semi-buildable, then $\mathsf{DNSC}_n^2(\mathsf{A})$ is semi-buildable.

Proof. We show that $\Sigma_{n,r} = \mathsf{DNSC}_n^2(1, \ldots, r)$ is nonevasive whenever $0 \leq r < n$. By Lemma 19.11, this will imply the desired result.

$\Sigma_{n,r}$ is trivially nonevasive if $n = 3$; thus assume that $n \geq 4$. If $r \leq 1$, then the edge $1n$ is a cone point; thus assume that $r \geq 2$.

Let E_n be the set of edges incident to the vertex n. If B contains two edges an and nb and hence a path from a to b, then ab is a cone point in $\Sigma_{n,r}(B, E_n \setminus B)$. If B contains no edges ending in n or no edges starting in n, then any edge in K_{n-1} is a cone point. The remaining case is that $B = \{kn, nk\}$ for some $k \in [n-1]$. One readily verifies that $\Sigma_{n,r}(\{kn, nk\}, E_n \setminus \{kn, nk\})$ coincides with $\{\{kn, nk\}\} * \Sigma_{n-1,r}$ if $k > r$ and with $\{\{kn, nk\}\} * \mathsf{DNSC}_{n-1}^2(1, \ldots, k-1, k+1, \ldots, r) \cong \{\{kn, nk\}\} * \Sigma_{n-1,r-1}$ if $k \leq r$. By induction, we are done. $\square$

Theorem 22.6 suggests the following conjecture, which implies Conjecture 22.5.

Conjecture 22.7. *For $n \geq 3$, $\mathsf{DNSC}_n^2(1, \ldots, n)$ is homotopy equivalent to a wedge of $(n-2) \cdot (n-2)!$ spheres of dimension $2n - 4$.*

22.3 Non-spanning Digraphs

Another variant is DNSp_n, the complex of non-spanning digraphs on n vertices. We consider the generalized complex $\mathsf{DNSp}_{n,k}$ of digraphs D such that $P(D)$ has at least $k+1$ atoms. Let us refer to a directed forest with k connected components as a k-*spanning* directed forest. $\mathsf{DNSp}_{n,k}$ is exactly the complex of digraphs that do not contain a k-spanning directed forest.

Theorem 22.8. *For $1 \leq k \leq n-1$, $\mathsf{DNSp}_{n,k} \sim d_{n,k} \cdot t^{2n-2k-3}$, where*

$$Q_n(x) := \sum_{k=1}^{n-1} d_{n,k} x^k = x(1+x)(2+x) \cdots (n-2+x) = \prod_{i=0}^{n-2} (i+x).$$

In particular, DNSp_n is homotopy equivalent to a wedge of $(n-2)!$ spheres of dimension $2n - 5$.

Proof. We use induction on n. First, note that $\mathsf{DNSp}_{n,n-1} = \{\emptyset\}$. Thus we may assume that $1 \leq k \leq n-2$; in particular, $n \geq 3$. Throughout the proof, we will frequently apply the fact that if a digraph D contains two edges ab and bc such that a, b, c are all different, then $D + ac \in \mathsf{DNSp}_{n,k}$ if and only if $D - ac \in \mathsf{DNSp}_{n,k}$.

Let $Y = \{in, ni : i \in [2, n-1]\}$ and consider the family $\Sigma_A = \mathsf{DNSp}_{n,k}(A, Y \setminus A)$ for each $A \subseteq Y$. We identify four cases:

- A contains edges in and nj such that $i \neq j$. Then ij is a cone point in Σ_A.

- A contains at least one edge in ending in n but no edge starting in n. Decompose with respect to $n1$. We have that $i1$ is a cone point in $\Sigma_A(n1, \emptyset)$. Moreover, $1n$ is a cone point in $\Sigma_A(\emptyset, n1)$. Namely, suppose that $D \in \Sigma_A(\emptyset, n1)$ and that $D + 1n$ contains a k-spanning directed forest F; obviously, $1n \in F$. We obtain a new k-spanning directed forest by replacing $1n$ with in. Since this forest is contained in D, we have a contradiction; $D \in \mathsf{DNSp}_{n,k}$.

- A does not contain any edge ending in n. Decompose with respect to $1n$. We have that $n1$ is a cone point in $\Sigma_A(1n, \emptyset)$. Namely, suppose that $D \in \Sigma_A(1n, \emptyset)$ and that $D + n1$ contains a k-spanning directed forest F; obviously, $n1 \in F$. Removing $n1$ from F and adding $1n$, we obtain another k-spanning directed forest contained in $D - n1$; this is a contradiction.

  It remains to consider $\Sigma_A(\emptyset, 1n)$; digraphs in this family have the property that no edges end in n. Decompose $\Sigma_A(\emptyset, 1n)$ with respect to $n1$. We claim that 21 is a cone point in $\Sigma_A(n1, 1n)$. Namely, if $D \in \Sigma_A(n1, 1n)$ and F is a k-spanning forest contained in $D + 21$, then $(F - 21) + n1$ is another k-spanning forest contained in D, a contradiction. The remaining family is $\Sigma_A(\emptyset, \{1n, n1\})$. If A contains some edge ni, then $1i$ is a cone point by the same argument as before. For $A = \emptyset$, we have that n is isolated in every digraph in $\Sigma_\emptyset(\emptyset, \{1n, n1\})$. As a consequence, this complex equals $\mathsf{DNSp}_{n-1,k-1}$. By induction, $\mathsf{DNSp}_{n-1,k-1} \sim d_{n-1,k-1} \cdot t^{2n-2k-3}$.

  To summarize, we have that $\Sigma_A \sim 0$ if $A \neq \emptyset$ and $\Sigma_\emptyset \sim d_{n-1,k-1} \cdot t^{2n-2k-3}$.

- $A = \{in, ni\}$ for some $i \in [2, n-1]$. Decompose with respect to $1n$ and $n1$. For any nonempty $I \subseteq \{1n, n1\}$, we have that $\Sigma_{\{in,ni\}}(I, \{1n, n1\} \setminus I)$ is a cone. Namely, $1i$ is a cone point if $1n \in I$ and $i1$ is a cone point if $n1 \in I$. The remaining case is $\Sigma_{\{in,ni\}}(\emptyset, \{1n, n1\})$. However, a digraph D containing $\{in, ni\}$ but no other edges incident to n belongs to $\Sigma_{\{in,ni\}}(\emptyset, \{1n, n1\})$ if and only if the induced subdigraph $D([n-1])$ belongs to $\mathsf{DNSp}_{n-1,k}$. Namely, we can extend a k-spanning directed forest contained in $D([n-1])$ to a k-spanning directed forest contained in D by adding the edge in. Conversely, if $D([n-1])$ does not contain a k-spanning directed forest, then neither does D: If n were a root of such a forest, then there would be a k-spanning directed forest in $D([n-1])$ with i among its roots, whereas a k-spanning directed forest such that n is not a root (i.e., the forest contains in) would obviously yield a k-spanning directed forest in $D([n-1])$. By induction, $\mathsf{DNSp}_{n-1,k} \sim d_{n-1,k} \cdot t^{2n-2k-5}$. Thus $\Sigma_{\{1n,n1\}} \sim d_{n-1,k} \cdot t^{2n-2k-3}$.

We conclude that $\mathsf{DNSp}_{n,k} \sim (d_{n-1,k-1} + (n-2)d_{n-1,k}) \cdot t^{2n-2k-3}$. As in the proof of Theorem 22.1, this implies by induction on n that $\sum_{k=1}^{n-1} |\tilde{\chi}(\mathsf{DNSp}_{n,k})| \cdot x^k = Q_n(x)$. $\square$

23

Not 2-edge-connected Graphs

We consider the complex NEC_n^k of not k-edge-connected graphs on n vertices. Such graphs have the property that we can make them disconnected by removing an edge set of size at most $k - 1$. For example, $G \in \mathsf{NEC}_n^2$ if and only if G is disconnected or $G - e$ is disconnected for some edge $e \in G$. Note that $\mathsf{NEC}_n^1 = \mathsf{NC}_n$, which has a very attractive structure by the results in Chapter 18; all homology appears in dimension $n - 3$. Moreover, NEC_n^2 is exactly the complex obtained from NEC_n^1 by adding $G + e$ for each $G \in \mathsf{NEC}_n^1$ and $e \in \binom{[n]}{2}$. While this suggests that the two complexes are combinatorially closely related, it does not at all indicate that they should have anything to do with each other topologically. Specifically, when adding the faces of $\mathsf{NEC}_n^2 \setminus \mathsf{NEC}_n^1$ to NEC_n^1, we kill all existing homology in NEC_n^1.

Indeed, in Section 23.1, we show that the homology of NEC_n^2 is nonvanishing below dimension $\lceil \frac{3n-7}{2} \rceil$, a bound way above the depth $n - 3$ of NC_n. Moreover, while there is homology in dimension $\lfloor \frac{5n-11}{3} \rfloor$ (see Section 23.4), there is no homology above this dimension. This rules out any deeper connection to the homology of NC_n^2, which is concentrated in dimension $2n - 5$; see Chapter 19.

However, there is an interesting connection between NEC_{2k-1}^2 and the complex NFC_{2k-1} of not factor-critical graphs. By a result of Linusson, Shareshian, and Welker [95], NFC_{2k-1} is homotopy equivalent to a wedge of $((2k - 3)!!)^2$ spheres of dimension $3k - 5$. In Section 23.3, we show that $\tilde{H}_{3k-5}(\mathsf{NEC}_{2k-1}^2; \mathbb{Z})$ is isomorphic to $\tilde{H}_{3k-5}(\mathsf{NFC}_{2k-1}; \mathbb{Z})$, thereby establishing that the shifted connectivity degree of NEC_{2k-1}^2 equals $3k - 5$. We have not been able to compute the shifted connectivity degree of NEC_n^2 for even n, but we conjecture it to be $\lceil \frac{3n-7}{2} \rceil = \frac{3n}{2} - 3$.

As usual, we use discrete Morse theory in our analysis. Similarly to the analysis of NC_n^3 in Chapter 20, our acyclic matching is explicit rather than defined in terms of a decision tree. In fact, the decision tree method does not appear to be useful in this case. Specifically, while NEC_5^2 is semi-collapsible and admits an acyclic matching with nine critical graphs of dimension four, the complex is not semi-nonevasive; the link of NEC_5^2 with respect to any element has homology in two dimensions.

Table 23.1. The integral homology of the complex NEC_n^2 of not 2-edge-connected graphs for $3 \leq n \leq 9$.

$\tilde{H}_i(\mathsf{NEC}_n^2)$	$i = 0$	1	2	3	4	5	6	7	8	9	10	11
$n = 3$	$-$	$\mathbb{Z}$	$-$	$-$	$-$	$-$	$-$	$-$	$-$	$-$	$-$	$-$
4	$-$	$-$	$-$	$\mathbb{Z}^2$	$-$	$-$	$-$	$-$	$-$	$-$	$-$	$-$
5	$-$	$-$	$-$	$-$	$\mathbb{Z}^9$	$-$	$-$	$-$	$-$	$-$	$-$	$-$
6	$-$	$-$	$-$	$-$	$-$	$-$	$\mathbb{Z}^{96}$	$-$	$-$	$-$	$-$	$-$
7	$-$	$-$	$-$	$-$	$-$	$-$	$-$	$\mathbb{Z}^{225}$	$\mathbb{Z}^{280}$	$-$	$-$	$-$
8	$-$	$-$	$-$	$-$	$-$	$-$	$-$	$-$	$-$	$\mathbb{Z}^{6768}$	$-$	$-$
9	$-$	$-$	$-$	$-$	$-$	$-$	$-$	$-$	$-$	$-$	$\mathbb{Z}^{11025}$	$\mathbb{Z}^{66528}$

In Table 23.1, we present the homology of NEC_n^2 for $n \leq 9$. All values are consequences of results in this chapter. Our acyclic matching is optimal for $n \leq 9$ and for $n = 11$. For $n \leq 6$ and $n = 8$, this is immediate from the fact that all critical graphs are of the same dimension. The situation is much less obvious for $n \in \{7, 9, 11\}$, because then we have critical graphs of two different adjacent dimensions. Still, the correspondence between NEC_{2k-1}^1 and NFC_{2k-1} makes it possible to settle the desired optimality. We do not know whether our matching is optimal for $n = 10$ and $n \geq 12$. Moreover, the homotopy type of NEC_n^2 remains an open problem for all $n \geq 7$ except $n = 8$.

The acyclic matching gives rise to an upper bound on the Betti numbers; we present an implicit formula for this bound in Section 23.2. Moreover, we obtain an implicit formula for the reduced Euler characteristic of NEC_n^2.

23.1 An Acyclic Matching

We prove that the quotient complex $\mathsf{EC}_n^2 = 2^{K_n}/\mathsf{NEC}_n^2$ admits an acyclic matching such that the dimension i of each critical graph is in the range

$$\left\lceil \frac{3n - 5}{2} \right\rceil \leq i \leq \left\lfloor \frac{5n - 8}{3} \right\rfloor . \tag{23.1}$$

Our acyclic matching turns out to be straightforward to translate into an acyclic matching on NEC_n^2; compare to the discussion at the beginning of Section 17.1.

While probably a coincidence, it might be worth noting that the size of the interval in (23.1) is exactly the same as the size of the corresponding interval for the matching complex; see Theorem 11.6 and Corollary 11.23. Specifically, the inequalities in (23.1) are equivalent to

$$\left\lceil \frac{n-4}{3} \right\rceil \leq 2n - i - 4 \leq \left\lfloor \frac{n-3}{2} \right\rfloor.$$

To prove (23.1), first note that $\mathsf{EC}_1^2 = \{\emptyset\}$ (by convention) and that $\mathsf{EC}_2^2 = \emptyset$. Moreover, by Corollary 19.5, $\mathsf{EC}_3^2 = \mathsf{C}_3^2$ admits an acyclic matching with one critical face of dimension 2, whereas $\mathsf{EC}_4^2 = \mathsf{C}_4^2$ admits an acyclic matching with two critical faces of dimension 4.

Assume that $n \geq 5$. We proceed in steps as follows.

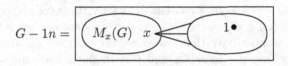

$$G - 1n = \boxed{\quad M_x(G) \quad x \quad\quad\quad 1 \bullet \quad}$$

Fig. 23.1. The set $M_x(G)$ introduced in Step 1; the two bounded regions represent connected induced subgraphs.

Step 1: *Defining the set $M_x(G)$ and partitioning EC_n^2 into subfamilies $\mathsf{EC}_{n,A}^2$.*

Our starting point is the edge between 1 and n. Before proceeding, we divide EC_n^2 into smaller families as follows. For any $x \in [2, n]$ and $G \in \mathsf{EC}_n^2$, let $M_x(G)$ be the set of vertices v such that every path from 1 to v in $G - 1n$ contains the vertex x; see Figure 23.1. We obtain $M_x(G)$ from $[n]$ by removing the vertices in the connected component containing 1 in $(G - 1n)([n] \setminus \{x\})$. Note that $x \in M_x(G)$. For each $A \subset [2, n]$ such that $n \in A$, let $\mathsf{EC}_{n,A}^2$ be the family of graphs $G \in \mathsf{EC}_n^2$ such that $A = M_n(G)$. Clearly, the families $\mathsf{EC}_{n,A}^2$ satisfy the Cluster Lemma 4.2; the set $M_n(G)$ can only increase when we remove edges from G.

Step 2: *Matching with the edge $1n$ to obtain the family $\Lambda_n(A)$.*

In $\mathsf{EC}_{n,A}^2$, match with $1n$ whenever possible, meaning that we pair $G - 1n$ and $G + 1n$ whenever $G - 1n$ is contained in $\mathsf{EC}_{n,A}^2$. Let $\Lambda_n(A)$ be the family of critical graphs with respect to this matching. By Lemma 4.1, any acyclic matching on $\Lambda_n(A)$ together with the matching just defined yields an acyclic matching on $\mathsf{EC}_{n,A}^2$. Let Λ_n be the union of all $\Lambda_n(A)$.

Step 3: *Defining the set $X(G)$ and the edge $e(G) = a(G)b(G)$.*

For any graph G, let $X(G)$ be the set of edges $e \in G$ such that $G - e$ is

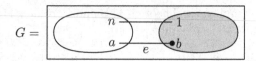

$$G =$$

Fig. 23.2. The vertices $a = a(G)$ and $b = b(G)$ and the edge $e = e(G)$ in Step 3; the shaded region represents a 2-edge-connected graph, whereas the white region represents a connected graph such that any edge separating it separates a from n.

disconnected. Let $G \in \Lambda_n$. Since G is 2-edge-connected, the set $X(G)$ is empty. This implies that any $e \in X(G - 1n)$ separates 1 and n in G. For any $G \in \Lambda_n(A)$ and $e_1, e_n \in X(G - 1n)$, since each of e_1 and e_n separates $G - 1n$, the subgraph $G - \{1n, e_1, e_n\}$ consists of exactly three connected components C_0, C_1, C_n, where $1 \in C_1$ and $n \in C_n$. Moreover, since neither of the two edges separates G, we must have that one edge, say e_1, joins C_0 and C_1, whereas the other edge, say e_n, joins C_0 and C_n. In particular, the component in $G - \{1n, e_1\}$ containing 1, which is C_1, is a proper subset of the component in $G - \{1n, e_n\}$ containing 1, which is $C_0 \cup C_1$. As a consequence, there is a unique edge $e = e(G)$ such that the component in $G - \{1n, e(G)\}$ containing 1 is minimal. Write $e(G) = a(G)b(G)$, where $a(G)$ belongs to the same component as n and $b(G)$ belongs to the same component as 1 in $G - \{1n, e(G)\}$. Note that we might have $a = n$ or $b = 1$ (but not both). See Figure 23.2.

Step 4: *Partitioning $\Lambda_n(A)$ into subfamilies $\Lambda_n(A, M, a, b)$, $\Pi_n(A, M, x)$, and $\Pi'_n(A, M, x)$.*

We have the following possibilities for a graph G in Λ_n:

$$a(G) \neq n : \begin{cases} G - a(G)n \in \Lambda_n. \text{ This is case A.} \\ \\ G - a(G)n \notin \Lambda_n : \begin{cases} b(G) \neq 1. \text{ This is case B1.} \\ \\ b(G) = 1. \text{ This is case C.} \end{cases} \end{cases}$$

$a(G) = n$. This is case B2.

Let $A \subset [2, n]$ be such that $n \in A$. We divide $\Lambda_n(A)$ into subfamilies according to the three cases A, B = B1+B2, and C:

Step 4A: *Defining the subfamilies $\Lambda_n(A, M, a, b)$.*

We consider all case A graphs. Let $M \subset [n]$ be a set containing A but not 1. For any $e = ab$ such that $a \in M \setminus A$ and $b \in [n-1] \setminus M$, define

$$\Lambda_n(A, M, a, b) = \{G : G - an \in \Lambda_n(A), a = a(G), b = b(G), M = M_a(G)\}.$$

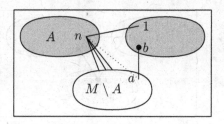

Fig. 23.3. A graph in $\Lambda_n(A, M, a, b)$ in Step 4A. The shaded regions represent 2-edge-connected graphs, whereas the white region represents a connected graph such that any edge separating it separates a from all vertices adjacent to n in G. There is at least one edge from n to $M \setminus A$ in $G - an$ (in the figure, there are three such edges). G may or may not contain the edge an.

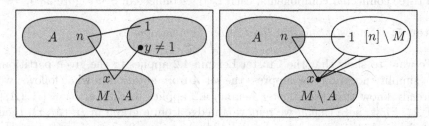

Fig. 23.4. The two kinds of graphs, B1 on the left and B2 on the right, in $\Pi_n(A, M, x)$ in Step 4B; the shaded regions represent 2-edge-connected graphs. In a graph G of the kind illustrated on the right, there are at least two edges from x to $[n] \setminus M$, and the induced subgraph $G([n] \setminus A)$ is 2-edge-connected. Note that G may contain the edge $1x$.

See Figure 23.3. Note that $a \neq n$, because $n \in A$.

Step 4B: *Defining the subfamilies $\Pi_n(A, M, x)$.*

We consider all case B graphs. Let M be a set as in Step 4A. For any $x \in M \setminus A$, define

$$\Pi_n(A, M, x) = \Lambda_n(A) \cap (\{G : x = a(G), xn = e(G + 1x), M = M_x(G)\}$$
$$\cup \{G : n = a(G), x = b(G), M = M_x(G)\}).$$

See Figure 23.4. Note that the condition $xn = e(G + 1x)$ means exactly that $G - xn \notin \Lambda_n$ and $1 \neq b(G)$.

Step 4C: *Defining the subfamilies $\Pi'_n(A, M, x)$.*

We consider all case C graphs. For each M containing A but not 1 and each $x \in M \setminus A$, define $\Pi'_n(A, M, x)$ as the family of graphs $G \in \Lambda_n(A)$ such that

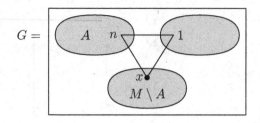

Fig. 23.5. A graph in $\Pi'_n(A, M, x)$ in Step 4C; the shaded regions represent 2-edge-connected graphs.

$M_x(G) = M$, such that $1x, xn, 1n \in G$, and such that $G - \{1x, xn, 1n\}$ consists of three connected components, each 2-edge-connected. See Figure 23.5.

Step 5: *Demonstrating that the Cluster Lemma applies.*

We want to show that the Cluster Lemma 4.2 applies to the given partition. To simplify notation, we suppress the set A from notation in what follows; we already know that the Cluster Lemma 4.2 applies to the partition $\{\Lambda_n(A)\}$ of Λ_n. First, note that if we remove an edge from a graph in $\Pi'_n(M, x)$, then we either obtain another graph in $\Pi'_n(M, x)$ or a not 2-edge-connected graph. Thus we may concentrate on $\Lambda_n(M, a, b)$ and $\Pi_n(M, x)$. For $M \subsetneq M'$, we consider each of $\Lambda_n(M, a, b)$ and $\Pi_n(M, x)$ as being above each of $\Lambda_n(M', a', b')$ and $\Pi_n(M', x')$ (note the direction) for every a, b, a', b', x, x'. Moreover, we consider $\Lambda_n(M, a, b)$ as being above $\Pi_n(M, x)$ for every M, a, b, x. Let G be a graph and let f be an edge in G. We want to show that G belongs to a class either above or equal to the class of $G - f$.

- $G \in \Lambda_n(M, a, b)$ and $G - f \in \Pi_n(N, x)$. The only possibility is that $x = a$ and that $e((G - f) + 1a) = an$; hence $M = N$. In particular, $\Lambda_n(M, a, b)$ is above $\Pi_n(N, x)$.
- $G \in \Lambda_n(M, a, b)$ and $G - f \in \Lambda_n(M', a', b')$. If $a = a'$, then clearly $b = b'$ and $M = M'$. If $a \neq a'$, then $a' \notin M = M_a(G)$ and hence $M \subsetneq M'$. Thus $\Lambda_n(M, a, b)$ is above or equal to $\Lambda_n(M', a', b')$.
- $G \in \Pi_n(N, x)$ and $G - f \in \Lambda_n(M, a, b)$. By construction, $(G - f) - an \in \Lambda_n$, which implies that $a \neq x$. In particular, $a \notin N = M_x(G)$ and hence $N \subsetneq M$; thus $\Pi_n(N, x)$ is above $\Lambda_n(M, a, b)$.
- $G \in \Pi_n(N, x)$ and $G - f \in \Pi_n(N', x')$. Then we must have $x = x'$ and $N \subseteq N'$; hence $\Pi_n(N, x)$ is equal to $\Pi_n(N', x')$.

As a consequence, the Cluster Lemma 4.2 applies.

Step 6: *Getting rid of $\Lambda_n(A, M, a, b)$ and defining a matching on $\Pi'_n(A, M, x)$.*

First, let us conclude that $\Lambda_n(A, M, a, b)$ admits a perfect acyclic matching.

Namely, we may pair $G - an$ and $G + an$; $e(G + an) = e(G)$ and $M_a(G + an) = M_a(G)$.

Next, consider $\Pi'_n(A, M, x)$. With $A_1 = [n] \setminus M$, $A_x = M \setminus A$, and $A_n = A$, $\Pi'_n(A, M, x)$ equals

$$\{\{1x, 1n, xn\}\} * \mathsf{EC}^2_{A_1} * \mathsf{EC}^2_{A_x} * \mathsf{EC}^2_{A_n},$$

where $\mathsf{EC}^2_{A_i}$ is the quotient complex of 2-edge-connected graphs on the set A_i. By induction on n, for $k < n$, EC^2_k admits an acyclic matching such that all unmatched graphs G satisfy

$$\left\lceil \frac{3k - 3}{2} \right\rceil \le |G| \le \left\lfloor \frac{5k - 5}{3} \right\rfloor. \tag{23.2}$$

By Theorem 5.29, this implies that $\Pi'_n(A, M, x)$ admits an acyclic matching with all unmatched graphs G satisfying

$$|G| \ge 3 + \frac{3|A_1| - 3}{2} + \frac{3|A_x| - 3}{2} + \frac{3|A_n| - 3}{2} = \frac{3n - 3}{2};$$

$$|G| \le 3 + \frac{5|A_1| - 5}{3} + \frac{5|A_x| - 5}{3} + \frac{5|A_n| - 5}{3} = \frac{5n - 6}{2} < \frac{5n - 5}{2}.$$

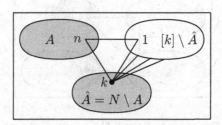

Fig. 23.6. A graph in $\mathcal{Q}_n(A, N, k)$ in Step 7; the shaded regions represent 2-edge-connected graphs. The induced subgraph $H = G([k])$ is 2-edge-connected, whereas $H - 1k$ is not. Moreover, $a(H) \ne k$.

Step 7: *Reducing $\Pi_n(A, N, k)$ to the family $\mathcal{Q}_n(A, N, k)$ and partitioning into the families $\Lambda_k(\hat{A}, \hat{M}, a, b)$, $\hat{\Pi}_k(\hat{A}, \hat{M}, y)$, and $\Pi'_k(\hat{A}, \hat{M}, y)$.*

It remains to consider $\Pi_n(A, N, x)$. For simplicity, assume that $A = [k + 1, n]$ and $x = k$. In $\Pi_n(A, N, k)$, match with $1k$ whenever possible and let $\mathcal{Q}_n(A, N, k)$ be the family of unmatched graphs. It is clear that we can always add the edge $1k$ without ending up outside $\Pi_n(A, N, k)$; see Figure 23.4. When removing the edge $1k$ from a graph G in $\Pi_n(A, N, k)$, we will end up inside $\Pi_n(A, N, k)$ if and only if either of the following is true:

- $G - 1k$ is a case B1 graph. This is equivalent to the induced subgraph $G([k])$ being a case B2 graph.
- $G - 1k$ is a case B2 graph. This is equivalent to $G([k]) - 1k$ being 2-edge-connected.

The conclusion is that we end up outside $\Pi_n(A, N, k)$ if and only if $G([k]) - 1k$ is a case A, case B1, or case C graph. Write $\hat{A} = N \setminus A$. Let $\hat{\Lambda}_k(\hat{A})$ be the subfamily of $\Lambda_k(\hat{A})$ obtained by removing all case B2 graphs. We obtain that $\mathcal{Q}_n(A, N, k)$ equals

$$\{\{1n, kn\}\} * \mathsf{EC}_A^2 * \hat{\Lambda}_k(\hat{A}).$$

The situation is illustrated in Figure 23.6. By induction on k, we have an acyclic matching on EC_A^2 such that the unmatched graphs G satisfy (23.2) for $k = |A|$. For $\hat{\Lambda}_k(\hat{A})$, take the same partition as in Step 4, the obvious exception being that we need to replace $\Pi_k(\hat{A}, \hat{M}, y)$ with

$$\hat{\Pi}_k(\hat{A}, \hat{M}, y) = \hat{\Lambda}_k(\hat{A}) \cap \{G : y = a(G), yk = e(G + 1y), \hat{M} = M_y(G)\}.$$

The other families $\Lambda_k(\hat{A}, \hat{M}, a, b)$ and $\Pi'_k(\hat{A}, \hat{M}, y)$ are defined as before. Mimicking the procedure in Step 5, we obtain that this partition of $\hat{\Lambda}_k(\hat{A})$ satisfies the Cluster Lemma 4.2.

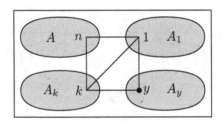

Fig. 23.7. A graph G in $\mathcal{Q}_n(A, A_k \cup A, k)$ in Step 8 such that $G([k])$ belongs to $\Pi'_k(A_k, A_y, y)$; the shaded regions represent 2-edge-connected graphs.

Step 8: *Getting rid of $\Lambda_k(\hat{A}, \hat{M}, a, b)$ and defining a matching on $\Pi'_k(\hat{A}, \hat{M}, y)$.*

As in Step 6, we obtain that $\Lambda_k(\hat{A}, \hat{M}, a, b)$ admits a perfect acyclic matching and that $\Pi'_k(\hat{A}, \hat{M}, y)$ equals

$$\{\{1y, 1k, yk\}\} * \mathsf{EC}_{A_1}^2 * \mathsf{EC}_{A_y}^2 * \mathsf{EC}_{A_k}^2,$$

where $A_1 = [k] \setminus \hat{M}$, $A_y = \hat{M} \setminus \hat{A}$, and $A_k = \hat{A}$. As a consequence, $\Sigma = \{\{1n, kn\}\} * \mathsf{EC}_A^2 * \Pi'_k(\hat{A}, \hat{M}, x)$ equals

$$\{\{1y, 1k, ky, 1n, kn\}\} * \mathsf{EC}_A^2 * \mathsf{EC}_{A_1}^2 * \mathsf{EC}_{A_y}^2 * \mathsf{EC}_{A_k}^2.$$

See Figure 23.7 for an illustration. Applying Theorem 5.29, (23.2), and induction on n, this implies that Σ admits an acyclic matching such that all unmatched graphs G satisfy

$$|G| \geq 5 + \frac{3|A| - 3}{2} + \frac{3|A_1| - 3}{2} + \frac{3|A_y| - 3}{2} + \frac{3|A_k| - 3}{2} > \frac{3n - 3}{2};$$

$$|G| \leq 5 + \frac{5|A| - 5}{5} + \frac{5|A_1| - 5}{3} + \frac{5|A_x| - 5}{3} + \frac{5|A_k| - 5}{3} = \frac{5n - 5}{3}.$$

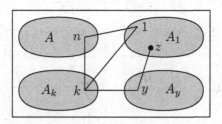

Fig. 23.8. A graph G in $\mathcal{Q}_n(A, A_k \cup A, k)$ in Step 9 such that $G([k])$ belongs to $\hat{\Pi}_k(A_k, A_y, y)$; the shaded regions represent 2-edge-connected graphs.

Step 9: *Defining a matching on $\hat{\Pi}_k(\hat{A}, \hat{M}, y)$.*

The one remaining family is $\hat{\Pi}_k(\hat{A}, \hat{M}, y)$. A graph G belongs to this family if and only if G contains the edges $1k, ky, yz$ for some $z \neq 1$ and has the property that $G - \{1k, ky, yz\}$ contains three connected components, each 2-edge-connected, one containing $\{1, z\}$, one containing k, and one containing y. Let $A_1 = [k] \setminus \hat{M}$, $A_y = \hat{M} \setminus \hat{A}$, and $A_k = \hat{A}$. Using exactly the same approach as in Step 8, we conclude that $\{\{1k, ky, yz, 1n, kn\}\} * \mathsf{EC}_A^2 * \hat{\Pi}_k(\hat{A}, \hat{M}, y)$ equals

$$\{\{1k, ky, yz, 1n, kn\}\} * \mathsf{EC}_A^2 * \mathsf{EC}_{A_1}^2 * \mathsf{EC}_{A_y}^2 * \mathsf{EC}_{A_k}^2$$

and hence admits an acyclic matching satisfying the desired bounds. See Figure 23.8 for an illustration.

Conclusion.

We have established the following result:

Theorem 23.1. *For $n \geq 3$, NEC_n^2 is $\lceil \frac{3(n-3)}{2} \rceil$-connected with the property that the homology group $\tilde{H}_i(\mathrm{NEC}_n^2, \mathbb{Z})$ is zero unless*

$$\left\lceil \frac{3n - 7}{2} \right\rceil \leq i \leq \left\lfloor \frac{5n - 11}{3} \right\rfloor. \quad \square$$

23.2 Enumerative Properties of the Given Matching

Throughout this section, generating functions are of the form $\sum_n r_n x^n/(n-1)!$; this is to obtain as simple formulas as possible. The acyclic matching in the previous section induces an upper bound on the Betti numbers:

Theorem 23.2. *For $n \geq 1$, let $\{f_n(t) = \sum_{i \geq 0} f_{n,i} t^i\}$ be the unique sequence of polynomials with the property that $F = F(t, x) = \sum_{n \geq 1} f_n(t) x^n/(n-1)!$ satisfies*

$$x \cdot \frac{\partial F}{\partial x} = \frac{F(1 + t^3 F^2)}{1 - t^5 F^3} \qquad (23.3)$$

and $f_1(t) = 1$. Then EC_n^2 admits an acyclic matching with $f_{n,i}$ unmatched sets of size i. In particular, $H = H(x) = F(-1, x) = -\sum_{n \geq 1} \tilde{\chi}(\mathsf{EC}_n^2) x^n/(n-1)!$ satisfies

$$xH' = \frac{H - H^2}{1 - H + H^2} \iff H = \ln \frac{H}{x(1 - H)}.$$

As a consequence,

$$\tilde{\chi}(\mathsf{EC}_n^2) = -\sum_{k=0}^{n-1}(-1)^k \binom{n}{k} \frac{n^{n-k-3} \cdot n!}{(n-k-1)!}.$$

Remark. Note that $\tilde{\chi}(\mathsf{EC}_n^2) = -\tilde{\chi}(\mathsf{NEC}_n^2)$ for $n \geq 2$.

Proof. From the proof of Theorem 23.1, we deduce that there are three types of critical graphs with respect to the given matching:

- With notation as in Step 6, we obtain graphs of the form

$$\{1x, 1n, xn\} \cup G_1 \cup G_x \cup G_n,$$

where G_i is a critical graph with respect to an acyclic matching on $\mathsf{EC}_{A_i}^2$. We may choose x in $(n-2)$ ways and A_1, A_x, A_n of size a_1, a_x, a_n (with $\sum a_i = n$) in $\dfrac{(n-3)!}{(a_1 - 1)!(a_x - 1)!(a_n - 1)!}$ ways.

- With notation as in Step 8, we obtain graphs of the form

$$\{1y, 1x, xy, 1n, xn\} \cup G_1 \cup G_x \cup G_y \cup G_n,$$

where G_i is a critical graph with respect to an acyclic matching on $\mathsf{EC}_{A_i}^2$. Note that we fixed $x = k$ and $A_n = [k+1, n]$ in Step 7, but that was only to simplify notation. We may choose x and y in $(n-2)(n-3)$ ways and A_1, A_x, A_y, A_n of size a_1, a_x, a_y, a_n (with $\sum a_i = n$) in

$$\frac{(n-4)!}{(a_1 - 1)!(a_x - 1)!(a_y - 1)!(a_n - 1)!}$$

ways.

- With notation as in Step 9, we obtain graphs of the form

$$\{1x, xy, yz, 1n, xn\} \cup G_1 \cup G_x \cup G_y \cup G_n,$$

where G_i is a critical graph with respect to an acyclic matching on $\mathsf{EC}^2_{A_i}$. Again we fixed $x = k$ and $A_n = [k+1, n]$, which was only for simplicity. We may choose x, y, and z in $(n-2)(n-3)(n-4)$ ways and A_1, A_x, A_y, A_n of size a_1, a_x, a_y, a_n (with $\sum a_i = n$) in $\dfrac{(n-5)!}{(a_1-2)!(a_x-1)!(a_y-1)!(a_n-1)!}$ ways.

As a conclusion, with $f_n = f_n(t)$ and $n \geq 3$, we obtain that

$$\frac{f_n}{(n-2)!} = \sum_{\sum a_i = n} t^3 \prod_{i \in \{1, x, n\}} \frac{f_{a_i}}{(a_i - 1)!} + \sum_{\sum a_i = n} t^5 \prod_{i \in \{1, x, y, n\}} \frac{f_{a_i}}{(a_i - 1)!}$$
$$+ \sum_{\sum a_i = n} t^5 \frac{f_{a_1}}{(a_1 - 2)!} \cdot \prod_{i \in \{x, y, n\}} \frac{f_{a_i}}{(a_i - 1)!}$$

and hence that

$$\partial_x \left(\frac{F}{x} \right) = \frac{t^3 F^3}{x^2} + \frac{t^5 F^4}{x^2} + t^5 F^3 \partial_x \left(\frac{F}{x} \right) \iff$$
$$x \partial_x(F) - F = t^3 F^3 + t^5 F^4 + x t^5 F^3 \partial_x(F) - t^5 F^4,$$

and we are done.

To compute the Euler characteristic of EC^2_n, we note that the inverse of $H(x)$ equals $G(y) = ye^{-y}/(1-y)$. By the Lagrange inversion formula (see Stanley [134]), the coefficient of x^n in $H(x)$ is equal to the coefficient of y^{n-1} in $(y/G(y))^n/n = (1-y)^n e^{yn}/n$, which is

$$\sum_{k=0}^{n-1} (-1)^k \binom{n}{k} \cdot \frac{n^{n-2-k}}{(n-1-k)!}.$$

Since the coefficient of x^n in $H(x)$ is also equal to $-\tilde{\chi}(\mathsf{EC}^2_n)/(n-1)!$, the final claim in the theorem follows. $\square$

As a side note, let us mention that the coefficient of $y^n/n!$ in $G(y)$ is the number of permutations on $n+1$ elements with exactly one fixed point; see sequence A000240 in Sloane's Encyclopedia [127].

For $k \geq 1$, note that there are no critical graphs in EC^2_{2k-1} with fewer than $3k - 3$ edges.

Corollary 23.3. *For $k \geq 1$, $f_{2k-1, 3k-3}$ is equal to $((2k-3)!!)^2$; $f_{1,0} = 1$.*

Proof. To obtain this identity, ignore the $t^5 F^3$ term in the equation (23.3); this yields the equation

$$\partial_x \left(\frac{F}{x} \right) = xt^3 \cdot \left(\frac{F}{x} \right)^3 \iff \frac{F}{x} = \frac{1}{\sqrt{1 - t^3 x^2}}.$$

In particular,

$$\sum_{k \geq 0} f_{2k+1,3k-3} \frac{x^{2k}}{(2k)!} = \frac{1}{\sqrt{1 - x^2}} = \sum_{k \geq 0} ((2k-1)!!)^2 \frac{x^{2k}}{(2k)!};$$

hence we are done. □

Let us also examine $f_{2k,3k-1}$ for $k \geq 1$; there are no critical graphs in EC_{2k}^2 with fewer than $3k - 1$ edges.

Theorem 23.4. *The integers $f_{2k,3k-1}$ satisfy the identity*

$$\frac{H(x)}{x} := \sum_{k \geq 1} f_{2k,3k-1} \frac{x^{2k-1}}{(2k-1)!} = \frac{x - \arcsin(x)\sqrt{1 - x^2}}{(1 - x^2)^2}.$$

Proof. Let $k \geq 2$ and write $n = 2k$. As in the proof of Theorem 23.2, there are three cases:

- In the first case, we have graphs of the form

$$G = \{1x, 1n, xn\} \cup G_1 \cup G_x \cup G_n$$

  and the vertex set of G_i is A_i. If A_1, A_x, and A_n are all of even size, then the number of edges in G is at least $3 + \sum_i \frac{3|A_i| - 2}{2} = 3k$. Thus we must have that two of the vertex sets have odd size; in this case it is possible to obtain exactly $3k - 1$ edges.
- In the second and third cases, we obtain graphs of the form

$$G = E_0 \cup G_1 \cup G_x \cup G_y \cup G_n,$$

  where E_0 is an edge set of size five. This time, all four graphs G_i must have an vertex set of odd size; otherwise, G will end up with more than $3k - 1$ edges.

Write $f_n = f_{n, \lceil \frac{3n-3}{2} \rceil}$. Analogously to the proof of Theorem 23.2, one may conclude that

$$\frac{f_{2k}}{(2k-2)!} = \sum_{\sum a_i = n} \prod_{i \in \{1,x,n\}} \frac{f_{a_i}}{(a_i - 1)!} + \sum_{\sum a_i = n} \prod_{i \in \{1,x,y,n\}} \frac{f_{a_i}}{(a_i - 1)!}$$

$$+ \sum_{\sum a_i = n} \frac{f_{a_1}}{(a_1 - 2)!} \cdot \prod_{i \in \{x,y,n\}} \frac{f_{a_i}}{(a_i - 1)!}.$$

In the first sum, we require that exactly two a_i be odd, whereas in the other sums, all a_i must be odd. Hence with F defined as in the proof of Corollary 23.3, fixing t to one, we obtain that

$$\frac{d}{dx}\left(\frac{H}{x}\right) = \frac{3F^2H}{x^2} + \frac{F^4}{x^2} + F^3\frac{d}{dx}\left(\frac{F}{x}\right) \iff$$
$$xH' = (1 + 3F^2)H + xF^3F'.$$

Substituting $F = x/\sqrt{1 - x^2}$, we derive the equation

$$xH' = \frac{2x^2 + 1}{1 - x^2}H + \frac{x^4}{(1 - x^2)^3} \iff$$
$$\frac{d}{dx}\left(\frac{(1 - x^2)^{3/2}}{x}H\right) = \frac{x^2}{(1 - x^2)^{3/2}} \iff$$
$$\frac{(1 - x^2)^{3/2}}{x}H = \frac{x}{(1 - x^2)^{1/2}} - \arcsin(x),$$

which concludes the proof. $\Box$

23.3 Bottom Nonvanishing Homology Group

We prove that the lower bound in Theorem 23.1 is sharp for odd n. Specifically, we demonstrate that the homology group $\tilde{H}_{3k-5}(\mathsf{NEC}^2_{2k-1}, \mathbb{Z})$ is nonzero for $k \geq 2$.

Recall that NFC_{2k-1} is the complex of not factor-critical graphs on $2k - 1$ vertices; a graph $G = (V, E)$ is factor-critical if $G(V \setminus \{v\})$ contains a perfect matching for each $v \in V$. Let FC_{2k-1} be the quotient complex of factor-critical graphs; $\mathsf{FC}_{2k-1} = 2^{K_{2k-1}}/\mathsf{NFC}_{2k-1}$. This complex is of importance in the analysis of the complex $\mathsf{NM}_{n,k}$ of graphs that do not contain any k-matching; see Chapter 24.

Theorem 23.5 (Linusson et al. [95]). *For $k \geq 2$, NFC_{2k-1} is homotopy equivalent to a wedge of $((2k - 3)!!)^2$ spheres of dimension $3k - 5$. As a consequence, FC_{2k-1} is homotopy equivalent to a wedge of $((2k - 3)!!)^2$ spheres of dimension $3k - 4$. $\Box$*

Linusson et al. [95] defined a *tree of triangles* to be a graph G on a linearly ordered vertex set $V = \{v_1, \ldots, v_n\}$ with $v_i < v_{i+1}$ satisfying either of the following properties:

- G consists of a single vertex.
- The edge v_1v_2 belongs to G, and there is a unique vertex v_i such that v_1v_i and v_2v_i belong to G. Moreover, the graph obtained from G by removing the three edges v_1v_2, v_1v_i, and v_2v_i has three connected components, each of which is a tree of triangles.

Any tree of triangles on the set $[2k - 1]$ is easily seen to be a minimal nonface of NFC_{2k-1} [95].

Theorem 23.6 (Linusson et al. [95]). *For $k \geq 2$, the set*

$$\mathcal{C}_k = \{[G] : G \text{ is a tree of triangles on } [2k-1]\}$$

is a basis for $\tilde{H}_{3k-4}(\mathsf{FC}_{2k-1}; \mathbb{Z})$ and the set

$$\mathcal{C}'_k = \{\partial([G]) : G \text{ is a tree of triangles on } [2k-1]\}$$

is a basis for $\tilde{H}_{3k-5}(\mathsf{NFC}_{2k-1}; \mathbb{Z})$. □

By the exact sequence for the pair $(2^{K_{2k-1}}, \mathsf{NFC}_{2k-1})$ and the remark before the theorem, the second statement is an immediate consequence of the first.

Theorem 23.7. *For $k \geq 2$, the sets $\mathcal{C}_k$ and $\mathcal{C}'_k$ in Theorem 23.6 form bases for the groups $\tilde{H}_{3k-4}(\mathsf{EC}^2_{2k-1}; \mathbb{Z})$ and $\tilde{H}_{3k-5}(\mathsf{NEC}^2_{2k-1}; \mathbb{Z})$, respectively. As a consequence, $\tilde{H}_{3k-4}(\mathsf{EC}^2_{2k-1}, \mathbb{Z})$ and $\tilde{H}_{3k-5}(\mathsf{NEC}^2_{2k-1}, \mathbb{Z})$ are both free of rank $((2k-3)!!)^2$.*

Proof. One easily adapts our acyclic matching on EC^2_{2k-1} such that a graph with $3k-3$ edges is critical if and only if the graph is a tree of triangles on the set $[2k-1]$. Specifically, switch the roles of the vertices 2 and n in our construction and proceed recursively in the natural manner on a graph as in Figure 23.5. Moreover, if H is a subgraph obtained from a tree of triangles G by removing an edge e, then H is not 2-edge-connected; each of f and g separates H, where $\{e, f, g\}$ is the unique triangle in G containing e. As a consequence, $\mathcal{C}_k$ generates $\tilde{H}_{3k-4}(\mathsf{EC}^2_{2k-1}; \mathbb{Z})$ and $\mathcal{C}'_k$ generates $\tilde{H}_{3k-5}(\mathsf{NEC}^2_{2k-1}; \mathbb{Z})$. By Theorems 23.5 and 23.6, we are done if we can prove that $\mathcal{C}'_k$ is a basis for $\tilde{H}_{3k-5}(\mathsf{NEC}^2_{2k-1}; \mathbb{Z})$.

Now, NFC_{2k-1} contains NEC^2_{2k-1}. Namely, suppose that G is not 2-edge-connected. If G is not connected, then let v be a vertex such that some connected component not containing v is odd (i.e., contains an odd number of vertices). Clearly, $H = G([2k-1] \setminus \{v\})$ does not contain a perfect matching, because H contains an odd component. If G is connected, let $e = ab$ be an edge separating G into two components; assume that the component containing a is odd. Then $H = G([2k-1] \setminus \{b\})$ does not contain a perfect matching, again because H contains an odd component.

We conclude that $\mathcal{C}'_k$ constitutes an independent set in $\tilde{H}_{3k-5}(\mathsf{NEC}^2_{2k-1}, \mathbb{Z})$ and hence forms a basis for this group. Namely, since NEC^2_{2k-1} is a subcomplex of NFC_{2k-1}, any boundary in the chain complex of NEC^2_{2k-1} would also be a boundary in the chain complex of NFC_{2k-1}. □

Corollary 23.8. *For $k \geq 2$, the shifted connectivity degree of NEC^2_{2k-1} equals $3k-5$.* □

The analogous problem for even n remains unsettled:

Conjecture 23.9. *For $k \geq 1$, $\tilde{H}_{3k-3}(\mathsf{NEC}^2_{2k}, \mathbb{Z})$ is free of rank $f_{2k,3k-1}$, where $f_{2k,3k-1}$ satisfies the identity in Theorem 23.4.*

The acyclic matching in Theorem 23.1 is optimal for $n \leq 11$, except possibly for $n = 10$. Namely, for $n = 8$ and $n \leq 6$, all critical faces have the same dimension. Moreover, for $n \in \{7, 9, 11\}$, all critical faces are concentrated in two dimensions. Since the matching is optimal in the lower of these two dimensions by Corollary 23.3 and Theorem 23.7, the matching must be optimal in both dimensions.

Problem 23.10. Is the acyclic matching in Section 23.1 optimal for $n = 10$ and $n \geq 12$? In particular, is NEC_n^2 semi-collapsible over $\mathbb{Z}$?

As Figure 23.9 exemplifies, not all critical graphs G and H have the property that $G \not\longmapsto H$. In particular, we cannot obtain a solution to Problem 23.10 simply by referring to Corollary 4.13.

Let $k, p \geq 1$ and let $\Pi_{kp+1}^{1 \bmod p}$ be the subposet of Π_{kp+1} consisting of all nontrivial proper partitions in which the size of each part is congruent to 1 modulo p. Björner [7] proved that $\Delta(\Pi_{kp+1}^{1 \bmod p})$ is shellable of dimension $k - 2$; hence the complex has homology only in its top dimension $k - 2$.

Shareshian and Wachs [121] recently proved that $\tilde{H}_{k-2}(\Delta(\Pi_{2k+1}^{1 \bmod 2}); \mathbb{Z})$ is isomorphic to each of $\tilde{H}_{3k-1}(\mathsf{FC}_{2k+1}; \mathbb{Z})$ and $\tilde{H}_{3k-2}(\mathsf{NEC}_{2k+1}^2; \mathbb{Z})$ as $\mathfrak{S}_{2k+1}$-modules. Based on this observation, they conjectured that the two groups $\tilde{H}_{k-2}(\Delta(\Pi_{kp+1}^{1 \bmod p}); \mathbb{Z})$ and $\tilde{H}_{k\binom{p+1}{2}-2}(\mathsf{NEC}_{kp+1}^p; \mathbb{Z})$ are isomorphic for all $p, k \geq 1$. The conjecture is known to be true for $p = 1$ and $p = 2$; in the former case, we obtain the complex NC_{k+1} of disconnected graphs and the proper part of the full partition lattice Π_{k+1} (see Section 18.1). The conjecture is also true in the trivial case $k = 1$. However, for $k = 2$ and $p = 3$, we obtain NEC_7^3, and via a computer calculation we have been able to prove that this complex is homotopy equivalent to a wedge of 310 spheres of dimension 10. Clearly, $\Delta(\Pi_7^{1 \bmod 3})$ is a discrete point set of size 35, which implies that the homology in the relevant degrees is not the same for the two complexes. Yet, there is still some hope that the following conjecture holds:

Conjecture 23.11. *For $p \geq 1$ and $n \geq p + 1$, the shifted connectivity degree of NEC_n^p equals $\left\lceil \frac{(n-1)(p+1)}{2} - 2 \right\rceil$. Moreover, for each $p, k \geq 1$, there is an embedding from $\tilde{H}_{k-2}(\Delta(\Pi_{kp+1}^{1 \bmod p}); \mathbb{Z})$ to $\tilde{H}_{k\binom{p+1}{2}-2}(\mathsf{NEC}_{kp+1}^p; \mathbb{Z})$.*

23.4 Top Nonvanishing Homology Group

Finally, we show that the upper bound in Theorem 23.1 is sharp.

Lemma 23.12. *Let G be a critical graph with respect to the given acyclic matching on EC_n^2 and let $\hat{u}_G$ be the corresponding group element in the resulting chain complex U in Theorem 4.16; see Corollary 4.17. If all 2-connected components in G contain either three or four vertices, then $\partial(\hat{u}_G) = 0$.*

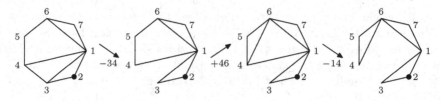

Fig. 23.9. A directed path between two critical graphs in the digraph corresponding to the given matching. Removing 14 in the first step and 34 in the third step, we obtain another directed path that "cancels out" this path; see Forman [49] for information about how paths cancel out.

Proof. All 2-connected components having vertex size three or four implies that G is either as in Figure 23.5 or as in Figure 23.7; we thus have two cases. We use induction on n to prove the desired result.

In the first case, the component containing the vertices 1 and n is a triangle; let x be the third vertex in this triangle. G is the union of this triangle and three 2-edge-connected graphs G_1, G_x, and G_n. By induction on n, each G_i corresponds to a group element $\hat{u}_i$ such that the boundary of $\hat{u}_i$ vanishes in the chain complex of $\mathsf{EC}^2_{V(G_i)}$. Moreover, the boundary of the element $\hat{u}_0 = [1x] \wedge [1n] \wedge [xn]$ in the chain complex of $\mathsf{EC}^2_{\{1,x,n\}}$ is zero. It follows that the boundary of the element $\hat{u} = \hat{u}_0 \wedge \hat{u}_1 \wedge \hat{u}_x \wedge \hat{u}_n$ in the chain complex of EC^2_n is zero. Using an induction argument, one easily concludes that all graphs appearing in the sum $\hat{u}$ are matched with smaller graphs. Thus $\hat{u} = \hat{u}_G$.

In the second case, the edge set of the component containing the vertices 1 and n is of the form $\{1y, 1k, ky, 1n, kn\}$ for some vertices k and y. G is the union of this component and four 2-edge-connected graphs G_1, G_k, G_y, and G_n. As above, each G_i corresponds to a group element $\hat{u}_i$ such that $\partial(\hat{u}_i) = 0$. Moreover, the boundary of the element $\hat{u}_0 = [1y] \wedge [1n] \wedge [ky] \wedge [kn] \wedge ([1k] - [yn])$ in the chain complex of $\mathsf{EC}^2_{\{1,k,y,n\}}$ is zero. As in the first case, we obtain an element $\hat{u}$ in the chain complex of EC^2_n such that the boundary vanishes. Now, consider a graph G' appearing in the sum $\hat{u}$ such that $yn \in G'$ and $1k \notin G'$. Then $\{1y, yn, ky, 1n, kn\}$ forms a 2-connected component in G', which implies that G' belongs to $\Lambda_n(A, M, y, 1)$ for some sets A and M. As a consequence, G' is matched with the smaller graph $G' - yn$. An induction argument yields that all noncritical graphs G' in $\hat{u}$ such that $1k \in G'$ and $yn \notin G'$ are also matched with smaller graphs. Again, we obtain that $\hat{u} = \hat{u}_G$. $\square$

Theorem 23.13. *For $n \geq 3$, the complex NEC^2_n has nonvanishing homology in dimension $\lfloor \frac{5n-11}{3} \rfloor$ but no homology above this dimension.*

Proof. By Lemma 23.12, we need only prove that there are critical graphs in EC^2_n with the maximum number $\lfloor \frac{5n-5}{3} \rfloor$ of edges such that all 2-connected components contain three or four vertices. This is clear if $n = 3$ or $n = 4$. Assume that $n \geq 5$ and write $n = 3k + r$, where $k \geq 0$ and $r \in \{5, 6, 7\}$.

- If $r = 5$, then take the wedge of k copies of the graph G_4 on four vertices with edge set $\{12, 13, 14, 23, 24\}$ and two copies of the triangle graph G_3 with edge set $\{12, 13, 23\}$.
- If $r = 6$, then take the wedge of $k + 1$ copies of G_4 and one copy of G_3.
- If $r = 7$, then take the wedge of $k + 2$ copies of G_4.

The number of vertices in the resulting graph is one more than the sum of the number of vertices in the separate graphs minus the number of graphs. This is n in each of the three cases. One easily checks that we obtain a graph isomorphic to some critical graph in all three cases. Moreover, the number of edges is easily seen to be $5k + 2r - 4 = \frac{5n}{3} + \frac{r-12}{3} = \frac{5n-5}{3}$. $\square$

Part VII

Cliques and Stable Sets

Graphs Avoiding k-matchings

We review the main known results about the complex $\mathsf{NM}_{n,k}$ of graphs on n vertices that do not contain a k-matching; these results are due to Linusson, Shareshian, and Welker [95]. Their most prominent achievement is the result that $\mathsf{NM}_{n,k}$ is homotopy equivalent to a wedge of spheres of dimension $3k - 4$; the number of spheres in the wedge is a polynomial in n for each k. They proved a similar formula for the complex $\mathsf{NM}_{m+n,k}(K_{m,n})$ of subgraphs of $K_{m,n}$ that do not contain a k-matching.

Theorem 24.1 (Linusson et al. [95]). *Let $\Pi^1_{n-1}(k)$ be the family of partitions* $\mathsf{U} = \{U_1, \ldots, U_{n-2k+1}\}$ *of $[n-1]$ such that $|U_i|$ is odd for each U_i. Then $\mathsf{NM}_{n,k}$ is homotopy equivalent to a wedge of $g_k(n)$ spheres of dimension $3k - 4$, where*

$$g_k(n) = \sum_{\mathsf{U} \in \Pi^1_{n-1}(k)} \prod_{i=1}^{n-2k+1} \left((|U_i| - 2)!! \right)^2. \tag{24.1}$$

In particular, $g_k(n)$ is a polynomial in n of degree $3k - 3$ such that $g_k(2k) = ((2k-3)!!)^2$ and $g_k(i) = 0$ for $1 \le i \le 2k - 1$. □

The proof of Linusson et al. relies on discrete Morse theory and the Gallai-Edmonds structure theorem; see Lovász and Plummer [97] for details about this theorem.

To see that $g_k(n)$ is indeed a polynomial with the given properties, let us investigate the formula (24.1) in greater detail. Let $\lambda = (\lambda_1, \ldots, \lambda_r)$ be a weakly decreasing sequence of odd integers such that $\lambda_i \ge 3$ and such that $\sum_i \lambda_i = 2k - 2 + r$. There are finitely many such sequences, because $2k - 2 + r = \sum_i \lambda_i \ge 3r$, which implies that $r \le k - 1$. Write $l(\lambda) = r$. The number of partitions $\mathsf{U} = \{U_1, \ldots, U_{n-2k+1}\}$ of $[n-1]$ such that $|U_i| = \lambda_i$ for $1 \le i \le l(\lambda)$ and $|U_i| = 1$ otherwise is

$$c_\lambda \cdot \binom{n-1}{2k-2+l(\lambda)}, \tag{24.2}$$

where c_λ is the number of partitions $U = \{U_1, \ldots, U_{l(\lambda)}\}$ of $[2k - 2 + l(\lambda)]$ such that $|U_i| = \lambda_i$ for all i. This implies that $g_k(n)$ is a polynomial, being a finite sum of expressions of the form (24.2). More precisely,

$$g_k(n) = \sum_\lambda \binom{n-1}{2k - 2 + l(\lambda)} \cdot c_\lambda \cdot \prod_{i=1}^{l(\lambda)} ((\lambda_i - 2)!!)^2, \qquad (24.3)$$

where the sum is over all λ with properties as above. As a consequence, we have the following corollary:

Corollary 24.2. *The coefficient of the highest-degree term n^{3k-3} in $g_k(n)$ equals $\frac{1}{(k-1)!6^{k-1}}$. In particular,*

$$g_k(n) \sim \frac{1}{(k-1)!} \cdot \left(\frac{n^3}{6} \right)^{k-1}.$$

Proof. The only weakly decreasing sequence λ such that $2k - 2 + l(\lambda) \geq 3k - 3$ and $\lambda_i \geq 3$ for all i is the sequence consisting of $k - 1$ threes. Clearly, $c_\lambda = \frac{(3k-3)!}{(k-1)!(3!)^{k-1}}$, which concludes the proof. $\square$

We give an alternative proof that g_k is a polynomial in a separate manuscript [73], where we also deduce that $g_k(0) = (-1)^{k-1}$. Equivalently, the polynomial counting the reduced Euler characteristic equals -1 at 0. For small values of k, one may easily compute an exact formula for $g_k(n)$:

Proposition 24.3. *We have that*

$$g_2(n) = \binom{n-1}{3};$$

$$g_3(n) = \binom{n-1}{5} \cdot \frac{5n-3}{3};$$

$$g_4(n) = \binom{n-1}{7} \cdot \frac{35n^2 - 28n + 9}{9};$$

$$g_5(n) = \binom{n-1}{9} \cdot \frac{(5n-1)(35n^2 - 14n + 15)}{15};$$

$$g_6(n) = \binom{n-1}{11} \cdot \frac{385n^4 + 374n^2 + 9}{9};$$

$$g_7(n) = \binom{n-1}{13} \cdot \frac{1001n^2(175n^3 + 175n^2 + 410n + 218) + 83343n - 945}{945};$$

$$g_8(n) = \binom{n-1}{15} \cdot \frac{1001n^3(25n^3 + 60n^2 + 145n + 168) + 120107n^2 + 34428n + 27}{27}.$$

Proof. This is just a matter of applying formula (24.3). For example, for $k = 4$, there are three relevant sequences λ; these are $(7), (5, 3)$ and $(3, 3, 3)$. Observing that $c_{(7)} = 1$, $c_{(5,3)} = 56$, and $c_{(3,3,3)} = 280$, we immediately obtain the desired formula for $g_4(n)$. $\square$

The coefficients in $g_8(n)$ are not alternating; the coefficients of 1, n, and n^2 are all negative. In fact, g_8 has two real and negative roots. g_4 and g_5 have two nonreal roots, whereas g_6, g_7, and g_8 have four nonreal roots. For $k \leq 6$, the real part of each root of $h_k(n) = g_k(n)/\binom{n-1}{2k-1}$ equals $(6 - k)/5$. This intriguing property does not hold for $h_7(n)$ and $h_8(n)$, but it does hold for the slightly modified polynomial $h_7(n) + \frac{55296}{4375}$.

For n odd, recall that a graph G on n vertices is factor-critical if $G([n]\setminus\{v\})$ contains a perfect matching for each $v \in [n]$. In Section 23.3, we discussed the complex FC_{2k-1} of factor-critical graphs. By Theorems 23.5 and 24.1, FC_{2k-1} and $\mathsf{NM}_{2k,k}$ are homotopy equivalent. Indeed, this is an observation of fundamental importance in the work of Linusson et al. [95]. One may establish a homotopy equivalence in the following manner:

Let Δ be the subcomplex of $\mathsf{NM}_{2k,k}$ consisting of all graphs G such that $G([2k - 1])$ is not factor-critical. For $G \in \Delta$, let $x = x_G$ be minimal such that $G([2k - 1] \setminus \{x\})$ does not contain a perfect matching. It is easy to see that we obtain a perfect acyclic matching on Δ by pairing $G - x_G(2k)$ and $G + x_G(2k)$. Thus $\mathsf{NM}_{2k,k}$ is homotopy equivalent to $\mathsf{NM}_{2k,k}/\Delta$ by the Contractible Subcomplex Lemma 3.16. Now, a graph G belonging to $\mathsf{NM}_{2k,k}/\Delta$ has the property that $G([2k-1])$ is factor-critical. As a consequence, $G+x(2k)$ contains a perfect matching for each $x \in [2k - 1]$. This implies that the vertex $2k$ must be isolated in G. Hence we have an isomorphism from $\mathsf{NM}_{2k,k}/\Delta$ to FC_{2k-1} defined by removing the vertex $2k$.

In this context, it is worth mentioning that there is an intriguing homological connection between FC_{2k-1} and a certain sublattice of the partition lattice Π_{2k-1}; this sublattice consists of all partitions in which all sets are odd. We refer the reader to Linusson et al. [95] for more information and references.

Finally, let us mention a beautiful result about the complex of subgraphs of a complete bipartite graph that do not contain a k-matching.

Theorem 24.4 (Linusson et al. [95]). *For $n, m, k \geq 1$, $\mathsf{NM}_{m+n,k}(K_{m,n})$ is homotopy equivalent to a wedge of $\binom{m-1}{k-1}\binom{n-1}{k-1}$ spheres of dimension $2k - 3$.* $\square$

t-colorable Graphs

We consider the complex Col_n^t of t-colorable graphs on n vertices, summarizing the results of Linusson and Shareshian [94] about the homotopy type and Euler characteristic of Col_n^t for $t \in \{2, n-3, n-2\}$. In addition, we present a conjecture about $\tilde{\chi}(\mathsf{Col}_n^{n-4})$ based on the intriguing observation that this value fits a certain polynomial of degree seven for nine different values of n.

First, note that $t = 2$ yields the complex B_n of bipartite graphs, which we examined in detail in Chapter 14. For easy reference, let us restate the main result (see Theorem 14.1) about this complex:

For $n \geq 1$, Col_n^2 is homotopy equivalent to a wedge of spheres of dimension $n-2$.

Next, consider $t = n - 2$. As Linusson and Shareshian observed [94], Col_n^{n-2} is the Alexander dual of the complex of star graphs considered in Proposition 14.16. This complex is semi-nonevasive and homotopy equivalent to a wedge of $\binom{n-1}{2}$ spheres of dimension 1 by Theorem 14.12 and Proposition 14.16.

Corollary 25.1. *For $n \geq 3$, Col_n^{n-2} is homotopy equivalent to a wedge of $\binom{n-1}{2}$ spheres of dimension $\binom{n}{2} - 4$.*

Proof. Use Proposition 5.36. $\square$

Finally, the case $t = n - 3$ is as follows:

Theorem 25.2 (Linusson and Shareshian [94]). *For $n \geq 4$, Col_n^{n-3} is homotopy equivalent to a wedge of $\binom{n-1}{3} \cdot \frac{3n^2 - 12n + 5}{5}$ spheres of dimension $\binom{n}{2} - 7$. $\square$*

The proof of Theorem 25.2 relies on the following general lemma:

Lemma 25.3 (Linusson and Shareshian [94]). *For $1 \leq t \leq n$, Col_n^t admits a decision tree such that G is evasive if and only if $G([n-1]) \notin \mathsf{Col}_{n-1}^{t-1}$ and the degree of every vertex in $G([n-1])$ is at least $t - 1$.*

Proof. Let $Y = \binom{[n-1]}{2}$ and $E_n = \{1n, \ldots, (n-1)n\}$. Consider the lifted complex $\Sigma_B = \mathrm{Col}_n^t(B, Y \setminus B)$ for each subset B of Y. Let H be the graph on the vertex set $[n-1]$ with edge set B. If H belongs to Col_{n-1}^{t-1}, then every edge in E_n is a cone point in Σ_B. If some vertex v has degree at most $t - 2$ in H, then vn is a cone point in Σ_B. Namely, for every graph G in Σ_B, v has fewer than t neighbors in $G + vn$, which implies that we may extend any t-coloring of $G([n] \setminus \{v\})$ to a t-coloring of $G + vn$. $\square$

In their proof of Theorem 25.2, Linusson and Shareshian applied Lemma 25.3 to Col_n^{n-3} and then defined an optimal acyclic matching on the remaining quotient complex of evasive graphs.

The following immediate consequence of Lemma 25.3 is worth mentioning.

Corollary 25.4 (Linusson and Shareshian [94]). *For $1 \leq t \leq n$, the shifted connectivity degree of Col_n^t is at least $\frac{(n-1)(t-1)}{2} - 1$.*

Proof. Every evasive graph with respect to the decision tree in Lemma 25.3 contains at least $\frac{(n-1)(t-1)}{2}$ edges, which implies that all faces of Col_n^t of dimension at most $\frac{(n-1)(t-1)}{2} - 2$ are contained in a collapsible subcomplex of Col_n^t. By Corollary 3.10, we are done. $\square$

The bound in Corollary 25.4 on the shifted connectivity degree of Col_n^t is not sharp. In fact, for $t \in \{1, 2, n - 3, n - 2, n - 1\}$, the actual value is $n(t - 1) - \binom{t}{2} - 1$; apply the results in this section. Computations by Linusson and Shareshian [94] yield some evidence that this value might be a bound on the shifted connectivity degree in general. More precisely, Linusson and Shareshian used computer to show that Col_7^3 is homotopy equivalent to a wedge of 1535 spheres of dimension ten. Moreover, they showed that

$$\tilde{H}_i(\mathrm{Col}_8^4; \mathbb{Z}) \cong \begin{cases} \mathbb{Z}^{9396} & \text{if } i = 17; \\ \mathbb{Z} & \text{if } i = 19; \\ 0 & \text{otherwise.} \end{cases}$$

Let p be fixed. In a separate manuscript [73], we prove that the Euler characteristic of the Alexander dual $\mathrm{NQP}_{n,n-p-1}$ of Col_n^{n-p-1} is a polynomial $g_p(n)$ such that $g_p(0) = -1$ and $g_p(k) = 0$ if $1 \leq k \leq p + 1$; see Section 26.8 for some more discussion. By Corollary 25.1 and Theorem 25.2, we know that

$$g_1(n) = -\binom{n-1}{2};$$

$$g_2(n) = \binom{n-1}{3} \cdot \frac{3n^2 - 12n + 5}{5}.$$

Consider $p = 3$. As already mentioned, $g_3(0) = -1$ and $g_3(1) = g_3(2) = g_3(3) = g_3(4) = 0$. Moreover, computations by Linusson and Shareshian yield that $g_3(5) = 1$, $g_3(6) = 105$, $g_3(7) = 1535$, and $g_3(8) = 9397$. Let $\hat{g}_3$ be the unique polynomial of degree at most seven with the property that $\hat{g}_3(k) = g_3(k)$ for $0 \leq k \leq 7$. A straightforward calculation yields that $\hat{g}_3(8) = g_3(8) = 9397$; hence we have quite strong evidence for the following conjecture:

Conjecture 25.5. *We have that* $g_3(n) = \tilde{\chi}(\mathsf{NQP}_{n,n-4})$ *is equal to*

$$\hat{g}_3(n) = \binom{n-1}{4} \cdot \frac{411n^3 - 4178n^2 + 10657n - 105}{105}.$$

Note that $\hat{g}_3(n)$ has two nonreal roots. To prove this conjecture, it suffices to demonstrate that $g_3(n)$ is of degree at most eight; we know the value of $g_3(n)$ at nine distinct points n.

For $2 \leq n \leq 7$, there is homology only in dimension eight. For $n = 8$, we have that $\dim \tilde{H}_8(\mathsf{NQP}_{8,4}) = 9396$ and $\dim \tilde{H}_6(\mathsf{NQP}_{8,4}) = 1$. This is very little evidence to base any conjecture on, but if each Betti number were given by a polynomial of degree at most seven, then we would obtain that

$$\dim \tilde{H}_8(\mathsf{NQP}_{n,n-4}) = \binom{n}{5} \cdot \frac{821n^2 - 8338n + 21207}{42};$$

$$\dim \tilde{H}_6(\mathsf{NQP}_{n,n-4}) = \binom{n-1}{7},$$

whereas all the other Betti numbers would vanish. Note that we do not know whether the Betti numbers are given by polynomials.

Let $t \geq 2$ and $p \geq 1$, and let $\mathsf{c} = (c_1, \ldots, c_{t-1})$ be a sequence of positive integers. Let $\mathsf{Col}^t_{n,\mathsf{c},p}$ be the complex of graphs admitting a t-coloring $\gamma : V \to [t]$ such that the following hold:

- $|\gamma^{-1}(i)| \leq c_i$ for $i \in [1, t-1]$.
- $|\gamma^{-1}(t)| \geq n - p$.

For $t = 2$ and $c_1 = p$, we obtain the complex $\mathsf{B}_{n,p}$ of graphs with balance number at most p. For $t = p+1$ and $c_1 = \ldots c_p = 1$, we obtain the complex $\mathsf{Cov}_{n,p}$ of graphs with covering number at most p; see Chapter 26. Note that if $\mathsf{c} = (p, \ldots, p)$, then $\mathsf{Col}^t_{n,\mathsf{c},p}$ is the complex of t-colorable graphs such that the color t is used on at least $n-p$ vertices. In a separate manuscript [73], we show that the Euler characteristic of $\mathsf{Col}^t_{n,\mathsf{c},p}$ is a polynomial in n for sufficiently large n for each fixed t, p, and c. This generalizes the corresponding results about $\mathsf{B}_{n,p}$ and $\mathsf{Cov}_{n,p}$ in Sections 14.3.3 above and 26.4 below, respectively.

Graphs and Hypergraphs with Bounded Covering Number

[1] A hypergraph H is p-*coverable* if there is a vertex set W of size at most p such that every edge in H contains at least one vertex from W. We refer to W as a $|W|$-*cover* of H. The *covering number* $\tau(H)$ of a hypergraph H is the smallest integer p such that H has a p-cover. For $1 \leq p \leq n$ and $1 \leq r \leq n$, let $\mathsf{HCov}_{n,p,r}$ be the simplicial complex of r-uniform hypergraphs on the vertex set $[n]$ with covering number at most p. Note that $\mathsf{HCov}_{n,p,2}$ coincides with the complex $\mathsf{Cov}_{n,p}$ of p-coverable graphs.

The main results of this chapter are as follows:

- In Sections 26.3 and 26.4, we show, for any fixed p and r, that the Betti numbers of $\mathsf{HCov}_{n,p,r}$ over any field $\mathbb{F}$ are polynomials in n. Specifically,

$$\dim \tilde{H}_i(\mathsf{HCov}_{n,p,r}, \mathbb{F}) = \sum_{k=p+r}^{\gamma+1} (-1)^{\gamma+1-k} f_{k,\gamma}(n) \dim \tilde{H}_i(\mathsf{HCov}_{k,p,r}, \mathbb{F}),$$

  where each $f_{k,\gamma}(n)$ is a polynomial in n and $\gamma = \gamma(p,r)$ is an integer. For $r = 2$, we have that $\gamma = 2p$, which turns out to imply that the degree of $f_{k,\gamma}(n)$ is at most $2p$ in this case.
- In Section 26.5, we give explicit formulas for the homology of $\mathsf{Cov}_{n,p} = \mathsf{HCov}_{n,p,2}$ for $p \leq 3$; for $p = 2$ and $p = 3$, our results are based on computer calculations with the program homology [42]. Notably, there is 2-torsion in dimension six in the homology of $\mathsf{Cov}_{n,3}$ for $n \geq 6$.
- In Section 26.6, we demonstrate, for any $p \geq 1$, that the $(2p-1)$-skeleton of $\mathsf{Cov}_{n,p}$ is vertex-decomposable and hence shellable. As a consequence, $\mathsf{Cov}_{n,p}$ is $(2p-2)$-connected and has no homology in dimension $i \leq 2p-2$. For $p \leq 3$ and $n \geq 2p+1$, we have detected nonzero homology in dimension $2p-1$. We have not been able to find meaningful counterparts of these results for $\mathsf{HCov}_{n,p,r}$ when $r \geq 3$.

---

[1] This chapter is a revised and extended version of a paper [72] published in *SIAM Journal of Discrete Mathematics*.

In Section 26.2, we introduce a complex $\mathsf{HCov}_{n,p,r}^{\#}$ with the same homotopy type as $\mathsf{HCov}_{n,p,r}$ and with certain nice properties that allow for a smooth analysis. We apply discrete Morse theory to $\mathsf{HCov}_{n,p,r}^{\#}$ in Section 26.3 and derive the polynomial property of the Betti numbers in Section 26.4.

The graph theory presented in Section 26.1 is crucial for our theorems and is used throughout the chapter. This is classical theory – basically Chapter 13 in Berge [6] – about graphs with the property that each vertex is contained in the complement of a cover of minimum size.

26.1 Solid Hypergraphs

Let us say that a hypergraph $H = (V, E)$ with covering number p is (p, r)-solid if, for every vertex set U of size at most $r - 1$, there is a p-cover W of H such that $U \cap W = \emptyset$. In this section, we present some useful results about (p, r)-solid $[r]$-hypergraphs; recall that a hypergraph is an S-hypergraph if all edges are of size an integer in S.

Lemma 26.1. *If an $[r]$-hypergraph H is (p, r)-solid, then H is r-uniform. Moreover, every covered vertex is contained in a p-cover of H.*

Proof. For the first statement, since H is (p, r)-solid, a vertex set of size at most $r - 1$ cannot form an edge in H. For the second statement, let v be a covered vertex and let e be an edge in H containing v; clearly, $|e \setminus \{v\}| = r - 1$. H being (p, r)-solid means that some p-cover does not intersect $e \setminus \{v\}$. Since this cover must then contain v, we are done. $\square$

By Lemma 26.1, we may restrict our attention to r-uniform hypergraphs. First, a simple observation:

Lemma 26.2. *If H is r-uniform with covering number p, then the vertex set of H has size at least $p + r - 1$. In particular, this is true if H is (p, r)-solid.*

Proof. Any r-uniform hypergraph on at most $p + r - 2$ vertices has covering number at most $p - 1$. $\square$

The bound in Lemma 26.2 is tight; the complete r-uniform hypergraph on $p + r - 1$ vertices is (p, r)-solid.

We will use the following lemma in Section 26.4 to prove that the Betti numbers of $\mathsf{HCov}_{n,p,r}$ are polynomials in n for each fixed p and r.

Lemma 26.3. *For every $p, r \geq 1$, there is a positive integer $\gamma(p, r)$ such that if H is a (p, r)-solid and r-uniform hypergraph with no uncovered vertices, then the number of vertices in H is at most $\gamma(p, r)$.*

Proof. Let H be (p, r)-solid without uncovered vertices. If we remove an edge e such that that $\tau(H) = \tau(H - e)$, then $H - e$ is again (p, r)-solid with no uncovered vertices. Namely, assume to the contrary that some vertex $v \in e$ is uncovered in $H - e$. By Lemma 26.1, there is a p-cover W of H containing v. However, this implies that $W \setminus \{v\}$ is a $(p-1)$-cover of $H - e$, which is a contradiction.

Starting with H, remove edges not affecting the covering number until we have a τ-*critical* hypergraph H', meaning that the removal of any edge in H' decreases the covering number of H'.[2] By a result of Bollobás [18], the number of edges in a τ-critical r-uniform hypergraph with covering number p is at most $\binom{p+r-1}{r}$; see Lovász [96, Ex. 13.32]. This implies that the number of vertices in H' is at most $r \cdot \binom{p+r-1}{r}$ (this is a very loose bound), and the lemma follows. $\square$

For $r = 2$, we can establish a tight bound on $\gamma(p, r)$:

Theorem 26.4 (Berge [6, Th. 13.13]). *If G is a simple graph with $\tau(G) = p$ such that G contains no uncovered vertices and such that every vertex is contained in a p-cover, then the number of vertices in G is at most $2p$. As a consequence, if G is $(p, 2)$-solid with no uncovered vertices, then the number of vertices in G is at most $2p$.* $\square$

The bound $2p$ is tight, as the $2p$-cycle is $(p, 2)$-solid. The first statement in the theorem is basically a consequence of some results due to Hajnal [58]; see Berge [6, Th. 13.8-9]. Unfortunately, these results seem hard to generalize to hypergraphs. By Lemma 26.1, the second statement in the theorem is a consequence of the first.

Finally, we state and prove a few results that we will use in Section 26.6 to prove that the $(2p - 1)$-skeleton of $\mathsf{Cov}_{n,p}$ is vertex-decomposable; again, we restrict our attention to graphs.

Lemma 26.5. *Let H be a graph with covering number p and with connected components $C_1, \ldots, C_k$. Then H is $(p, 2)$-solid if and only if there are integers $p_1, \ldots, p_k$ summing up to p such that C_i is $(p_i, 2)$-solid for each i.*

Proof. With $p_i = \tau(C_i)$, it is clear that $\sum_i p_i = \tau(H) = p$. Suppose that some vertex $v \in C_i$ is contained in every p_i-cover of C_i. Then v is contained in every p-cover of H; we cannot cover $H \setminus C_i$ with fewer than $p - p_i$ vertices. Conversely, if v is *not* contained in a given p_i-cover of C_i, then we can extend this cover to a p-cover of H not containing v by picking an arbitrary p_j-cover of every other C_j. $\square$

Lemma 26.6. *A $(p, 2)$-solid graph H contains at least $2p - k$ edges, where k is the number of connected components in H with at least two vertices.*

Proof. The lemma is clear for $p = 1$; assume that $p \geq 2$. We may assume that H contains no uncovered vertices. Let the connected components of H be $C_1, \ldots, C_k$. With $p_i = \tau(C_i)$, we have that C_i is $(p_i, 2)$-solid for each i

---

[2] This is equivalent to H' being α-*critical* as defined by Berge [6, Sec. 13.3].

by Lemma 26.5. In particular, if $k \geq 2$, then we may use induction on p to conclude that C_i contains at least $2p_i - 1$ edges. Summing over i and using the fact that $\sum_i p_i = p$, we obtain that H contains at least $2p - k$ edges.

Thus assume that H is connected. As in the proof of Lemma 26.3, note that if we remove an edge that does not affect the covering number of H, then the resulting graph is again $(p, 2)$-solid with no uncovered vertices. Remove such edges from H until we have a τ-critical graph H'; the removal of any edge from H' decreases the covering number.

If the obtained graph H' is disconnected with k components, then we removed at least $k - 1$ edges, and by the same induction argument as above, H' contains at least $2p - k$ edges. Hence H contains at least $2p - 1$ edges as desired.

Assume that H' is connected; for simplicity, let us write H instead of H'. Berge [6, Th. 13.6] proved that a τ-critical and connected graph is 2-connected. We want to find a vertex x in H such that the induced subgraph K obtained by removing x from H is $(p - 1, 2)$-solid. By induction, this will imply that K contains at least $2(p - 1) - 1$ edges, which in turn will imply that H contains at least $2(p - 1) - 1 + 2 = 2p - 1$ edges as desired. Namely, we get rid of at least two edges when we remove x, and the resulting graph K is connected, as H is 2-connected.

To find the vertex x, let $y \leq z$ mean that any p-cover of H containing y also contains z. This defines a partial order. Namely, since H is τ-critical, we have, for each y, w such that $yw \in H$, that the graph $H - yw$ has a $(p - 1)$-cover Q with the property that $y, w \notin Q$. If $z \notin Q$, then $y \not\leq z$, as $Q \cup \{y\}$ is a p-cover of H not containing z. If $z \in Q$, then $z \not\leq y$, as $Q \cup \{w\}$ is a p-cover of H containing z but not y. Now, pick x maximal with respect to the given partial order. This means, for any $y \neq x$, that there is a p-cover of H containing x but not y. In particular, there is a $(p - 1)$-cover not containing y of the induced subgraph K obtained by removing x from H. However, this means exactly that K is $(p - 1, 2)$-solid, and we are done. $\square$

The bound in Lemma 26.6 is tight: Let G be the graph consisting of a path of vertex length $2(p - k + 1)$ and $k - 1$ additional components, each of vertex size two. Then G is $(p, 2)$-solid and contains $k - 1 + 2p - 2k + 1 = 2p - k$ edges.

26.2 A Related Simplicial Complex

For $n, p, r \geq 1$, let $\mathsf{HCov}^{\#}_{n,p,r}$ be the simplicial complex of $[r]$-hypergraphs on the vertex set $[n]$ with covering number at most p. Hence $\mathsf{HCov}^{\#}_{n,p,r}$ consists of hypergraphs with edges of size between 1 and r, whereas $\mathsf{HCov}_{n,p,r}$ consists of r-uniform hypergraphs. As it turns out, $\mathsf{HCov}^{\#}_{n,p,r}$ has several attractive properties that make the complex easier to handle than the original complex $\mathsf{HCov}_{n,p,r}$. We will write $\mathsf{Cov}^{\#}_{n,p} = \mathsf{HCov}^{\#}_{n,p,2}$.

Lemma 26.7. *For $p \geq 1$ and $1 \leq r \leq n$, $\mathsf{HCov}_{n,p,r} \simeq \mathsf{HCov}_{n,p,r}^{\#}$.*

Proof. We show how to collapse $\mathsf{HCov}_{n,p,r}^{\#}$ down to $\mathsf{HCov}_{n,p,r}$. Fix a linear order on $\binom{[n]}{r}$; this is the family of edges of maximum size r. For a hypergraph $H \in \mathsf{HCov}_{n,p,r}^{\#} \setminus \mathsf{HCov}_{n,p,r}$, let $e = e(H)$ be maximal with respect to this linear order such that e contains an edge $e' \in H$ of size at most $r - 1$; e itself is not necessarily contained in H. For each e of size r, let $\mathcal{F}(e)$ be the family of hypergraphs $H \in \mathsf{HCov}_{n,p,r}^{\#} \setminus \mathsf{HCov}_{n,p,r}$ such that $e(H) = e$. It is clear that the families $\mathcal{F}(e)$ satisfy the Cluster Lemma 4.2. Namely, $H \mapsto e(H) \in \binom{[n]}{r}$ is a poset map with the given linear order on $\binom{[n]}{r}$. Now, we obtain a perfect matching on $\mathcal{F}(e)$ by pairing $H + e$ with $H - e$ for each $H \in \mathcal{F}(e)$. Namely, adding or deleting e does not affect $e(H)$. Also, the covering number remains the same when e is added or deleted, as H already contains an edge $e' \subsetneqq e$. By the Cluster Lemma 4.2, we are done. $\square$

Next, we prove that $\mathsf{HCov}_{n,p,r}^{\#}$ and $\mathsf{HCov}_{n,r,p}^{\#}$ are homotopy equivalent; we may hence swap p and r without affecting the homotopy type. One may view this result as an analogue of the result about the complex $\mathsf{HB}_{n,p,t}$ in Proposition 14.21.

Proposition 26.8. *For $n, p, r \geq 1$, we have that $\mathsf{HCov}_{n,p,r}^{\#} \simeq \mathsf{HCov}_{n,r,p}^{\#}$. In particular, $\mathsf{HCov}_{n,p,r} \simeq \mathsf{HCov}_{n,r,p}$ whenever $n \geq \max\{p, r\}$.*

Proof. For $1 \leq n \leq p + r - 1$, $\mathsf{HCov}_{n,p,r}^{\#}$ and $\mathsf{HCov}_{n,r,p}^{\#}$ are both cones and hence collapsible; every edge of maximum size is a cone point. Assume that $n \geq p + r$. Consider the nerve complex $\mathsf{N}_{n,p,r} = \mathsf{N}(\mathsf{HCov}_{n,p,r}^{\#})$; see the Nerve Theorem 6.2. We may identify the 0-cells in $\mathsf{N}_{n,p,r}$ with subsets of $[n]$ of size p. Namely, every maximal hypergraph $H \in \mathsf{HCov}_{n,p,r}^{\#}$ has a unique p-cover consisting of those x with the property that the singleton edge x belongs to H.

For a set U of size p, let H_U be the maximal hypergraph in $\mathsf{HCov}_{n,p,r}^{\#}$ with unique p-cover U. A family $\mathcal{W}$ of 0-cells in $\mathsf{N}_{n,p,r}$ forms a face of $\mathsf{N}_{n,p,r}$ if and only if the intersection $\bigcap_{W \in \mathcal{W}} H_W$ is nonempty. This means that there is a set S of size at most r such that $|W \cap S| \geq 1$ for each $W \in \mathcal{W}$. However, this is exactly the condition that the hypergraph $([n], \mathcal{W})$ admits a cover of size at most r. As a consequence, we may identify $\mathsf{N}_{n,p,r}$ with $\mathsf{HCov}_{n,r,p}$. Thus

$$\mathsf{HCov}_{n,p,r}^{\#} \simeq \mathsf{N}_{n,p,r} \cong \mathsf{HCov}_{n,r,p} \simeq \mathsf{HCov}_{n,r,p}^{\#};$$

the first equivalence follows from the Nerve Theorem 6.2, whereas the last equivalence follows from Lemma 26.7. $\square$

26.3 An Acyclic Matching

The purpose of this section is to present an acyclic matching on $\mathsf{HCov}_{n,p,r}^{\#}$ such that the unmatched graphs have certain rather strong properties. Observant readers may note that our matching is quite similar in nature to the matching that Linusson and Shareshian provided for complexes of t-colorable graphs; see Lemma 25.3.

For an $[r]$-hypergraph H on the vertex set $[n]$, let $\mathcal{X}(H)$ be the family of all subsets of $[n-1]$ of size at most $r-1$ that have nonempty intersection with every p-cover of $H([n-1])$. Note that if $H \in \mathsf{HCov}^{\#}_{n,p,r}$, then we may add the edge $X \cup \{n\}$ to H for any $X \in \mathcal{X}(H)$ without ending up outside $\mathsf{HCov}^{\#}_{n,p,r}$.

Define

$$\mathcal{A}_{n,p,r} = \{H \in \mathsf{HCov}^{\#}_{n,p,r} : H([n-1]) \in \mathsf{HCov}^{\#}_{n-1,p-1,r}\};$$
$$\mathcal{B}_{n,p,r} = \{H \in \mathsf{HCov}^{\#}_{n,p,r} : H([n-1]) \notin \mathsf{HCov}^{\#}_{n-1,p-1,r} \text{ and } \mathcal{X}(H) \neq \emptyset\};$$
$$\mathcal{C}_{n,p,r} = \{H \in \mathsf{HCov}^{\#}_{n,p,r} : H([n-1]) \notin \mathsf{HCov}^{\#}_{n-1,p-1,r} \text{ and } \mathcal{X}(H) = \emptyset\}.$$

It is clear that $\mathsf{HCov}^{\#}_{n,p,r}$ is the disjoint union of $\mathcal{A}_{n,p,r}$, $\mathcal{B}_{n,p,r}$, and $\mathcal{C}_{n,p,r}$ and that $\mathcal{A}_{n,p,r}$ and $\mathcal{A}_{n,p,r} \cup \mathcal{C}_{n,p,r}$ are both simplicial complexes. This implies that the three families satisfy the Cluster Lemma 4.2. We want to prove that there are perfect acyclic matchings on $\mathcal{A}_{n,p,r}$ and $\mathcal{B}_{n,p,r}$. The remaining family $\mathcal{C}_{n,p,r}$ is the family of all $[r]$-hypergraphs H such that $H([n-1])$ has covering number p and such that every subset of $[n-1]$ of size at most $r-1$ is disjoint from some p-cover of $H([n-1])$. This means exactly that $H([n-1])$ is (p,r)-solid.

We obtain a perfect acyclic matching on $\mathcal{A}_{n,p,r}$ by pairing $H - n$ with $H + n$; we match with the singleton edge n. Namely, for any cover W of $H([n-1])$, $W \cup \{n\}$ is a cover of H.

For a family $\mathcal{X}$ of subsets of $[n-1]$, let $\mathcal{B}_{n,p,r}(\mathcal{X})$ be the family of hypergraphs $H \in \mathcal{B}_{n,p,r}$ such that $\mathcal{X}(H) = \mathcal{X}$. It is clear that the families $\mathcal{B}_{n,p,r}(\mathcal{X})$ satisfy the Cluster Lemma 4.2. Namely, $H \mapsto \mathcal{X}(H)$ is a poset map; $\mathcal{X}(H)$ cannot grow when we delete edges from H. Let $X(H)$ be minimal in $\mathcal{X}(H)$ with respect to some fixed linear order. If $H \in \mathcal{B}_{n,p,r}(\mathcal{X})$, then the same is true for $H + X(H)n$; every p-cover of H contains an element from $X(H)$, and $X(H)$ has size at most $r-1$. $\mathcal{X}(H)$ does not depend on the set of edges containing n, which means that we obtain a perfect matching on $\mathcal{B}_{n,p,r}(\mathcal{X})$ by pairing $H - X(H)n$ with $H + X(H)n$. Taking the union over all $\mathcal{X}$, we get a perfect acyclic matching on $\mathcal{B}_{n,p,r}$.

Combining our two perfect acyclic matchings on $\mathcal{A}_{n,p,r}$ and $\mathcal{B}_{n,p,r}$, we obtain an acyclic matching on $\mathsf{HCov}^{\#}_{n,p,r}$ with $\mathcal{C}_{n,p,r}$ as the set of critical graphs. Theorem 4.11 yields the following:

Proposition 26.9. *With notation as above and as in (4.2) in Section 4.3,*

$$\mathsf{HCov}^{\#}_{n,p,r} \simeq (\mathsf{HCov}^{\#}_{n,p,r})_{\mathcal{C}_{n,p,r}}.$$

Also, given an acyclic matching on $\mathcal{C}_{n,p,r}$ with c_i critical sets of dimension i for each i, $\mathsf{HCov}^{\#}_{n,p,r}$ is homotopy equivalent to a CW complex with c_i cells of dimension i for each i and one additional 0-cell. □

26.4 Homotopy Type and Homology

Before proceeding, let us examine some special cases. First of all, note that $\mathsf{HCov}_{n,p,r}^{\#}$ is a cone and hence collapsible whenever $1 \le n \le p+r-1$. Also, by Lemma 26.7, $\mathsf{HCov}_{p+r,p,r}^{\#}$ is homotopy equivalent to $\mathsf{HCov}_{p+r,p,r}$, which contains all r-uniform hypergraphs on the vertex set $[p+r]$ except the complete hypergraph. This implies that

$$\mathsf{HCov}_{p+r,p,r}^{\#} \simeq S^{C(p+r,r)-2}, \tag{26.1}$$

where $C(m,k) = \binom{m}{k}$.

Next, consider $p = 1$. The complex $\mathsf{HCov}_{n,1,r}^{\#}$ consists of *star hypergraphs*, which are hypergraphs covered by a single vertex. By Proposition 26.8, $\mathsf{HCov}_{n,1,r}^{\#}$ is homotopy equivalent to $\mathsf{HCov}_{n,r,1}^{\#} = \mathsf{HCov}_{n,r,1}$. Now, the latter complex is obviously the $(r-1)$-skeleton of an $(n-1)$-simplex. As a consequence, we have the following simple result.

Proposition 26.10. *For $n, r \ge 1$, $\mathsf{HCov}_{n,1,r}^{\#}$ and $\mathsf{HCov}_{n,r,1}^{\#}$ are both homotopy equivalent to a wedge of $\binom{n-1}{r}$ spheres of dimension $r-1$.* $\square$

Note that $\mathsf{Cov}_{n,1} = \mathsf{HCov}_{n,1,2}$ coincides with the complex $\mathsf{B}_{n,1}$ of star graphs considered in Proposition 14.16. Since $\mathsf{Cov}_{n,1}$ is homotopy equivalent to $\mathsf{HCov}_{n,1,2}^{\#}$ by Lemma 26.7, Proposition 26.10 is equivalent to Proposition 14.16 for the special case $r = 2$.

Now, proceed with general n, p, r. Recall that $\mathcal{C}_{n,p,r}$ is the set of critical hypergraphs in Proposition 26.9 and that a hypergraph H in $\mathsf{HCov}_{n,p,r}^{\#}$ belongs to $\mathcal{C}_{n,p,r}$ if and only if $H([n-1])$ is (p,r)-solid. For a nonempty vertex set $J \subseteq [n-1]$, let $\mathcal{C}_{n,p,r}(J)$ be the family of hypergraphs H in $\mathcal{C}_{n,p,r}$ such that J is the set of vertices that are covered in $H([n-1])$. Write

$$\Lambda_{n,p,r}(J) = (\mathsf{HCov}_{n,p,r}^{\#})_{\mathcal{C}_{n,p,r}(J)} \quad \text{(notation as in (4.2))};$$
$$\Lambda_{k,p,r} = \Lambda_{k,p,r}([k-1]).$$

Lemma 26.11. *Let $n, p, r \ge 1$. For any nonempty vertex set $J \subseteq [n-1]$, $\Lambda_{n,p,r}(J)$ is the union of $\mathcal{C}_{n,p,r}(J)$ and a collapsible subcomplex of the complex $\mathcal{A}_{n,p,r}$ defined in Section 26.3. Moreover, we have that*

$$\Lambda_{n,p,r}(J) \simeq \Lambda_{|J|+1,p,r}. \tag{26.2}$$

Proof. For the first claim, let H be a hypergraph in $\mathcal{C}_{n,p,r}(J)$. We want to prove that $H \not\longmapsto \mathcal{C}_{n,p,r}(I)$ if $I \ne J$. Note that if we remove an edge e from H, then we obtain a hypergraph in $\mathcal{C}_{n,p,r}(I)$ for some $I \subseteq J$ or a hypergraph in $\mathcal{A}_{n,p,r}$. If $n \in e$, then $H - e \in \mathcal{C}_{n,p,r}(J)$; thus assume that $n \notin e$. It is clear that $\mathcal{A}_{n,p,r} \not\longmapsto \mathcal{C}_{n,p,r}$, which means that we only have to prove that if the new hypergraph $G = H - e$ belongs to $\mathcal{C}_{n,p,r}(I)$, then $I = J$.

Assume the opposite. Then some $x \in e$ is uncovered in $G([n-1])$. Since $H \in \mathcal{C}_{n,p,r}$, there is a p-cover W of $H([n-1])$ such that $(e \setminus \{x\}) \cap W = \emptyset$. Since $e \in H([n-1])$, we must have that $x \in W$. However, since x is uncovered in $G([n-1])$, $W \setminus \{x\}$ covers $G([n-1])$, which implies that $G \in \mathcal{A}_{n,p,r}$, contradictory to assumption. Thus our claim is proved.

For the second claim, we have that the first claim implies that $\Lambda_{n,p,r}(J)$ is the union of $\mathcal{C}_{n,p,r}(J)$ and a collapsible subcomplex $\mathcal{T}$ of $\mathcal{A}_{n,p,r}$. To see that $\mathcal{T}$ is collapsible, just note that $H - n \in \mathcal{T}$ if and only if $H + n \in \mathcal{T}$; this is by definition of $\Lambda_{n,p,r}(J)$ and Lemma 4.10. In particular, $\mathcal{T}$ is a cone with cone point the singleton edge n. $\mathcal{C}_{n,p,r}(J) \cup \mathcal{T}$ is easily seen to be homotopy equivalent to $\mathcal{C}_{n,p,r}(J) \cup \mathcal{A}_{n,p,r}$. Namely, we obtain a perfect acyclic matching on $\mathcal{A}_{n,p,r} \setminus \mathcal{T}$ by pairing $H - n$ with $H + n$ whenever $H \in \mathcal{A}_{n,p,r} \setminus \mathcal{T}$; $\mathcal{A}_{n,p,r}$ and $\mathcal{T}$ are both cones with cone point n.

Let $\mathcal{C}'_{n,p,r}(J)$ be the subfamily of $\mathcal{C}_{n,p,r}(J)$ consisting of those H with the property that all vertices in $[n-1] \setminus J$ are uncovered in H (not only in $H([n-1])$). We obtain a perfect acyclic matching on $\mathcal{C}_{n,p,r}(J) \setminus \mathcal{C}'_{n,p,r}(J)$ in the following manner. In a hypergraph $H \in \mathcal{C}_{n,p,r}(J) \setminus \mathcal{C}'_{n,p,r}(J)$, define $e(H)$ as the maximal edge in H with respect to some fixed linear order such that $e(H)$ contains some vertex in $[n-1] \setminus J$. Let $\mathcal{C}_{n,p,r}(J, e)$ be the subfamily of $\mathcal{C}_{n,p,r}(J) \setminus \mathcal{C}'_{n,p,r}(J)$ consisting of those H satisfying $e(H) = e$.

It is clear that the families $\mathcal{C}_{n,p,r}(J, e)$ satisfy the Cluster Lemma 4.2. Namely, $H \mapsto e(H)$ is a poset map; $e(H)$ cannot increase when edges are removed from H. Write $e'(H) = (e(H) \cap J) \cup \{n\}$. We claim that we may define a perfect matching on $\mathcal{C}_{n,p,r}(J, e)$ by pairing $H - e'(H)$ with $H + e'(H)$ whenever $H \in \mathcal{C}_{n,p,r}(J, e)$; note that $e'(H)$ is the same for all $H \in \mathcal{C}_{n,p,r}(J, e)$. To prove the claim, it suffices to prove that $H - e'(H) \in \mathcal{C}_{n,p,r}(J)$ if and only if $H + e'(H) \in \mathcal{C}_{n,p,r}(J)$; $e(H)$ does not depend on whether the edge $e'(H)$ is present in H. To prove this, we need only show that $H + e'(H) \in \mathsf{HCov}^{\#}_{n,p,r}$ whenever $H \in \mathcal{C}_{n,p,r}(J)$. Now, every p-cover W of H is contained in J by assumption; otherwise, we would have a $(p-1)$-cover of $H([n-1])$. This implies that W must contain an element from $e(H) \cap J = e'(H) \setminus \{n\}$. Thus W intersects e', and we are done.

The conclusion is that the simplicial complex $\mathcal{C}_{n,p,r}(J) \cup \mathcal{A}_{n,p,r}$ is homotopy equivalent to $\mathcal{C}'_{n,p,r}(J) \cup \mathcal{A}_{n,p,r}$. Now, $\mathcal{C}'_{n,p,r}(J) \cup \mathcal{A}_{n,p,r}(J)$ is a simplicial complex, where $\mathcal{A}_{n,p,r}(J)$ is the set of all graphs in $\mathcal{A}_{n,p,r}$ such that all vertices in $[n-1] \setminus J$ are uncovered. We may collapse $\mathcal{C}'_{n,p,r}(J) \cup \mathcal{A}_{n,p,r}$ down to $\mathcal{C}'_{n,p,r}(J) \cup \mathcal{A}_{n,p,r}(J)$ by matching $H - n$ with $H + n$ whenever $H \in \mathcal{A}_{n,p,r} \setminus \mathcal{A}_{n,p,r}(J)$. The resulting complex is clearly isomorphic to $\mathcal{C}_{|J|+1,p,r}([|J|]) \cup \mathcal{A}_{|J|+1,p,r}$. By the proof above, we may collapse this complex down to $\Lambda_{|J|+1,p,r}$, and we are done. $\square$

Define
$$\gamma(p, r) = \min\{\gamma : \mathcal{C}_{n,p,r}(J) = \emptyset \text{ whenever } |J| > \gamma\}. \tag{26.3}$$

Such a $\gamma(p, r)$ exists by Lemma 26.3, and $\gamma(p, 2) = 2p$ by Theorem 26.4.

Theorem 26.12. *Let $n, p, r \geq 1$. With notation as above,*

$$\mathsf{HCov}_{n,p,r}^{\#} \simeq \bigvee_{k=p+r}^{\gamma(p,r)+1} \bigvee_{\binom{n-1}{k-1}} \Lambda_{k,p,r} = \bigvee_{k=p+r}^{\min\{\gamma(p,r)+1,n\}} \bigvee_{\binom{n-1}{k-1}} \Lambda_{k,p,r}, \qquad (26.4)$$

where $\gamma = \gamma(p,r)$ is defined as in (26.3); $\gamma(p,r) = pr$ for $1 \leq r \leq 2$.

Remark. Since the 1-skeleton of $\mathsf{HCov}_{n,p,r}^{\#}$ is full as soon as $p \geq 2$, the right-hand side in (26.4) is unambiguous from a homotopy point of view. For $p = 1$, $\mathsf{HCov}_{n,p,r}^{\#}$ is homotopy equivalent to a wedge of spheres in a fixed dimension by Proposition 26.10, which immediately yields unambiguity.

Proof. First, note that Lemma 26.11 implies that

$$\mathsf{HCov}_{n,p,r}^{\#} \simeq \bigvee_{J \subseteq [n-1]} \Lambda_{n,p,r}(J) \simeq \bigvee_{k=1}^{n} \bigvee_{\binom{n-1}{k-1}} \Lambda_{k,p,r}.$$

Namely, by the proof of the lemma, $\mathcal{C}_{n,p,r}(J) \not\longmapsto \mathcal{C}_{n,p,r}(I)$ if $I \neq J$; hence Theorem 4.11 yields the desired result. It remains to prove that $\mathcal{C}_{n,p,r}(J)$ is void unless $p + r \leq |J| + 1 \leq \gamma(p,r) + 1$. The lower bound follows by Lemma 26.2, whereas the upper bound is by definition of $\gamma(p,r)$; see (26.3). $\square$

Corollary 26.13. *Let $p, r \geq 1$ and $n \geq \gamma(p,r)+1$. For any field $\mathbb{F}$ and any integer $i \geq -1$, $\tilde{H}_i(\mathsf{HCov}_{n,p,r}^{\#}, \mathbb{F})$ is nonzero if and only if $\tilde{H}_i(\mathsf{HCov}_{\gamma(p,r)+1,p,r}^{\#}, \mathbb{F})$ is nonzero. Moreover, the connectivity degrees of the complexes $\mathsf{HCov}_{n,p,r}^{\#}$ and $\mathsf{HCov}_{\gamma(p,r)+1,p,r}^{\#}$ are the same. In particular, for $n \geq 2p+1$, $\tilde{H}_i(\mathsf{Cov}_{n,p}^{\#}, \mathbb{F}) \neq 0$ if and only if $\tilde{H}_i(\mathsf{Cov}_{2p+1,p}^{\#}, \mathbb{F}) \neq 0$, and the connectivity degrees of $\mathsf{Cov}_{n,p}^{\#}$ and $\mathsf{Cov}_{2p+1,p}^{\#}$ are the same.*

Proof. Whenever $n \geq \gamma(p,r) + 1$, $\tilde{H}_i(\mathsf{HCov}_{n,p,r}^{\#}, \mathbb{F})$ is nonzero if and only if $\tilde{H}_i(\Lambda_{k,p,r}, \mathbb{F})$ is nonzero for some k such that $p + r \leq k \leq \gamma(p,r) + 1$; use Theorem 26.12. By the same theorem, the connectivity degree of $\mathsf{HCov}_{n,p,r}^{\#}$ is the minimum of the connectivity degrees of $\Lambda_{k,p,r}$ for $p+r \leq k \leq \gamma(p,r) + 1$. Since these conditions do not depend on n, we are done. For the last claim, apply Theorem 26.4. $\square$

Corollary 26.14. *Let $n, p, r \geq 1$. For any field $\mathbb{F}$ and any integer $i \geq -1$, the Betti number $\beta_i(\mathsf{HCov}_{n,p,r}^{\#}, \mathbb{F}) = \dim \tilde{H}_i(\mathsf{HCov}_{n,p,r}^{\#}, \mathbb{F})$ satisfies*

$$\beta_i(\mathsf{HCov}_{n,p,r}^{\#}, \mathbb{F}) = \sum_{k=p+r}^{\gamma+1} (-1)^{\gamma+1-k} \binom{n-1}{k-1}\binom{n-1-k}{\gamma+1-k} \beta_i(\mathsf{HCov}_{k,p,r}^{\#}, \mathbb{F});$$

$\gamma = \gamma(p,r)$. In particular, $\beta_i(\mathsf{HCov}_{n,p,r}^{\#}, \mathbb{F})$ is a polynomial in n of degree at most $\gamma(p,r)$.

Remark. Since $\gamma(p,2) = 2p$ by Theorem 26.4, we have that

$$\beta_i(\mathsf{Cov}_{n,p}^{\#}, \mathbb{F}) = \sum_{k=p+2}^{2p+1} (-1)^{k-1} \binom{n-1}{k-1} \binom{n-1-k}{2p+1-k} \beta_i(\mathsf{Cov}_{k,p}^{\#}, \mathbb{F}).$$

By Proposition 26.8, we may choose $\gamma(p,r) = pr$ in the corollary whenever $p \leq 2$.

Proof. By Theorem 26.12, we know that $f_{p,r,i}(n) = \beta_i(\mathsf{HCov}_{n,p,r}^{\#}, \mathbb{F})$ defines a polynomial in n of degree at most $\gamma(p,r)$ such that $f_{p,r,i}(k) = 0$ for $1 \leq k \leq p+r-1$. By Proposition 6.13 with $s = 1$, we are done. $\square$

For the remainder of this section, we confine ourselves to the case $r = 2$.

Corollary 26.15. *Let $\mathbb{F}$ be a field or $\mathbb{Z}$. For $1 \leq p \leq n-2$, $\tilde{H}_i(\mathsf{Cov}_{n,p}, \mathbb{F}) = \tilde{H}_i(\mathsf{Cov}_{n,p}^{\#}, \mathbb{F})$ is zero unless $i \leq p \cdot \min\{p+1, \frac{n+1}{2}\} - 1$. Hence, for $2 \leq q \leq n-1$, $\tilde{H}_i(\mathsf{Cov}_{n,n-q}, \mathbb{F})$ is zero unless $i \leq \lfloor \frac{(n+1)(n-q)}{2} \rfloor - 1$, which implies that the Alexander dual of $\mathsf{Cov}_{n,n-q}$ has no homology strictly below dimension $\lceil \frac{(q-2)(n+1)}{2} \rceil - 1$.*

Proof. It is clear that all hypergraphs $G \in \mathcal{C}_{k,p,2}([k-1])$ are ordinary graphs; since $G([k-1])$ is $(p,2)$-solid, $G([k-1])$ has this property (apply Lemma 26.1), and the singleton edge k cannot be present in G. We claim that a graph $G \in \mathcal{C}_{k,p,2}([k-1])$ has at most $p \cdot \frac{k+1}{2}$ edges; inserting $k = \min\{2p+1, n\}$ yields the desired bound. Now, by construction, the degree of each vertex in $G([k-1])$ is at most p; otherwise some vertices would necessarily be part of every p-cover of $G([k-1])$. Also, the vertex k is not part of any p-cover, which implies that the degree of k is at most p. Summing, we get $p \cdot \frac{k-1}{2} + p = p \cdot \frac{k+1}{2}$ as claimed. The last statement follows by Alexander duality; $\binom{n}{2} - (\frac{(n+1)(n-q)}{2} - 1) - 3 = \frac{(q-2)(n+1)}{2} - 1$. $\square$

Remark. In Section 26.6, we show that $\tilde{H}_i(\mathsf{Cov}_{n,p})$ is zero unless $i \geq 2p-1$.

One may compare the last statement in the corollary to Linusson and Shareshian's Corollary 25.4, which states that the complex Col_n^t of t-colorable graphs on n vertices is $(\lceil \frac{(t-1)(n-1)}{2} \rceil - 2)$-connected. In this context, it might be worth noting that Col_n^t is contained in the Alexander dual of $\mathsf{Cov}_{n,n-(t+1)}$; a t-colorable graph does not contain any $(t+1)$-cliques. Since our acyclic matching is closely related to the acyclic matching of Linusson and Shareshian (see Lemma 25.3), it is therefore not too surprising that our bound is only slightly different from theirs. See Section 26.8 for a potential improvement of this bound.

Finally, we prove a minor result about the Euler characteristic of $\mathsf{Cov}_{n,p}^{\#}$. Note that Corollaries 26.14 and 26.15 imply that $\tilde{\chi}(\mathsf{Cov}_{n,p}^{\#})$ defines a polynomial in n of degree at most $2p$ for each fixed p.

Proposition 26.16. *Let $p \geq 1$ and let f_p be the polynomial with the property that $f_p(n) = \tilde{\chi}(\mathsf{Cov}_{n,p}^{\#})$ for $n \geq 1$. Then $f_p(0) = -1$. Moreover, let $\mathcal{Y}_{n,p}$ be the family of hypergraphs in $\mathsf{Cov}_{n,p}^{\#}$ with no uncovered vertices. Then $\tilde{\chi}(\mathcal{Y}_{n,p}) = 0$ whenever $n > 2p$.*

Proof. Define $\mathsf{Cov}_{0,p}^{\#} = \mathcal{Y}_{0,p} = \{\emptyset\}$. Clearly,

$$\tilde{\chi}(\mathsf{Cov}_{n,p}^{\#}) = \sum_{k=0}^{n} \binom{n}{k} \tilde{\chi}(\mathcal{Y}_{k,p}) \tag{26.5}$$

for all $n \geq 0$. Moreover, for $n \geq 1$,

$$\tilde{\chi}(\mathsf{Cov}_{n,p}^{\#}) = f_p(n) = \sum_{k \geq 0} \binom{n}{k} y_k, \tag{26.6}$$

where $y_k = 0$ for $k > 2p$; the degree of f_p is at most $2p$. One easily derives from (26.5) and (26.6) that

$$y_n - \tilde{\chi}(\mathcal{Y}_{n,p}) = (-1)^n (y_0 - \tilde{\chi}(\mathcal{Y}_{0,p})) = (-1)^n (y_0 + 1)$$

for $n \geq 0$. Thus it suffices to prove that $\tilde{\chi}(\mathcal{Y}_{n,p}) = 0$ for some $n > 2p$; this will imply that $y_0 = \tilde{\chi}(\mathcal{Y}_{0,p}) = -1$ and hence that $f_p(0) = \tilde{\chi}(\mathsf{Cov}_{0,p}^{\#}) = -1$ as desired. As a byproduct, we will also obtain that $\tilde{\chi}(\mathcal{Y}_{n,p}) = 0$ for all $n > 2p$.

Now, for a given hypergraph $H \in \mathcal{Y}_{n,p}$, let H^* be the graph obtained from H by removing all singleton edges. Let $\mathcal{X}_{n,p}$ be the subfamily of $\mathcal{Y}_{n,p}$ consisting of all hypergraphs H such that some vertex x is contained in every p-cover of the underlying graph H^*. For each $H \in \mathcal{X}_{n,p}$, let $x(H)$ be minimal with this property. We obtain a perfect element matching on $\mathcal{X}_{n,p}$ by pairing $H - \{x(H)\}$ and $H + \{x(H)\}$.

Let $H \in \mathcal{Y}_{n,p} \setminus \mathcal{X}_{n,p}$ and let W be a p-cover of H. By assumption, for each $w \in W$, there is a p-cover of H^* not containing w, which implies that w is adjacent to at most p vertices in H. It follows that there are at most $p + p^2$ covered vertices in H; hence $\mathcal{Y}_{n,p} \setminus \mathcal{X}_{n,p} = \emptyset$ whenever $n > p + p^2$. As a consequence, $\tilde{\chi}(\mathcal{Y}_{n,p}) = 0$ whenever $n > p + p^2$, and we are done. $\square$

26.5 Computations

Corollary 26.14 reduces the problem of determining the homology of the complex $\mathsf{HCov}_{n,p,r} \simeq \mathsf{HCov}_{n,p,r}^{\#}$ for general $n \geq p + r$ to the special cases $p + r \leq n \leq \gamma(p, r) + 1$. For $r = 2$, we know by Theorem 26.4 that it suffices to consider $p + 2 \leq n \leq 2p + 1$. Using the computer program homology [42], we have been able to compute the homology of $\mathsf{Cov}_{n,p} = \mathsf{HCov}_{n,p,2}$ for $p = 2, 3$; the results are presented in Theorems 26.17 and 26.18 below.

For integers m, r, define $C(m, r) = \binom{m}{r}$.

Theorem 26.17. *For $n \geq 4$, the kth homology group of $\mathsf{Cov}_{n,2}$ is zero unless $3 \leq k \leq 4$, in which case we have that*

$$\tilde{H}_3(\mathsf{Cov}_{n,2}) \cong \mathbb{Z}^{C(n-1,4)};$$
$$\tilde{H}_4(\mathsf{Cov}_{n,2}) \cong \mathbb{Z}^{C(n,4)}.$$

In particular, the reduced Euler characteristic of $\mathsf{Cov}_{n,2}$ is $\binom{n-1}{3}$.

Proof. Running `homology` [42] on the complex $\mathsf{Cov}_{5,2}$, we obtain that

$$\tilde{H}_3(\mathsf{Cov}_{5,2}) \cong \mathbb{Z};$$
$$\tilde{H}_4(\mathsf{Cov}_{5,2}) \cong \mathbb{Z}^5.$$

By (26.1), $\mathsf{Cov}_{4,2} \simeq S^4$. Thus Corollary 26.14 yields that the homology of $\mathsf{Cov}_{n,2}$ is torsion-free and that

$$\dim \tilde{H}_3(\mathsf{Cov}_{n,2}, \mathbb{Q}) = \binom{n-1}{5-1}\binom{n-5-1}{4-5+1} = \binom{n-1}{4};$$
$$\dim \tilde{H}_4(\mathsf{Cov}_{n,2}, \mathbb{Q}) = -\binom{n-1}{4-1}\binom{n-4-1}{4-4+1} + 5\binom{n-1}{5-1}\binom{n-5-1}{4-5+1} = \binom{n}{4}. \quad \square$$

Remark. We have not been able to determine the homotopy type of $\mathsf{Cov}_{n,2}$.

Theorem 26.18. *For $n \geq 5$, the kth homology group of $\mathsf{Cov}_{n,3}$ is zero unless $5 \leq k \leq 8$, in which case we have that*

$$\tilde{H}_5(\mathsf{Cov}_{n,3}) \cong \mathbb{Z}^{C(n-1,6)};$$
$$\tilde{H}_6(\mathsf{Cov}_{n,3}) \cong (\mathbb{Z}_2)^{C(n,6)};$$
$$\tilde{H}_7(\mathsf{Cov}_{n,3}) \cong \mathbb{Z}^{9C(n,6)};$$
$$\tilde{H}_8(\mathsf{Cov}_{n,3}) \cong \mathbb{Z}^{C(n,5)}.$$

In particular, the reduced Euler characteristic of $\mathsf{Cov}_{n,3}$ is $-\binom{n-1}{4} \cdot \frac{5n^2 - 31n + 15}{15}$. By Proposition 26.8, the same holds for the complex $\mathsf{Cov}_{n,2,3}$.

Proof. Computations with `homology` [42] yield that

$$\begin{cases} \tilde{H}_6(\mathsf{Cov}_{6,3}) \cong \mathbb{Z}_2; \\ \tilde{H}_7(\mathsf{Cov}_{6,3}) \cong \mathbb{Z}^9; \\ \tilde{H}_8(\mathsf{Cov}_{6,3}) \cong \mathbb{Z}^6 \end{cases} \quad \text{and} \quad \begin{cases} \tilde{H}_5(\mathsf{Cov}_{7,3}) \cong \mathbb{Z}; \\ \tilde{H}_6(\mathsf{Cov}_{7,3}) \cong (\mathbb{Z}_2)^7; \\ \tilde{H}_7(\mathsf{Cov}_{7,3}) \cong \mathbb{Z}^{63}; \\ \tilde{H}_8(\mathsf{Cov}_{7,3}) \cong \mathbb{Z}^{21}. \end{cases}$$

By (26.1), we know that $\tilde{H}_i(\mathsf{Cov}_{5,3}) = \mathbb{Z}$ if $i = 8$ and 0 otherwise. By Corollary 26.14, there is no torsion in $\tilde{H}_i(\mathsf{Cov}_{n,3}, \mathbb{Z})$ unless $i = 6$, in which case there is 2-torsion but no free homology. Corollary 26.14 yields that

Table 26.1. The reduced Euler characteristic of $\mathsf{HCov}^{\#}_{n,p,2}$ for small values on n and p. Recall that $\tilde{\chi}(\mathsf{HCov}^{\#}_{n,p,2}) = \tilde{\chi}(\mathsf{Cov}_{n,p})$ whenever $n \geq 2$.

$\tilde{\chi}(\mathsf{HCov}^{\#}_{n,p,2})$	$n = 0$	1	2	3	4	5	6	7	8	9	10
$p = 1$	-1	0	0	-1	-3	-6	-10	-15	-21	-28	-36
2	-1	0	0	0	1	4	10	20	35	56	84
3	-1	0	0	0	0	1	-3	-43	-203	-658	-1722
4	-1	0	0	0	0	0	-1	-61	?	?	?

Table 26.2. The homology of $\Lambda_{k,p,2}$ for all interesting (k,p) such that $2 \leq p \leq 3$ and for $(k,p) = (6,4),(7,4)$ (we obtained the latter homology via a computer calculation of the homology of $\mathsf{Cov}_{7,4}$).

$\tilde{H}_i(\Lambda_{k,p,2},\mathbb{Z})$	$i = 3$	4	5	6	7	8	9	10	11	12	13
$(k,p) = (4,2)$	-	$\mathbb{Z}$	-	-	-	-	-	-	-	-	-
$(5,2)$	$\mathbb{Z}$	$\mathbb{Z}$	-	-	-	-	-	-	-	-	-
$(5,3)$	-	-	-	-	-	$\mathbb{Z}$	-	-	-	-	-
$(6,3)$	-	-	-	$\mathbb{Z}_2$	$\mathbb{Z}^9$	$\mathbb{Z}$	-	-	-	-	-
$(7,3)$	-	-	$\mathbb{Z}$	$\mathbb{Z}_2$	$\mathbb{Z}^9$	-	-	-	-	-	-
$(6,4)$	-	-	-	-	-	-	-	-	-	-	$\mathbb{Z}$
$(7,4)$	-	-	-	-	-	-	-	$\mathbb{Z}$	$\mathbb{Z}^{55} \oplus \mathbb{Z}_2$	-	$\mathbb{Z}$

$$\dim \tilde{H}_5(\mathsf{Cov}_{n,3},\mathbb{Q}) = \binom{n-1}{6}\binom{n-8}{0} = \binom{n-1}{6};$$

$$\dim \tilde{H}_6(\mathsf{Cov}_{n,3},\mathbb{Z}_2) = -\binom{n-1}{5}\binom{n-7}{1} + 7\binom{n-1}{6}\binom{n-8}{0} = \binom{n}{6};$$

$$\dim \tilde{H}_7(\mathsf{Cov}_{n,3},\mathbb{Q}) = -9\binom{n-1}{5}\binom{n-7}{1} + 63\binom{n-1}{6}\binom{n-8}{0} = 9\binom{n}{6};$$

$$\dim \tilde{H}_8(\mathsf{Cov}_{n,3},\mathbb{Q}) = \binom{n-1}{4}\binom{n-6}{2} - 6\binom{n-1}{5}\binom{n-7}{1} + 21\binom{n-1}{6}\binom{n-8}{0} = \binom{n}{5}.$$

$\square$

See Table 26.1 for the Euler characteristic of $\mathsf{Cov}_{n,2}$ for small n and p.

Remark. Note that all Betti numbers are integer multiples of binomial coefficients. This is due to Theorem 26.12 and the simple structure of the homology of $\Lambda_{k,p,2}$; see Table 26.2. We would be surprised if this property held in general; see Proposition 26.22 (e) for a potential conjecture that might be a bit more realistic.

26.6 Homotopical Depth

We prove that $\mathsf{Cov}_{n,p}$ has a vertex-decomposable $(2p-1)$-skeleton for $n \geq p+2$. Note that we consider the graph complex $\mathsf{Cov}_{n,p}$, not the hypergraph complex $\mathsf{HCov}_{n,p,2}^{\#}$ We have not been able to prove anything of interest about the depth or the connectivity degree of $\mathsf{HCov}_{n,p,r}$ for $r \geq 3$.

Theorem 26.19. *For $1 \leq p \leq n-2$, $\mathsf{Cov}_{n,p}$ is $VD(2p-1)$. In particular, the homotopical depth of $\mathsf{Cov}_{n,p}$ is at least $(2p-1)$.*

Proof. Let $Y = \binom{[n-1]}{2}$ and $E_n = \{1n, \ldots, (n-1)n\}$. For any subset B of Y, let

$$d_p(B) = p + \min\{p - 1, |Y \setminus B|\};$$

$d_p(B) = 2p - 1$ if $|Y \setminus B| \geq p - 1$. We claim that $\mathsf{Cov}_{n,p}(A, B)$ is $VD(d_p(B))$ for any disjoint subsets A and B of Y. The special case $A = B = \emptyset$ yields the theorem, since $|Y| \geq p - 1$.

To prove the claim, we use induction on $|Y \setminus B|$. We distinguish three cases:

(i) $|Y \setminus B| \leq p - 1$. Then the covering number of the graph with edge set $Y \setminus B$ is at most $p - 1$. As a consequence, the graph with edge set $E_n \cup (Y \setminus B)$ has covering number at most p, which implies that all edges in $E_n \cup (Y \setminus (A \cup B))$ are cone points in $\mathsf{Cov}_{n,p}(A, B)$. In particular, $\mathsf{Cov}_{n,p}(A, B)$ is the join of $\{A\}$ and the full simplex on

$$|E_n| + |Y \setminus B| - |A| = n - 1 + |Y \setminus B| - |A| \geq d_p(B) - |A| + 1$$

elements ($n - 1 \geq p + 1$). This implies that $\mathsf{Cov}_{n,p}(A, B)$ is $VD(d_p(B))$ as desired.

(ii) $|Y \setminus B| \geq p$ and $A \subsetneq Y \setminus B$. Then let $e \in Y \setminus (A \cup B)$. We have by induction on $|Y \setminus (A \cup B)|$ that $\mathsf{Cov}_{n,p}(A+e, B)$ and $\mathsf{Cov}_{n,p}(A, B+e)$ are $VD(2p-1)$. As a consequence, $\mathsf{Cov}_{n,p}(A, B)$ is $VD(2p-1)$ by Lemma 6.9.

(iii) $|Y \setminus B| = |A| \geq p$ and $A = Y \setminus B$. In this case, we consider complexes $\mathsf{Cov}_{n,p}(A, Y \setminus A)$ such that $|A| \geq p$. Note that all faces of $\mathsf{Cov}_{n,p}(A, Y \setminus A)$ are of the form $A \cup C$ for some set $C \subseteq E_n$. Let H be the graph with edge set A. We identify three subcases:

  (a) $\tau(H) \leq p - 1$. Then all $n - 1$ edges in E_n are cone points in $\mathsf{Cov}_{n,p}(A, B)$, and we are done; $|E_n| - 1 = n - 2 \geq p \geq 2p - |A| > d_p(Y \setminus A) - |A|$.

  (b) $\tau(H) = p$ and some vertex x is contained in every p-cover of H. Since every p-cover contains x, the edge xn is a cone point in $\mathsf{Cov}_{n,p}(A, B)$. In particular, $\mathsf{Cov}_{n,p}(A, Y \setminus A)$ is $VD(2p-1)$ if and only if $\mathsf{Cov}_{n,p}(A+xn, Y \setminus A)$ is $VD(2p - 1)$.

  Define A_0 and Y_0 as the sets obtained from A and Y by removing all edges containing x; hence $Y_0 = \binom{[n-1] \setminus \{x\}}{2}$. We have that $\mathsf{Cov}_{n,p}(A+xn, Y \setminus A)$ coincides with $\mathsf{Cov}_{n-1,p-1}(A_0, Y_0 \setminus A_0) * \{(A+xn) \setminus A_0\}$, where we remove the vertex x (rather than n) to obtain $\mathsf{Cov}_{n-1,p-1}$.

Namely, a graph G containing H and being contained in $H+E_n$ has a p-cover if and only if $G([n]\setminus\{x\})$ has a $(p-1)$-cover. By induction on n, $\mathsf{Cov}_{n-1,p-1}(A_0, Y_0\setminus A_0)$ is $VD(d_{p-1})$, where $d_{p-1} = d_{p-1}(Y_0\setminus A_0)$. We need to prove that

$$d_{p-1} \geq 2p - 1 - |(A + xn) - A_0| = 2p - 2 - |A| + |A_0|.$$

Now,

$$d_{p-1} = p - 1 + \min\{p - 2, |A_0|\}.$$

If $p - 2 \geq |A_0|$, then $d_{p-1} = p - 1 + |A_0|$, which is at least $2p - 1 - |A| + |A_0|$, as $|A| \geq p$. If $p - 2 < |A_0|$, then $d_{p-1} = 2p - 3$, which is at least $2p - 2 + |A_0| - |A|$, as $|A \setminus A_0| \geq 1$. In fact, we must have $|A \setminus A_0| > 1$, because x is contained in every p-cover. Thus we are done.

(c) $\tau(H) = p$ and no vertex is contained in every p-cover of H. This means that H is $(p, 2)$-solid. As a consequence, Lemma 26.6 yields that $|A| \geq 2p - k$, where k is the number of connected components of H with at least two vertices. Thus it suffices to prove that $\Delta = \mathsf{Cov}_{n,p}(A, Y \setminus A)$ is $VD(|A|+k-1)$. Let $C_1, \ldots, C_k$ be the connected components of H (uncovered vertices excluded); by Lemma 26.5, each C_i is $(p_i, 2)$-solid for some $p_i \geq 1$ satisfying $\sum_i p_i = p$. Let T_i be the set of edges $xn \in E_n$ with one endpoint x in C_i. Let Δ_i be the induced subcomplex of Δ on the set T_i. It is clear that $\Delta = \{A\} * \Delta_1 * \cdots * \Delta_k$; we can add a subset Q of E_n to H without increasing $p = \tau(H)$ if and only if we can add the corresponding subsets $Q \cap T_i$ without increasing $p_i = \tau(C_i)$.

Now, each vertex in C_i is contained in a p_i-cover of C_i by Lemma 26.1, and $p_i \geq 1$ for each i. As a consequence, Δ_i is $VD(0)$ for each i, which implies by Lemma 6.11 that Δ is $VD(|A| + k - 1)$. Thus we are done. $\square$

We conjecture that there is nonvanishing homology in dimension $2p - 1$ for $n \geq 2p + 1$; this would imply that the shifted connectivity degree of $\mathsf{Cov}_{n,p}$ equals $2p - 1$ whenever $n \geq 2p + 1$. See Section 26.8 for further discussion.

26.7 Triangle-Free Graphs

Note that $\mathsf{Cov}_{n,p}$ is the Alexander dual of the complex of graphs on n vertices that do not contain a clique of size $n - p$. For $p = n - 3$, we obtain the complex $\not{\triangleright}_n$ of triangle-free graphs on n vertices. In this section, we summarize our humble results for this very important graph property.

Corollary 26.20. *For $n \geq r + 2$,*

$$\mathsf{HCov}_{n,n-r-1,r} \simeq \Lambda_{n,n-r-1,r} \vee \bigvee_{n-1} \Lambda_{n-1,n-r-1,r}$$

$$\simeq \Lambda_{n,n-r-1,r} \vee \bigvee_{n-1} S^{C(n-1,r)-2},$$

where $C(m,k) = \binom{m}{k}$. In particular, for $r = 2$, the dual complex $\not\!\!{\triangleright}_n$ of triangle-free graphs on n vertices has the property that $\tilde{H}_{n-2}(\not\!\!{\triangleright}_n, \mathbb{Z})$ contains $\mathbb{Z}^{n-1}$ as a free subgroup.

Proof. The first equivalence is Theorem 26.12. The second equivalence follows from the fact that

$$\Lambda_{p+r,p,r} \simeq \mathsf{HCov}_{p+r,p,r} \simeq S^{C(p+r,r)-2};$$

use Theorem 26.12 and (26.1) with $p = n - r - 1$. For the final statement, use Theorem 3.4. $\square$

Corollary 26.21. *For $n \geq 3$, $\not\!\!{\triangleright}_n$ is $VD(n-2)$ and has homotopical depth $n-2$.*

Proof. $\not\!\!{\triangleright}_n$ is $VD(n-2)$ by Corollary 13.8; all minimal nonfaces are triangles, which are isthmus-free. However, the $(n-1)$-skeleton of $\not\!\!{\triangleright}_n$ is not even Cohen-Macaulay, as there is homology in dimension $n-2$; use Theorem 13.9 or Corollary 26.20. $\square$

Table 26.3. The homology of $\not\!\!{\triangleright}_n$ for $4 \leq n \leq 7$. The figures are collected from Table 26.2 and translated via Alexander duality.

$\tilde{H}_i(\not\!\!{\triangleright}_n, \mathbb{Z})$	$i=2$	3	4	5	6	7	8
$n=4$	$\mathbb{Z}^3$	-	-	-	-	-	-
5	-	$\mathbb{Z}^5$	$\mathbb{Z}$	-	-	-	-
6	-	-	$\mathbb{Z}^6$	$\mathbb{Z}^9 \oplus \mathbb{Z}_2$	-	-	-
7	-	-	-	$\mathbb{Z}^7$	$\mathbb{Z}_2$	$\mathbb{Z}^{55}$	$\mathbb{Z}$

We have no complete description of the homology of $\not\!\!{\triangleright}_n$ except for $n \leq 7$; see Table 26.3.

26.8 Concluding Remarks and Open Problems

We have not been able to compute the homology of $\mathsf{HCov}_{n,p,r}$ for general n, p, and r, and we have very little hope to ever see this being achieved; see the complexity-theoretic remark below for some further discussion. Nevertheless, the homology of $\mathsf{Cov}_{n,p} = \mathsf{HCov}_{n,p,2}$ certainly has plenty of structure, and our computations for small values of n and p suggest that there is quite some more structure to be found. In the following proposition, note that we restrict our attention to $p \leq 3$.

Proposition 26.22. *The following hold for* $1 \leq p \leq 3$:

(a) $\mathsf{Cov}_{n,p}$ *has no homology over any field strictly above dimension* $\binom{p+2}{2} - 2$. *Equivalently, the Alexander dual of* $\mathsf{Cov}_{n,n-r}$ *has no homology strictly below dimension* $d_{n,r} = n(r-2) - \binom{r-1}{2} - 1$.

(b) *For* $p + 2 \leq n \leq 2p + 1$, $\mathsf{Cov}_{n,p}$ *has no homology strictly below dimension* $2p - 1 + \binom{2p-n+2}{2}$.

(c) *For* $p = 2$ *and* $p = 3$, $\tilde{H}_{\binom{p+2}{2}-2}(\mathsf{Cov}_{n,p}, \mathbb{Z})$ *is free of rank* $\binom{n}{p+2}$.

(d) $\tilde{H}_{2p-1}(\mathsf{Cov}_{n,p}, \mathbb{Z})$ *is free of rank* $\binom{n-1}{2p}$.

(e) *For* $i \geq 2p$ *and for any field* $\mathbb{F}$,

$$\sum_{k=p+2}^{2p+1} (-1)^k \beta_i(\Lambda_{k,p,2}, \mathbb{F}) = 0.$$

Equivalently, for $i \geq 2p$, *the polynomial* $f_{p,i}(n) = \beta_i(\mathsf{Cov}_{n,p}, \mathbb{F})$ *vanishes at zero.*

(f) *All roots of the polynomial* $f_{p,i}$ *are real and nonnegative (they are indeed integers). Moreover, the Euler characteristic of* $\mathsf{Cov}_{n,p}$ *is a polynomial in* n *with only real and positive roots.* $\square$

Note that properties (a)-(d) are also true for $p = 4$ and $n \leq 7$; see Table 26.2. Moreover, by Corollary 26.21, property (a) is true whenever $p = n - 3$.

Proposition 26.22 suggests the following problem.

Question 2. Among the six properties listed in Proposition 26.22, which of them hold for general p?

We are particularly interested in knowing whether property (a) remains true in general. First, this relates to the important problem of determining the connectivity degree of the complex of K_{n-p}-free graphs; this is the Alexander dual $\mathsf{Cov}_{n,p}^*$ of $\mathsf{Cov}_{n,p}$. Second, we would like to know more about connections between the complex Col_n^t of t-colorable graphs and $\mathsf{Cov}_{n,n-(t+1)}^*$; recall that the latter complex contains the former. As Linusson and Shareshian [94] observed (see Chapter 25 for discussion), most homology of Col_n^t is concentrated in dimension $n(t-1) - \binom{t}{2} - 1 = d_{n,t+1}$ for all known examples, and so far no homology below this dimension has been found.

Regarding property (b), one may also ask whether the corresponding skeleton is vertex-decomposable or at least Cohen-Macaulay. Regarding property (c), we know that $\tilde{H}_{\binom{p+2}{2}-2}(\mathsf{Cov}_{n,p}, \mathbb{Z})$ contains a free subgroup of rank $\binom{n-1}{p+1}$; use (26.1) and Corollary 26.14.

Since we do not have much data, it may well turn out that several of the properties in Proposition 26.22 do not generalize to larger values of p. We are particularly skeptical about properties (b) and (f). Regarding property (f), one may recall Conjecture 14.20 about the real-rootedness of $\tilde{\chi}(\mathsf{B}_{n,p})$; see Section 14.3.3. We justified this conjecture by reducing it to a conjecture about the homotopy type of certain simplicial complexes. In the present case, we have no such justification.

For hypergraphs, the situation is even worse, as we have almost *no* data. Still, regarding property (a), one may ask whether it is true that $\mathsf{HCov}_{n,p,r}$ has no homology over any field strictly above dimension $\binom{p+r}{r} - 2$.

In our opinion however, the most important open problem for $r \geq 3$ is to determine the maximum integer k for which $\Lambda_{k,p,r}$ has nonvanishing homology. This would give an upper bound on the degree of the polynomials $f_{p,r,i}$. Our hope is that the answer is pr, but we have no evidence whatsoever for this guess when $p, r \geq 3$. In particular, pr is not an upper bound on $\gamma(p, r)$; $\gamma(3,3) \geq 10$, as the hypergraph on the vertex set $\{0, 1, \ldots, 9\}$ with edges $012, 234, 456, 678, 890$ is $(3, 3)$-solid.

Recall that $\mathsf{NM}_{n,k}$ is the complex of graphs on n vertices that do not contain a k-matching and that $\mathsf{NQP}_{n,t}$ is the complex of graphs on n vertices that do not admit a clique partition into t parts; see Chapter 25 for more information. In a separate manuscript [73], we give a unified proof that $\mathsf{NM}_{n,p+1}$, $\mathsf{NQP}_{n,n-p-1}$, and $\mathsf{Cov}_{n,p}$ have Euler characteristics given by polynomials. These graph properties have in common that they all avoid $(p+1)$-matchings while admitting p-matchings. Moreover, if Σ_n is any of the properties, then there is a d such that the following holds:

- If $G \in \Sigma_n$ and $x \in G$ has the property that $\deg_G(x) \geq d$, then $G + E_n(x) \in \Sigma_n$.

We show that the polynomial property holds for any class of graph properties satisfying these two conditions, provided that the union of all properties in the class is closed under addition and deletion of isolated vertices.

Complexity-theoretic remark. The (VERTEX) COVER problem on input a pair (G, p) is to determine whether $G \in \mathsf{Cov}_{n,p}$; n is the number of vertices in G. This is the *containment problem* for the family $\{\mathsf{Cov}_{n,p} : n, p \geq 1\}$. COVER is well-known to be NP-complete [81, 33]. A potentially interesting question is whether there is any deeper connection between this fact and the fact that the homology of $\mathsf{Cov}_{n,p}$ seems difficult to compute for general n and p. One may raise the analogous question for any family of monotone graph properties with an NP-complete containment problem. One example is the NP-complete HAMILTONIAN problem: On input a graph G, determine whether $G \in \mathsf{Ham}_n$, where Ham_n is the quotient complex of Hamiltonian graphs discussed in Chapter 17.

Part VIII

Open Problems

27

Open Problems

We collect different open problems related to the complexes discussed in this book. Some of these problems have already been discussed in earlier chapters, but we restate them for completeness. Whenever we make a statement about the connectivity degree of a complex, we ignore special cases for which the complex happens to be contractible.

Problems on Chapter 5

Throughout this section, Δ and Γ are simplicial complexes.

Problem 27.1 (cf. Propositions 5.17 and 5.19). Is it true that every contractible complex is buildable? Is it true that every homotopically Cohen-Macaulay complex is semi-buildable?
Conjecture: False; we suggest the dunce hat as a potential counterexample.

Problem 27.2 (Welker [146]; cf. Theorem 5.27). If $\Delta * \Gamma$ is collapsible, is it true that at least one of Δ and Γ is collapsible?

Problem 27.3 (cf. Theorem 5.28). If $\Delta * \Gamma$ is semi-collapsible but not collapsible, is it true that each of Δ and Γ is semi-collapsible?

Problem 27.4 (Welker [146]; cf. Theorem 5.30). If the barycentric subdivision of Δ is nonevasive, is it true that Δ is collapsible?

Problem 27.5 (cf. Theorem 5.31). If the barycentric subdivision of Δ is semi-nonevasive, is it true that Δ is semi-collapsible?

Problem 27.6 (Welker [146]; cf. Theorem 5.30). If the barycentric subdivision of Δ is collapsible, is it true that this subdivision is in fact nonevasive?

Problem 27.7 (cf. Theorem 5.31). If the barycentric subdivision of Δ is semi-collapsible, is it true that this subdivision is in fact semi-nonevasive?

Problem 27.8 (Welker [146]; cf. Theorem 5.33). If $\Delta(P \times Q)$ is nonevasive, is it true that $\Delta(P)$ and $\Delta(Q)$ are both nonevasive?

Problem 27.9 (cf. Theorem 5.34). If $\Delta(P)$ and $\Delta(Q)$ are semi-nonevasive and evasive, is it true that $\Delta(P \times Q)$ is semi-nonevasive?

Problems on Chapter 11

Problem 27.10 (cf. Theorems 11.16 and 11.26). Find the rank of the elementary 3-group $\tilde{H}_{k-1+r}(M_{2k+1+3r}; \mathbb{Z})$ for $k \in \{1,2\}$ and $r \geq k + 2$.

Problem 27.11 (cf. Corollary 11.23). Is it true that $\tilde{H}_d(M_n; \mathbb{Z})$ is torsion-free if and only if $d < \frac{n-4}{3}$ or $d > \frac{n-5}{2}$?
Conjecture: True. Open for n odd and $d = \frac{n-5}{2}$.

Problem 27.12 (cf. Section 11.2.3). For which integers p and n is there a d such that $\tilde{H}_d(M_n; \mathbb{Z})$ contains p-torsion?
Remark. For almost all n, the homology of M_n contains 3-torsion. A daring conjecture would be that this is true not only for $p = 3$ but for all odd p.

Problem 27.13 (cf. Section 11.2.4). For each n and d, is it true that

$$\tilde{H}_d(M_n; \mathbb{Z}) \cong \tilde{H}_d(\mathrm{del}_{M_n}(e); \mathbb{Z}) \oplus \tilde{H}_{d-1}(M_{n-2}; \mathbb{Z}),$$

where e is any edge in the complete graph K_n?

Problem 27.14 (Shareshian and Wachs [122]; cf. Theorem 11.32). For $1 \leq m \leq n$, with $\nu_{m,n}$ defined as in (11.6) in Section 11.3.1, is it true that $\tilde{H}_{\nu_{m,n}}(M_{m,n}; \mathbb{Z})$ is torsion-free if and only if $n \geq 2m - 4$?
Conjecture: True. Open for $n \in \{2m - 4, 2m - 3\}$.

Problem 27.15 (Shareshian and Wachs [122]; cf. Theorem 11.37). Is it true that $\tilde{H}_d(M_{m,n}; \mathbb{Z})$ is torsion-free if and only if $d < \nu_{m,n}$ or $d > m - 3$?
Conjecture: True. Open for $d = m - 2$ when $m + 2 \leq n \leq 2m - 3$ and for $d = m - 3$ when $8 \leq m = n$.

Problems on Chapter 12

Problem 27.16 (cf. Theorem 12.2). For $n \geq 3$, Is it true that the shifted connectivity degree and the homotopical depth of BD_n^2 are equal to $\lceil \frac{7n-13}{9} \rceil$?

Problem 27.17 (cf. Theorem 12.17). For $n \geq 3$, is it true that the shifted connectivity degree and the homotopical depth of $\overline{\mathrm{BD}}_n^2$ are equal to $\lceil \frac{7n-13}{9} \rceil$?

Problem 27.18 (cf. Theorems 12.8, Theorem 12.10, and 12.17). Determine the shifted connectivity degree and the homotopical depth of BD_n^d and $\overline{\mathrm{BD}}_n^d$ for general d.

Problems on Chapter 13

Problem 27.19 (cf. Proposition 13.16). A necessary condition for a simplicial complex to be SPI over a matroid M is that each minimal nonface is a circuit. Find a necessary *and* sufficient condition in terms of the set of nonfaces.

Problem 27.20 (cf. Section 13.3). Find axioms for the class of SPI complexes (or a strictly larger class with all nice topological properties preserved) without referring to any underlying matroid.

Problems on Chapter 14

Problem 27.21 (cf. Theorem 14.8). Let $\mathcal{T}_n$ be the family of spanning trees T on the vertex set $[n]$ with the property that each simple path $(1 = a_1, a_2, \ldots, a_k)$ in T starting at the vertex 1 has the property that $a_i < a_{i+2}$ for $1 \leq i \leq k - 2$. Find a bijection between $\mathcal{T}_n$ and the family of ordered partitions of the set $[n - 1]$.

Problem 27.22 (cf. Corollaries 14.18 and 14.19). For $p \geq 1$, is it true that the polynomial $f_p(n) = \tilde{\chi}(\mathsf{B}_{n,p})$ is of degree exactly $2p$ with only real and positive roots?
Conjecture: True.

Problem 27.23 (cf. Section 14.3.4). Is it true that the hypergraph complex $\mathsf{HB}_{n,p,t}$ is homotopy equivalent to a wedge of spheres of dimension $pt - 1$ whenever $n \geq pt + 1$?

Problems on Chapter 15

Problem 27.24. Analyze the topology of $\mathsf{DF}_n(D)$ in cases where Kozlov's Corollary 15.2 does not apply.

Problem 27.25. Analyze the topology of $\mathsf{DAcy}_n(D)$ in cases where Hultman's Theorem 15.5 does not apply.

Problem 27.26 (cf. Section 15.5). Compute the Euler characteristic of DOAC_n.

Problem 27.27 (cf. Theorem 15.15). Is it true that DNOCy_n is $VD(2n-3)$ or at least has a Cohen-Macaulay $(2n - 3)$-skeleton?
Conjecture: True.

Problems on Chapter 16

Problem 27.28 (cf. Theorem 16.8). Is it true that the rank of the group $\tilde{H}_{k-1}(\mathsf{NXM}_{3k+1}; \mathbb{Z})$ is $\frac{1}{k+1}\binom{4k+2}{k} = \frac{1}{3k+2}\binom{4k+2}{k+1}$?
Conjecture: True.

Problem 27.29 (cf. Theorem 16.8). Find an explicit formula for the homology of NXM_n in any given degree d.

Problem 27.30 (cf. Theorem 16.15). For $n \geq 2p+1$, is it true that the shifted connectivity degree and the homotopical depth of $\mathsf{NXB}_{n,p}$ are equal to $2p-1$?
Conjecture: True.

Problem 27.31 (cf. Theorem 16.15). Compute the homology of $\mathsf{NXB}_{n,p}$.

Problems on Chapter 17

Problem 27.32 (cf. Theorem 17.2 and Corollary 17.3). Compute the shifted connectivity degree and depth of NHam_n.

Problem 27.33 (cf. Section 17.3). Is it true that $\tilde{H}_{3(n-2)}(\mathsf{DNHam}_n, \mathbb{Z})$ contains a free subgroup isomorphic to $\mathbb{Z}^{(n-2)!}$ for all $n \geq 2$? What is the shifted connectivity degree of DNHam_n?

Problems on Chapter 18

Problem 27.34 (cf. Theorems 18.6 and 18.7 and Section 18.2.2). Let $k, n \geq 1$. Write $r = (n-1) \bmod (k+1)$. Is it true that the shifted connectivity degree of $\mathsf{NLC}_{n,k}$ equals $\lceil \alpha_{n,k} \rceil$, where $\alpha_{n,k} := \frac{(k-1)(n-1+r/k)}{k+1} - 1$?
Conjecture: True. Open for $k \geq 4$ unless $k+2 \leq n \leq 2k+2$, $n = 3k+2$, or $n = t(k+1)/2$ for $t \geq 2$ and k odd.

Problem 27.35 (cf. Theorem 18.11). With notation as in the previous Problem, is it true that the $\alpha_{n,k}$-skeleton of $\mathsf{NLC}_{n,k}$ is VD or at least shellable?
Conjecture: True. Open for $k \geq 3$.

Problem 27.36 (cf. Theorem 18.6 and Section 18.2.2). Compute the homology of $\mathsf{NLC}_{n,k}$.
Open for $n \geq 2k+3$, $n \neq 3k+2$.

Problems on Chapter 19

Problem 27.37 (cf. Theorem 19.4). Is it true that the $(2n-5)$-skeleton of NC_n^2 is VD or at least shellable?
Conjecture: True.

Problem 27.38. Let M be a matroid on the set E with rank function ρ. A set $\sigma \subseteq E$ is *connected* if σ has full rank and if there is no partition $\sigma = \sigma_1 \cup \sigma_2$ such that $\rho(\sigma) = \rho(\sigma_1) + \rho(\sigma_2)$. NC_n^2 is the complex of disconnected sets in the graphic matroid $M(K_n)$. Examine the topology of this complex for other matroids.

Problems on Chapter 20

Problem 27.39 (cf. Section 20.1). Is it true that NC_n^3 is semi-nonevasive?

Problem 27.40 (cf. Section 20.1). Is it true that the $(2n-4)$-skeleton of NC_n^3 is Cohen-Macaulay?
Conjecture: True.

Problem 27.41 (cf. Section 20.4). Prove *any* nontrivial and general result about NC_n^k for $k \notin \{1,2,3,n-2\}$.

Problems on Chapter 21

Problem 27.42 (cf. Theorem 21.3). Is it true that the $(2n-5)$-skeleton of $NCR_n^{1,0}$ is VD or at least shellable?
Conjecture: True.

Problem 27.43 (cf. Theorem 21.8). Is it true that the order complex of $\overline{NX\Pi_n^2}$ is homeomorphic to a sphere or even the boundary complex of a polytope?
Conjecture: True.

Problem 27.44 (cf. Theorem 21.9). Is it true that $\overline{NCR}_n^{(1,1)}$ is semi-nonevasive?

Problem 27.45 (cf. Theorem 21.9). Is it true that the $(2n-4)$-skeleton of $NCR_n^{(1,1)}$ (equivalently, the $(n-4)$-skeleton of $\overline{NCR}_n^{1,1}$) is Cohen-Macaulay?
Conjecture: True.

Problems on Chapter 22

Problem 27.46 (cf. Theorem 22.4). Is it true that the $(2n-4)$-skeleton of $DNSC_n$ is VD or at least shellable?
Conjecture: True.

Problem 27.47 (cf. Section 22.2). Is it true that $DNSC_n^2$ is homotopy equivalent to a wedge of $(n-2)\cdot(n-2)!$ spheres of dimension $3n-5$?
Conjecture: True.

Problem 27.48 (cf. Theorem 22.8). Is it true that the $(2n-5)$-skeleton of $DNSp_n$ is Cohen-Macaulay?
Conjecture: True.

Problems on Chapter 23

Problem 27.49 (cf. Theorems 23.1 and 23.7). Is it true that the shifted connectivity degree of NEC_n^2 equals $\lceil \frac{3n-7}{2} \rceil$?
Conjecture: True. Open for even n.

Problem 27.50 (cf. Theorem 23.1). Is the homology of NEC_n^2 torsion-free? More generally, is NEC_n^2 semi-collapsible over $\mathbb{Z}$?
Conjecture: True.

Problem 27.51 (cf. Theorem 23.4). Is it true that $\tilde{H}_{3k-5}(\mathsf{NEC}_{2k-1}^2, \mathbb{Z})$ is free of rank $((2k-3)!!)^2$?
Conjecture: True.

Problem 27.52 (cf. Section 23.3). For $p \geq 1$ and $n \geq p+1$, is it true that the shifted connectivity degree of NEC_n^p equals $\left\lceil \frac{(n-1)(p+1)}{2} - 2 \right\rceil$?

Problems on Chapter 25

Problem 27.53. Find the degree of the polynomial $g_p = \tilde{\chi}(\mathsf{NQP}_{n,n-p-1})$.

Problems on Chapter 26

See Section 26.8 for discussion.

Problem 27.54. Is it true that $\mathsf{Cov}_{n,p}$ has no homology over any field strictly above dimension $\binom{p+2}{2} - 2$ and that $\tilde{H}_{\binom{p+2}{2}-2}(\mathsf{Cov}_{n,p}, \mathbb{Z})$ is free of rank $\binom{n}{p+2}$?
Conjecture: True.

Problem 27.55. For $p+2 \leq n \leq 2p+1$, is it true that $\mathsf{Cov}_{n,p}$ has no homology strictly below dimension $2p - 1 + \binom{2p-n+2}{2}$?

Problem 27.56. Is it true that $\tilde{H}_{2p-1}(\mathsf{Cov}_{n,p}, \mathbb{Z})$ is free of rank $\binom{n-1}{2p}$?
Conjecture: True.

Problem 27.57. For $i \geq 2p$, is it true that the polynomial $f_{p,i}(n) = \beta_i(\mathsf{Cov}_{n,p}, \mathbb{F})$ satisfies $f_{p,i}(0) = 0$?
Conjecture: True.

Problem 27.58. With notation as in the previous problem, is it true that all roots of the polynomial $f_{p,i}$ are real and nonnegative?
Conjecture: False.

References

1. J. L. Andersen. *Determinantal Rings Associated with Matrices: a Counterexample*. PhD thesis, University of Minnesota, 1992.
2. C. A. Athanasiadis. Decompositions and connectivity of matching and chessboard complexes. *Discrete Comput. Geom.*, 31(3):395–403, 2004.
3. E. Babson, A. Björner, S. Linusson, J. Shareshian, and V. Welker. Complexes of not i-connected graphs. *Topology*, 38(2):271–299, 1999.
4. E. Babson and P. Hersh. Discrete Morse functions from lexicographic orders. *Trans. Amer. Math. Soc.*, 357:509–534, 2005.
5. E. Babson and D. M. Kozlov. Complexes of graph homomorphisms. *Israel J. Math.*, 152:285–312, 2006.
6. C. Berge. *Graphs and Hypergraphs*. North-Holland, Amsterdam, 2nd revised edition, 1976.
7. A. Björner. Shellable and Cohen-Macaulay partially ordered sets. *Trans. Amer. Math. Soc.*, 260:159–183, 1980.
8. A. Björner. The homology and shellability of matroids and geometric lattices. In N. White, editor, *Matroid Applications*, pages 226–283. Cambridge University Press, 1992.
9. A. Björner. Topological methods. In R. L. Graham, M. Grötschel, and L. Lovász, editors, *Handbook of Combinatorics*, volume II, pages 1819–1872, Cambridge, MA, 1996. The MIT Press.
10. A. Björner and J. D. Farley. Chain polynomials of distributive lattices are 75% unimodal. *Electronic J. Combin.*, 12(1):N4, 2005.
11. A. Björner, L. Lovász, S. T. Vrećica, and R. T. Živaljević. Chessboard complexes and matching complexes. *J. London Math. Soc. (2)*, 49:25–39, 1994.
12. A. Björner and M. Wachs. On lexicographically shellable posets. *Trans. Amer. Math. Soc.*, 277:323–341, 1983.
13. A. Björner and M. Wachs. Shellable nonpure complexes and posets. I. *Trans. Amer. Math. Soc.*, 348:1299–1327, 1996.
14. A. Björner and M. Wachs. Shellable nonpure complexes and posets. II. *Trans. Amer. Math. Soc.*, 349:3945–3975, 1997.
15. A. Björner, M. Wachs, and V. Welker. On sequentially Cohen-Macaulay complexes and posets. Preprint, 2007.
16. A. Björner and V. Welker. Homology of the "k-equal" manifolds and related partition lattices. *Adv. Math.*, 110(2):277–313, 1995.

17. A. Björner and V. Welker. Complexes of directed graphs. *SIAM J. Discrete Math.*, 12(4):413–424, 1999.
18. B. Bollobás. On generalized graphs. *Acta Math. Acad. Sci. Hungar.*, 16:447–452, 1965.
19. B. Bollobás. Extremal graph theory. In R. L. Graham, M. Grötschel, and L. Lovász, editors, *Handbook of Combinatorics*, volume II, pages 1231–1292, Cambridge, MA, 1996. The MIT Press.
20. R. Bott and C. Taubes. On the self-linking of knots. *J. Math. Phys.*, 35(10):5247–5287, 1994.
21. S. Bouc. Homologie de certains ensembles de 2-sous-groupes des groupes symétriques. *J. Algebra*, 150:187–205, 1992.
22. K.S. Brown. Euler characteristics of discrete groups and G-spaces. *Invent. Math.*, 27:229–264, 1974.
23. K.S. Brown. Euler characteristics of groups: the p-fractional part. *Invent. Math.*, 29:1–5, 1975.
24. H. Bruggesser and P. Mani. Shellable decompositions of cells and spheres. *Math. Scand.*, 29:197–205, 1971.
25. W. Bruns and A. Conca. Gröbner bases and determinantal ideals. In J. Herzog and V. Vuletescu, editors, *Commutative Algebra, Singularities and Computer Algebra*, pages 9–66, Dordrecht, 2003. Kluwer Academic.
26. W. Bruns and J. Herzog. *Cohen-Macaulay rings.* Cambridge University Press, 1993.
27. V. Capoyleas and J. Pach. A Turán-type theorem on chords of a convex polygon. *J. Combin. Theory, Ser. B*, 56:9–15, 1992.
28. A. Chakrabarti, S. Khot, , and Y. Shi. Evasiveness of subgraph containment and related properties. *SIAM J. Comput.*, 31(3):866–875, 2001.
29. F. Chapoton, S. Fomin, and A.V. Zelevinsky. Polytopal realizations of generalized associahedra. *Canad. Math. Bull.*, 45:537–566, 2002.
30. H. Charalambous. Pointed simplicial complexes. *Illinois J. Math.*, 41(1):1–9, 1997.
31. M. K. Chari. Manuscript, 1999.
32. M. K. Chari. On discrete Morse functions and combinatorial decompositions. *Discrete Math.*, 217(1–3):101–113, 2000.
33. S. Cook. The complexity of theorem proving procedures. In *Proceedings of Third Annual ACM Symposium on Theory of Computing*, pages 151–158, 1971.
34. K. Crowley. *Discrete Morse Theory and the Geometry of Nonpositively Curved Simplicial Complexes.* PhD thesis, Rice University, May 2001.
35. E. Deutsch. Dyck path enumeration. *Discrete Math.*, 204:167–202, 1999.
36. X. Dong. *The Topology of Bounded Degree Graph Complexes and Finite Free Resolutions.* PhD thesis, University of Minnesota, August 2001.
37. X. Dong and M. L. Wachs. Combinatorial Laplacian of the matching complex. *Electronic J. Combin.*, 9(1):R17, 2002.
38. A. Dress, M. Klucznik, J. Koolen, and V. Moulton. $2kn - \binom{2k+1}{2}$: A note on extremal combinatorics of cyclic split systems. *Seminaire Lotharingien de Combinatoire*, 47, 2001.
39. A. Dress, J. Koolen, and V. Moulton. On line arrangements in the hyperbolic plane. *European J. Combin.*, 23:549–557, 2002.
40. A. Dress, J. Koolen, and V. Moulton. $4n - 10$. *Ann. Combin.*, 8:463–471, 2004.
41. S. Dulucq and J. G. Penaud. Cordes, arbres et permutations. *Discrete Math.*, 117:89–105, 1993.

42. J.-G. Dumas, F. Heckenbach, B. D. Saunders, and V. Welker. *Simplicial Homology, a share package for GAP*, 2000.

43. D. Eisenbud. *Commutative Algebra with a View Toward Algebraic Geometry.* Springer-Verlag New York, 1995.

44. T. Fine. Extrapolation when very little is known about the source. *Inform. Control*, 16:331–359, 1970.

45. P. Flajolet and M. Noy. Analytic combinatorics of non-crossing configurations. *Discrete Math.*, 204:203–229, 1999.

46. J. Folkman. The homology groups of a lattice. *J. Math. Mech.*, 15:631–636, 1966.

47. S. Fomin and A. Zelevinsky. Y-systems and generalized associahedra. *Ann. Math*, 158:977–1018, 2003.

48. R. Forman. A discrete Morse theory for cell complexes. In S. T. Yau, editor, *Geometry, Topology & Physics for Raoul Bott.* International Press, 1995.

49. R. Forman. Morse theory for cell complexes. *Adv. Math.*, 134:90–145, 1998.

50. R. Forman. Morse theory and evasiveness. *Combinatorica*, 20:489–504, 2000.

51. J. S. Frame, G. de B. Robinson, and R. M. Thrall. The hook graphs of the symmetric group. *Canad. J. Math.*, 6:316–325, 1954.

52. J. Friedman and P. Hanlon. On the Betti numbers of chessboard complexes. *J. Algebraic Combin.*, 8:193–203, 1998.

53. P. F. Garst. *Cohen-Macaulay complexes and group actions.* PhD thesis, University of Wisconsin-Madison, 1979.

54. M. Goresky and R. D. MacPherson. *Stratified Morse Theory, Volume 14 of Ergebnisse der Mathematik und ihrer Grenzgebiete.* Springer-Verlag, Berlin, Heidelberg, New York, 1988.

55. O. A. Gross. Preferential arrangements. *Amer. Math. Monthly*, 69:4–8, 1962.

56. M. Hachimori. *Combinatorics of constructible complexes.* PhD thesis, University of Tokyo, 2000.

57. M. Haiman. *Constructing the associahedron.* MIT, 1984. Handwritten manuscript published electronically at math.berkeley.edu/~mhaiman.

58. A. Hajnal. A theorem on k-saturated graphs. *Canad. Math. J.*, 17:720–724, 1965.

59. A. Hatcher. *Algebraic Topology.* Cambridge University Press, 2002.

60. P. Hersh. On optimizing discrete Morse functions. *Adv. in Appl. Math.*, 35:294–322, 2005.

61. P. Hersh and V. Welker. Gröbner basis degree bounds on $\mathrm{Tor}_\bullet^{k[\Lambda]}(k,k)_\bullet$ and Discrete Morse Theory for posets. In *Proceedings of the Summer Research Conference on Integer Points in Polyhedra*, 2003.

62. J. Herzog and N. V. Trung. Gröbner bases and multiplicity of determinantal and Pfaffian ideals. *Adv. Math.*, 96:1–37, 1992.

63. M. Hochster. Rings of invariants of tori, Cohen-Macaulay rings generated by monomials, and polytopes. *Ann. Math.*, 96:318–337, 1972.

64. A. Hultman. Directed subgraph complexes. *Electronic Journal of Combinatorics*, 11(1):R75, 2004.

65. I. M. Isaacs. *Algebra: A Graduate Course.* Brook/Coles Publishing Company, Pacific Grove, CA, 1994.

66. M. Jöllenbeck and V. Welker. Resolution of the residue field via algebraic discrete Morse theory. *Mem. Amer. Math. Soc.* to appear.

67. J. Jonsson. On the topology of simplicial complexes related to 3-connected and Hamiltonian graphs. *J. Combin. Theory, Ser. A*, 104(1):169–199, 2003.

68. J. Jonsson. Classifying monotone digraph properties that are strong pseudo-independence complexes. Manuscript, 2005.

69. J. Jonsson. Matching complexes on grids. Manuscript, 2005.

70. J. Jonsson. Optimal decision trees on simplicial complexes. *Electronic J. Combin.*, 12(1):R3, 2005.

71. J. Jonsson. *Simplicial Complexes of Graphs*. PhD thesis, Royal Institute of Technology, Stockholm, 2005.

72. J. Jonsson. Simplicial complexes of graphs and hypergraphs with a bounded covering number. *SIAM J. Discrete Math.*, 19(3):633–650, 2005.

73. J. Jonsson. Some classes of simplicial complexes with Euler characteristics given by polynomials. Manuscript, 2005.

74. J. Jonsson. Exact sequences for the homology of the matching complex. Submitted, 2007.

75. J. Jonsson. Five-torsion in the homology of the matching complex on 14 vertices. Submitted, 2007.

76. J. Jonsson. On the 3-torsion part of the homology of the chessboard complex. Submitted, 2007.

77. T. Józefiak and J. Weyman. Representation-theoretic interpretation of a formula of D. E. Littlewood. *Math. Proc. Cambridge Philos. Soc.*, 103:193–196, 1988.

78. J. Kahn, M. Saks, and D. Sturtevant. A topological approach to evasiveness. *Combinatorica*, 4:297–306, 1984.

79. M. M. Kapranov. The permutoassociahedron, Mac Lane's coherence theorem and asymptotic zones for the KZ equation. *J. Pure Appl. Algebra*, 85(2):119–142, 1993.

80. D. B. Karaguezian. *Homology of complexes of degree one graphs*. PhD thesis, Stanford University, 1994.

81. R. Karp. Reducibility among combinatorial problems. In R. Miller and J. Thatcher, editors, *Complexity of Computer Computations*, pages 85–104. Plenum Press, 1972.

82. D. J. Kleitman and D. J. Kwiatkowski. Further results on the Aanderaa-Rosenberg conjecture. *J. Combin. Theory, Ser. B*, 28:85–95, 1980.

83. M. Kontsevich. Formal (non)commutative symplectic geometry. In *The Gelfand Mathematical Seminars, 1990–1992*, pages 173–187. Birkhauser Boston, 1993.

84. M. Kontsevich. Vassiliev's knot invariants. *Adv. Soviet Math.*, 16(2):137–150, 1993.

85. M. Kontsevich. Feynman diagrams and low-dimensional topology. In *First European Congress of Mathematics, Vol. II (Paris, 1992)*, pages 97–121. Birkhäuser Basel, 1994.

86. D. M. Kozlov. Complexes of directed trees. *J. Combin. Theory, Ser. A*, 88(1):112–122, 1999.

87. G. Kreweras. Sur les partitions noncrois ees d'un cycle. *Discrete Math.*, 1:333–350, 1972.

88. R. Ksontini. *Propriétés homotopiques du complexe de Quillen du groupe symétrique*. PhD thesis, Université de Lausanne, 2000.

89. R. Ksontini. Simple connectivity of the Quillen complex of the symmetric group. *J. Combin. Theory, Ser. A*, 103:257–279, 2003.

90. C. W. Lee. The associahedron and triangulations of an *n*-gon. *European J. Combin.*, 10:551–560, 1989.

91. T. Lewiner, H. Lopes, and G. Tavares. Optimal discrete Morse functions for 2-manifolds. *Comput. Geom.*, 26(3):221–233, 2003.

92. T. Lewiner, H. Lopes, and G. Tavares. Towards optimality in discrete Morse theory. *Experiment. Math.*, 12(3):271–285, 2003.

93. T. Lewiner, H. Lopes, and G. Tavares. Visualizing Forman's discrete vector field. In H.-C. Hege and K. Polthier, editors, *Visualization and Mathematics III*, pages 95–112. Springer Berlin, 2003.

94. S. Linusson and J. Shareshian. Complexes of t-colorable graphs. *SIAM J. Discrete Math.*, 16(3):371–389, 2003.

95. S. Linusson, J. Shareshian, and V. Welker. Complexes of graphs with bounded matching size. *J. Algebraic Combin.* to appear.

96. L. Lovász. *Combinatorial Problems and Exercises*. North-Holland, Amsterdam, 2nd edition, 1993.

97. L. Lovász and M.D. Plummer. *Matching Theory*. North-Holland, Amsterdam, 1986.

98. F. H. Lutz. Examples of $\mathbb{Z}$-acyclic and contractible vertex-homogeneous simplicial complexes. *Discrete Comput. Geom.*, 27:137–154, 2002.

99. M. Markl. Simplex, associahedron, and cyclohedron. *Contemp. Math.*, 227:235–265, 1999.

100. S. Moriyama and F. Takeuchi. Incremental construction properties in dimension two – shellability, extendable shellability and vertex decomposability. In *Proceedings of the 12th Canadian conference on computational geometry*, pages 65–72. University of New Brunswick, 2000.

101. J. R. Munkres. *Elements of Algebraic Topology*. Perseus Books Publishing, 1984.

102. T. Nakamigawa. A generalization of diagonal flips in a convex polygon. *Theor. Comput. Sci.*, 235(2):271–282, 2000.

103. I. Novik, A. Postnikov, and B. Sturmfels. Syzygies of oriented matroids. *Duke Mathematical Journal*, 111(2):287–317, 2002.

104. M. Noy. Enumeration of non-crossing trees on a circle. *Discrete Math.*, 180:301–313, 1998.

105. J.G. Oxley. *Matroid Theory*. Oxford University Press, 1992.

106. H. Poincaré. Cinquième Complément à l'Analysis Situs. *Rend. Circ. Mat. Palermo*, 18:277–308, 1904.

107. J. S. Provan. *Decompositions, shellings and diameters of simplicial complexes and convex polytopes*. PhD thesis, Cornell University, 1977.

108. J. S. Provan and L. J. Billera. Decompositions of simplicial complexes related to diameters of convex polyhedra. *Math. Oper. Res.*, 5:576–594, 1980.

109. J. Przytycki and A. Sikora. Polygon dissections and Euler, Fuss, Kirkman and Cayley numbers. *J. Combin. Theory, Ser. A*, 92:68–76, 2000.

110. D. Quillen. Homotopy properties of the poset of nontrivial p-subgroups of a group. *Adv. Math.*, 28:101–128, 1978.

111. V. Reiner and J. Roberts. Minimal resolutions and homology of chessboard and matching complexes. *J. Algebraic Combin.*, 11:135–154, 2000.

112. V. Reiner and G. M. Ziegler. Coxeter-associahedra. *Mathematika*, 41:364–393, 1994.

113. G. Reisner. Cohen-Macaulay quotients of polynomial rings. *Adv. Math.*, 21:30–49, 1976.

114. R. L. Rivest and J. Vuillemin. A generalization and proof of the Aanderaa-Rosenberg conjecture. In *Proceedings of the 7th Annual ACM Symposium on Theory of Computing*, pages 6–12. ACM press, 1975.

115. R. L. Rivest and J. Vuillemin. On recognizing graph properties from adjacency matrices. *Theoret. Comput. Sci.*, 3(3):371–384, 1976.

116. G.-C. Rota. On the foundations of combinatorial theory I. Theory of Möbius functions. *Z. Wahrscheinlichkeitstheorie*, 2:340–368, 1964.

117. J. Shareshian. Links in the complex of separable graphs. *J. Combin. Theory, Ser. A*, 88(1):54–65, 1999.

118. J. Shareshian. Discrete Morse theory for complexes of 2-connected graphs. *Topology*, 40(4):681–701, 2001.

119. J. Shareshian. Hypergraph matching complexes and Quillen complexes of symmetric groups. *J. Combin. Theory, Ser. A*, 106:299–314, 2004.

120. J. Shareshian and M. L. Wachs. Homology of matching and chessboard complexes. In *Formal Power Series and Algebraic Combinatorics, 13th International Conference*. Arizona State University, 2001. Extended abstract.

121. J. Shareshian and M. L. Wachs. On the 1 mod k partition poset and graph complexes. Manuscript, 2005.

122. J. Shareshian and M. L. Wachs. Torsion in the matching complex and chessboard complex. *Adv. Math.*, 212(2):525–570, 2007.

123. S. Sigg. Laplacian and homology of free two-step nilpotent Lie algebras. *J. Algebra*, 185:144–161, 1996.

124. R. Simion. Noncrossing partitions. *Discrete Math.*, 217:367–409, 2000.

125. R. Simion. A type-B associahedron. *Adv. in Appl. Math.*, 30:2–25, 2003.

126. E. Sköldberg. Combinatorial discrete Morse theory from an algebraic viewpoint. Preprint, 2003.

127. N. J. A. Sloane. *The On-Line Encyclopedia of Integer Sequences*. Published electronically at www.research.att.com/~njas/sequences, 2007.

128. D. E. Smith. On the Cohen-Macaulay property in commutative algebra and simplicial topology. *Pacific J. Math.*, 14:165–196, 1990.

129. D. Soll. Evasive Simpliziale Komplexe und Diskrete Morse Theorie. Master's thesis, Philipps-Universität Marburg, 2002.

130. E. H. Spanier. *Algebraic Topology*. McGraw-Hill, 1966.

131. R. P. Stanley. Some aspects of groups acting on finite posets. *J. Combin. Theory, Ser. A*, 32(2):132–161, 1982.

132. R. P. Stanley. *Combinatorics and Commutative Algebra*. Birkhauser Boston, 2nd edition, 1996.

133. R. P. Stanley. *Enumerative Combinatorics, Vol. 1*. Cambridge University Press, 1997.

134. R. P. Stanley. *Enumerative Combinatorics, Vol. 2*. Cambridge University Press, 1999.

135. R. P. Stanley. *Catalan Addendum*. Published electronically at www-math.mit.edu/~rstan/ec, version of 14 March 2005.

136. J. Stasheff. Homotopy associativity of h-spaces I, II. *Trans. Amer. Math. Soc.*, 108:275–312, 1963.

137. S. Sundaram. On the topology of two partition posets with forbidden block sizes. *J. Pure. Appl. Algebra*, 155:271–304, 2001.

138. S. Sundaram and M. Wachs. The homology representations of the k-equal partition lattice. *Trans. AMS*, 349:935–954, 1997.

139. V. Turchin. Homologies of complexes of doubly connected graphs. *Russian Math. Surveys (Uspekhi)*, 52:426–427, 1997.

140. E. Tzanaki. Polygon dissections and some generalizations of cluster complexes. *J. Combin. Theory, Ser. A*, 113(6):1189–1198, 2006.

141. V. A. Vassiliev. Cohomology of knot spaces. In V. I. Arnold, editor, *Adv. in Sov. Math.; Theory of Singularities and its Applications*, pages 23–69, Providence, R.I., 1990.

142. V. A. Vassiliev. *Complements of Discriminants of Smooth Maps: Applications (Transl. Math. Monogr., vol. 98)*. Amer. Math. Soc., revised edition, 1994.

143. V. A. Vassiliev. Topology of two-connected graphs and homology of spaces of knots. In S. Tabachnikov, editor, *Differential and Symplectic Topology of Knots and Curves (Amer. Math. Soc. Transl. Ser. 2, vol. 190)*, pages 253–286. Amer. Math. Soc., 1999.

144. M. L. Wachs. Whitney homology of semipure shellable posets. *J Algebraic Combin.*, 9:173–207, 1999.

145. M. L. Wachs. Topology of matching, chessboard and general bounded degree graph complexes. *Alg. Universalis*, 49(4):345–385, 2003.

146. V. Welker. Constructions preserving evasiveness and collapsibility. *Discrete Math.*, 207:243–255, 1999.

147. D. J. A. Welsh. *Matroid Theory*. Academic Press, London, 1976.

148. H. S. Wilf. *generatingfunctionology*. Academic Press, NY, 2nd edition, 1994.

149. A. C.-C. Yao. Monotone bipartite graph properties are evasive. *SIAM J. Comput.*, 17(3):517–520, 1988.

150. E. C. Zeeman. On the dunce hat. *Topology*, 2:341–358, 1964.

151. G. M. Ziegler. Shellability of chessboard complexes. *Israel J. Math.*, 87, 1994.

152. G. M. Ziegler. *Lectures on Polytopes*. Springer-Verlag New York, 2nd edition, 1998.

153. R. T. Zivaljević and S.T. Vrećica. The colored Tverberg problem and complexes of injective functions. *J. Combin. Theory, Ser. A*, 61, 1992.

Index

Lecture Notes in Mathematics

For information about earlier volumes
please contact your bookseller or Springer
LNM Online archive: springerlink.com

Vol. 1835: O.T. Izhboldin, B. Kahn, N.A. Karpenko, A. Vishik, Geometric Methods in the Algebraic Theory of Quadratic Forms. Summer School, Lens, 2000. Editor: J.-P. Tignol (2004)

Vol. 1836: C. Nastasescu, F. Van Oystaeyen, Methods of Graded Rings. XIII, 304 p, 2004.

Vol. 1837: S. Tavaré, O. Zeitouni, Lectures on Probability Theory and Statistics. Ecole d Eté de Probabilités de Saint-Flour XXXI-2001. Editor: J. Picard (2004)

Vol. 1838: A.J. Ganesh, N.W. O Connell, D.J. Wischik, Big Queues. XII, 254 p, 2004.

Vol. 1839: R. Gohm, Noncommutative Stationary Processes. VIII, 170 p, 2004.

Vol. 1840: B. Tsirelson, W. Werner, Lectures on Probability Theory and Statistics. Ecole d Eté de Probabilités de Saint-Flour XXXII-2002. Editor: J. Picard (2004)

Vol. 1841: W. Reichel, Uniqueness Theorems for Variational Problems by the Method of Transformation Groups (2004)

Vol. 1842: T. Johnsen, A. L. Knutsen, K_3 Projective Models in Scrolls (2004)

Vol. 1843: B. Jefferies, Spectral Properties of Noncommuting Operators (2004)

Vol. 1844: K.F. Siburg, The Principle of Least Action in Geometry and Dynamics (2004)

Vol. 1845: Min Ho Lee, Mixed Automorphic Forms, Torus Bundles, and Jacobi Forms (2004)

Vol. 1846: H. Ammari, H. Kang, Reconstruction of Small Inhomogeneities from Boundary Measurements (2004)

Vol. 1847: T.R. Bielecki, T. Björk, M. Jeanblanc, M. Rutkowski, J.A. Scheinkman, W. Xiong, Paris-Princeton Lectures on Mathematical Finance 2003 (2004)

Vol. 1848: M. Abate, J. E. Fornaess, X. Huang, J. P. Rosay, A. Tumanov, Real Methods in Complex and CR Geometry, Martina Franca, Italy 2002. Editors: D. Zaitsev, G. Zampieri (2004)

Vol. 1849: Martin L. Brown, Heegner Modules and Elliptic Curves (2004)

Vol. 1850: V. D. Milman, G. Schechtman (Eds.), Geometric Aspects of Functional Analysis. Israel Seminar 2002-2003 (2004)

Vol. 1851: O. Catoni, Statistical Learning Theory and Stochastic Optimization (2004)

Vol. 1852: A.S. Kechris, B.D. Miller, Topics in Orbit Equivalence (2004)

Vol. 1853: Ch. Favre, M. Jonsson, The Valuative Tree (2004)

Vol. 1854: O. Saeki, Topology of Singular Fibers of Differential Maps (2004)

Vol. 1855: G. Da Prato, P.C. Kunstmann, I. Lasiecka, A. Lunardi, R. Schnaubelt, L. Weis, Functional Analytic Methods for Evolution Equations. Editors: M. Iannelli, R. Nagel, S. Piazzera (2004)

Vol. 1856: K. Back, T.R. Bielecki, C. Hipp, S. Peng, W. Schachermayer, Stochastic Methods in Finance, Bressanone/Brixen, Italy, 2003. Editors: M. Fritelli, W. Runggaldier (2004)

Vol. 1857: M. Émery, M. Ledoux, M. Yor (Eds.), Séminaire de Probabilités XXXVIII (2005)

Vol. 1858: A.S. Cherny, H.-J. Engelbert, Singular Stochastic Differential Equations (2005)

Vol. 1859: E. Letellier, Fourier Transforms of Invariant Functions on Finite Reductive Lie Algebras (2005)

Vol. 1860: A. Borisyuk, G.B. Ermentrout, A. Friedman, D. Terman, Tutorials in Mathematical Biosciences I. Mathematical Neurosciences (2005)

Vol. 1861: G. Benettin, J. Henrard, S. Kuksin, Hamiltonian Dynamics   Theory and Applications, Cetraro, Italy, 1999. Editor: A. Giorgilli (2005)

Vol. 1862: B. Helffer, F. Nier, Hypoelliptic Estimates and Spectral Theory for Fokker-Planck Operators and Witten Laplacians (2005)

Vol. 1863: H. Führ, Abstract Harmonic Analysis of Continuous Wavelet Transforms (2005)

Vol. 1864: K. Efstathiou, Metamorphoses of Hamiltonian Systems with Symmetries (2005)

Vol. 1865: D. Applebaum, B.V. R. Bhat, J. Kustermans, J. M. Lindsay, Quantum Independent Increment Processes I. From Classical Probability to Quantum Stochastic Calculus. Editors: M. Schürmann, U. Franz (2005)

Vol. 1866: O.E. Barndorff-Nielsen, U. Franz, R. Gohm, B. Kümmerer, S. Thorbjønsen, Quantum Independent Increment Processes II. Structure of Quantum Lévy Processes, Classical Probability, and Physics. Editors: M. Schürmann, U. Franz, (2005)

Vol. 1867: J. Sneyd (Ed.), Tutorials in Mathematical Biosciences II. Mathematical Modeling of Calcium Dynamics and Signal Transduction. (2005)

Vol. 1868: J. Jorgenson, S. Lang, $Pos_n(R)$ and Eisenstein Series. (2005)

Vol. 1869: A. Dembo, T. Funaki, Lectures on Probability Theory and Statistics. Ecole d Eté de Probabilités de Saint-Flour XXXIII-2003. Editor: J. Picard (2005)

Vol. 1870: V.I. Gurariy, W. Lusky, Geometry of Müntz Spaces and Related Questions. (2005)

Vol. 1871: P. Constantin, G. Gallavotti, A.V. Kazhikhov, Y. Meyer, S. Ukai, Mathematical Foundation of Turbulent Viscous Flows, Martina Franca, Italy, 2003. Editors: M. Cannone, T. Miyakawa (2006)

Vol. 1872: A. Friedman (Ed.), Tutorials in Mathematical Biosciences III. Cell Cycle, Proliferation, and Cancer (2006)

Vol. 1873: R. Mansuy, M. Yor, Random Times and Enlargements of Filtrations in a Brownian Setting (2006)

Vol. 1874: M. Yor, M. Émery (Eds.), In Memoriam Paul-André Meyer - Séminaire de Probabilités XXXIX (2006)

Vol. 1875: J. Pitman, Combinatorial Stochastic Processes. Ecole d Eté de Probabilités de Saint-Flour XXXII-2002. Editor: J. Picard (2006)

Vol. 1876: H. Herrlich, Axiom of Choice (2006)

Vol. 1877: J. Steuding, Value Distributions of L-Functions (2007)

Vol. 1878: R. Cerf, The Wulff Crystal in Ising and Percolation Models, Ecole d Eté de Probabilités de Saint-Flour XXXIV-2004. Editor: Jean Picard (2006)

Vol. 1879: G. Slade, The Lace Expansion and its Applications, Ecole d Eté de Probabilités de Saint-Flour XXXIV-2004. Editor: Jean Picard (2006)

Vol. 1880: S. Attal, A. Joye, C.-A. Pillet, Open Quantum Systems I, The Hamiltonian Approach (2006)

Vol. 1881: S. Attal, A. Joye, C.-A. Pillet, Open Quantum Systems II, The Markovian Approach (2006)

Vol. 1882: S. Attal, A. Joye, C.-A. Pillet, Open Quantum Systems III, Recent Developments (2006)

Vol. 1883: W. Van Assche, F. Marcellàn (Eds.), Orthogonal Polynomials and Special Functions, Computation and Application (2006)

Vol. 1884: N. Hayashi, E.I. Kaikina, P.I. Naumkin, I.A. Shishmarev, Asymptotics for Dissipative Nonlinear Equations (2006)

Vol. 1885: A. Telcs, The Art of Random Walks (2006)

Vol. 1886: S. Takamura, Splitting Deformations of Degenerations of Complex Curves (2006)

Recent Reprints and New Editions